Arc-Search Techniques for Interior-Point Methods

Yaguang Yang
Office of Research
US Nuclear Regulatory Commission
Rockville, Maryland, USA

T0262327

CRC Press
Taylor & Francis Group
Boca Raton London New York

CRC Press is an imprint of the
Taylor & Francis Group, an **informa** business
A SCIENCE PUBLISHERS BOOK

CRC Press
Taylor & Francis Group
6000 Broken Sound Parkway NW, Suite 300
Boca Raton, FL 33487-2742

Version Date: 20200415

International Standard Book Number-13: 978-0-367-48728-7 (Hardback)

Visit the Taylor & Francis Web site at
http://www.taylorandfrancis.com

and the CRC Press Web site at
http://www.routledge.com

To my wife, Rong Cai
my son, Minghui Yang
and my daughter, Maria M Yang

Preface

My interest in optimization dates back to the time of my graduate studies between 1982–1985 in China. My master thesis was on the solution of linear programming with all parameters (**A**, **b**, **c**) being in some intervals. A paper based on my thesis entitled "A new approach to uncertain parameter linear programming" was published in the European Journal of Operational Research in 1991, volume 54, pp. 95–114.

After I came to the US in 1991, I studied controls in the Electrical Engineering department at the University of Maryland and worked as a control engineer after graduation. However, my interest in optimization never faded. I used optimization techniques to solve my engineering problems, and studied optimization books and read papers in my spare time. In 2007, while reading a paper by Megiddo ("Pathways to the Optimal Set in Linear Programming") published in Progress in Mathematical Programming, I realized that using arc-search to estimate the central path (the pathway suggested by Megiddo) may be a better method than line search, although the latter has been widely used in the optimization community. This thought inspired me to systematically study the interior-point methods, including some excellent books like Wright's "Primal-Dual Interior-Point Methods", Ye's "Interior-Point Algorithm: Theory and Analysis", and Roos, Terlaky and Vial's book "Interior Point Methods for Linear Optimization".

At first, I was amazed by the beautiful convergence analysis which proved that many interior-point algorithms are convergent in polynomial time while there was no simplex algorithm that was proved to be convergent in polynomial time (this is still a fact today). This means that, as the problem size becomes larger, the time to find a solution may become exponentially longer if a simplex algorithm is used. However, if a convergent interior-point algorithm is used, the time to find a solution increases moderately as the problem size increases. Computational experience shows that Mehrotra's algorithm, one of the most successful interior-point algorithms, which has been implemented in the state-of-the-art interior-point software packages, is competitive with the simplex algorithms for large linear programming problems.

Then, to my surprise, behind the huge success of the interior-point method, there were also noticeable gaps between the theoretical analysis of the interior-point algorithms and the computational experiences of these interior-point algorithms. Researchers noted that, in order to make an interior-point algorithm efficient, three strategies are very important: (a) use higher order derivatives to approximate the central path, (b) search for the optimizer in a wide neighborhood of the central path, and (c) start from an infeasible initial point. However, convergence analysis shows that using any of these proven strategies will increase the polynomial bound of these algorithms. This clearly contradicts the definition of the polynomial bound. In addition, using a combination of these proven strategies will increase the polynomial bound even more (making things worse). To further exacerbate the problem, Mehrotra's predictor-corrector (MPC) algorithm (the most popular and most efficient interior-point algorithm around 2010) uses all of the strategies listed above but does not have a convergence proof. Since it fails to converge, it doesn't have polynomiality, a critical issue with the simplex method (see Klee and Minty "How good is the simplex algorithm?" in Inequalities, O Shisha, eds., Academic Press, Providence, RI, 1972, pp. 159–175"). This lack of polynomiality motivated Khachiyan's ellipsoid method for linear programming and was the main argument for the development of interior-point algorithms.

This observation motivated me to work on the interior-point method using the arc-search technique. I hoped that by using some kind of arc-search technique I would be able to solve the dilemma. My intuition was to use a part of an ellipse arc to approximate the central path because by changing the major and minor axes and the center, the shape and location of the ellipse changes. I figured out some necessary details and posted my first result on the website Optimization Online in 2009, http://www.optimization-online.org/DB HTML/2009/08/2375.html.

The early support I received was from my Ph.D adviser, Professor Andre L Tits, and Professor Tamas Terlaky. They both read my posted article and provided valuable feedback.

It took some time for journals to accept my first two papers on polynomial arc-search interior-point algorithms for linear programming and convex quadratic programming. Due to the different paths in the review processes, the paper on arc-search interior-point algorithms for convex quadratic programming was published before the paper on arc-search interior-point algorithms for linear programming. These two papers showed that, by using the arc-search technique, the higher order interior-point method results in the best polynomial bound for any interior-point algorithms for linear programming. This solved part of the dilemma that we discussed earlier.

The main criticism of the reviewers for these two papers was the lack of extensive numerical testing to provide support for the aforementioned theoretical analysis. Therefore, I decided to test the arc-search technique using Netlib benchmark linear programming problems. Since the majority of the Netlib benchmark problems do not have an interior-point, the infeasible starting point method is the only choice. I decided, therefore, to write two codes, one implements Mehrotra's method and the other implements a similar algorithm. The only difference between the two algorithms is that the former uses line search while the latter uses arc-search. I got support from Dr. Chris Hoxie, at the Office of Research in the U.S. NRC, who provided the computational environment for this research. The extensive testing on the Netlib benchmark problems was very promising, it showed that arc-search algorithm is more efficient and more robust than Mehrotra's predictor-corrector (MPC) method. This result motivated me to devise some arc-search algorithms that use all three proven strategies and to show their convergence. Not long after I finished the test on the Netlib benchmark problems, I developed two algorithms, both are more efficient and more robust than Mehrotra's method and both converge in polynomial time. In particular, one of the algorithms achieves the best polynomial bound for all existing interior-point algorithms for linear programming (feasible or infeasible). This solves all the dilemmas mentioned above. I didn't submit these results until the benchmark test paper was accepted.

While I was focused on arc-search techniques for linear and convex quadratic programming, researchers in several groups extended the arc-search technique to more general problems, such as linear complementarity problem (LCP), symmetric cone optimization problem (SCP), and positive semi-definite programming problem (SDP). Unsurprisingly, all algorithms in these extensions have obtained better polynomial bounds than their counterparts that use the traditional line search method.

However, algorithms in these extensions have not considered all three proven strategies at the same time and have not achieved the best polynomial bound that the best algorithm for linear programming has achieved (in my opinion, they should be able to achieve the best polynomial bound that the best algorithm for linear programming has achieved). In addition, these algorithms for LCP, SCP, and SDP have not been tested as extensively as similar algorithms designed for linear programming. I also believe that there will be more meaningful extensions of the arc-search techniques to other optimization areas, such as general convex programming and nonlinear programming. This motivated me to write a book about the arc-search technique; to discuss the merits of the arc-search in linear programming, LCP, SCP, and SDP; to introduce some of the latest developments in applying arc-search technique to other optimization problems; to encourage more researchers to identify new applications based on existing results. If some readers decide to work on this new area after reading this book, I will have achieved my goal.

Like other similar projects, I have received support from people at different stages of development. First, I am indebted to Professor Tits at University of Maryland, and Professor Terlaky at Lehigh University, for reading my very first result and providing useful comments. I would like to thank Dr. Hoxie of U.S. NRC, for providing the computational environment for this research. I also would like to thank Professor Yamashita at Tokyo Institute of Technology, who co-authored two papers on arc-search interior-point algorithms, one of them is included in this book. I am grateful to Professor Zhang at China Three Gorges University, who kindly sent me his very recent papers which are included in this book as well.

Contents

I

LINE SEARCH INTERIOR-POINT METHODS FOR LINEAR PROGRAMMING

Chapter 1

Introduction

Most numerical optimization algorithms are composed of two major steps in every iteration: (a) from the current iterate, find a direction that will improve the objective function, and (b) determine a step-size in this direction to obtain the next iterate that can actually improve the objective function. However, people later realized that searching along a straight line may not be a good idea in some applications, because many applications are nonlinear. Therefore, arc-search methods were developed, analyzed, and tested. For example, G. P. McCormick [87] proposed a method that searches for the optimizer along an arc for unconstrained optimization problems. For equality constrained optimization problems, if both objective function and equality constraints are smooth, noticing that the smooth equality constraints form a Riemannian manifold, Luenberger realized in 1972 [79] that the constraints set is similar to a surface in three dimensional space and search for the optimizer might be carried out along the geodesic on the Riemannian manifold. This idea was then fully developed by Gabay [39] and extended by Smith [127] and Yang [150] among others. Some good resources on this topic are [1, 31]. However, in this book, the proposed arc-search uses a different technique which is specifically developed for interior-point method of numerical optimization.

Interior-point method was originally developed for linear programming (LP) problems after Karmarkar published his seminal work in 1984 [57][1]. The tremendous success of the interior-point method is on two folds: Many interior-point algorithms are proved to be polynomial for linear programming problems, an attractive feature that simplex algorithms do not have; some interior-point algorithms, such as Mehrotra's predictor-corrector (MPC) method, are computationally competitive compared to the simplex algorithms, at least for large linear programming problems.

[1] The interior-point method is related to the log barrier function which was proposed probably as early as 1955 by Frisch [38] and to the affine scaling used by Dikin [27], but their works were not widely noticed by researchers until interior-point method became a popular tool in numerical optimization in the 1980s.

By the middle of the 1990s, interior-point method was viewed as a mature discipline in optimization when three monographs were published [10, 144, 164]. However, there was a big gap for interior-point method between the convergence theory and the efficiency of practical computation. In theory, the efficiency of an optimization algorithm can be measured by the worst case iteration bounds. Although the majority interior-point algorithms have polynomial bounds, these bounds are in different orders. The lowest polynomial bounds increase (become worse) in the following order[2]: (1) the first-order method with feasible starting point in narrow neighborhood, $O(\sqrt{n}L)$; (2a) the second-order method with feasible starting point in narrow neighborhood, $O(n^{3/4}L^{3/2})$, which may be comparable to (2b) the first-order method with feasible starting point in wide neighborhood, $O(nL)$, which has the same complexity as (2c) the first-order method with infeasible starting point in narrow neighborhood, $O(nL)$; (3) the first-order method with infeasible starting point in wide neighborhood, $O(n^2L)$; and (4) the second-order method with infeasible starting point in wide neighborhood (Mehrotra's method), no polynomial bound (no convergence result). However, the observed computational efficiency is exactly in the opposite direction, i.e., Mehrotra's method is the most efficient, and the efficiency decreases from (4) to (3) to (2) and to (1). This dilemma has been known since the very beginning of the development of the interior-point methods [144]. Todd, in 2002 [133], raised two questions to the researchers in the optimization community: "Can we give a theoretical explanation for the difference between worst-case bounds and observed practical performance? Can we find a theoretically and practically efficient way to reoptimize?". However, this gap was not closed until the 2010s when this author published a series of papers to resolve the dilemma. The main idea involved in these papers is a new arc-search technique.

This book has three major parts. The first part, including Chapters 2, 3, and 4, introduces some important algorithms during the development of the interior-point method around the 1990s, most of them being widely known. The main purpose of this part is (a) to explain the dilemma described above by analyzing these algorithms' polynomial bounds and providing the computational experience associated with these algorithms, and (b) summarize three techniques that were proved to enhance the computational efficiency for the interior-point algorithms. These techniques are (1) using higher-order derivative to approximate central path, (2) starting from infeasible initial point, and (3) searching for the optimizer in a wide neighborhood.

The second part, including Chapters 5, 6, 7, and 8, discusses how to solve the dilemma step by step. Chapter 5 considers a higher-order interior-point algorithm that uses arc-search technique and shows that higher-order interior-point algorithm can achieve the same polynomial complexity as the first-order interior-point algorithms if an arc-search technique is used. Chapter 6 goes beyond by considering an algorithm that uses both higher-order derivatives and starts from infeasible initial point and proves that the algorithm still can achieve a good polynomial bound if the

[2]This complexity order will be discussed in Chapters 2, 3, 4, and summarized in the last section of Chapter 4.

arc-search technique is used. To verify that the arc-search strategy can indeed enhance the computational efficiency, Chapter 7 describes an algorithm that is almost identical to the Mehrotra's algorithm, except that one uses arc-search and the other uses line search. These two algorithms are tested against each other using the widely accepted Netlib benchmark problems. To be fair, both algorithms start with the same infeasible initial point, use the same pre-process and post-process method, use the same parameters, and terminate with the same stopping criterion. Extensive numerical result shows that the arc-search algorithm is indeed more efficient and robust than Mehrotra's algorithm. Chapter 8 introduces two new arc-search infeasible interior-point algorithms and their convergence is analyzed. These algorithms have all the features that have been demonstrated to be good for computational efficiency: (a) use second-order derivative, (b) start from an infeasible starting point, and (c) search in the widest neighborhood. Yet, we show that these algorithms are convergent in polynomial time and more robust and efficient than Mehrotra's method. In addition, one algorithm achieves the best polynomial bound for all interior-point algorithms, feasible or infeasible, for the linear programming problem.

Because of the success of the interior-point method in linear programming, there are many extensions of the interior-point method to solve convex quadratic programming (CQP), linear complementarity problem (LCP), semidefinite programming problem (SDP), general convex programming problem (CP), and nonlinear programming problem using interior-point method. Part three of this book, including Chapters 9, 10, 11, and 12, extends arc-search techniques to these problems. Convergence analysis shows that these algorithms are superior in complexity bounds to their counterparts, the interior-point algorithms that use traditional line search. Unlike the interior-point algorithms for linear programming, these arc-search algorithms have not been extensively tested and their computationally superiority to their counterparts has not been demonstrated. The author believes that there is a lot of room for researchers to work on arc-search algorithms in these areas. Part of the purpose of this book is to provide the readers the latest development in this direction so that they will be familiar with this area and will be able to contribute in this direction.

To have a big picture on the optimization research, we provide some fundamentals in the next few sections and point out where the interior-point method may apply.

1.1 About the Notations

Throughout the book, normal letters are used for scalars, bold capital letters are used for matrices, and bold small letters are used for vectors. For real scalar a, $\lceil a \rceil$ is used to represent the smallest integer that is larger than a and $\lfloor a \rfloor$ is used to represent the largest integer that is smaller than a. The set of all real numbers is denoted as $\mathbb{R}$, The set of all vectors in n-dimensional space is denoted as $\mathbb{R}^n$. The set of all matrices in $m \times n$ dimension is denoted as $\mathbb{R}^{m \times n}$. For two vectors $\mathbf{x}$ and $\mathbf{s}$, their Hadamard (element-wise) product is denoted by $\mathbf{x} \circ \mathbf{s}$, the ith component of $\mathbf{x}$ is denoted by x_i, the element-wise inverse of $\mathbf{x}$ is denoted by $\mathbf{x}^{-1}$ if $\min |x_i| > 0$, the element-wise division of the two vectors is denoted by $\mathbf{s}^{-1} \circ \mathbf{x}$, or $\mathbf{x} \circ \mathbf{s}^{-1}$, or $\frac{\mathbf{x}}{\mathbf{s}}$ if $\min |s_i| > 0$, the Euclidean norm of $\mathbf{x}$ is denoted by $\|\mathbf{x}\|$ or explicitly by $\|\mathbf{x}\|_2$, the 1-norm of $\mathbf{x}$ is denoted by $\|\mathbf{x}\|_1$, the ∞-norm of $\mathbf{x}$ is denoted by $\|\mathbf{x}\|_\infty$. For $p \geq 1$, $\|\mathbf{x}\|_1$, $\|\mathbf{x}\|_p$, and $\|\mathbf{x}\|_\infty$ are defined as

$$\|\mathbf{x}\|_1 = \sum_{i=1,\dots,n} |x_i|, \tag{1.1a}$$

$$\|\mathbf{x}\|_p = \left(|x_1|^p + |x_2|^p + \ldots + |x_n|^p\right)^{1/p}, \tag{1.1b}$$

$$\|\mathbf{x}\|_\infty = \max |x_i|. \tag{1.1c}$$

The identity matrix of any dimension is denoted by $\mathbf{I}$, the vector of all ones with appropriate dimension is denoted by $\mathbf{e}$, the transpose of matrix $\mathbf{A}$ is denoted by $\mathbf{A}^{\mathrm{T}}$, a basis for the null space of $\mathbf{A}$ is denoted by $\hat{\mathbf{A}}$, the ith row of $\mathbf{A}$ is denoted by $\mathbf{A}_{i,\cdot}$, the jth column of $\mathbf{A}$ is denoted by $\mathbf{A}_{\cdot,j}$. The matrix norms $\|\mathbf{A}\|_1$, $\|\mathbf{A}\|_2$, and $\|\mathbf{A}\|_\infty$ are all induced norms which are defined as

$$\|\mathbf{A}\|_p = \sup_{\mathbf{x}\neq\mathbf{0}} \frac{\|\mathbf{Ax}\|_p}{\|\mathbf{x}\|_p}, \tag{1.2}$$

where $1 \leq p \leq \infty$. It is straightforward to see from the definition that

$$\|\mathbf{Ax}\|_p \leq \|\mathbf{A}\|_p \|\mathbf{x}\|_p. \tag{1.3}$$

where $\mathbf{A} \in \mathbb{R}^{m\times n}$. For a matrix $\mathbf{H} \in \mathbb{R}^{n\times n}$, $Tr(\mathbf{H}) = \sum_{\mathbf{i=1}}^{\mathbf{n}} \mathbf{H}_{ii}$ is named the trace of $\mathbf{H}$, if $\mathbf{H}$ is positive semidefinite, it is denoted by $\mathbf{H} \succeq \mathbf{0}$, and if $\mathbf{H}$ is positive definite, then it is denoted by $\mathbf{H} \succ \mathbf{0}$. To make the notation simple for block column vectors, $[\mathbf{x}^{\mathrm{T}}, \mathbf{s}^{\mathrm{T}}]^{\mathrm{T}}$ is denoted by $(\mathbf{x}, \mathbf{s})$. For $\mathbf{x} \in \mathbb{R}^n$, a related diagonal matrix is denoted by $\mathbf{X} \in \mathbb{R}^{n\times n}$ whose diagonal elements are components of the vector $\mathbf{x}$. For any vector $\mathbf{x}$ in any iterative algorithm, its initial point is denoted by $\mathbf{x}^0$, the point at the kth iteration is denoted by $\mathbf{x}^k$; for any scalar a in any iterative algorithm, its initial point is denoted by a_0, the point at the kth iteration is denoted by a_k. Finally, the empty set is denoted by $\emptyset$.

1.2 Linear Programming

We consider the linear programming (LP) in the standard form:

$$\min\ \mathbf{c}^{\mathrm{T}}\mathbf{x}, \quad \text{subject to} \quad \mathbf{Ax} = \mathbf{b}, \quad \mathbf{x} \geq \mathbf{0}, \tag{1.4}$$

where $\mathbf{A} \in \mathbb{R}^{m\times n}$, $\mathbf{b} \in \mathbb{R}^m$, $\mathbf{c} \in \mathbb{R}^n$ are given, and $\mathbf{x} \in \mathbb{R}^n$ is the vector to be optimized. The dual programming (DP) of (1.4) is:

$$\max\ \mathbf{b}^{\mathrm{T}}\lambda, \quad \text{subject to} \quad \mathbf{A}^{\mathrm{T}}\lambda + \mathbf{s} = \mathbf{c}, \quad \mathbf{s} \geq \mathbf{0}, \tag{1.5}$$

with dual variable vector $\lambda \in \mathbb{R}^m$, and dual slack vector $\mathbf{s} \in \mathbb{R}^n$.

A different presentation for linear programming is the canonical form which can be presented as follows:

$$\min\ \mathbf{c}^{\mathrm{T}}\mathbf{x}, \quad \text{subject to} \quad \mathbf{Ax} \geq \mathbf{b}, \quad \mathbf{x} \geq \mathbf{0}, \tag{1.6}$$

where $\mathbf{A} \in \mathbb{R}^{m \times d}$, $\mathbf{b} \in \mathbb{R}^m$, $\mathbf{c} \in \mathbb{R}^d$ are given, and $\mathbf{x} \in \mathbb{R}^d$ is the vector to be optimized. The dual of (1.6) is:

$$\max \ \mathbf{b}^{\mathrm{T}}\mathbf{y}, \quad \text{subject to} \ \ \mathbf{A}^{\mathrm{T}}\mathbf{y} \leq \mathbf{c}, \ \ \mathbf{y} \geq \mathbf{0}, \tag{1.7}$$

with dual variable vector $\mathbf{y} \in \mathbb{R}^m$. By introducing slack variables, the canonical form can easily be converted into the standard form. While most computational algorithms are based on standard form, canonical form is convenient for discrete mathematics and computational geometry, such as the study for polytopes, which is closely related to the complexity of the simplex method.

Linear programming has been one of the most extensively studied problems in contemporary mathematics since it was formulated and solved by Dantzig in 1940s [24]. Dantzig's simplex method is not only mathematically elegant but also computationally very efficient in practice. Noticing that the simplex method searches for optimizers from a vertex to vertex along the edges of the polytope, Hirsch conjectured that the upper bound of the diameter of the polytopes, which is also the upper bound for simplex methods to find the optimizer in the worst case[3], is $m-d$ for $m > d \geq 2$ [25]. After a 50-year effort by many researchers, Santos [121] showed that this conjecture is incorrect. Most experts now believe that the best upper bound for simplex method to find the optimizer in the worst case may be bounded by a polynomial. However, Klee and Minty in 1972 [65] showed that Dantzig's pivot rule (that governs how to move iterate from a vertex to the next vertex) needs exponentially many iterations to find the optimizer in the worst case. Different pivot rules, such as Zadeh's rule [169], were proposed with the hope to achieve a polynomial upper bound of iterations in the worst case. Unfortunately, people have shown that all popular pivot rules in simplex method cannot find the optimizer in polynomial iterations in the worst case [108], including Zadeh's rule [37]. Researches realized that solving this dilemma can be a very difficult problem (see [126]).

Another direction is to find other methods that are good in theory, meaning that their worst case computational complexities are bounded above by some polynomial. There was a surprising breakthrough in 1979: Khachiyan proved that the ellipsoid method for linear programming finds the optimizers in polynomial time [59]. However, researchers quickly realized that Khachiyan's ellipsoid algorithm is not computationally competitive in practice [11]. Finding an algorithm polynomial in theory and competitive in practice became important. A different method, the interior-point method, was then developed by Karmarkar [57]. Karmarkar's work inspired many researchers and the interior-point method has achieved tremendous success. However, as it was pointed out early, there was still a major gap that had to be closed up. A major task of the book is to present some recent algorithms that are not only polynomial but also computationally efficient.

We start with some basic and important results. Let $(\mathbf{x}^*, \lambda^*, \mathbf{s}^*)$ be the a primal-dual optimal solution of (1.4) and (1.5), by applying the Karush-Kuhn-Tucker (KKT)

[3]Hirsch's conjecture actually states that the diameter of the polytope described by (1.6) is less than $m-d$.

conditions[4] to LP, we have

$$\mathbf{A}\mathbf{x}^* = \mathbf{b} \tag{1.8a}$$
$$\mathbf{A}^{\mathrm{T}}\boldsymbol{\lambda}^* + \mathbf{s}^* = \mathbf{c} \tag{1.8b}$$
$$x_i^* s_i^* = 0, \quad i = 1, \ldots, n \tag{1.8c}$$
$$(\mathbf{x}^*, \mathbf{s}^*) \geq \mathbf{0}. \tag{1.8d}$$

Since (1.4) is a convex problem , $(\mathbf{x}^*, \boldsymbol{\lambda}^*, \mathbf{s}^*)$ is the optimal solution of (1.4) if and only if all the KKT conditions (1.8) hold. Let P^* be the set of all optimal primal solutions of $\mathbf{x}^*$ and D^* be the set of all optimal dual solutions of $(\boldsymbol{\lambda}^*, \mathbf{s}^*)$. Denote the feasible set $\mathcal{F}$ as a collection of all points that meet the constraints of LP and DP.

$$\mathcal{F} = \{(\mathbf{x}, \boldsymbol{\lambda}, \mathbf{s}) \mid \mathbf{A}\mathbf{x} = \mathbf{b}, \mathbf{A}^{\mathrm{T}}\boldsymbol{\lambda} + \mathbf{s} = \mathbf{c}, (\mathbf{x}, \mathbf{s}) \geq \mathbf{0}\}, \tag{1.9}$$

and the strictly feasible set $\mathcal{F}^o$ as a collection of all points that meet the constraints of LP and DP and are strictly positive

$$\mathcal{F}^o = \{(\mathbf{x}, \boldsymbol{\lambda}, \mathbf{s}) \mid \mathbf{A}\mathbf{x} = \mathbf{b}, \mathbf{A}^{\mathrm{T}}\boldsymbol{\lambda} + \mathbf{s} = \mathbf{c}, (\mathbf{x}, \mathbf{s}) > \mathbf{0}\}. \tag{1.10}$$

Similarly, $\mathbf{x} \geq \mathbf{0}$ is said to be a primal feasible solution if $\mathbf{A}\mathbf{x} = \mathbf{b}$; $\mathbf{x} > \mathbf{0}$ is said to be a strictly primal feasible solution if $\mathbf{A}\mathbf{x} = \mathbf{b}$; $\mathbf{s} \geq \mathbf{0}$ is said to be a dual feasible solution if $\mathbf{A}^{\mathrm{T}}\boldsymbol{\lambda} + \mathbf{s} = \mathbf{c}$ for some $\boldsymbol{\lambda}$; and $\mathbf{s} > \mathbf{0}$ is said to be a strictly dual feasible solution if $\mathbf{A}^{\mathrm{T}}\boldsymbol{\lambda} + \mathbf{s} = \mathbf{c}$ for some $\boldsymbol{\lambda}$. The following result is about the existence and boundedness of the solution set of P^* and D^*.

Theorem 1.1
Suppose that the primal and dual problems are feasible; that is, $\mathcal{F} \neq \emptyset$. If the dual problem has a strictly feasible point, then the primal optimal solution set P^* is not empty and bounded. Similarly, if the primal problem has a strictly feasible point, then the set

$$\{\mathbf{s}^* \mid (\boldsymbol{\lambda}^*, \mathbf{s}^*) \in D^* \textit{ for some } \boldsymbol{\lambda}^* \in \mathbb{R}^m\}$$

is not empty and bounded.

Proof 1.1 The first claim is proved in [144]. The second part is proved as follows. Let $(\bar{\mathbf{s}}, \bar{\boldsymbol{\lambda}})$ with $\bar{\mathbf{s}} \geq \mathbf{0}$ be a feasible dual solution and $\hat{\mathbf{x}} > 0$ be a strictly primal solution. Then,

$$0 \leq \bar{\mathbf{s}}^{\mathrm{T}}\hat{\mathbf{x}} = (\mathbf{c} - \mathbf{A}^{\mathrm{T}}\bar{\boldsymbol{\lambda}})^{\mathrm{T}}\hat{\mathbf{x}} = \mathbf{c}^{\mathrm{T}}\hat{\mathbf{x}} - \bar{\boldsymbol{\lambda}}^{\mathrm{T}}\mathbf{b}. \tag{1.11}$$

Consider the set $\mathcal{T}$ defined by

$$\mathcal{T} = \{\mathbf{s} \mid \mathbf{A}^{\mathrm{T}}\boldsymbol{\lambda} + \mathbf{s} = \mathbf{c}, \quad \mathbf{s} \geq \mathbf{0}, \quad \boldsymbol{\lambda}^{\mathrm{T}}\mathbf{b} \geq \bar{\boldsymbol{\lambda}}^{\mathrm{T}}\mathbf{b}\}. \tag{1.12}$$

[4] KKT conditions are named after Karush, Kuhn, and Tucker. This result will be presented in Section 1.4.

The set $\mathcal{T}$ is not empty because $\bar{\mathbf{s}} \in \mathcal{T}$. $\mathcal{T}$ is also closed by definition. For any $\mathbf{s} \in \mathcal{T}$, using (1.11), we have

$$\sum_{i=1}^{n} \hat{x}_i s_i = \hat{\mathbf{x}}^{\mathrm{T}}\mathbf{s} = \hat{\mathbf{x}}^{\mathrm{T}}(\mathbf{c}-\mathbf{A}^{\mathrm{T}}\lambda) = \mathbf{c}^{\mathrm{T}}\hat{\mathbf{x}} - \lambda^{\mathrm{T}}\mathbf{b} \le \mathbf{c}^{\mathrm{T}}\hat{\mathbf{x}} - \bar{\lambda}^{\mathrm{T}}\mathbf{b} = \bar{\mathbf{s}}^{\mathrm{T}}\hat{\mathbf{x}} \tag{1.13}$$

Therefore,

$$s_i \le \frac{1}{\hat{x}_i}\bar{\mathbf{s}}^{\mathrm{T}}\hat{\mathbf{x}}, \quad i = 1,2,\ldots,n,$$

which means that

$$\|\mathbf{s}\|_\infty \le \big(\max_i \frac{1}{\hat{x}_i}\big)\bar{\mathbf{s}}^{\mathrm{T}}\hat{\mathbf{x}}$$

Since $\mathbf{s}$ is an arbitrary element of $\mathcal{T}$, it can be concluded that $\mathcal{T}$ is bounded. By definition, it is clear that $(\lambda^*, \mathbf{s}^*) \in \mathcal{T}$. Since $\mathcal{T}$ is bounded and closed, the set $\{\mathbf{s}^* \mid (\lambda^*, \mathbf{s}^*) \in D^* \text{ for some } \lambda^* \in \mathbb{R}^m\}$ is not empty and bounded. This completes the proof. ■

Corollary 1.1
If $\mathcal{F}^o \neq \emptyset$, then the set $\{(\mathbf{x}^*, \mathbf{s}^*) \mid x^* \in P^*, (\lambda^*, \mathbf{s}^*) \in D^* \text{ for some } \lambda^* \in \mathbb{R}^m\}$ is bounded.

In view of condition (1.8c), it follows that

$$\mathbf{x}^{*\mathrm{T}}\mathbf{s}^* = 0. \tag{1.14}$$

This condition is called complementarity condition. If all conditions of (1.8) except (1.8c) hold, then we have

$$0 \le \mathbf{x}^{\mathrm{T}}\mathbf{s} = \mathbf{x}^{\mathrm{T}}(\mathbf{c}-\mathbf{A}^{\mathrm{T}}\lambda) = \mathbf{x}^{\mathrm{T}}\mathbf{c} - \mathbf{b}^{\mathrm{T}}\lambda.$$

Hence, $\mathbf{x}^{\mathrm{T}}\mathbf{c} \ge \mathbf{b}^{\mathrm{T}}\lambda$ and $\mathbf{x}^{\mathrm{T}}\mathbf{s}$ is called the duality gap. In the discussion of the rest of the book, along with duality gap, we will use duality measure which is defined as

$$\mu = \frac{\mathbf{x}^{\mathrm{T}}\mathbf{s}}{n} \tag{1.15}$$

for linear programming problem (1.4).

When conditions (1.8a), (1.8b), and (1.8d) hold, and the duality gap is zero, an optimal solution is found. The key idea of "feasible interior-point method" is to maintain the conditions (1.8a), (1.8b), and (1.8d) and reduce the duality gap in all iterations until an optimizer is found. The term "feasible" is used in the method because the conditions (1.8a) and (1.8b) are maintained in all iterations, while the term "interior-point" is used in the method because the strict inequality of (1.8d) is maintained in all iterations. If the conditions of (1.8a) and (1.8b) are not enforced but only strict inequality is maintained in all iterations before the optimal solution is found, then the method is called “infeasible interior-point method”. Both strategies will be discussed in this book.

It will be seen later that the convergence properties of the interior-point algorithms are closely related to a classical result, Goldman-Tucker Theorem. Before we state and provide a proof for the Goldman-Tucker Theorem, we introduce the separating hyperplane theorem and Farkas Lemma.

Theorem 1.2
Let $\mathcal{C}$ and $\mathcal{D}$ be two closed convex sets in $\mathbb{R}^n$ with at least one of them bounded, and assume that $\mathcal{C} \cap \mathcal{D} = \emptyset$. Then, there exist $\mathbf{a} \in \mathbb{R}^n$, and $b \in \mathbb{R}$ such that

$$\mathbf{a}^{\mathrm{T}}\mathbf{x} > b, \quad \forall \mathbf{x} \in \mathcal{D} \quad and \quad \mathbf{a}^{\mathrm{T}}\mathbf{x} < b, \quad \forall \mathbf{x} \in \mathcal{C}. \tag{1.16}$$

Proof 1.2 Define

$$Dist(\mathcal{C},\mathcal{D}) = \inf \|\mathbf{u}-\mathbf{v}\|, \quad \text{s.t.} \quad \mathbf{u} \in \mathcal{C}, \quad ,\mathbf{v} \in \mathcal{D}.$$

The infimum is achievable and is positive. Let $\mathbf{c} \in \mathcal{C}$, $\mathbf{d} \in \mathcal{D}$ be the points that achieve the infimum, and

$$\mathbf{a} = \mathbf{d} - \mathbf{c}, \quad b = \frac{\|\mathbf{d}\|^2 - \|\mathbf{c}\|^2}{2}.$$

Note that $\mathbf{a} \neq \mathbf{0}$ because $\mathcal{C} \cap \mathcal{D} = \emptyset$. We show that the separating hyperplane is the function $f(\mathbf{x}) = \mathbf{a}^{\mathrm{T}}\mathbf{x} - b$, and

$$f(\mathbf{x}) > 0, \quad \forall \mathbf{x} \in \mathcal{D} \quad \text{and} \quad f(\mathbf{x}) < 0, \quad \forall \mathbf{x} \in \mathcal{C} \tag{1.17}$$

Note that

$$f\left(\frac{\mathbf{c}+\mathbf{d}}{2}\right) = (\mathbf{d}-\mathbf{c})^{\mathrm{T}}\frac{\mathbf{c}+\mathbf{d}}{2} - \frac{\|\mathbf{d}\|^2 - \|\mathbf{c}\|^2}{2} = 0.$$

We show only that $f(\mathbf{x}) > 0, \ \ \forall \mathbf{x} \in \mathcal{D}$ because the proof for $f(\mathbf{x}) < 0, \ \ \forall \mathbf{x} \in \mathcal{C}$ is identical. Denoting $g(\mathbf{x}) = \|\mathbf{x}-\mathbf{c}\|^2$, we have $\nabla g(\mathbf{x}) = 2(\mathbf{x}-\mathbf{c})$. Suppose by contradiction, there exists a $\bar{\mathbf{d}} \in \mathbf{D}$ with $f(\bar{\mathbf{d}}) \leq 0$. Then,

$$(\mathbf{d}-\mathbf{c})^{\mathrm{T}}\bar{\mathbf{d}} - \frac{\|\mathbf{d}\|^2 - \|\mathbf{c}\|^2}{2} \leq 0.$$

Using this formula, we have

$$\begin{aligned}
\nabla g(\mathbf{d})^{\mathrm{T}}(\bar{\mathbf{d}}-\mathbf{d}) &= 2(\mathbf{d}-\mathbf{c})^{\mathrm{T}}(\bar{\mathbf{d}}-\mathbf{d}) \\
&= 2(-\|\mathbf{d}\|^2 + (\mathbf{d}-\mathbf{c})^{\mathrm{T}}\bar{\mathbf{d}} + \mathbf{c}^{\mathrm{T}}\mathbf{d}) \\
&\leq 2(-\|\mathbf{d}\|^2 + \tfrac{\|\mathbf{d}\|^2 - \|\mathbf{c}\|^2}{2} + \mathbf{c}^{\mathrm{T}}\mathbf{d}) \\
[\text{Since } \mathbf{d} \neq \mathbf{c}] \quad &= -\|\mathbf{d}-\mathbf{c}\|^2 < 0.
\end{aligned}$$

This means that $\bar{\mathbf{d}} - \mathbf{d}$ is a strictly decent direction for g at $\mathbf{d}$. Hence, there is a $\bar{\alpha} > 0$ such that for all $\alpha \in (0, \bar{\alpha})$,

$$g(\mathbf{d} + \alpha(\bar{\mathbf{d}} - \mathbf{d})) < g(\mathbf{d})$$

i.e.,

$$\|\mathbf{d}+\alpha(\bar{\mathbf{d}}-\mathbf{d})-\mathbf{c}\|^2 < \|\mathbf{d}-\mathbf{c}\|^2.$$

This contradicts to the assumption that **d** is the closest point to **c**. ■

The following corollary will be used in the proof for the Farxas Lemma.

Corollary 1.2
Let $\mathcal{C}$ be a closed convex set and $\mathbf{x} \in \mathbb{R}^n$ be a point not in $\mathcal{C}$. Then, $\mathbf{x}$ and $\mathcal{C}$ can be separated by a hyperplane.

Lemma 1.1
Let $\mathbf{A}$ be a $m \times n$ matrix, and $\mathbf{b} \in \mathbb{R}^m$. Then exactly one of the following statements holds, but not both:

$$\begin{array}{lll} (I) & \exists \mathbf{y} \in \mathbb{R}^m : & \mathbf{y}^{\mathrm{T}}\mathbf{A} \geq \mathbf{0},\ \mathbf{y}^{\mathrm{T}}\mathbf{b} < 0,\ or \\ (II) & \exists \mathbf{x} \in \mathbb{R}^n : & \mathbf{A}\mathbf{x} = \mathbf{b},\ \mathbf{x} \geq \mathbf{0}. \end{array} \tag{1.18}$$

Proof 1.3 We will first show that if (II) is true, then (I) necessarily false. Assume $\mathbf{Ax} = \mathbf{b}$ for some $\mathbf{x} \geq \mathbf{0}$. If $\mathbf{y}^{\mathrm{T}}\mathbf{A} \geq \mathbf{0}$, then for $\mathbf{x} \geq \mathbf{0}$, we have that $\mathbf{y}^{\mathrm{T}}\mathbf{Ax} \geq 0$. Since $\mathbf{Ax} = \mathbf{b}$, this implies that $\mathbf{y}^{\mathrm{T}}\mathbf{b} \geq 0$, and thus, it cannot be that both $\mathbf{y}^{\mathrm{T}}\mathbf{Ax} \geq 0$ and $\mathbf{y}^{\mathrm{T}}\mathbf{b} < 0$.

Now we will prove that if (II) is false then (I) is necessarily true. Define:

$$\mathcal{C} = \{\mathbf{q} \in \mathbb{R}^m \mid \exists \mathbf{x} \geq \mathbf{0}, \mathbf{Ax} = \mathbf{q}\}.$$

Notice that $\mathcal{C}$ is a convex set: for $\mathbf{q}_1, \mathbf{q}_2 \in \mathcal{C}$, there exist $\mathbf{x}_1, \mathbf{x}_2$ such that $\mathbf{q}_1 = \mathbf{Ax}_1$ and $\mathbf{q}_2 = \mathbf{Ax}_2$, and for any $\lambda \in [0,1]$ we have that

$$\lambda\mathbf{q}_1 + (1-\lambda)\mathbf{q}_2 = \lambda\mathbf{Ax}_1 + (1-\lambda)\mathbf{Ax}_2 = \mathbf{A}(\lambda\mathbf{x}_1 + (1-\lambda)\mathbf{x}_2),$$

and hence, $\mathbf{q}_1 + (1-\lambda)\mathbf{q}_2 \in \mathcal{C}$. Since (II) is false $\mathbf{b} \notin \mathcal{C}$. From the separating hyperplane theorem, we know there exists $\mathbf{y} \in \mathbb{R}^m \setminus \{\mathbf{0}\}$ s.t. $\mathbf{y}^{\mathrm{T}}\mathbf{q} \geq 0$ and $\mathbf{y}^{\mathrm{T}}\mathbf{b} < 0$, for all $\mathbf{q} \in \mathcal{C}$. Since $\mathbf{q} = \mathbf{Ax}$ that implies that $\forall \mathbf{x} \geq \mathbf{0}$ we have that $\mathbf{y}^{\mathrm{T}}\mathbf{A} \geq \mathbf{0}$ and $\mathbf{y}^{\mathrm{T}}\mathbf{b} < 0$, as required. ■

A simple extension of this lemma is provided in [144].

Corollary 1.3
For each pair of matrices $\mathbf{G} \in \mathbb{R}^{p\times n}$ and $\mathbf{H} \in \mathbb{R}^{\mathbf{q}\times\mathbf{n}}$ and each vector $\mathbf{d} \in \mathbb{R}^n$, either

$$\begin{array}{lll} (I) & \exists \mathbf{x} \in \mathbb{R}^n : & \mathbf{Gx} \geq \mathbf{0},\ \mathbf{Hx} = \mathbf{0},\ \mathbf{d}^{\mathrm{T}}\mathbf{x} < 0\ or \\ (II) & \exists \mathbf{y} \in \mathbb{R}^p, \mathbf{z} \in \mathbb{R}^q : & \mathbf{G}^{\mathrm{T}}\mathbf{y} + \mathbf{H}^{\mathrm{T}}\mathbf{z} = \mathbf{d},\ \mathbf{y} \geq \mathbf{0}. \end{array} \tag{1.19}$$

Let $(\mathbf{x}^*, \lambda^*, \mathbf{s}^*)$ be any solution of (1.8). Let index sets $\mathcal{B}, \mathcal{N}$ be defined as

$$\mathcal{B} = \{j \in \{1, \ldots, n\} \mid x_j^* \neq 0\}. \tag{1.20}$$

$$\mathcal{N} = \{j \in \{1,\ldots,n\} \mid s_j^* \neq 0\}. \tag{1.21}$$

Goldman-Tucker [40] proved the following result.

Theorem 1.3

There exists at least one primal solution $\mathbf{x}^*$ and one dual solution of $(\lambda^*, \mathbf{s}^*)$, for the corresponding index sets $\mathcal{B}$ and $\mathcal{N}$, the following conditions hold simultaneously: $\mathcal{B} \cup \mathcal{N} = \{1,\ldots,n\}$ and $\mathcal{B} \cap \mathcal{N} = \emptyset$; i.e., $\mathbf{x}^* + \mathbf{s}^* > \mathbf{0}$ and $\mathbf{x}^* \circ \mathbf{s}^* = \mathbf{0}$.

Proof 1.4 It is easy to see that $\mathcal{B} \cap \mathcal{N} = \emptyset$. Otherwise, there is a pair of $x_i^* > 0$ and $s_i^* > 0$ which is contradict to the condition of (1.8c).

Let $\mathcal{T}$ be the set of indices in $\{1,2,\ldots,n\}$ that do not intersect to either $\mathcal{B}$ or $\mathcal{N}$. The proof is to show that $\mathcal{T} = \emptyset$. Since $\mathcal{B} \cap \mathcal{N} = \emptyset$, $\mathcal{B} \cup \mathcal{N} \cup \mathcal{T}$ is a partition of $\{1,2,\ldots,n\}$. Let $\mathbf{A}_{\mathcal{B}}$ and $\mathbf{A}_{\mathcal{T}}$ denotes the submatrices of columns of $\mathbf{A}$ that correspond to $\mathcal{B}$ and $\mathcal{T}$. Select any index $i \in \mathcal{T}$. The goal is show that i must be in either $\mathcal{B}$ or $\mathcal{N}$, depending on whether there exists a vector $\mathbf{w}$ satisfying the following relations:

$$\begin{aligned} \mathbf{A}_{\cdot,i}^{\mathrm{T}}\mathbf{w} &< 0, \\ -\mathbf{A}_{\cdot,j}^{\mathrm{T}}\mathbf{w} &\geq 0, \quad \text{for } j \in \mathcal{T} \setminus \{i\} \\ \mathbf{A}_{\mathcal{B}}^{\mathrm{T}}\mathbf{w} &= 0, \end{aligned} \tag{1.22}$$

where $\mathbf{A}_{\cdot,i}$ denotes the ith column of $\mathbf{A}$. Suppose that $\mathbf{w}$ satisfying (1.22) exists. Let $(\mathbf{x}^*, \lambda^*, \mathbf{s}^*)$ be a primal-dual solution for which $\mathbf{s}_{\mathcal{N}}^* > \mathbf{0}$, define the vector $(\bar{\lambda}, \bar{\mathbf{s}})$ as

$$\bar{\lambda} = \lambda^* + \varepsilon\mathbf{w}, \quad \bar{\mathbf{s}} = \mathbf{c} - \mathbf{A}^{\mathrm{T}}\bar{\lambda} = \mathbf{s}^* - \varepsilon\mathbf{A}^{\mathrm{T}}\mathbf{w},$$

and choose $\varepsilon > 0$ small enough that

$$\begin{aligned} \bar{s}_i = s_i^* - \varepsilon\mathbf{A}_{\cdot,i}^{\mathrm{T}}\mathbf{w} &> 0, \\ \bar{s}_j = s_j^* - \varepsilon\mathbf{A}_{\cdot,j}^{\mathrm{T}}\mathbf{w} \geq 0, \quad \text{for } j &\in \mathcal{T} \setminus \{i\} \\ \bar{\mathbf{s}}_{\mathcal{B}} = \mathbf{s}_{\mathcal{B}}^* &= \mathbf{0}, \\ \bar{\mathbf{s}}_{\mathcal{N}} = \mathbf{s}_{\mathcal{N}}^* - \varepsilon\mathbf{A}_{\mathcal{N}}^{\mathrm{T}}\mathbf{w} &> \mathbf{0}. \end{aligned}$$

It follows from these relations that $(\bar{\lambda}, \bar{\mathbf{s}})$ is feasible for the dual problem. In fact, it is also optimal, since any primal solution vector $\mathbf{x}^*$ must have $x_i^* = 0$ for all $i \notin \mathcal{B}$ and $\bar{\mathbf{s}}^{\mathrm{T}}\mathbf{x}^* = 0$. Therefore, by the definition of $\mathcal{N}$, we must have $i \in \mathcal{N}$.

Suppose, alternatively, that no vector $\mathbf{w}$ satisfies (1.22). Denote $\mathbf{G} = -\mathbf{A}_{\cdot,j}^{\mathrm{T}}$, $j \in \mathcal{T} \setminus \{i\}$, $\mathbf{H} = \mathbf{A}_{\mathrm{B}}^{\mathrm{T}}$, $\mathbf{d} = \mathbf{A}_{\cdot,i}$, and $\mathbf{x} = \mathbf{w}$. In view of Corollary 1.3, we may claim that the following system must have a solution:

$$-\sum_{j \in \mathcal{T}} \mathbf{A}_{\cdot,j} y_j + \mathbf{A}_{\mathcal{B}}\mathbf{z} = \mathbf{A}_{\cdot,i}, \quad y_j \geq 0, \text{ for all } j \in \mathcal{T} \setminus \{i\}. \tag{1.23}$$

Let a vector $\mathbf{v} \in \mathbb{R}^{|\mathcal{T}|}$ be defined as

$$v_i = 1, \quad v_j = y_j, \text{ for all } j \in \mathcal{T} \setminus \{i\},$$

the formula of (1.23) can be rewritten as

$$\mathbf{A}_{\mathcal{T}}\mathbf{v} = \mathbf{A}_{B}\mathbf{z}, \quad \mathbf{v} \geq \mathbf{0}, \quad v_i > 0. \tag{1.24}$$

Let $\mathbf{x}^*$ be a primal solution for which $\mathbf{x}_B^* > 0$ and define $\bar{\mathbf{x}}$ by

$$\bar{\mathbf{x}}_B = \mathbf{x}_B^* - \varepsilon\mathbf{z}, \quad \bar{\mathbf{x}}_{\mathcal{T}} = \varepsilon\mathbf{v}, \quad \bar{\mathbf{x}}_N = \mathbf{0}. \tag{1.25}$$

Using (1.24) and (1.25), we have

$$\begin{aligned} \mathbf{A}\bar{\mathbf{x}} &= \mathbf{A}_B\bar{\mathbf{x}}_B + \mathbf{A}_{\mathcal{T}}\bar{\mathbf{x}}_{\mathcal{T}} + \mathbf{A}_N\bar{\mathbf{x}}_N \\ &= \mathbf{A}_B(\mathbf{x}_B^* - \varepsilon\mathbf{z}) + \mathbf{A}_{\mathcal{T}}\varepsilon\mathbf{v} \\ &= \mathbf{A}_B\mathbf{x}_B^* = \mathbf{b}. \end{aligned} \tag{1.26}$$

Since for ε small enough, $\bar{\mathbf{x}} \geq \mathbf{0}$, $\bar{\mathbf{x}}$ is a feasible solution and, in fact, an optimal solution, because $\bar{\mathbf{x}}_N = \mathbf{0}$. Since $v_i = 1$ and $\bar{x}_i = \varepsilon > 0$, it must have $i \in \mathcal{B}$ by (1.20). Therefore, by the definition of $\mathcal{T}$, it must have $\mathcal{T} = \emptyset$. ∎

An optimal solution with the properties described in the Goldman-Tucker theorem is strictly complementary. The Goldman-Tucker theorem declares that there is always a strictly complementary optimal solution for linear programming, but this is not the case for general optimization problems, even for quadratic programming problem [49].

Before we state an important theorem, we provide a simple lemma here.

Lemma 1.2
For all $\beta > -1$, we have

$$\log(1+\beta) \leq \beta$$

with equality if and only if $\beta = 0$.

Proof 1.5 For $\beta \geq 1$, clearly, the inequality holds. For $-1 < \beta < 1$, it follows from Taylor expansion that

$$\log(1+\beta) = \beta - \frac{\beta^2}{2} + \frac{\beta^3}{3} - \frac{\beta^4}{4} + \ldots \leq \beta$$

with equality if and only if $\beta = 0$. ∎

The following theorem will be used repeatedly to prove the polynomiality for many algorithms in this book.

Theorem 1.4
Let $\varepsilon \in (0,1)$ be given. Suppose that an algorithm for solving (1.8) generates a sequence of iterations that satisfies

$$\mu_{k+1} \leq \left(1 - \frac{\delta}{n^{\omega}}\right)\mu_k, \quad k = 0, 1, 2, \ldots, \tag{1.27}$$

for some positive constants δ and ω. Suppose that the starting point $(\mathbf{x}^0, \lambda^0, \mathbf{s}^0)$ satisfies

$$\mu_0 \leq 1/\varepsilon. \tag{1.28}$$

Then there exists an index K with

$$K = O(n^{\omega} \log(1/\varepsilon))$$

such that

$$\mu_k \leq \varepsilon \quad \text{for} \quad \forall k \geq K.$$

Proof 1.6 By taking logarithms on both sides of (1.27), we have

$$\log(\mu_{k+1}) \leq \log\left(1 - \frac{\delta}{n^{\omega}}\right) + \log(\mu_k).$$

Repeatedly using this formula and using (1.28) yields

$$\log \mu_k \leq k \log\left(1 - \frac{\delta}{n^{\omega}}\right) + \log \mu_0 \leq k \log\left(1 - \frac{\delta}{n^{\omega}}\right) + \log(1/\varepsilon).$$

Using Lemma 1.2 yields

$$\log \mu_k \leq k\left(-\frac{\delta}{n^{\omega}}\right) + \log(1/\varepsilon).$$

Therefore, $\mu_k \leq \varepsilon$ is met if we have

$$k\left(-\frac{\delta}{n^{\omega}}\right) + \log(1/\varepsilon) \leq \log(\varepsilon) = -\log(1/\varepsilon).$$

This inequality holds for all k that satisfy

$$k \geq K = \frac{2}{\delta} n^{\omega} \log \frac{1}{\varepsilon}.$$

This finishes the proof. ∎

1.3 Convex Optimization

Convex optimization was studied at least as early as the 1950s, when Markowitz [86] formulated portfolio selection problem as a convex quadratic optimization problem. In the 1960s, Rosen considered pattern separation by using convex programming [119]. A very nice book on convex analysis was published in the 1970s by Rockafellar [115]. The discipline became popular in the 1990s, when Boyd, EI Ghaoui, Feron, and Balakrishnan indicated [14] that many control system design problems are related to convex optimization problems, Alizadeh, Nesterov and Nemirovski [2, 101] found out that the very successful interior-point method for linear programming can be a powerful tool for convex optimization problems. This book, however, does not discuss general convex optimization problems as some excellent books [15, 100] that cover this topic are already available. This book will discuss some special convex optimization problems, including convex quadratic programming, monotone linear complementarity problem, and positive semidefinite programming, which are closely related to linear programming problem and very efficient algorithms are available.

1.3.1 Convex sets, functions, and optimization

The set $\mathcal{U} \in \mathbb{R}^n$ is convex if

$$\mathbf{u},\mathbf{v} \in \mathcal{U} \Rightarrow t\mathbf{u}+(1-t)\mathbf{v} \in \mathcal{U}, \qquad \text{for all } t \in [0,1]. \tag{1.29}$$

Given a convex set $\mathcal{U}$ and a function f, we say that f is convex if

$$\mathbf{u},\mathbf{v} \in \mathcal{U} \Rightarrow f(\alpha\mathbf{u}+(1-\alpha)\mathbf{v}) \le \alpha f(\mathbf{u})+(1-\alpha)f(\mathbf{v}), \qquad \text{for all } \alpha \in [0,1]. \tag{1.30}$$

We say that f is strict convex if the inequality in (1.30) is strict if $\mathbf{u} \neq \mathbf{v}$ and $\alpha \in (0,1)$. Special examples are (a) linear functions are convex; (b) if $\mathcal{U}$ is an open convex set and $f : \mathcal{U} \to \mathbb{R}$ is twice continuously differentiable on $\mathcal{U}$ with a positive semidefinite Hessian for all $\mathbf{u} \in \mathcal{U}$, then the function f is convex; if the Hessian is positive definite for all $\mathbf{u} \in \mathcal{U}$, then the function f is strictly convex.

In this book, we will consider a linear constrained convex optimization problem:

$$\min_{\mathbf{x}} f(\mathbf{x}) \quad \text{subject to} \quad \mathbf{Gx} = \mathbf{c}, \quad \mathbf{Hx} \ge \mathbf{d}, \quad \mathbf{x} \in \mathcal{U}, \tag{1.31}$$

where f is a smooth convex function, $\mathcal{U} \in \mathbb{R}^n$ is convex set, $\mathbf{G}$ and $\mathbf{H}$ are matrices, and $\mathbf{c}$ and $\mathbf{d}$ are vectors. Denote the convex constraint set

$$\Omega = \{\mathbf{x} \mid \mathbf{Gx} = \mathbf{c}, \quad \mathbf{Hx} \ge \mathbf{d}, \quad \mathbf{x} \in \mathcal{U}\}.$$

A vector $\mathbf{x}^*$ is a local solution of the problem (1.31) if $\mathbf{x}^* \in \Omega$ and there is a neighborhood $\mathcal{N}$ of $\mathbf{x}^*$ such that $f(\mathbf{x}) \ge f(\mathbf{x}^*)$ for $\mathbf{x} \in \mathcal{N} \cap \Omega$. A vector $\mathbf{x}^*$ is a strictly local solution of the problem (1.31) if $\mathbf{x}^* \in \Omega$ and there is a neighborhood $\mathcal{N}$ of $\mathbf{x}^*$ such that $f(\mathbf{x}) > f(\mathbf{x}^*)$ for $\mathbf{x} \in \mathcal{N} \cap \Omega$ with $\mathbf{x} \neq \mathbf{x}^*$. A vector $\mathbf{x}^*$ is a global solution of the problem (1.31) if $\mathbf{x}^* \in \Omega$ such that $f(\mathbf{x}) \ge f(\mathbf{x}^*)$. A vector $\mathbf{x}^*$ is a strictly global solution of the problem (1.31) if $\mathbf{x}^* \in \Omega$ such that $f(\mathbf{x}) > f(\mathbf{x}^*)$ for $\mathbf{x} \in \Omega$ with $\mathbf{x} \neq \mathbf{x}^*$. Applying the KKT conditions (to be discussed at the end of the chapter) to the problem (1.31) gives the necessary conditions for the optimal solution(s) $\mathbf{x}^*$ of the problem (1.31):

$$\nabla f(\mathbf{x}^*) - \mathbf{G}^{\mathrm{T}}\mathbf{y} - \mathbf{H}^{\mathrm{T}}\mathbf{z} = \mathbf{0}, \tag{1.32a}$$

$$\mathbf{Gx}^* = \mathbf{c}, \tag{1.32b}$$

$$\mathbf{Hx}^* \ge \mathbf{d}, \tag{1.32c}$$

$$\mathbf{z} \ge \mathbf{0}, \tag{1.32d}$$

$$\mathbf{z}^{\mathrm{T}}(\mathbf{Hx}^* - \mathbf{d}) = 0. \tag{1.32e}$$

One of the main results for the convex optimization problem (1.31) is the following theorem:

Theorem 1.5
For the convex programming problem (1.31), the KKT conditions (1.32) are sufficient for $\mathbf{x}^*$ to be a global solution. That is, if there are vectors $\mathbf{y}$ and $\mathbf{z}$ such that (1.32) hold, then $\mathbf{x}^*$ is a global solution of (1.31). If, in addition, the function f is strictly convex on its convex region, then any local solution is a uniquely defined global solution.

Proof 1.7 We prove the first claim and leave the second claims for readers. Given $\mathbf{x}^*$ together with vectors $\mathbf{y}$ and $\mathbf{z}$ that satisfy the KKT conditions, we show that

$$f(\mathbf{x}^*+\mathbf{v}) \ge f(\mathbf{x}^*) \tag{1.33}$$

for all $\mathbf{v}$ such that $\mathbf{x}^*+\mathbf{v}$ is feasible. Because f is convex, we have for any vector $\mathbf{v}$ with $\mathbf{x}^*+\mathbf{v}\in\mathcal{U}$ that

$$f(\mathbf{x}^*+\alpha\mathbf{v}) = f((1-\alpha)\mathbf{x}^*+\alpha(\mathbf{x}^*+\mathbf{v})) \le (1-\alpha)f(\mathbf{x}^*)+\alpha f(\mathbf{x}^*+\mathbf{v}),$$

which is equivalent to

$$\frac{1}{\alpha}[f(\mathbf{x}^*+\alpha\mathbf{v})-f(\mathbf{x}^*)] \le f(\mathbf{x}^*+\mathbf{v})-f(\mathbf{x}^*).$$

By taking the limit as $\alpha\to 0^+$ and using the differentiability of f, we have

$$f(\mathbf{x}^*+\mathbf{v}) \ge f(\mathbf{x}^*)+\mathbf{v}^{\mathrm{T}}\nabla f(\mathbf{x}^*). \tag{1.34}$$

Since the feasible set is convex, we can express any feasible set as $\mathbf{x}^*+\mathbf{v}$. Since $\mathbf{G}\mathbf{x}^*=\mathbf{c}$ and $\mathbf{G}(\mathbf{x}^*+\mathbf{v})=\mathbf{c}$, the vector $\mathbf{v}$ must satisfy $\mathbf{G}\mathbf{v}=\mathbf{0}$. Consider the active components of the inequality constraints $\mathbf{H}\mathbf{x}^*=\mathbf{d}$, that is, the index i for which $\mathbf{H}_{i,\cdot}\mathbf{x}^*=\mathbf{d_i}$, because $\mathbf{x}^*+\mathbf{v}$ is feasible, we must have $\mathbf{H}_{i,\cdot}\mathbf{v}\ge\mathbf{0}$. For the inactive components, the rows i for which $\mathbf{H}_{i,\cdot}\mathbf{x}^*>\mathbf{d_i}$, we have from the conditions (1.32d) and (1.32e) that $z_i=0$. Inserting the vector $\mathbf{v}$ into the formula (1.34), from (1.32a), we obtain

$$\begin{aligned}
\mathbf{v}^{\mathrm{T}}\nabla f(\mathbf{x}^*) &= \mathbf{v}^{\mathrm{T}}(\mathbf{G}^{\mathrm{T}}\mathbf{y}+\mathbf{H}^{\mathrm{T}}\mathbf{z}) \\
&= \mathbf{y}^{\mathrm{T}}\mathbf{G}\mathbf{v}+\sum_{i\text{ active}} z_i\mathbf{H}_{i,\cdot}\mathbf{v}+\sum_{i\text{ inactive}} z_i\mathbf{H}_{i,\cdot}\mathbf{v}.
\end{aligned}$$

Since $\mathbf{G}\mathbf{v}=\mathbf{0}$ and $z_i=0$ for i inactive components, the first and the third terms on the right-hand side are zeros. The middle term is non-negative, since z_i and $\mathbf{H}_{i,\cdot}\mathbf{v}$ are both non-negative when i is an active index. Hence, $\mathbf{v}^{\mathrm{T}}\nabla f(\mathbf{x}^*)\ge 0$, we have from (1.34) that (1.33) holds. This shows that $\mathbf{x}^*$ is a global solution, as claimed. ∎

1.3.2 Convex quadratic programming

In Chapters 9 and 10, we will consider the convex quadratic programming (QP) in the standard form:

$$\begin{aligned} (QP) \qquad & \min \qquad \frac{1}{2}\mathbf{x}^{\mathrm{T}}\mathbf{H}\mathbf{x}+\mathbf{c}^{\mathrm{T}}\mathbf{x}, \\ & \text{subject to} \qquad \mathbf{A}\mathbf{x}=\mathbf{b}, \quad \mathbf{x} \geq \mathbf{0}, \end{aligned} \tag{1.35}$$

where $\mathbf{0} \preceq \mathbf{H} \in \mathbb{R}^{n\times n}$ is a positive semidefinite matrix, $\mathbf{A} \in \mathbb{R}^{m\times n}$, $\mathbf{b} \in \mathbb{R}^m$, $\mathbf{c} \in \mathbb{R}^n$ are given, and $\mathbf{x} \in \mathbb{R}^n$ is the vector to be optimized. Associated with the quadratic programming is the dual programming (DQP) that is also presented in the standard form:

$$\begin{aligned} (DQP) \qquad & \max \qquad -\frac{1}{2}\mathbf{x}^{\mathrm{T}}\mathbf{H}\mathbf{x}+\mathbf{b}^{\mathrm{T}}\lambda, \\ & \text{subject to} \quad -\mathbf{H}\mathbf{x}+\mathbf{A}^{\mathrm{T}}\lambda+\mathbf{s}=\mathbf{c}, \quad \mathbf{s} \geq \mathbf{0}, \quad \mathbf{x} \geq \mathbf{0}, \end{aligned} \tag{1.36}$$

where $\lambda \in \mathbb{R}^m$ is the dual variable vector, and $\mathbf{s} \in \mathbb{R}^n$ is the dual slack vector.

It is well-known that this problem can be solved by simplex method [143] and other alternatives [106]. However, in this book, we will apply the arc-search interior-point method to this problem and show that arc-search interior-point algorithms can achieve the best polynomial bound as the interior-point algorithms can achieve for linear program problem.

Again, the interior-point algorithms for quadratic programming are based on the KKT conditions (1.49) applied to the QP, which is given as follows:

$$\mathbf{A}\mathbf{x}=\mathbf{b} \tag{1.37a}$$

$$\mathbf{A}^{\mathrm{T}}\lambda+\mathbf{s}-\mathbf{H}\mathbf{x}=\mathbf{c} \tag{1.37b}$$

$$x_i s_i=0, \quad i=1,\ldots,n \tag{1.37c}$$

$$(\mathbf{x},\mathbf{s}) \geq \mathbf{0}. \tag{1.37d}$$

It is worthwhile to point out that the linear programming is a special case of the convex quadratic programming. Indeed, setting $\mathbf{H}=\mathbf{0}$ in (1.35) and (1.36), the QP and its dual DQP reduce to LP and its dual DP. In addition, setting $\mathbf{H}=\mathbf{0}$ in (1.37), the KKT conditions for QP problem reduce to the KKT conditions for LP problem.

1.3.3 Monotone and mixed linear complementarity problem

The monotone linear complementarity problem (LCP) [21] is to solve the following systems:

$$\mathbf{s}=\mathbf{M}\mathbf{x}+\mathbf{q}, \quad (\mathbf{x},\mathbf{s}) \geq \mathbf{0}, \quad \mathbf{x}^{\mathrm{T}}\mathbf{s}=0, \tag{1.38}$$

where $\mathbf{0} \preceq \mathbf{M} \in \mathbb{R}^{n\times n}$, $\mathbf{q} \in \mathbb{R}^n$ are given, $\mathbf{x} \in \mathbb{R}^n$ and $\mathbf{s} \in \mathbb{R}^n$ are variables to be determined.

A more general LCP is the mixed LCP [69] which is given as

$$\begin{bmatrix} \mathbf{s} \\ \mathbf{0} \end{bmatrix} = \begin{bmatrix} \mathbf{M}_{11} & \mathbf{M}_{12} \\ \mathbf{M}_{21} & \mathbf{M}_{22} \end{bmatrix} \begin{bmatrix} \mathbf{x} \\ \lambda \end{bmatrix} + \begin{bmatrix} \mathbf{q}_1 \\ \mathbf{q}_2 \end{bmatrix}, \quad (\mathbf{x},\mathbf{s}) \geq \mathbf{0}, \quad \mathbf{x}^{\mathrm{T}}\mathbf{s}=0, \tag{1.39}$$

where $\mathbf{x},\mathbf{s} \in \mathbb{R}^n$ and $\lambda \in \mathbb{R}^m$ are vectors to determined. It is easy to see that the monotone LCP is a special case of mixed LCP.

From the KKT conditions (1.37) for QP, one can rewrite it as follows:

$$\begin{bmatrix} \mathbf{s} \\ \mathbf{0} \end{bmatrix} = \begin{bmatrix} \mathbf{H} & -\mathbf{A}^{\mathrm{T}} \\ \mathbf{A} & \mathbf{0} \end{bmatrix} \begin{bmatrix} \mathbf{x} \\ \lambda \end{bmatrix} + \begin{bmatrix} \mathbf{c} \\ -\mathbf{b} \end{bmatrix}, \quad (\mathbf{x},\mathbf{s}) \geq \mathbf{0}, \quad \mathbf{x}^{\mathrm{T}}\mathbf{s} = 0, \tag{1.40}$$

which is the mixed LCP problem . Therefore, convex quadratic programming is also a special case of mixed LCP problem.

1.3.4 Positive semidefinite programming

Another special case of convex programming is the positive semidefinite programming (SDP), which has many interesting applications to physical problems, to control system design problems, and to other areas of mathematical programming [3, 15, 107].

Let $\mathcal{S}^n$ be the set of real symmetric $n \times n$ matrices. For $\mathbf{G} \in \mathcal{S}^n$ and $\mathbf{H} \in \mathcal{S}^n$, their inner product on $\mathcal{S}^n$ is defined by

$$\mathbf{G} \bullet \mathbf{H} = Tr(\mathbf{G}^{\mathrm{T}}\mathbf{H}), \tag{1.41}$$

where $Tr(\mathbf{M})$ is the trace of the matrix $\mathbf{M}$. Using this notation, we can define the SDP as follows:

$$\min_{\mathbf{X} \in \mathcal{S}^n} \mathbf{C} \bullet \mathbf{X}, \quad \text{s.t.} \quad \mathbf{A}_i \bullet \mathbf{X} = b_i, \quad i = 1, \ldots, m, \quad \mathbf{X} \succeq \mathbf{0}, \tag{1.42}$$

where $\mathbf{C} \in \mathcal{S}^n$, $\mathbf{A}_i \in \mathcal{S}^n$ for $k = 1, \ldots, m$, and $\mathbf{b} = (b_1, \ldots, b_m) \in \mathbb{R}^n$ are given, and $\mathbf{0} \preceq \mathbf{X} \in \mathcal{S}^n$ is the primal variables to be optimized. Due to the symmetric restriction, there are $n(n+1)/2$ variables (not n^2 variables) in $\mathbf{X}$. The dual problem of (1.42) is given by

$$\min_{\mathbf{X} \in \mathcal{S}^n} \mathbf{b}^{\mathrm{T}}\mathbf{y}, \quad \text{s.t.} \quad \sum_{i=1}^{m} y_i \mathbf{A}_i + \mathbf{S} = \mathbf{C}, \quad \mathbf{S} \succeq \mathbf{0}, \tag{1.43}$$

where $\mathbf{y} = [y_1, \ldots, y_m]^{\mathrm{T}} \in \mathbb{R}^m$ and $\mathbf{S} \in \mathcal{S}^n$ are the dual variables. Again, there are $n(n+1)/2$ variables in $\mathbf{S}$ because of the symmetric restriction. Without loss of generality, we assume that $\mathbf{A}_i$ is linear independent.

Applying the KKT conditions (1.49) to (1.42) gives the KKT conditions for the positive semidefinite problem:

$$\mathbf{A}_i \bullet \mathbf{X} = Tr(\mathbf{A}_i\mathbf{X}) = b_i, \quad i = 1, \ldots, m, \quad \mathbf{X} \succeq \mathbf{0} \tag{1.44a}$$

$$\sum_{i=1}^{m} y_i \mathbf{A}_i + \mathbf{S} = \mathbf{C}, \quad \mathbf{S} \succeq \mathbf{0}, \tag{1.44b}$$

$$\mathbf{X}\mathbf{S} = \mathbf{0}. \tag{1.44c}$$

The KKT system has $n(n+1)+m$ equations and exact $n(n+1)+m$ variables.

1.4 Nonlinear Programming

The first order optimality conditions are probably the most important result for the general constrained optimization problems. These conditions are applicable to many optimization problems, such as a linear optimization problem which has linear objective function and linear constraints, a convex quadratic optimization problem which has a convex quadratic objective function and linear constraints, and a general nonlinear optimization problem which has general nonlinear objective function and nonlinear constraints. Although the first order optimality conditions for the general constrained optimization problems are necessary conditions, these conditions are necessary and sufficient conditions for a linear optimization problem, a convex quadratic optimization problem, and other convex optimization problems which are considered extensively in this book.

1.4.1 *Problem description*

Consider the general optimization problem:

$$\begin{array}{ll} \min_{\mathbf{x}\in\mathbb{R}^n} & f(\mathbf{x}) \\ \text{subject to} & c_i(\mathbf{x}) = 0, \quad i \in \mathcal{E} \\ & c_i(\mathbf{x}) \geq 0, \quad i \in \mathcal{I} \end{array}$$

where f is the objective function and c_i are the constraint functions; these functions are all smooth, real-valued on a subset of $\mathbb{R}^n$, and $\mathcal{E}$ and $\mathcal{I}$ are two finite sets of indices for equality constraints and inequality constraints, respectively. The feasible set Ω is defined as the set of all points $\mathbf{x}$ that satisfy all the constraints; i.e.,

$$\Omega = \{\mathbf{x} \mid c_i(\mathbf{x}) = 0, \quad i \in \mathcal{E}; \quad c_i(\mathbf{x}) \geq 0, \quad i \in \mathcal{I}\} \tag{1.45}$$

so that one can rewrite (1.45) as

$$\min_{\Omega} \; f(\mathbf{x}). \tag{1.46}$$

A vector $\mathbf{x}^*$ is a local solution of the problem (1.45) if $\mathbf{x}^* \in \Omega$ and there is a neighborhood $\mathcal{N}$ of $\mathbf{x}^*$ such that $f(\mathbf{x}) \geq f(\mathbf{x}^*)$ for $\mathbf{x} \in \mathcal{N} \cap \Omega$. A vector $\mathbf{x}^*$ is a strictly local solution of the problem (1.45) if $\mathbf{x}^* \in \Omega$ and there is a neighborhood $\mathcal{N}$ of $\mathbf{x}^*$ such that $f(\mathbf{x}) > f(\mathbf{x}^*)$ for $\mathbf{x} \in \mathcal{N} \cap \Omega$ with $\mathbf{x} \neq \mathbf{x}^*$. A vector $\mathbf{x}^*$ is a global solution of the problem (1.45) if $\mathbf{x}^* \in \Omega$ such that $f(\mathbf{x}) \geq f(\mathbf{x}^*)$. A vector $\mathbf{x}^*$ is a strictly global solution of the problem (1.45) if $\mathbf{x}^* \in \Omega$ such that $f(\mathbf{x}) > f(\mathbf{x}^*)$ for $\mathbf{x} \in \Omega$ with $\mathbf{x} \neq \mathbf{x}^*$.

1.4.2 *Karush-Kuhn-Tucker conditions*

To state the first order optimality conditions, we introduce the Lagrangian function for the constrained optimization problem (1.45) which is defined as

$$\mathcal{L}(\mathbf{x}, \lambda) = f(\mathbf{x}) - \sum_{i \in \mathcal{E} \cup \mathcal{I}} \lambda_i c_i(\mathbf{x}). \tag{1.47}$$

The active set at any feasible $\mathbf{x}$ is the union of the set $\mathcal{E}$ and a subset of $\mathcal{I}$ with the indices of the active inequality constraints given by

$$\mathcal{A}(\mathbf{x}) = \mathcal{E} \cup \{c_i(\mathbf{x}) = 0 \mid i \in \mathcal{I}\}. \tag{1.48}$$

The first order optimality conditions are directly related to the linearly independent constraint qualification (LICQ), which is defined as follows:

Definition 1.1 Given the point $\mathbf{x}^*$ and the active set $\mathcal{A}(\mathbf{x}^*)$ defined by (1.48), the linear independent constraint qualification is said to be held if the set of active constraint gradients $\{\nabla c_i(\mathbf{x}^*), i \in \mathcal{A}(\mathbf{x}^*)\}$ is linearly independent.

Note, that if this condition holds, none of the active constraint gradients can be zero. Now we are ready to state the first-order necessary conditions.

Theorem 1.6
Suppose that $\mathbf{x}^*$ is a local solution of (1.45) and that the LICQ holds at $\mathbf{x}^*$. Then, there is a Lagrange multiplier vector λ^*, with components $\lambda_i^*, i \in \mathcal{E} \cup \mathcal{I}$ such that the following conditions are satisfied at $(\mathbf{x}^*, \lambda^*)$

$$\nabla_x \mathcal{L}(\mathbf{x}^*, \lambda^*) = \mathbf{0}, \tag{1.49a}$$
$$c_i(\mathbf{x}^*) = 0, \quad \forall i \in \mathcal{E}, \tag{1.49b}$$
$$c_i(\mathbf{x}^*) \geq 0, \quad \forall i \in \mathcal{I}, \tag{1.49c}$$
$$\lambda_i^* \geq 0, \quad \forall i \in \mathcal{I}, \tag{1.49d}$$
$$\lambda_i^* c_i(\mathbf{x}^*) = 0,, \quad \forall i \in \mathcal{E} \cup \mathcal{I}. \tag{1.49e}$$

The proof of Theorem 1.6 is very technical, therefore, is omitted. Readers who are interested in the proof are referred to [106]. The conditions of (1.49) are widely known as the Karush-Kuhn-Tucker conditions or KKT conditions for short. The KKT conditions were first proved by Karush in his master thesis in 1939 [58] and rediscovered by Kuhn and Tucker in 1951 [75]. A special solution is important and deserves its own definition:

Definition 1.2 Given a local solution $\mathbf{x}^*$ of (1.45) and a vector λ^* satisfying (1.49), we say that the solution is strictly complementary if exactly one of λ_i^* and $c_i(\mathbf{x}^*)$ is zero for each index $i \in \mathcal{I}$. In other words, $\lambda_i^* > 0$ for each $i \in \mathcal{I} \cap \mathcal{A}(\mathbf{x}^*)$.

Chapter 2

A Potential-Reduction Algorithm for LP

Potential-reduction method is closely related to the logarithmic barrier function, which is usually attributed to Frisch [38]. Logarithmic barrier function was extensively studied by Fiacco and McCormick in their book [36] for the nonlinear programming. Since the calculation of logarithmic barrier function becomes increasingly difficulty as the barrier parameter approaches to zero, the method of using logarithmic barrier function was not widely accepted. However, after Karmarkar announced his first polynomial interior-point algorithm for linear programming [57] which uses a logarithmic barrier function, research of interior-point methods using some logarithmic barrier functions became popular. Two classes of potential-reduction methods, primal potential-reduction method and primal-dual potential-reduction method, were proposed. Since primal-dual potential functions have a more balanced consideration on both primal and dual variables, they are now considered to be better than their counterpart. Therefore, this chapter discusses only a potential-reduction method that uses a logarithmic barrier function involving both primal and dual variables [130, 135, 72, 163].

2.1 Preliminaries

One of the main ideas of an interior-point algorithm is to search for the optimizer from inside of $\mathcal{F}^0$, which reduces the duality gap in every iteration. For this reason, we need to know if the iterate stays in a bounded set.

Theorem 2.1
Suppose that $\mathcal{F}^0 \neq \emptyset$. Then, for each $K \geq 0$, the set

$$\{(\mathbf{x},\mathbf{s}) \mid (\mathbf{x},\boldsymbol{\lambda},\mathbf{s}) \in \mathcal{F} \text{ for some } \boldsymbol{\lambda}, \text{ and } \mathbf{x}^{\mathrm{T}}\mathbf{s} \leq K\} \tag{2.1}$$

is bounded.

Proof 2.1 Let $(\bar{\mathbf{x}},\bar{\boldsymbol{\lambda}},\bar{\mathbf{s}})$ be any vector in $\mathcal{F}^o$ and $(\mathbf{x},\boldsymbol{\lambda},\mathbf{s})$ be any vector in $\mathcal{F}$ with $\mathbf{x}^{\mathrm{T}}\mathbf{s} \leq K$. Since $\mathbf{A}(\bar{\mathbf{x}}-\mathbf{x}) = \mathbf{0}$, and $\mathbf{A}^{\mathrm{T}}(\bar{\boldsymbol{\lambda}}-\boldsymbol{\lambda})+(\bar{\mathbf{s}}-\mathbf{s}) = \mathbf{0}$, we have

$$(\bar{\mathbf{x}}-\mathbf{x})^{\mathrm{T}}(\bar{\mathbf{s}}-\mathbf{s}) = -(\bar{\mathbf{x}}-\mathbf{x})^{\mathrm{T}}\mathbf{A}^{\mathrm{T}}(\bar{\boldsymbol{\lambda}}-\boldsymbol{\lambda}) = 0.$$

Using $\mathbf{x}^{\mathrm{T}}\mathbf{s} \leq K$ yields

$$\bar{\mathbf{x}}^{\mathrm{T}}\mathbf{s}+\bar{\mathbf{s}}^{\mathrm{T}}\mathbf{x} \leq K+\bar{\mathbf{x}}^{\mathrm{T}}\bar{\mathbf{s}}. \tag{2.2}$$

Since $(\bar{\mathbf{x}},\bar{\mathbf{s}}) > \mathbf{0}$, denoting $c = \min_{i=1,2,\ldots,n}\min(\bar{x}_i,\bar{s}_i) > 0$ and using (2.2), we have

$$c\mathbf{e}^{\mathrm{T}}(\mathbf{x}+\mathbf{s}) \leq K+\bar{\mathbf{x}}^{\mathrm{T}}\bar{\mathbf{s}}.$$

This implies that for $i = 1,2,\ldots,n$,

$$\begin{aligned} 0 &\leq x_i \leq \frac{1}{c}(K+\bar{\mathbf{x}}^{\mathrm{T}}\bar{\mathbf{s}}), \\ 0 &\leq s_i \leq \frac{1}{c}(K+\bar{\mathbf{x}}^{\mathrm{T}}\bar{\mathbf{s}}). \end{aligned} \tag{2.3}$$

This completes the proof. ∎

When the strict feasible set of the linear programming is not empty, we know that the level set defined by (2.1) is bounded. Starting from an interior-point, the best way to search for the next iterate is to follow a curve named central path [88] because this strategy avoids the numerical difficulty when the iterate moves too close to the boundary before an optimizer is approached. The central path C is an arc of the points $(\mathbf{x}(t),\boldsymbol{\lambda}(t),\mathbf{s}(t)) \in \mathcal{F}^o$ parametrized by a positive scalar t. All points on C satisfy the perturbed KKT system for some $t > 0$:

$$\mathbf{A}\mathbf{x} = \mathbf{b}, \tag{2.4a}$$

$$\mathbf{A}^{\mathrm{T}}\mathbf{y}+\mathbf{s} = \mathbf{c}, \tag{2.4b}$$

$$\mathbf{x}\circ\mathbf{s} = t\mathbf{e}, \tag{2.4c}$$

$$(\mathbf{x},\mathbf{s}) > \mathbf{0}. \tag{2.4d}$$

Remark 2.1 The important feature of the points on central path C is that the pair-wise product $x_i s_i$ are identical for all i. ∎

A linear system of equations related to the perturbed KKT system is as follows:

$$\begin{bmatrix} \mathbf{A} & \mathbf{0} & \mathbf{0} \\ \mathbf{0} & \mathbf{A}^{\mathrm{T}} & \mathbf{I} \\ \mathbf{S} & \mathbf{0} & \mathbf{X} \end{bmatrix} \begin{bmatrix} \Delta\mathbf{x}^k \\ \Delta\boldsymbol{\lambda}^k \\ \Delta\mathbf{s}^k \end{bmatrix} = \begin{bmatrix} \mathbf{0} \\ \mathbf{0} \\ \sigma_k\mu_k\mathbf{e}-\mathbf{x}^k\circ\mathbf{s}^k \end{bmatrix}, \tag{2.5}$$

where σ_k is the centering parameter[1] and μ_k is the duality measure. Quite a few algorithms are related to the solution of this linear system of equations.

2.2 A Potential-Reduction Algorithm

A popular primal and dual potential function is defined by

$$f_\rho(\mathbf{x},\mathbf{s}) = \rho\log(\mathbf{x}^{\mathrm{T}}\mathbf{s}) - \sum_{i=1}^{n}\log(x_i s_i) \tag{2.6}$$

where $\rho > n$ is a weight parameter. It is easy to see that the first part is related to the duality gap. When an optimizer is found, this term goes to zero. The second term is a barrier which prevents the iterate from going to zero too fast before an optimizer is approached. The second term is also related to centrality, which can easily be seen if we rewrite $f_\rho(\mathbf{x},\mathbf{s})$ as follows:

$$\begin{aligned} f_\rho(\mathbf{x},\mathbf{s}) &= \rho\log(\mathbf{x}^{\mathrm{T}}\mathbf{s}) - \sum_{i=1}^{n}\log(x_i s_i) \\ &= (\rho-n)\log(\mathbf{x}^{\mathrm{T}}\mathbf{s}) - \sum_{i=1}^{n}\log\frac{x_i s_i}{\mathbf{x}^{\mathrm{T}}\mathbf{s}/n} + n\log n. \end{aligned} \tag{2.7}$$

Notice that the first term in (2.7) is the objective with the weight of $(\rho - n)$. The second term is the penalty for $x_i s_i$ deviating from the duality measure $\mathbf{x}^{\mathrm{T}}\mathbf{s}/n$ (better if no $x_i s_i$ is much smaller than the average of $\mu = \mathbf{x}^{\mathrm{T}}\mathbf{s}/n$). If $\mathbf{x}^{\mathrm{T}}\mathbf{s} > 0$ but $x_i s_i = 0$, then $f_\rho(\mathbf{x},\mathbf{s})$ will blow up. The following lemma shows that the summation of the second and third terms has a low bound of $n\log n$. Therefore, $f_\rho(\mathbf{x},\mathbf{s}) \to -\infty$ only if the first term goes to $-\infty$, which means that duality gap is reduced to zero.

Lemma 2.1
If the vector pair $(\mathbf{x},\mathbf{s}) > \mathbf{0}$, then we have

$$-\sum_{i=1}^{n}\log\frac{x_i s_i}{\mathbf{x}^{\mathrm{T}}\mathbf{s}/n} \geq 0. \tag{2.8}$$

The equality holds if and only if $\frac{s_i x_i}{\mu} = 1$.

Proof 2.2 Let $\beta = \frac{s_i x_i}{\mu} - 1$ or equivalently $1+\beta = \frac{s_i x_i}{\mu}$. In view of Lemma 1.2, we have

$$-\sum_{i=1}^{n}\log\frac{x_i s_i}{\mathbf{x}^{\mathrm{T}}\mathbf{s}/n} \geq -\sum_{i=1}^{n}\left(\frac{s_i x_i}{\mu} - 1\right) = -(n-n) = 0. \tag{2.9}$$

[1]The role of the centering parameter has been fully discussed in [144].

The equality holds if and only if $0 = \beta = \frac{s_i x_i}{\mu} - 1$, i.e., $\frac{s_i x_i}{\mu} = 1$. ■

This lemma is useful to derive an important relation between $f_\rho(\mathbf{x}, \mathbf{s})$ and duality measure $\mu = \mathbf{x}^{\mathrm{T}}\mathbf{s}/n$.

Lemma 2.2
For any $(\mathbf{x}, \lambda, \mathbf{s}) \in \mathcal{F}^o$, it must have

$$\mu \le exp\left(\frac{f_\rho(\mathbf{x}, \mathbf{s})}{\rho - n}\right). \tag{2.10}$$

Proof 2.3 By the definition of $f_\rho(\mathbf{x}, \mathbf{s})$, we have

$$\begin{aligned} f_\rho(\mathbf{x}, \mathbf{s}) &= (\rho - n)\log(\mathbf{x}^{\mathrm{T}}\mathbf{s}) - \sum_{i=1}^{n} \log \frac{x_i s_i}{\mathbf{x}^{\mathrm{T}}\mathbf{s}/n} + n\log n \\ \text{use Lemma 2.1} \quad &\ge (\rho - n)\log\mu + (\rho - n)\log n + n\log n \\ &\ge (\rho - n)\log\mu. \end{aligned} \tag{2.11}$$

Inequality (2.10) follows immediately from the above inequality. ■

Lemma 2.2 indicates that if $f_\rho(\mathbf{x}, \mathbf{s}) \to -\infty$, then $\mu \to 0$. For $(\mathbf{x}, \lambda, \mathbf{s}) \in \mathcal{F}$, when $\mu \to 0$, in view of KKT conditions (1.8) (which are necessary and sufficient for linear programming), an optimal solution of the linear programming problem (1.4) is found. Now we are ready to present the potential-reduction algorithm.

Algorithm 2.1
Data: $\varepsilon > 0$, $\rho > n$, $\sigma = n/\rho$, initial point $(\mathbf{x}^0, \lambda^0, \mathbf{s}^0) \in \mathcal{F}^o$, and $\mu_0 = \frac{\mathbf{x}^{0^{\mathrm{T}}}\mathbf{s}^0}{n}$.
for *iteration $k = 0, 1, 2, \ldots$*

Step 1: If $\mu_k < \varepsilon$ and $(\mathbf{x}^k, \mathbf{s}^k) > \mathbf{0}$ hold, stop. Otherwise continue.

Step 2: Solve the linear system of equations (2.5) for $(\Delta\mathbf{x}^k, \Delta\lambda^k, \Delta\mathbf{s}^k)$.

Step 3: Calculate

$$\alpha_{\max} = \sup\{\alpha \in [0, 1] \mid (\mathbf{x}, \mathbf{s}) + \alpha(\Delta\mathbf{x}^k, \Delta\mathbf{s}^k) \ge 0\}$$

Step 4: Select α_k such that

$$\alpha_k = \arg \min_{\alpha \in (0, \alpha_{\max})} f_\rho(\mathbf{x}^k + \alpha\Delta\mathbf{x}^k, \mathbf{x}^k + \alpha\Delta\mathbf{x}^k).$$

Step 5: Set $(\mathbf{x}^{k+1}, \lambda^{k+1}, \mathbf{s}^{k+1}) = (\mathbf{x}^k, \lambda^k, \mathbf{s}^k) + \alpha_k(\Delta\mathbf{x}^k, \Delta\lambda^k, \Delta\mathbf{s}^k)$.

end *(for)*

2.3 Convergence Analysis

We first show that if $(\mathbf{x}^0, \lambda^0, \mathbf{s}^0) \in \mathcal{F}^o$, then, all iterates will stay in $\mathcal{F}^o$.

Lemma 2.3
Let $(\mathbf{x}^k, \lambda^k, \mathbf{s}^k) \in \mathcal{F}^o$, then the next iterate generated by Algorithm 2.1 satisfies $(\mathbf{x}^{k+1}, \lambda^{k+1}, \mathbf{s}^{k+1}) \in \mathcal{F}^o$.

Proof 2.4 From Step 5 of Algorithm 2.1, we have

$$(\mathbf{x}^{k+1}, \lambda^{k+1}, \mathbf{s}^{k+1}) = (\mathbf{x}^k, \lambda^k, \mathbf{s}^k) + \alpha_k(\Delta\mathbf{x}^k, \Delta\lambda^k, \Delta\mathbf{s}^k).$$

In view of the first row of (2.5), it follows

$$\mathbf{A}\mathbf{x}^{k+1} = \mathbf{A}(\mathbf{x}^k + \alpha_k\Delta\mathbf{x}^k) = \mathbf{A}\mathbf{x}^k = \mathbf{b}.$$

In view of the second row of (2.5), it follows

$$\begin{aligned}\mathbf{A}^{\mathrm{T}}\lambda^{k+1} + \mathbf{s}^{k+1} &= \mathbf{A}^{\mathrm{T}}(\lambda^k + \alpha_k\Delta\lambda^k) + (\mathbf{s}^k + \alpha_k\Delta\mathbf{s}^k) \\ &= (\mathbf{A}^{\mathrm{T}}\lambda^k + \mathbf{s}^k) + \alpha_k(\mathbf{A}^{\mathrm{T}}\Delta\lambda^k + \Delta\mathbf{s}^k) \\ &= (\mathbf{A}^{\mathrm{T}}\lambda^k + \mathbf{s}^k) = \mathbf{c}. \end{aligned} \tag{2.12}$$

Steps 3 and 4 of Algorithm 2.1 guarantees that $(\mathbf{x}^{k+1}, \mathbf{s}^{k+1}) > \mathbf{0}$. This completes the proof. ■

Next we show that $f_\rho(\mathbf{x}, \mathbf{s})$ is reduced by at least a fix amount of $\delta > 0$, which is independent of n, at every iteration of Algorithm 2.1. First, we introduce a technical result.

Lemma 2.4
Let $\mathbf{t} \in \mathbb{R}^n$ and $\|\mathbf{t}\|_\infty \le \tau < 1$. We have

$$-\sum_{i=1}^{n} \log(1 + t_i) \le -\mathbf{e}^{\mathrm{T}}\mathbf{t} + \frac{\|\mathbf{t}\|^2}{2(1-\tau)}. \tag{2.13}$$

If $\tau = 0.5$, it is reduced to

$$-\sum_{i=1}^{n} \log(1 + t_i) \le -\mathbf{e}^{\mathrm{T}}\mathbf{t} + \|\mathbf{t}\|^2. \tag{2.14}$$

Proof 2.5 Let $f(\mathbf{t}) = -\sum_{i=1}^{n} \log(1 + t_i)$, which is well defined and smooth for $\|\mathbf{t}\|_\infty \le \tau < 1$. Then, it follows

$$\mathbf{f}'(\mathbf{t}) = \left(-\frac{1}{1+t_1}, \ldots, -\frac{1}{1+t_n}\right),$$

and

$$\mathbf{f}''(\mathbf{t}) = \mathrm{diag}\left(\frac{1}{(1+t_i)^2}\right).$$

Using Taylor's theorem, we have

$$f(\mathbf{t}) = f(\mathbf{0}) + \mathbf{t}^{\mathrm{T}}\mathbf{f}'(\mathbf{0}) + \frac{1}{2}\int_0^1 \mathbf{t}^{\mathrm{T}}\mathbf{f}''(\alpha\mathbf{t})\mathbf{t}\,d\alpha.$$

Note, that $f(\mathbf{0}) = 0$ and $\mathbf{t}^{\mathrm{T}}\mathbf{f}'(\mathbf{0}) = -\mathbf{e}^{\mathrm{T}}\mathbf{t}$, we need only to estimate the third term. Since

$$\mathbf{t}^{\mathrm{T}}\mathbf{f}''(\alpha\mathbf{t})\mathbf{t} = \sum_{i=1}^{n} \frac{t_i^2}{(1+\alpha t_i)^2} \leq \sum_{i=1}^{n} \frac{t_i^2}{(1-\alpha\tau)^2} = \frac{\|\mathbf{t}\|^2}{(1-\alpha\tau)^2},$$

we have

$$\int_0^1 \mathbf{t}^{\mathrm{T}}\mathbf{f}''(\alpha\mathbf{t})\mathbf{t}\,d\alpha \leq \|\mathbf{t}\|^2 \int_0^1 \frac{d\alpha}{(1-\alpha\tau)^2} = \|\mathbf{t}\|^2 \frac{1}{\tau}\frac{1}{1-\alpha\tau}\Big]_{\alpha=0}^{\alpha=1} = \frac{\|\mathbf{t}\|^2}{1-\tau}$$

Combining formulas above yields the inequality of (2.13). Substituting $\tau = 0.5$ into (2.13) gives (2.14). ∎

To simplify the notation, we will drop the superscript k. Some formulas derived from (2.5) are also needed. Pre-multiplying $\Delta\mathbf{x}^{\mathrm{T}}$ in the second row of (2.5) yields

$$\Delta\mathbf{x}^{\mathrm{T}}\Delta\mathbf{s} = 0. \tag{2.15}$$

By examining the last row of (2.5), it is easy to see

$$\mathbf{s}^{\mathrm{T}}\Delta\mathbf{x} + \mathbf{x}^{\mathrm{T}}\Delta\mathbf{s} = -\mathbf{x}^{\mathrm{T}}\mathbf{s} + \sigma\mu n = -\mathbf{x}^{\mathrm{T}}\mathbf{s}(1-\sigma). \tag{2.16}$$

Lemma 2.5
Let α_c *satisfy* $\alpha_c \max(\|\mathbf{X}^{-1}\Delta\mathbf{x}\|_\infty, \|\mathbf{S}^{-1}\Delta\mathbf{s}\|_\infty) = 0.5$. *For* $\alpha \in (0, \alpha_c] \subset (0, \alpha_{\max}) \subset (0, 1]$, *we have*

$$f_\rho(\mathbf{x}+\alpha\Delta\mathbf{x}, \mathbf{s}+\alpha\Delta\mathbf{s}) - f_\rho(\mathbf{x},\mathbf{s}) \leq \alpha f_1 + \alpha^2 f_2$$

where

$$f_1 := \rho\frac{\mathbf{s}^{\mathrm{T}}\Delta\mathbf{x} + \mathbf{x}^{\mathrm{T}}\Delta\mathbf{s}}{\mathbf{x}^{\mathrm{T}}\mathbf{s}} - \mathbf{e}^{\mathrm{T}}(\mathbf{X}^{-1}\Delta\mathbf{x} + \mathbf{S}^{-1}\Delta\mathbf{s}),$$
$$f_2 := \|\mathbf{X}^{-1}\Delta\mathbf{x}\|^2 + \|\mathbf{S}^{-1}\Delta\mathbf{s}\|^2.$$

Proof 2.6 Using (2.7) and (2.15), we have

$$\begin{aligned} & f_\rho(\mathbf{x}+\alpha\Delta\mathbf{x}, \mathbf{s}+\alpha\Delta\mathbf{s}) - f_\rho(\mathbf{x},\mathbf{s}) \\ = \; & \rho\log(\mathbf{x}+\alpha\Delta\mathbf{x})^{\mathrm{T}}(\mathbf{s}+\alpha\Delta\mathbf{s}) - \sum_{i=1}^{n}\log(x_i+\alpha\Delta x_i)(s_i+\alpha\Delta s_i) \end{aligned}$$

$$
\begin{aligned}
& -\rho \log(\mathbf{x}^{\mathrm{T}}\mathbf{s}) + \sum_{i=1}^{n} \log(x_i s_i) \\
= \;& \rho \log \frac{(\mathbf{x}+\alpha\Delta\mathbf{x})^{\mathrm{T}}(\mathbf{s}+\alpha\Delta\mathbf{s})}{\mathbf{x}^{\mathrm{T}}\mathbf{s}} - \sum_{i=1}^{n} \log \frac{x_i+\alpha\Delta x_i}{x_i} - \sum_{i=1}^{n} \log \frac{s_i+\alpha\Delta s_i}{s_i} \\
= \;& \rho \log\left(1+\alpha \frac{\mathbf{s}^{\mathrm{T}}\Delta\mathbf{x}+\mathbf{x}^{\mathrm{T}}\Delta\mathbf{s}}{\mathbf{x}^{\mathrm{T}}\mathbf{s}}\right) \\
& -\sum_{i=1}^{n} \log\left(1+\alpha\frac{\Delta x_i}{x_i}\right) - \sum_{i=1}^{n} \log\left(1+\alpha\frac{\Delta s_i}{s_i}\right) \qquad (2.17)
\end{aligned}
$$

In view of (2.16), it has

$$
\beta := \alpha \frac{\mathbf{s}^{\mathrm{T}}\Delta\mathbf{x}+\mathbf{x}^{\mathrm{T}}\Delta\mathbf{s}}{\mathbf{x}^{\mathrm{T}}\mathbf{s}} = -\alpha(1-\sigma) > -1.
$$

Therefore, the first term meets the condition of Lemma 1.2. By the definition of α_c, we know that $\alpha\frac{|\Delta x_i|}{|x_i|} < 1$ and $\alpha\frac{|\Delta s_i|}{|s_i|} < 1$. Therefore, applying Lemmas 1.2 and 2.4 to (2.17) yields

$$
\begin{aligned}
& f_\rho(\mathbf{x}+\alpha\Delta\mathbf{x}, \mathbf{s}+\alpha\Delta\mathbf{s}) - f_\rho(\mathbf{x},\mathbf{s}) \\
\leq \;& \rho\alpha \frac{\mathbf{s}^{\mathrm{T}}\Delta\mathbf{x}+\mathbf{x}^{\mathrm{T}}\Delta\mathbf{s}}{\mathbf{x}^{\mathrm{T}}\mathbf{s}} - \alpha\mathbf{e}^{\mathrm{T}}\left(\mathbf{X}^{-1}\Delta\mathbf{x}+\mathbf{S}^{-1}\Delta\mathbf{s}\right) \\
& +\alpha^2\left(\|\mathbf{X}^{-1}\Delta\mathbf{x}\|^2 + \|\mathbf{S}^{-1}\Delta\mathbf{s}\|^2\right) \\
:= \;& \alpha f_1 + \alpha^2 f_2.
\end{aligned}
$$

This completes the proof. ■

The next task is to estimate the size of f_1 and f_2. To simplify the analysis and to get a good polynomial bound, we select

$$
\sigma = \frac{n}{\rho} = \frac{n}{n+\sqrt{n}} \approx 1 - \frac{1}{\sqrt{n}}, \tag{2.18}
$$

which was proposed in [72][2]. Therefore, the third row of (2.5) can be rewritten as

$$
\mathbf{S}\Delta\mathbf{x} + \mathbf{X}\Delta\mathbf{s} = -\mathbf{XSe} + \frac{n}{\rho}\mu\mathbf{e}. \tag{2.19}
$$

Denote

$$
\mathbf{V} = (\mathbf{XS})^{1/2}, \quad \mathbf{v} = \mathbf{Ve}, \quad v_{\min} = \min_{i=1,\ldots,n} v_i = \min_{i=1,\ldots,n} \sqrt{x_i s_i}. \tag{2.20}
$$

and

$$
\mathbf{D} = \mathbf{X}^{1/2}\mathbf{S}^{-1/2}, \quad \mathbf{u} = -\mathbf{v} + \frac{n\mu}{\rho}\mathbf{V}^{-1}\mathbf{e}. \tag{2.21}
$$

[2] Computational experience shows that using adaptively selected σ with small σ in later iterations will be much more efficient than this fixed one.

The following lemma provides the formulas which will be used in the remainder of this chapter.

Lemma 2.6
By the definitions of (2.20) and (2.21), the following formulas hold.

$$||\mathbf{v}||^2 = \mathbf{x}^{\mathrm{T}}\mathbf{s} = n\mu, \quad \mathbf{X} = \mathbf{VD}, \quad \mathbf{S} = \mathbf{VD}^{-1}, \tag{2.22}$$

$$\mathbf{S}\Delta\mathbf{x} + \mathbf{X}\Delta\mathbf{s} = \mathbf{Vu}, \tag{2.23}$$

$$\mathbf{D}^{-1}\Delta\mathbf{x} + \mathbf{D}\Delta\mathbf{s} = \mathbf{u}, \tag{2.24}$$

$$\mathbf{X}^{-1}\Delta\mathbf{x} + \mathbf{S}^{-1}\Delta\mathbf{s} = \mathbf{V}^{-1}\mathbf{u}, \tag{2.25}$$

$$||\mathbf{D}^{-1}\Delta\mathbf{x}||, ||\mathbf{D}\Delta\mathbf{s}|| \leq ||\mathbf{u}||. \tag{2.26}$$

Proof 2.7 Formulas in (2.22) are directly followed from the definitions of (2.20) and (2.21). The formula of (2.23) is easy to see from (2.19). Pre-multiplying $\mathbf{V}^{-1}$ on both sides of (2.19) gives (2.24). Pre-multiplying $(\mathbf{XS})^{-1}$ on both sides of (2.19) shows (2.25). To show (2.26), we use (2.15) and (2.24) to get the following relations.

$$\begin{aligned} ||\mathbf{u}||^2 &= ||\mathbf{D}^{-1}\Delta\mathbf{x} + \mathbf{D}\Delta\mathbf{s}||^2 \\ &= ||\mathbf{D}^{-1}\Delta\mathbf{x}||^2 + 2\Delta\mathbf{x}^{\mathrm{T}}\mathbf{D}^{-1}\mathbf{D}\Delta\mathbf{s} + ||\mathbf{D}\Delta\mathbf{s}||^2 \\ &= ||\mathbf{D}^{-1}\Delta\mathbf{x}||^2 + ||\mathbf{D}\Delta\mathbf{s}||^2 \\ &\geq \max\{||\mathbf{D}^{-1}\Delta\mathbf{x}||^2, ||\mathbf{D}\Delta\mathbf{s}||^2\}. \end{aligned} \tag{2.27}$$

Taking the square root on both side proves (2.26). ■

The estimation of f_1 and f_2 follows from Lemma 2.6.

Lemma 2.7
Let f_1 and f_2 be defined as in Lemma 2.5, then, the following relations hold.

$$f_1 \leq -\frac{\rho}{n\mu}||\mathbf{u}||^2, \qquad f_2 \leq \frac{||\mathbf{u}||^2}{v_{\min}^2}.$$

Proof 2.8 From Lemma 2.5, we have

$$\begin{aligned} f_1 &= \tfrac{\rho}{\mathbf{x}^{\mathrm{T}}\mathbf{s}}\mathbf{e}^{\mathrm{T}}(\mathbf{S}\Delta\mathbf{x} + \mathbf{X}\Delta\mathbf{s}) - \mathbf{e}^{\mathrm{T}}(\mathbf{X}^{-1}\Delta\mathbf{x} + \mathbf{S}^{-1}\Delta\mathbf{s}) \\ \text{use (2.23) and (2.25)} \quad &= \tfrac{\rho}{n\mu}\mathbf{e}^{\mathrm{T}}\mathbf{Vu} - \mathbf{e}^{\mathrm{T}}\mathbf{V}^{-1}\mathbf{u} \\ &= \tfrac{\rho}{n\mu}(\mathbf{e}^{\mathrm{T}}\mathbf{V} - \tfrac{n\mu}{\rho}\mathbf{e}^{\mathrm{T}}\mathbf{V}^{-1})\mathbf{u} \\ \text{use (2.20)} \quad &= \tfrac{\rho}{n\mu}(\mathbf{v} - \tfrac{n\mu}{\rho}\mathbf{V}^{-1}\mathbf{e})^{\mathrm{T}}\mathbf{u} \\ &= -\tfrac{\rho}{n\mu}(-\mathbf{v} + \tfrac{n\mu}{\rho}\mathbf{V}^{-1}\mathbf{e})^{\mathrm{T}}\mathbf{u} \\ \text{use (2.21)} \quad &= -\tfrac{\rho}{n\mu}||\mathbf{u}||^2 \end{aligned}$$

In view of Lemma 2.5, we have

$$
\begin{aligned}
f_2 &= \|\mathbf{X}^{-1}\Delta\mathbf{x}\|^2 + \|\mathbf{S}^{-1}\Delta\mathbf{s}\|^2 \\
\text{use (2.22)} \quad &= \|\mathbf{V}^{-1}\mathbf{D}^{-1}\Delta\mathbf{x}\|^2 + \|\mathbf{V}^{-1}\mathbf{D}\Delta\mathbf{s}\|^2 \\
&\le \|\mathbf{V}^{-1}\|^2(\|\mathbf{D}^{-1}\Delta\mathbf{x}\|^2 + \|\mathbf{D}\Delta\mathbf{s}\|^2) \\
\text{use (2.27) and (2.20)} \quad &\le \frac{\|\mathbf{u}\|^2}{v_{\min}^2}.
\end{aligned}
$$

This completes the proof. ■

Since $f_1 < 0$ and $f_2 > 0$, from Lemma 2.5, as α becomes small enough, it must have

$$f_\rho(\mathbf{x}+\alpha\Delta\mathbf{x}, \mathbf{s}+\alpha\Delta\mathbf{s}) \le f_\rho(\mathbf{x},\mathbf{s}).$$

We need to determine what is the value of the α for the potential function to decrease and what amount is the decrease. Since f_1 and f_2 depend on $\|\mathbf{u}\|$, we need to estimate the size of $\mathbf{u}$ in terms of some constants and $v_{\min}$.

Lemma 2.8

For any $(\mathbf{x},\lambda,\mathbf{s}) \in \mathcal{F}^o$ and for $\rho > n+\sqrt{n}$, it must have

$$\|\mathbf{u}\| \ge \frac{\sqrt{3}n\mu}{2v_{\min}\rho} \tag{2.28}$$

Proof 2.9 Using (2.22), we have

$$\mathbf{v}^{\mathrm{T}}(\mathbf{V}^{-1}\mathbf{e} - \mathbf{v}/\mu) = n - n = 0. \tag{2.29}$$

For the vector $\frac{\mathbf{e}}{\mathbf{v}} - \frac{1}{\mu}\mathbf{v}$, the norm

$$\|\mathbf{V}^{-1}\mathbf{e} - \frac{1}{\mu}\mathbf{v}\|^2$$

is bounded below by taking the square of any component of $\mathbf{v}$ that corresponds to $v_{\min}$, i.e.,

$$\|\mathbf{V}^{-1}\mathbf{e} - \frac{1}{\mu}\mathbf{v}\|^2 \ge \left(\frac{1}{v_{\min}} - \frac{v_{\min}}{\mu}\right)^2. \tag{2.30}$$

Also, the following equalities hold:

$$
\begin{aligned}
&\|\mathbf{V}^{-1}\mathbf{e} - \frac{n+\sqrt{n}}{n\mu}\mathbf{v}\|^2 \\
&= \|\mathbf{V}^{-1}\mathbf{e}\|^2 - 2\frac{n+\sqrt{n}}{n\mu}\mathbf{e}^{\mathrm{T}}\mathbf{V}^{-1}\mathbf{v} + \left(\frac{n+\sqrt{n}}{n\mu}\right)^2\mathbf{v}^{\mathrm{T}}\mathbf{v} \\
\text{use of (2.22)} \quad &= \|\mathbf{V}^{-1}\mathbf{e}\|^2 - 2\frac{n+\sqrt{n}}{n\mu}\mathbf{e}^{\mathrm{T}}\mathbf{e} + \left(\frac{n+\sqrt{n}}{n\mu}\right)^2 n\mu \\
&= \|\mathbf{V}^{-1}\mathbf{e}\|^2 - 2\frac{n+\sqrt{n}}{\mu} + \frac{n^2+2\sqrt{n}n+n}{n\mu}
\end{aligned}
$$

$$= \quad \|\mathbf{V}^{-1}\mathbf{e}\|^2 - \frac{n-1}{\mu} \tag{2.31}$$

By using (2.21) and $\mathbf{V}^{-1}\mathbf{v} = \mathbf{e}$, we have

$$\begin{array}{rcl}
\|\mathbf{u}\|^2 \frac{\rho^2}{n^2\mu^2} & = & \|\mathbf{V}^{-1}\mathbf{e} - \frac{\rho}{n\mu}\mathbf{v}\|^2 \\
& = & \|\mathbf{V}^{-1}\mathbf{e}\|^2 + \frac{\rho^2}{n^2\mu^2}\|\mathbf{v}\|^2 - 2\frac{\rho}{n\mu}\mathbf{e}^{\mathrm{T}}\mathbf{V}^{-1}\mathbf{v} \\
\text{use of (2.22)} & = & \|\mathbf{V}^{-1}\mathbf{e}\|^2 + \frac{\rho^2}{n\mu} - 2\frac{\rho}{\mu} \\
\text{use of elementary algebra} & = & \|\mathbf{V}^{-1}\mathbf{e}\|^2 - \frac{n-1}{\mu} + \frac{(\rho-n-\sqrt{n})^2 + 2\sqrt{n}(\rho-n-\sqrt{n})}{n\mu} \\
\rho > n+\sqrt{n} & \geq & \|\mathbf{V}^{-1}\mathbf{e}\|^2 - \frac{n-1}{\mu} \\
\text{use (2.31)} & = & \|\mathbf{V}^{-1}\mathbf{e} - \frac{n+\sqrt{n}}{n\mu}\mathbf{v}\|^2 \\
\text{use of (2.29)} & = & \|\mathbf{V}^{-1}\mathbf{e} - \frac{1}{\mu}\mathbf{v}\|^2 + \|\frac{1}{\sqrt{n\mu}}\mathbf{v}\|^2 \\
\text{use of (2.30)} & \geq & \left(\frac{1}{v_{\min}} - \frac{v_{\min}}{\mu}\right)^2 + \frac{\mathbf{v}^{\mathrm{T}}\mathbf{v}}{n\mu^2} \\
& = & \frac{1}{\mu^2}\left[\left(\frac{\mu}{v_{\min}} - v_{\min}\right)^2 + \mu\right] \\
& = & \frac{1}{\mu^2}\left[\left(\frac{\mu}{2v_{\min}} - v_{\min}\right)^2 + \frac{3}{4}\frac{\mu^2}{v_{\min}^2}\right] \\
& \geq & \frac{3}{4v_{\min}^2}.
\end{array}$$

Rearranging the inequality proves the lemma. ■

We are now ready to prove the convergence of the potential reduction algorithm.

Lemma 2.9
Let $\bar{\alpha} = \frac{v_{\min}}{2\|\mathbf{u}\|}$. Then, we have

$$f_\rho(\mathbf{x} + \bar{\alpha}\Delta\mathbf{x}, \mathbf{s} + \bar{\alpha}\Delta\mathbf{s}) - f_\rho(\mathbf{x}, \mathbf{s}) < -0.15 := \delta. \tag{2.32}$$

Proof 2.10 First, it is easy to see that $\bar{\alpha}$ satisfies the condition of Lemma 2.5 because of the following relation:

$$\begin{array}{rcl}
\bar{\alpha}\max(\|\mathbf{X}^{-1}\Delta\mathbf{x}\|_\infty, \|\mathbf{S}^{-1}\Delta\mathbf{s}\|_\infty) & \leq & \bar{\alpha}\max(\|\mathbf{X}^{-1}\Delta\mathbf{x}\|_2, \|\mathbf{S}^{-1}\Delta\mathbf{s}\|_2) \\
& \leq & \frac{v_{\min}}{2\|\mathbf{u}\|}\frac{\|\mathbf{u}\|}{v_{\min}} = 0.5,
\end{array}$$

where the last inequality follows from the proof of the last inequality in Lemma 2.7. Now, using Lemmas 2.5, 2.7, and 2.8, we have

$$\begin{array}{rl}
& f_\rho(\mathbf{x} + \bar{\alpha}\Delta\mathbf{x}, \mathbf{s} + \bar{\alpha}\Delta\mathbf{s}) - f_\rho(\mathbf{x}, \mathbf{s}) \\
\leq & \bar{\alpha}f_1 + \bar{\alpha}^2 f_2 \\
\leq & -\frac{v_{\min}}{2\|\mathbf{u}\|}\frac{\rho}{n\mu}\|\mathbf{u}\|^2 + \frac{v_{\min}^2}{4\|\mathbf{u}\|^2}\frac{\|\mathbf{u}\|^2}{v_{\min}^2}
\end{array}$$

$$\begin{aligned} &\leq -\frac{\rho v_{\min}}{2n\mu}\|\mathbf{u}\| + \frac{1}{4} \\ &\leq -\frac{\rho v_{\min}}{2n\mu}\frac{\sqrt{3}n\mu}{2v_{\min}\rho} + \frac{1}{4} < -0.15. \end{aligned}$$

This completes the proof. ■

Finally, we have the convergence result for the potential-reduction interior-point algorithm.

Theorem 2.2
Given a starting point $(\mathbf{x}^0, \lambda^0, \mathbf{s}^0) \in \mathcal{F}^o$*, all iterates will stay in* $\mathcal{F}^o$*, and there is a* $\delta = -0.15$ *such that (2.32) holds. Moreover, for* $\varepsilon \in (0,1)$ *and* $\rho - n > \sqrt{n}$*, we have an index K defined by*

$$K = \left\lceil \frac{f_\rho(\mathbf{x}^0, \mathbf{s}^0)}{\delta} + \frac{\sqrt{n}}{\delta}|\log \varepsilon| \right\rceil \tag{2.33}$$

such that

$$\mu_k \leq \varepsilon, \qquad \forall k \geq K. \tag{2.34}$$

Proof 2.11 The first claim is shown in Lemma 2.3. The second claim is proved in Lemma 2.9. The last claim follows the following argument. From (2.10), (2.32) and $\rho > n + \sqrt{n}$, it follows

$$\begin{aligned} \log \mu &\leq f_\rho(\mathbf{x}^k, \mathbf{s}^k)/(\rho - n) \\ &\leq f_\rho(\mathbf{x}^k, \mathbf{s}^k)/\sqrt{n} \\ &\leq (f_\rho(\mathbf{x}^0, \mathbf{s}^0) - k\delta)/\sqrt{n}. \end{aligned}$$

Clearly, if $(f_\rho(\mathbf{x}^0, \mathbf{s}^0) - k\delta)/\sqrt{n} \leq -|\log \varepsilon|$, then, it must have $\mu \leq \varepsilon$, i.e., for $k \geq K$ with K given in (2.33), it must have $\mu_k \leq \varepsilon$. ■

2.4 Concluding Remarks

The primal-dual potential-reduction interior-point algorithm discussed in this chapter is a feasible interior-point algorithm because it requires the initial point in $\mathcal{F}^o$. This may not be a reasonable requirement as it may take a lot of effort to find a point in $\mathcal{F}^o$. Also, primal-dual potential-reduction interior-point algorithms are not as efficient as primal-dual path-following interior-point algorithms which will be discussed in the next chapter. We will also discuss infeasible primal-dual path-following interior-point algorithms which do not require the initial point in $\mathcal{F}^o$, a very desired feature for practical computations.

However, potential-reduction interior-point method was found to be useful in the development of interior-point algorithms for nonlinear programming. This is the main reason that we have included the method in this book.

Chapter 3

Feasible Path-Following Algorithms for LP

Path-following is an excellent idea of Megiddo [88] in the redevelopment of interior-point method in the 1980's that has been studied extensively. The key idea of the path-following algorithms is to search for the optimizer approximately along a curve which is known as the central path. Following the central-path prevents the iterate from approaching the boundary too early before the iterate approaches the optimal solution. Many algorithms were proposed based on this idea. Central-path is an arc defined by (2.4) in Chapter 2. The potential-reduction algorithm discussed in Chapter 2 does not restrict its iterate to follow the central path explicitly, but the potential function includes an item that implicitly penalizes the iterate far away from the central-path. However, all path-following algorithms explicitly require the iterate to stay in a neighborhood of the central-path. Several neighborhoods are used by researchers. In this chapter, two mostly used neighborhoods are considered.

The first neighborhood is defined by using the 2-norm:

$$\mathcal{N}_2(\theta) = \{(\mathbf{x}, \lambda, \mathbf{s}) \in \mathcal{F}^o \mid \|\mathbf{x} \circ \mathbf{s} - \mu \mathbf{e}\|_2 \leq \theta \mu\} \tag{3.1}$$

where $\theta \in (0,1)$ is a constant and $\mathcal{F}^o$ is defined in (1.10). It is clear that this neighborhood restricts $x_i s_i$ to stay not too far away from the average

$$\mu = \mathbf{x}^{\mathrm{T}}\mathbf{s}/n = \frac{1}{n}\sum_{i=1}^{n} x_i s_i.$$

The second neighborhood is motivated from the one-side ∞-norm:

$$\mathcal{N}_{-\infty}(\gamma) = \{(\mathbf{x}, \lambda, \mathbf{s}) \in \mathcal{F}^o \mid x_i s_i \geq \gamma \mu\}. \tag{3.2}$$

where $\gamma \in (0,1)$ is a constant. This neighborhood does not restrict $x_i s_i$ from becoming too big, it cares only if some $x_i s_i$ are much smaller than the average characterized by duality measure μ.

Both neighborhoods require that the iterate stays inside the feasible set, therefore, $\mathbf{x} > \mathbf{0}$ and $\mathbf{s} > \mathbf{0}$, hence, $x_i s_i > 0$ hold for all $i = 1, 2, \ldots, n$ in all iterations. Note that the $\mathcal{N}_2(\theta)$ neighborhood can be expressed as

$$\sum_{i=1}^{n} \left(\frac{s_i x_i}{\mu} - 1 \right)^2 \leq \theta^2 < 1.$$

This means that each relative deviation of $x_i s_i$ from μ is very small. Therefore, the neighborhood is called narrow neighborhood. On the other hand, $\mathcal{N}_{-\infty}(\gamma)$ neighborhood requires only that $x_i s_i$ is not too close to zero, therefore, the neighborhood is called wide neighborhood. For algorithms that restrict the iterate inside the narrow neighborhood, since the central-path is a curve, the step-size is normally short when searching is along a straight line, otherwise the iterate will be outside the narrow neighborhood. Therefore, these algorithms are named as short step algorithms. For algorithms that restrict the iterate inside the wide neighborhood, although the central-path is a still the same curve, but the room to search for the new iterate is much larger, the step-size is normally longer. Therefore, these algorithms are known as long-step algorithms.

Another idea to make the iterate take a longer step is proposed by Mizuno, Todd, and Ye [93]. The initial iterate is inside a narrow neighborhood with a small parameter θ and the search for the next iterate is in a larger narrow neighborhood with a larger parameter 2θ, this will generate a longer step than a search in the neighborhood with a small θ. This step is called the predictor step. Since $\theta < 1$, to prevent θ becomes bigger and bigger, after the iterate moves to a neighborhood with a larger parameter 2θ, the algorithm brings the iterate back to the neighborhood with the original small θ but the objective function will not increase while moving the iterate back to the original narrow neighborhood. This is called the corrector step. This algorithm is known as the Mizuno-Todd-Ye algorithm, or MTY algorithm for short.

A different way to make a longer step for the iterate is proposed by Monteiro, Adler, and Resende [97]. As the central-path is an arc, searching along a straight line will keep the step-size small in order to prevent the iterate from moving to the outside of the neighborhood, an intuitive idea is to search for the optimizer along the central-path. However the calculation of the central-path defined by (2.4) will be too expansive to be feasible. Monteiro, Adler, and Resende proposed using power series to approximate the central-path. This method needs second-order derivatives to approximate the central-path, which is different from aforementioned algorithms which need only the first-order derivative to get the search direction. Therefore, algorithms that need more than first order derivatives are called higher-order algorithms.

It is worthwhile to note that all algorithms mentioned above assume that the initial point is either in $\mathcal{N}_2(\theta)$ or $\mathcal{N}_{-\infty}(\gamma)$. Therefore, they are called feasible interior-point algorithms. It has been known that, in general, feasible interior-point algorithms are not as efficient as the infeasible interior-point algorithms to be discussed in the next chapter. However, all the iterates in the feasible interior-point algorithm are feasible, which is desirable for some engineering problems that will be discussed in Chapter 10.

This chapter discusses only feasible interior-point algorithms. We start with the short-step path-following algorithm.

3.1 A Short-Step Path-Following Algorithm

A well-known short-step path-following algorithm was proposed by Kojima, Mizuno, and Yoshise [70]. The polynomial bound established for this algorithm is still the best among all interior-point algorithms for linear programming. Before we discuss the algorithm and its convergence, we state several technical lemmas that will be used to analyze this algorithm and some algorithms to be discussed in the rest of the book.

Lemma 3.1
For any vector $\mathbf{x} \in \mathbb{R}^n$, the following vector norm inequalities hold

$$\|\mathbf{x}\|_\infty \le \|\mathbf{x}\|_2 \le \|\mathbf{x}\|_1 \le \sqrt{n}\|\mathbf{x}\|_2 \le n\|\mathbf{x}\|_\infty.$$

Lemma 3.2
For any vector pairs of $\mathbf{u}$ and $\mathbf{v}$ with $\mathbf{u}^{\mathrm{T}}\mathbf{v} \ge 0$, we have

$$\|\mathbf{u} \circ \mathbf{v}\| \le 2^{-3/2}\|\mathbf{u}+\mathbf{v}\|^2. \tag{3.3}$$

Proof 3.1 Since $\mathbf{u}^{\mathrm{T}}\mathbf{v} \ge 0$, we have

$$0 \le \mathbf{u}^{\mathrm{T}}\mathbf{v} = \sum_{u_i v_i \ge 0} |u_i v_i| - \sum_{u_i v_i < 0} |u_i v_i|. \tag{3.4}$$

or equivalently

$$\sum_{u_i v_i \ge 0} |u_i v_i| \ge \sum_{u_i v_i < 0} |u_i v_i|. \tag{3.5}$$

For any two scalars a and b with $ab \ge 0$, in view of the algebraic-geometric mean inequality, we have

$$\sqrt{ab} \le \frac{1}{2}|a+b|. \tag{3.6}$$

Using these two relations and $\|\cdot\|_2 \le \|\cdot\|_1$, we have

$$\begin{array}{rcl}
\|\mathbf{u}\circ\mathbf{v}\| & = & \left(\|\sum_{u_iv_i\ge 0}[u_iv_i]\|_2^2+\|\sum_{u_iv_i<0}[u_iv_i]\|_2^2\right)^{1/2} \\
\text{use } \|\cdot\|_2\le\|\cdot\|_1 & \le & \left(\|\sum_{u_iv_i\ge 0}[u_iv_i]\|_1^2+\|\sum_{u_iv_i<0}[u_iv_i]\|_1^2\right)^{1/2} \\
\text{use (3.5)} & \le & \left(2\|\sum_{u_iv_i\ge 0}[u_iv_i]\|_1^2\right)^{1/2} \\
 & \le & 2^{1/2}\|\sum_{u_iv_i\ge 0}[u_iv_i]\|_1 \\
\text{use (3.6)} & \le & 2^{1/2}\left\|\frac{1}{4}\sum_{u_iv_i\ge 0}\left(u_i+v_i\right)^2\right\|_1 \\
 & \le & 2^{-3/2}\sum_{u_iv_i\ge 0}(u_i+v_i)^2 \\
 & \le & 2^{-3/2}\sum_1^n(u_i+v_i)^2 \\
 & \le & 2^{-3/2}\|\mathbf{u}+\mathbf{v}\|^2.
\end{array}$$

This finishes the proof. ■

Lemma 3.3
Let $\mathbf{S}$ *be the diagonal matrix whose diagonal elements are the vector* $\mathbf{s}$ *and* $\mathbf{X}$ *be the diagonal matrix whose diagonal elements are vector* $\mathbf{x}$*. Let* $\mathbf{e}$ *be a vector of all ones, and* σ *and* μ *be two constants. Denote* $\mathbf{D}=\mathbf{X}^{1/2}\mathbf{S}^{-1/2}$*. For two vectors* $\Delta\mathbf{x}$ *and* $\Delta\mathbf{s}$*, if*

$$\mathbf{S}\Delta\mathbf{x}+\mathbf{X}\Delta\mathbf{s}=-\mathbf{XSe}+\sigma\mu\mathbf{e}, \tag{3.7}$$

then we have

$$\begin{array}{rcl}
\|\Delta\mathbf{x}\circ\Delta\mathbf{s}\| & \le & 2^{-3/2}\|\mathbf{D}^{-1}\Delta\mathbf{x}+\mathbf{D}\Delta\mathbf{s}\|^2 \\
 & = & 2^{-3/2}\|(\mathbf{XS})^{-1/2}(-\mathbf{x}\circ\mathbf{s}+\sigma\mu\mathbf{e})\|^2.
\end{array} \tag{3.8}$$

Proof 3.2 Pre-multiplying $(\mathbf{XS})^{-1/2}$ in (3.7) gives

$$\mathbf{D}^{-1}\Delta\mathbf{x}+\mathbf{D}\Delta\mathbf{s}=(\mathbf{XS})^{-1/2}(-\mathbf{x}\circ\mathbf{s}+\sigma\mu\mathbf{e}).$$

Using Lemma 3.2 and this equality yields

$$\begin{array}{rcl}
\|\Delta\mathbf{x}\circ\Delta\mathbf{s}\| & = & \|(\mathbf{D}^{-1}\Delta\mathbf{x})\circ(\mathbf{D}\Delta\mathbf{s})\| \\
 & \le & 2^{-3/2}\|\mathbf{D}^{-1}\Delta\mathbf{x}+\mathbf{D}\Delta\mathbf{s}\|^2 \\
 & = & 2^{-3/2}\|(\mathbf{XS})^{-1/2}(-\mathbf{x}\circ\mathbf{s}+\sigma\mu\mathbf{e})\|^2.
\end{array} \tag{3.9}$$

This finishes the proof. ■

The short-step path-following algorithm uses several fixed parameters. Assume that the initial point is in $\mathcal{N}_2(\theta)$, the algorithm makes sure that all iterates stay in $\mathcal{N}_2(\theta)$. It is formally stated as below:

Algorithm 3.1
Data: $\mathbf{A}$, $\mathbf{b}$, $\mathbf{c}$.
Parameters: $\theta=0.4$, $\varepsilon\in(0,1)$, *and* $\sigma=1-0.4/\sqrt{n}$.
Initial point $(\mathbf{x}^0,\lambda^0,\mathbf{s}^0)\in\mathcal{N}_2(\theta)$ *and* $\mu_0=\frac{\mathbf{x}^{0\mathrm{T}}\mathbf{s}^0}{n}$.
for *iteration* $k=0,1,2,\ldots$

Step 1: If $\mu_k < \varepsilon$, stop. Otherwise continue.

Step 2: Solve the linear systems of equations (2.5) to get $(\Delta\mathbf{x}, \Delta\lambda, \Delta\mathbf{s})$.

Step 3: Set

$$(\mathbf{x}^{k+1}, \lambda^{k+1}, \mathbf{s}^{k+1}) = (\mathbf{x}^k, \lambda^k, \mathbf{s}^k) + (\Delta\mathbf{x}, \Delta\lambda, \Delta\mathbf{s}) \tag{3.10}$$

and $\mu_{k+1} = \frac{\mathbf{x}^{k+1^{\mathrm{T}}}\mathbf{s}^{k+1}}{n}$. *Set $k+1 \to k$. Go back to Step 1.*

end (for)

Following the proof of Lemma 2.3, it is easy to show the following relations.

$$\mathbf{A}\mathbf{x}^{k+1} = \mathbf{A}(\mathbf{x}^k + \Delta\mathbf{x}^k) = \mathbf{A}\mathbf{x}^k = \mathbf{b}. \tag{3.11}$$

and

$$\mathbf{A}^{\mathrm{T}}\lambda^{k+1} + \mathbf{s}^{k+1} = (\mathbf{A}^{\mathrm{T}}\lambda^k + s^k) = \mathbf{c}. \tag{3.12}$$

Hence, to prove polynomiality of the algorithm, we need to show (a) that the duality measure reduces at least at a constant rate in every iteration, and (b) that $(\mathbf{x}^{k+1}, \mathbf{s}^{k+1}) > \mathbf{0}$. The next lemma is stated in a slightly general form so that we can reuse it in other algorithms. For this purpose, we denote:

$$(\mathbf{x}(\alpha), \lambda(\alpha), \mathbf{s}(\alpha)) = (\mathbf{x}, \lambda, \mathbf{s}) + \alpha(\Delta\mathbf{x}, \Delta\lambda, \Delta\mathbf{s}), \tag{3.13}$$

$$\mu(\alpha) = \mathbf{x}(\alpha)^{\mathrm{T}}\mathbf{s}(\alpha)/n. \tag{3.14}$$

Lemma 3.4

Let $(\Delta\mathbf{x}, \Delta\lambda, \Delta\mathbf{s})$ be defined by (2.5). Then, we have

$$\Delta\mathbf{x}^{\mathrm{T}}\Delta\mathbf{s} = 0, \tag{3.15}$$

$$\mu(\alpha) = (1 - \alpha(1-\sigma))\mu. \tag{3.16}$$

Proof 3.3 The (3.15) is proved in (2.15). The proof of (3.16) uses (2.15) and (2.16).

$$\begin{aligned}
\mathbf{x}(\alpha)^{\mathrm{T}}\mathbf{s}(\alpha) &= \mathbf{x}^{\mathrm{T}}\mathbf{s} + \alpha(\mathbf{s}^{\mathrm{T}}\Delta\mathbf{x} + \mathbf{x}^{\mathrm{T}}\Delta\mathbf{s}) + \alpha^2\Delta\mathbf{x}^{\mathrm{T}}\Delta\mathbf{s} \\
\text{use (2.15)} &= \mathbf{x}^{\mathrm{T}}\mathbf{s} + \alpha(\mathbf{s}^{\mathrm{T}}\Delta\mathbf{x} + \mathbf{x}^{\mathrm{T}}\Delta\mathbf{s}) \\
\text{use (2.16)} &= \mathbf{x}^{\mathrm{T}}\mathbf{s}(1 - \alpha(1-\sigma)).
\end{aligned} \tag{3.17}$$

Dividing both sides by n proves the (3.16). ∎

Remark 3.1 If $\mu_{k+1} = \mu(\alpha)$ for a fixed α, equation (3.16) means that the duality measure reduces at a constant rate in every iteration. ∎

The next lemma estimates the size of $\Delta\mathbf{x} \circ \Delta\mathbf{s}$.

Lemma 3.5
Let $(\mathbf{x},\lambda,\mathbf{s})\in\mathcal{F}^o$*. Then, we have*

$$\|\Delta\mathbf{x}\circ\Delta\mathbf{s}\|\le\frac{\theta^2+n(1-\sigma)^2}{2^{3/2}(1-\theta)}\mu. \tag{3.18}$$

Proof 3.4 First, it is easy to see

$$\mathbf{e}^{\mathrm{T}}(\mathbf{x}\circ\mathbf{s}-\mu\mathbf{e})=\mathbf{x}^{\mathrm{T}}\mathbf{s}-\mu n=0. \tag{3.19}$$

Since $(\mathbf{x},\lambda,\mathbf{s})\in\mathcal{N}_2(\theta)$, we have

$$\min_i\{s_ix_i\}\ge(1-\theta)\mu. \tag{3.20}$$

This gives

$$\begin{aligned}\|\mathbf{x}\circ\mathbf{s}-\sigma\mu\mathbf{e}\|^2 &= \|(\mathbf{x}\circ\mathbf{s}-\mu\mathbf{e})+(1-\sigma)\mu\mathbf{e}\|^2\\ &= \|\mathbf{x}\circ\mathbf{s}-\mu\mathbf{e}\|^2+2(1-\sigma)\mu\mathbf{e}^{\mathrm{T}}(\mathbf{x}\circ\mathbf{s}-\mu\mathbf{e})+(1-\sigma)^2\mu^2n\\ \text{use (3.19)}\quad &\le \theta^2\mu^2+(1-\sigma)^2\mu^2n. \end{aligned}\tag{3.21}$$

By using Lemma 3.3, we have

$$\begin{aligned}\|\Delta\mathbf{x}\circ\Delta\mathbf{s}\| &\le 2^{-3/2}\|(\mathbf{XS})^{-1/2}(-\mathbf{x}\circ\mathbf{s}+\sigma\mu\mathbf{e})\|^2\\ &\le 2^{-3/2}\sum_{i=1}^{n}\frac{(-x_is_i+\sigma\mu)^2}{x_is_i}\\ &\le 2^{-3/2}\frac{\|\mathbf{x}\circ\mathbf{s}-\sigma\mu\mathbf{e}\|^2}{\min_i(x_is_i)}\\ \text{use (3.20) and (3.21)}\quad &\le 2^{-3/2}\frac{\theta^2\mu^2+(1-\sigma)^2\mu^2n}{(1-\theta)\mu}\\ &\le \frac{\theta^2+n(1-\sigma)^2}{2^{3/2}(1-\theta)}\mu \end{aligned}\tag{3.22}$$

This completes the proof. ■

The next estimation is based on Lemma 3.5.

Lemma 3.6
Let $(\mathbf{x},\lambda,\mathbf{s})\in\mathcal{N}_2(\theta)$*. Then, we have*

$$\begin{aligned}\|\mathbf{x}(\alpha)\circ\mathbf{s}(\alpha)-\mu(\alpha)\mathbf{e}\| &\le |1-\alpha|\|\mathbf{x}\circ\mathbf{s}-\mu\mathbf{e}\|+\alpha^2\|\Delta\mathbf{x}\circ\Delta\mathbf{s}\| \qquad (3.23)\\ &\le |1-\alpha|\theta\mu+\alpha^2\frac{\theta^2+n(1-\sigma)^2}{2^{3/2}(1-\theta)}\mu. \qquad (3.24)\end{aligned}$$

Proof 3.5 Formula (3.24) follows directly from the definition of $\mathcal{N}_2(\theta)$ and Lemma 3.5. Using (3.13) and Lemma 3.4, we have

$$\begin{aligned} & x_i(\alpha)s_i(\alpha) - \mu(\alpha) \\ = \; & x_is_i + \alpha(s_i\Delta x_i + x_i\Delta s_i) + \alpha^2\Delta x_i\Delta s_i - (1-\alpha(1-\sigma))\mu \\ \text{last row of (2.5)} \quad = \; & x_is_i(1-\alpha) + \alpha\sigma\mu + \alpha^2\Delta x_i\Delta s_i - (1-\alpha+\alpha\sigma)\mu \\ = \; & x_is_i(1-\alpha) + \alpha^2\Delta x_i\Delta s_i - (1-\alpha)\mu \end{aligned} \tag{3.25}$$

This can be rewritten in matrix form

$$\begin{aligned} & \|\mathbf{x}(\alpha)\circ\mathbf{s}(\alpha) - \mu(\alpha)\mathbf{e}\| \\ = \; & \|\mathbf{x}\circ\mathbf{s}(1-\alpha) + \alpha^2\Delta\mathbf{x}\circ\Delta\mathbf{s} - (1-\alpha)\mu\mathbf{e}\| \\ \le \; & |1-\alpha|\|\mathbf{x}\circ\mathbf{s} - \mu\mathbf{e}\| + \alpha^2\|\Delta\mathbf{x}\circ\Delta\mathbf{s}\| \\ \le \; & |1-\alpha|\theta\mu + \alpha^2\frac{\theta^2+n(1-\sigma)^2}{2^{3/2}(1-\theta)}\mu. \end{aligned} \tag{3.26}$$

The last inequality follows from the definition of $\mathcal{N}_2(\theta)$ and Lemma 3.5. ■

Finally, the following lemma shows that all iterate will stay in $\mathcal{N}_2(\theta)$.

Lemma 3.7
Let $\theta = 0.4$ and $\sigma = 1 - 0.4/\sqrt{n}$. Then, the following inequality

$$\frac{\theta^2+n(1-\sigma)^2}{2^{3/2}(1-\theta)} \le \sigma\theta \tag{3.27}$$

holds. Moreover, if $(\mathbf{x},\lambda,\mathbf{s}) \in \mathcal{N}_2(\theta)$, then, for all $\alpha \in [0,1]$, it must have

$$(\mathbf{x}(\alpha),\lambda(\alpha),\mathbf{s}(\alpha)) \in \mathcal{N}_2(\theta). \tag{3.28}$$

Proof 3.6 The first claim is easy to verify by substituting the values of θ and $\sigma = 1-0.4/\sqrt{n}$ into the first formula. Substituting (3.27) into (3.24) yields

$$\begin{aligned} \|\mathbf{x}(\alpha)\circ\mathbf{s}(\alpha) - \mu(\alpha)\mathbf{e}\| \; & \le \; |1-\alpha|\theta\mu + \alpha^2\sigma\theta\mu \\ \text{since } \alpha\in[0,1] \; & \le \; (1-\alpha+\sigma\alpha)\theta\mu \\ \text{use Lemma 3.4} \; & \le \; \theta\mu(\alpha), \end{aligned} \tag{3.29}$$

which is the proximity condition of $\mathcal{N}_2(\theta)$. Since it has been shown that (3.11) and (3.12) hold, to show that $(\mathbf{x}^{k+1},\mathbf{s}^{k+1}) \in \mathcal{F}^o$, we need to show $(\mathbf{x}^{k+1},\mathbf{s}^{k+1}) > 0$. It follows from (3.29) that

$$x_i(\alpha)s_i(\alpha) \le (1-\theta)\mu(\alpha) = (1-\theta)(1-\alpha(1-\sigma))\mu > 0$$

holds for all $\alpha \in [0,1]$ because $\theta \in (0,1)$ and $\sigma > 0$. Since $(\mathbf{x}(0),\mathbf{s}(0)) = (\mathbf{x}^k,\mathbf{s}^k) > \mathbf{0}$ and it has no chance for $x_i(\alpha) = 0$ or $s_i(\alpha) = 0$, it must have $(\mathbf{x}(1),\mathbf{s}(1)) = (\mathbf{x}^{k+1},\mathbf{s}^{k+1}) > \mathbf{0}$. This shows that $(\mathbf{x}^{k+1},\lambda^{k+1},\mathbf{s}^{k+1}) \in \mathcal{F}^o$. ■

Since all iterates stay in $\mathcal{N}_2(\theta)$, using Lemma 3.4, we have

Lemma 3.8
Given $\varepsilon > 0$, suppose that initial point $(\mathbf{x}^0, \lambda^0, \mathbf{s}^0) \in \mathcal{N}_2(0.4)$ and $\mu_0 \le 1/\varepsilon^\kappa$ for some positive constant κ. Then, there is a large $K = \mathcal{O}(\sqrt{n}\log(1/\varepsilon))$ such that

$$\mu_k \le \varepsilon \text{ for all } k \ge K.$$

Proof 3.7 Using Lemma 3.4 with $\alpha = 1$ and $\sigma = 1 - 0.4/\sqrt{n}$, we have

$$\mu_{k+1} = \mu(\alpha) = \left(1 - \frac{0.4}{\sqrt{n}}\right)\mu.$$

Using Theorem 1.4 with $\omega = 0.5$ and $\delta = 0.4$, the claim follows. ■

Remark 3.2 In interior-point method literature, people say that the polynomial complexity of the short step path-following algorithm is $\mathcal{O}(\sqrt{n}\log(1/\varepsilon))$. ■

3.2 A Long-Step Path-Following Algorithm

As pointed out earlier, the short step path-following algorithm is not computationally very efficient because the searching neighborhood is unnecessarily tight. Therefore, researchers developed long-step path-following algorithms. The first and a well-known long-step path-following algorithm is formally stated below:

Algorithm 3.2
Data: $\mathbf{A}$, $\mathbf{b}$, $\mathbf{c}$.
Parameters: $\gamma \in (0,1)$, $\varepsilon \in (0,1)$, *and* $0 < \sigma_{\min} < \sigma_{\max} < 1$.
Initial point $(\mathbf{x}^0, \lambda^0, \mathbf{s}^0) \in \mathcal{N}_{-\infty}(\gamma)$ *and* $\mu_0 = \frac{\mathbf{x}^{0^{\mathrm{T}}}\mathbf{s}^0}{n}$.
for *iteration* $k = 0, 1, 2, \ldots$

Step 1: If $\mu_k < \varepsilon$, stop. Otherwise continue.

Step 2: Select $\sigma \in [\sigma_{\min}, \sigma_{\max}]$ and solve the linear systems of equations (2.5) to get $(\Delta\mathbf{x}, \Delta\lambda, \Delta\mathbf{s})$.

Step 3: Choose the largest α_k in $[0,1]$ such that

$$(\mathbf{x}(\alpha), \lambda(\alpha), \mathbf{s}(\alpha)) = (\mathbf{x}^k, \lambda^k, \mathbf{s}^k) + \alpha(\Delta\mathbf{x}, \Delta\lambda, \Delta\mathbf{s}) \in \mathcal{N}_{-\infty}(\gamma). \tag{3.30}$$

Step 4: Set

$$(\mathbf{x}^{k+1}, \lambda^{k+1}, \mathbf{s}^{k+1}) = (\mathbf{x}(\alpha_k), \lambda(\alpha_k), \mathbf{s}(\alpha_k)),$$

and $\mu_{k+1} = \frac{\mathbf{x}^{{k+1}^{\mathrm{T}}}\mathbf{s}^{k+1}}{n}$. *Set $k+1 \to k$. Go back to Step 1.*

end (for)

The rest of this section is to prove the convergence of Algorithm 3.2 and estimate its corresponding polynomial bound. First, it is easy to verify that (3.11) and (3.12) hold for Algorithm 3.2. It will be shown that for all i, and $\alpha \in (0, \alpha_k]$, $x_i(\alpha)s_i(\alpha) \geq \gamma\mu(\alpha) > 0$ in Lemma 3.11, therefore, there is no chance for $x_i(\alpha) = 0$ or $s_i(\alpha) = 0$, i.e., $x_i(\alpha) > 0$ and $s_i(\alpha) > 0$. This shows

Lemma 3.9
If $(\mathbf{x}^0, \lambda^0, \mathbf{s}^0) \in \mathcal{F}^o$, then, $(\mathbf{x}^k, \lambda^k, \mathbf{s}^k) \in \mathcal{F}^o$.

Therefore, we need only to show that proximity condition defined by $\mathcal{N}_{-\infty}(\gamma)$ holds and μ_k decreases at least at a constant rate.

Lemma 3.10
Suppose that $(\mathbf{x}, \lambda, \mathbf{s}) \in \mathcal{N}_{-\infty}(\gamma)$, then

$$\|\Delta\mathbf{x} \circ \Delta\mathbf{s}\| = 2^{-3/2}\left(1 + \frac{1}{\gamma}\right)n\mu_k. \tag{3.31}$$

Proof 3.8 By using Lemma 3.3, it has

$$\begin{aligned}
\|\Delta\mathbf{x} \circ \Delta\mathbf{s}\| &\leq 2^{-3/2}\|(\mathbf{XS})^{-1/2}(-\mathbf{x} \circ \mathbf{s} + \sigma\mu_k\mathbf{e})\|^2 \\
&= 2^{-3/2}\| - (\mathbf{XS})^{1/2}\mathbf{e} + \sigma\mu_k(\mathbf{XS})^{-1/2}\mathbf{e}\|^2 \\
&= 2^{-3/2}\left(\mathbf{x}^{\mathrm{T}}\mathbf{s} - 2\sigma\mu_k\mathbf{e}^{\mathrm{T}}\mathbf{e} + \sigma^2\mu_k^2\sum_{i=1}^{n}\frac{1}{x_i s_i}\right) \\
\text{use } x_i s_i \geq \gamma\mu_k \quad &\leq 2^{-3/2}\left(\mathbf{x}^{\mathrm{T}}\mathbf{s} - 2\sigma\mu_k n + \sigma^2\mu_k^2\frac{n}{\gamma\mu_k}\right) \\
&\leq 2^{-3/2}\left(1 - 2\sigma + \frac{\sigma^2}{\gamma}\right)n\mu_k \\
\text{use } \sigma \in (0,1) \quad &\leq 2^{-3/2}\left(1 + \frac{1}{\gamma}\right)n\mu_k
\end{aligned}$$

This completes the proof. ∎

The following lemma provides the condition for $(\mathbf{x}^k, \lambda^k, \mathbf{s}^k)$ to stay in $\mathcal{N}_{-\infty}(\gamma)$.

Lemma 3.11
Let $(\mathbf{x}^k, \lambda^k, \mathbf{s}^k) \in \mathcal{N}_{-\infty}(\gamma)$. For

$$\alpha \in \left[0, 2^{3/2}\gamma\frac{(1-\gamma)\sigma}{(1+\gamma)n}\right], \tag{3.32}$$

it has $(\mathbf{x}(\alpha), \lambda(\alpha), \mathbf{s}(\alpha)) \in \mathcal{N}_{-\infty}(\gamma)$.

Proof 3.9 For $i = 1, \ldots, n$, using Lemma 3.10, we have

$$|\Delta x_i \Delta s_i| \leq \|\Delta \mathbf{x} \circ \Delta \mathbf{s}\|_2 \leq 2^{-3/2}\left(1+\frac{1}{\gamma}\right)n\mu_k. \tag{3.33}$$

Using (3.13), we have

$$\begin{aligned} x_i(\alpha)s_i(\alpha) &= x_is_i + \alpha(s_i\Delta x_i + x_i\Delta s_i) + \alpha^2\Delta x_i\Delta s_i \\ &\geq x_is_i(1-\alpha) + \alpha\sigma\mu_k - |\alpha^2\Delta x_i\Delta s_i| \\ \text{use (3.33)} \quad &\geq (1-\alpha)\gamma\,\mu_k + \alpha\sigma\mu_k - \alpha^2 2^{-3/2}\left(1+\frac{1}{\gamma}\right)n\mu_k. \end{aligned} \tag{3.34}$$

Therefore, in view of (3.16), the proximity condition

$$x_i(\alpha)s_i(\alpha) \geq \gamma\,\mu(\alpha) = \gamma(1-\alpha(1-\sigma))\mu_k > 0 \tag{3.35}$$

will be satisfied if the following condition holds:

$$(1-\alpha)\gamma\,\mu_k + \alpha\sigma\mu_k - \alpha^2 2^{-3/2}\left(1+\frac{1}{\gamma}\right)n\mu_k \geq \gamma(1-\alpha(1-\sigma))\mu_k$$

Simple manipulation gives

$$(1-\gamma)\alpha\sigma\mu_k \geq \alpha^2 2^{-3/2}\left(1+\frac{1}{\gamma}\right)n\mu_k,$$

which yields

$$\alpha \leq 2^{3/2}\gamma\frac{(1-\gamma)\sigma}{(1+\gamma)n},$$

the upper bound of (3.32). Therefore, the proximity condition (3.35) holds if (3.32) is satisfied. From (3.35), there is no chance for $x_i(\alpha) = 0$ or $s_i(\alpha) = 0$ if α meets the condition of (3.32). In view of Lemma 3.9, we have $(\mathbf{x}^k(\alpha), \lambda^k(\alpha), \mathbf{s}^k(\alpha)) \in \mathcal{N}_{-\infty}(\gamma)$. This completes the proof. ■

Now, we are ready to prove the convergence and provide the polynomial bound for Algorithm 3.2.

Theorem 3.1
Given the parameters γ, $\sigma_{\min}$, and $\sigma_{\max}$ in Algorithm 3.2, there exists a constant δ independent of n such that, for all $k \geq 0$,

$$\mu_{k+1} \leq \left(1-\frac{\delta}{n}\right)\mu_k.$$

Proof 3.10 From (3.16) and (3.32), it must have

$$\mu_{k+1} = (1-\alpha(1-\sigma))\mu_k$$

$$\leq \left(1 - 2^{3/2}\gamma\frac{(1-\gamma)\sigma}{(1+\gamma)n}(1-\sigma)\right)\mu_k. \tag{3.36}$$

Noticing that $\sigma(1-\sigma)$ is concave, its minimum is attained at one of its end points, i.e., $\sigma(1-\sigma) \geq \min\{\sigma_{\min}(1-\sigma_{\min}), \sigma_{\max}(1-\sigma_{\max})\}$ for all $\sigma \in [\sigma_{\min}, \sigma_{\max}]$. By setting

$$\delta = 2^{3/2}\gamma\frac{(1-\gamma)}{(1+\gamma)}\min\{\sigma_{\min}(1-\sigma_{\min}), \sigma_{\max}(1-\sigma_{\max})\},$$

the proof is completed. ■

Invoking Lemma 1.4, we have

Theorem 3.2
Given the parameters $\varepsilon > 0$, $\gamma \in (0,1)$, $\sigma_{\min}$, and $\sigma_{\max}$ in Algorithm 3.2, suppose that the initial point $(\mathbf{x}^0, \lambda^0, \mathbf{s}^0) \in \mathcal{N}_{-\infty}(\gamma)$ with

$$\mu_0 \leq 1/\varepsilon^{\kappa}$$

for some positive constant $\kappa = 1$. Then, there is an integer $K = \mathcal{O}(n\log(1/\varepsilon)$ such that for all $k \geq K$

$$\mu_k \leq \varepsilon.$$

Remark 3.3 Algorithm 3.2 has the polynomial bound of $\mathcal{O}(n\log(1/\varepsilon))$ which is worse than the polynomial bound $\mathcal{O}(\sqrt{n}\log(1/\varepsilon))$ of Algorithm 3.1. However, it is well-known that Algorithm 3.2 is much more efficient in numerical computation. This is the first dilemma between theoretical analysis and numerical experience. Improved algorithms will be provided in the later Chapters to fix this dilemma. ■

3.3 MTY Predictor-Corrector Algorithm

The dilemma that the short step path-following interior-point algorithm has better theoretical polynomial bound but the long step path-following interior-point algorithm has demonstrated better performance in numerical experience was known in the very beginning of the development of interior-point method. Researchers have been working on closing the gap. This section discusses the MTY predictor-correct algorithm, which is proposed for the purpose of improving the computational performance for the short step interior-point algorithm.

In the short step path-following algorithm, the parameters $\theta = 0.4$ and $\sigma = 1 - 0.4/\sqrt{n}$ are fixed, which balances the requirements of staying close to the centrality and reducing the duality measure. Noticing that reducing σ will reduce the weight of centrality and increasing θ will enlarge the neighborhood and possibly the step size to search for a smaller duality measure, the MTY predictor-corrector path-following algorithm separates the two tasks in two steps. In predictor step (when the iterate stay in a narrow neighborhood defined by $\mathcal{N}_2(0.25)$), its goal is to reduce the duality measure as much as it can, therefore, it reduces $\sigma = 0$ and increases $\theta = 0.5$ (the searching is carried out in a wider neighborhood defined by $\mathcal{N}_2(0.5)$). In the corrector step, it increases $\sigma = 1$ and reduces $\theta = 0.25$ in order to bring the iterate back to $\mathcal{N}_2(0.25)$, but does not increase the duality measure. The algorithm is formally stated as follows:

Algorithm 3.3
Data: **A**, **b**, **c**.
Parameters: $\varepsilon > 0$,
Initial point $(\mathbf{x}^0, \lambda^0, \mathbf{s}^0) \in \mathcal{N}_2(0.25)$, *and* $\mu_0 = \frac{\mathbf{x}^{0^T}\mathbf{s}^0}{n}$.
for *iteration* $k = 0, 1, 2, \ldots$

If $\mu_k < \varepsilon$, *stop. Otherwise continue.*

If k is even (predictor):

Set $\sigma = 0$ *and solve the linear systems of equations (2.5) to get* $(\Delta\mathbf{x}, \Delta\lambda, \Delta\mathbf{s})$.
Choose the largest $\alpha_k \in [0, 1]$ *such that*

$$(\mathbf{x}(\alpha), \lambda(\alpha), \mathbf{s}(\alpha)) = (\mathbf{x}^k, \lambda^k, \mathbf{s}^k) + \alpha(\Delta\mathbf{x}, \Delta\lambda, \Delta\mathbf{s}) \in \mathcal{N}_2(0.5). \quad (3.37)$$

Set

$$(\mathbf{x}^{k+1}, \lambda^{k+1}, \mathbf{s}^{k+1}) = (\mathbf{x}(\alpha_k), \lambda(\alpha_k), \mathbf{s}(\alpha_k)),$$

and $\mu_{k+1} = \frac{\mathbf{x}^{k+1^T}\mathbf{s}^{k+1}}{n}$.

Else (corrector)

Set $\sigma = 1$ *and solve the linear systems of equations (2.5) to get* $(\Delta\mathbf{x}, \Delta\lambda, \Delta\mathbf{s})$.
Set

$$(\mathbf{x}^{k+1}, \lambda^{k+1}, \mathbf{s}^{k+1}) = (\mathbf{x}^k, \lambda^k, \mathbf{s}^k) + (\Delta\mathbf{x}, \Delta\lambda, \Delta\mathbf{s}),$$

and $\mu_{k+1} = \frac{\mathbf{x}^{k+1^T}\mathbf{s}^{k+1}}{n}$.

Set $k + 1 \to k$.

end (for)

The convergence analysis can be simplified by using the results established earlier for short step and long step algorithms.

Lemma 3.12
Suppose that $(\mathbf{x}, \lambda, \mathbf{s}) \in \mathcal{N}_2(0.25)$ *and* $(\Delta\mathbf{x}, \Delta\lambda, \Delta\mathbf{s})$ *be calculated by solving the linear systems of equations (2.5) with* $\sigma = 0$. *Then,* $(\mathbf{x}(\alpha), \lambda(\alpha), \mathbf{s}(\alpha)) \in \mathcal{N}_2(0.5)$ *for all* $\alpha \in [0, \bar{\alpha}]$, *where*

$$\bar{\alpha} = \min\left(\frac{1}{2}, \left(\frac{\mu}{8\|\Delta\mathbf{x} \circ \Delta\mathbf{s}\|}\right)^{\frac{1}{2}}\right). \quad (3.38)$$

Proof 3.11 By using (3.23), $\alpha \le 0.5$, and $\sigma = 0$, we have

$$\begin{array}{lll}
\|\mathbf{x}(\alpha)\circ\mathbf{s}(\alpha)-\mu(\alpha)\mathbf{e}\| & \le & (1-\alpha)\|\mathbf{x}\circ\mathbf{s}-\mu\mathbf{e}\|+\alpha^2\|\Delta\mathbf{x}\circ\Delta\mathbf{s}\| \\
\text{[use (3.38)]} & \le & (1-\alpha)\|\mathbf{x}\circ\mathbf{s}-\mu\mathbf{e}\|+\frac{\mu\|\Delta\mathbf{x}\circ\Delta\mathbf{s}\|}{8\|\Delta\mathbf{x}\circ\Delta\mathbf{s}\|} \\
[(\mathbf{x},\lambda,\mathbf{s})\in\mathcal{N}_2(0.25)] & \le & (1-\alpha)\frac{\mu}{4}+\frac{\mu}{8} \\
 & \le & (1-\alpha)\frac{\mu}{4}+(1-\alpha)\frac{\mu}{4} \\
 & \le & 0.5\mu(\alpha).
\end{array}$$

This proves that $(\mathbf{x}(\alpha),\lambda(\alpha),\mathbf{s}(\alpha))$ meets the proximity condition of $\mathcal{N}_2(0.5)$. Therefore,

$$-0.5\mu(\alpha)\le x_i(\alpha)s_i(\alpha)-\mu(\alpha)\le 0.5\mu(\alpha),$$

hence, using Lemma 3.4 with $\sigma = 0$ gives

$$x_i(\alpha)s_i(\alpha)\ge 0.5\mu(\alpha)=0.5(1-\alpha)\mu>0.$$

There is no chance for $x_i(\alpha)=0$ or $x_i(\alpha)=0$ to occur. Positivity of $(\mathbf{x}(\alpha),\mathbf{s}(\alpha))>\mathbf{0}$ holds. Using the same argument used to prove Lemma 3.9, it is easy to verify that $\mathbf{A}\mathbf{x}(\alpha)=\mathbf{b}$ and $\mathbf{A}^{\mathrm{T}}\lambda(\alpha)+\mathbf{s}(\alpha)=\mathbf{c}$ hold. This proves $(\mathbf{x}(\alpha),\lambda(\alpha),\mathbf{s}(\alpha))\in\mathcal{N}_2(0.5)$ for all $\alpha\in[0,\bar{\alpha}]$. ∎

The next lemma estimates the reduction rate of $\mu(\alpha)$ in predictor step.

Lemma 3.13
The reduction rate of the duality measure for predictor step is given by

$$\mu_{k+1}\le\left(1+\frac{0.4}{\sqrt{n}}\right)\mu_k. \tag{3.39}$$

Proof 3.12 By using Lemma 3.5, $\theta = 0.25$, $\sigma = 0$, and $n \ge 1$, we have

$$\begin{array}{lll}
\dfrac{\mu}{8\|\Delta\mathbf{x}\circ\Delta\mathbf{s}\|} & \ge & \dfrac{\mu}{8}\dfrac{2^{3/2}(1-\theta)}{(\theta^2+n(1-\sigma)^2)\mu} \\
 & \ge & \dfrac{2^{3/2}(1-0.25)}{8(0.25^2+n)} \\
 & = & \dfrac{3\sqrt{2}}{1+16n} \\
 & \ge & \dfrac{0.16}{n}.
\end{array} \tag{3.40}$$

This shows

$$\bar{\alpha}=\min\left(\frac{1}{2},\left(\frac{\mu}{8\|\Delta\mathbf{x}\circ\Delta\mathbf{s}\|}\right)^{\frac{1}{2}}\right)\ge\min\left(\frac{1}{2},\left(\frac{0.16}{n}\right)^{\frac{1}{2}}\right)=\frac{0.4}{\sqrt{n}}.$$

Since $\sigma = 0$ in predictor step, from (3.16), we have

$$\mu_{k+1}\le\left(1-\frac{0.4}{\sqrt{n}}\right).$$

This completes the proof. ■

Then, we will show that the corrector step will bring the iterate from $\mathcal{N}_2(0.5)$ back to $\mathcal{N}_2(0.25)$ without changing the duality measure.

Lemma 3.14
Suppose that $(\mathbf{x},\lambda,\mathbf{s}) \in \mathcal{N}_2(0.5)$ *and let* $(\Delta\mathbf{x},\Delta\lambda,\Delta\mathbf{s})$ *be calculated by solving the linear systems of equations (2.5) with* $\sigma = 1$. *Then, we have*

$$(\mathbf{x}(1),\lambda(1),\mathbf{s}(1)) \in \mathcal{N}_2(0.25), \qquad \mu(1) = \mu. \tag{3.41}$$

Proof 3.13 Substituting $\sigma = 1$ into Lemma 3.4 yields $\mu(1) = \mu$. Substituting $\sigma = 1$, $\theta = 0.5$, and $\alpha = 1$ into (3.24) gives

$$\|\mathbf{X}(1)\mathbf{S}(1)\mathbf{e} - \mu(1)\mathbf{e}\| \leq 0.25\mu = 0.25\mu(1). \tag{3.42}$$

The proof is completed by verifying that $(\mathbf{x}(1),\lambda(1),\mathbf{s}(1)) \in \mathcal{F}^o$, which is very similar to the argument that was used to prove Lemma 3.9. ■

Invoking Lemma 1.4 and (3.39) and (3.41), we have

Theorem 3.3
Give $\varepsilon \in (0,1)$, *suppose that the initial point* $(\mathbf{x}^0,\lambda^0,\mathbf{s}^0) \subset \mathcal{N}_2(0.25)$ *and* $\mu_0 \leq 1/\varepsilon^\kappa$ *for some positive constant* $\kappa = 1$ *in Algorithm 3.3. Then, there is an integer* $K = \mathcal{O}(\sqrt{n}\log(1/\varepsilon))$ *such that*

$$\mu_k \leq \varepsilon \quad \textit{for all} \quad k \geq K.$$

Remark 3.4 MTY algorithm is proposed to improve the efficiency of the short step interior-point algorithm by enlarging the search neighborhood without sacrificing the convergence property. This idea has been used by many people to design computationally efficient algorithms. We will discuss some of the algorithms later in this book. ■

3.4 A Modified Short Step Path-Following Algorithm

Although interior-point method was viewed to be a mature discipline in mathematical programming after several research monographs were published in the 1990s [144, 164, 117], and some of the most cited books in mathematical programming [81, 106] included the method, there are still some fundamental problems that need to be answered [133]. For example, we have seen that the short step interior-point algorithm has better polynomial bound than the long step interior-point algorithm, but their performances are in the reversed order. One of the main reasons leading to this dilemma is that, for most interior-point algorithms, the selection of the centering parameter is based on heuristics. To develop algorithms with the best polynomial bound, some researches use the heuristics with the sole purpose in mind of devising algorithms wherein it is easy to show the low polynomial bound without considering the efficiency in practice. To develop efficient algorithms in practice, other researches focus on the heuristics which, by intuition, will generate good iterates but ignore the problem of proving a polynomial bound.

A widely used shortcut in developing interior-point algorithms is to separate the selection of the centering parameter σ from the selection of the line-search step size [120, 68, 97, 89, 82, 172, 152]. This strategy makes the problem simple to deal with but has to use heuristics in the selection of the centering parameter. Therefore, this is not an optimal strategy.

In this section, we propose a systematic way to optimally select the centering parameters and line-search step size at the same time, aiming at minimizing the duality gap in all iterations. We show that this algorithm will have the best-known polynomial bound, even though the estimation is extremely conservative. We use some Netlib test problems to demonstrate that the proposed algorithm may be very efficient compared to the very efficient algorithm MPC. The main purpose of this section is to show that optimally selecting σ_k and line-search step size at the same time will be feasible and will improve the computational efficiency significantly. This strategy will be used in Chapters 7 and 8 for arc-search algorithms. The materials presented in this section are based on [154].

3.4.1 A modified interior point algorithm for LP

Starting from any initial point $(\mathbf{x}^0, \lambda^0, \mathbf{s}^0)$ in a central-path neighborhood $\mathcal{N}_2(\theta)$ that satisfies $(\mathbf{x}^0, \mathbf{s}^0) > \mathbf{0}$ and $\|\mathbf{x}^0\mathbf{s}^0 - \mu_0\mathbf{e}\| \leq \theta\mu$, instead of searching along the central-path, which is difficult to find in practice, we consider searching along a line inside $\mathcal{F}^o(\theta)$, defined as follows:

$$(\mathbf{x}(\alpha,\sigma), \lambda(\alpha,\sigma), \mathbf{s}(\alpha,\sigma)) := (\mathbf{x}^k - \alpha\dot{\mathbf{x}}(\sigma), \lambda^k - \alpha\dot{\lambda}(\sigma), \mathbf{s}^k - \alpha\dot{\mathbf{s}}(\sigma)), \tag{3.43}$$

where $\alpha \in [0,1]$, $\sigma \in [0,1]$, and $(\dot{\mathbf{x}}(\sigma), \dot{\lambda}(\sigma), \dot{\mathbf{s}}(\sigma))$ is defined by

$$\begin{bmatrix} \mathbf{A} & \mathbf{0} & \mathbf{0} \\ \mathbf{0} & \mathbf{A}^{\mathrm{T}} & \mathbf{I} \\ \mathbf{S}^k & \mathbf{0} & \mathbf{X}^k \end{bmatrix} \begin{bmatrix} \dot{\mathbf{x}}(\sigma) \\ \dot{\lambda}(\sigma) \\ \dot{\mathbf{s}}(\sigma) \end{bmatrix} = \begin{bmatrix} \mathbf{0} \\ \mathbf{0} \\ \mathbf{x}^k \circ \mathbf{s}^k - \sigma\mu_k\mathbf{e} \end{bmatrix}. \tag{3.44}$$

We use $(\mathbf{x}(\alpha,\sigma), \lambda(\alpha,\sigma), \mathbf{s}(\alpha,\sigma))$ to emphasize that the updated variables are functions of both α and σ, which will be selected at the same time. Since the search stays in $\mathcal{F}^o(\theta)$, as $\mathbf{x}^k \circ \mathbf{s}^k \to \mathbf{0}$, (1.15) implies that $\mu_k \to 0$; hence, the iterates will approach to an optimal solution of (1.4) because the perturbed KKT system (2.4) reduces to KKT condition.

We will use several results that can easily be derived from (3.44). To simplify the notations, we will drop the superscript and subscript k unless a confusion may be introduced.

Lemma 3.15
Let $(\dot{\mathbf{x}}(\sigma), \dot{\lambda}(\sigma), \dot{\mathbf{s}}(\sigma))$ *be defined in (3.44). Then, the following relations hold.*

$$\mathbf{s}^{\mathrm{T}}\dot{\mathbf{x}}(\sigma) + \mathbf{x}^{\mathrm{T}}\dot{\mathbf{s}}(\sigma) = \mathbf{x}^{\mathrm{T}}\mathbf{s} - \sigma\mu n, \tag{3.45}$$

$$\mu(\alpha,\sigma) = \frac{\mathbf{x}(\alpha,\sigma)^{\mathrm{T}}\mathbf{s}(\alpha,\sigma)}{n} = \mu(1-\alpha(1-\sigma)). \tag{3.46}$$

Proof 3.14 The first relation is straightforward, using the third row of (3.44), therefore, the proof is omitted. For the second relation, Pre-multiplying $\dot{\mathbf{x}}(\sigma)^{\mathrm{T}}$ to both sides of the second row of (3.44) shows that $\dot{\mathbf{x}}(\sigma)^{\mathrm{T}}\dot{\mathbf{s}}(\sigma) = 0$. Using this relation, together with (3.43) and (3.45), we have

$$\begin{aligned}
\mu(\alpha,\sigma) &= \frac{\mathbf{x}(\alpha,\sigma)^{\mathrm{T}}\mathbf{s}(\alpha,\sigma)}{n} \\
&= \frac{(\mathbf{x}^k - \alpha\dot{\mathbf{x}}(\sigma))^{\mathrm{T}}(\mathbf{s}^k - \alpha\dot{\mathbf{s}}(\sigma))}{n} \\
&= \frac{\mathbf{x}^{k^{\mathrm{T}}}\mathbf{s}^k}{n} - \alpha\frac{\dot{\mathbf{s}}(\sigma)^{\mathrm{T}}\mathbf{x}^k + \dot{\mathbf{x}}(\sigma)^{\mathrm{T}}\mathbf{s}^k}{n} \\
&= \mu(1-\alpha(1-\sigma)).
\end{aligned}$$

This completes the proof. ■

Lemma 3.16
Let $(\dot{\mathbf{x}}(\sigma), \dot{\lambda}(\sigma), \dot{s}(\sigma))$ *be defined in (3.44). Assume that* $(\mathbf{x}, \lambda, \mathbf{s}) \in \mathcal{F}^o(\theta)$*. Then, the following relations hold.*

$$\mathbf{A}\mathbf{x}(\alpha,\sigma) = \mathbf{b}, \quad \mathbf{A}^{\mathrm{T}}\lambda(\alpha,\sigma) + \mathbf{s}(\alpha,\sigma) = \mathbf{c}. \tag{3.47}$$

Proof 3.15 The proof is straightforward and is, therefore, omitted. ■

Let $\hat{\mathbf{A}}$ be an orthonormal basis for the null space of $\mathbf{A}$. The following lemma will be needed later in this section.

Lemma 3.17
Let $(\dot{\mathbf{x}}(\sigma), \dot{\lambda}(\sigma), \dot{\mathbf{s}}(\sigma))$ *be defined in (3.44). Then, the following relations hold.*

$$\dot{\mathbf{x}}(\sigma) = \hat{\mathbf{A}}(\hat{\mathbf{A}}^{\mathrm{T}}\mathbf{S}\mathbf{X}^{-1}\hat{\mathbf{A}})^{-1}\hat{\mathbf{A}}^{\mathrm{T}}(\mathbf{S}\mathbf{e} - \sigma\mathbf{X}^{-1}\mu\mathbf{e}) := \mathbf{p}_x - \sigma\mathbf{q}_x, \tag{3.48a}$$

$$\dot{\mathbf{s}}(\sigma) = \mathbf{A}^{\mathrm{T}}(\mathbf{A}\mathbf{X}\mathbf{S}^{-1}\mathbf{A}^{\mathrm{T}})^{-1}\mathbf{A}(\mathbf{X}\mathbf{e} - \sigma\mathbf{S}^{-1}\mu\mathbf{e}) := \mathbf{p}_s - \sigma\mathbf{q}_s, \tag{3.48b}$$

$$\dot{\lambda}(\sigma) = -(\mathbf{A}\mathbf{X}\mathbf{S}^{-1}\mathbf{A}^{\mathrm{T}})^{-1}\mathbf{A}(\mathbf{X}\mathbf{e} - \sigma\mathbf{S}^{-1}\mu\mathbf{e}), \tag{3.48c}$$

where

$$\mathbf{p}_x = \hat{\mathbf{A}}(\hat{\mathbf{A}}^{\mathrm{T}}\mathbf{S}\mathbf{X}^{-1}\hat{\mathbf{A}})^{-1}\hat{\mathbf{A}}^{\mathrm{T}}\mathbf{S}\mathbf{e}, \quad \mathbf{q}_x = \mu\hat{\mathbf{A}}(\hat{\mathbf{A}}^{\mathrm{T}}\mathbf{S}\mathbf{X}^{-1}\hat{\mathbf{A}})^{-1}\hat{\mathbf{A}}^{\mathrm{T}}\mathbf{X}^{-1}\mathbf{e},$$

$$\mathbf{p}_s = \mathbf{A}^{\mathrm{T}}(\mathbf{AXS}^{-1}\mathbf{A}^{\mathrm{T}})^{-1}\mathbf{AXe}, \quad \mathbf{q}_s = \mu\mathbf{A}^{\mathrm{T}}(\mathbf{AXS}^{-1}\mathbf{A}^{\mathrm{T}})^{-1}\mathbf{AS}^{-1}\mathbf{e}.$$

Proof 3.16 In view of the first row of (3.44), it is clear that $\dot{\mathbf{x}}(\sigma)$ is a vector in the null space of $\mathbf{A}$. From the first two rows of (3.44), we have, for some vector $\mathbf{v}$,

$$\mathbf{X}^{-1}\dot{\mathbf{x}}(\sigma) = \mathbf{X}^{-1}\hat{\mathbf{A}}\mathbf{v}, \quad \mathbf{S}^{-1}\mathbf{A}^{\mathrm{T}}\dot{\lambda}(\sigma) + \mathbf{S}^{-1}\dot{\mathbf{s}}(\sigma) = \mathbf{0}. \tag{3.49}$$

From the third row of (3.44), we have,

$$\mathbf{X}^{-1}\dot{\mathbf{x}}(\sigma) + \mathbf{S}^{-1}\dot{\mathbf{s}}(\sigma) = \mathbf{e} - \sigma\mu\mathbf{X}^{-1}\mathbf{S}^{-1}\mathbf{e}.$$

Substituting (3.49) into the above equation and writing the result as a matrix form yields

$$[\mathbf{X}^{-1}\hat{\mathbf{A}}, -\mathbf{S}^{-1}\mathbf{A}^{\mathrm{T}}]\begin{bmatrix} \mathbf{v} \\ \dot{\lambda}(\sigma) \end{bmatrix} = \mathbf{e} - \sigma\mu\mathbf{X}^{-1}\mathbf{S}^{-1}\mathbf{e}.$$

Since $\mathbf{A}$ is full rank and $\hat{\mathbf{A}}$ is the orthonormal basis for the null space of $\mathbf{A}$, we have

$$\begin{bmatrix} (\hat{\mathbf{A}}^{\mathrm{T}}\mathbf{SX}^{-1}\hat{\mathbf{A}})^{-1}\hat{\mathbf{A}}^{\mathrm{T}}\mathbf{S} \\ -(\mathbf{AXS}^{-1}\mathbf{A}^{\mathrm{T}})^{-1}\mathbf{AX} \end{bmatrix} [\mathbf{X}^{-1}\hat{\mathbf{A}}, -\mathbf{S}^{-1}\mathbf{A}^{\mathrm{T}}] = \mathbf{I}.$$

This gives

$$\begin{bmatrix} \mathbf{v} \\ \dot{\lambda}(\sigma) \end{bmatrix} = \begin{bmatrix} (\hat{\mathbf{A}}^{\mathrm{T}}\mathbf{SX}^{-1}\hat{\mathbf{A}})^{-1}\hat{\mathbf{A}}^{\mathrm{T}}\mathbf{S} \\ -(\mathbf{AXS}^{-1}\mathbf{A}^{\mathrm{T}})^{-1}\mathbf{AX} \end{bmatrix} (\mathbf{e} - \sigma\mu\mathbf{X}^{-1}\mathbf{S}^{-1}\mathbf{e}).$$

Substituting this equation into (3.49) proves the result. ∎

Since

$$\begin{aligned} \dot{\mathbf{x}}(\sigma)\circ\dot{\mathbf{s}}(\sigma) &= (\mathbf{p}_x - \sigma\mathbf{q}_x)\circ(\mathbf{p}_s - \sigma\mathbf{q}_s) \\ &= \mathbf{p}_x\circ\mathbf{p}_s - \sigma(\mathbf{q}_x\circ\mathbf{p}_s + \mathbf{p}_x\circ\mathbf{q}_s) + \sigma^2\mathbf{q}_x\circ\mathbf{q}_s \\ &:= \mathbf{p} - \sigma\mathbf{q} + \sigma^2\mathbf{r}, \end{aligned}$$

where

$$\mathbf{p} = \mathbf{p}_x\circ\mathbf{p}_s, \quad \mathbf{q} = \mathbf{q}_x\circ\mathbf{p}_s + \mathbf{p}_x\circ\mathbf{q}_s, \quad \mathbf{r} = \mathbf{q}_x\circ\mathbf{q}_s, \tag{3.50}$$

to make sure that $(\mathbf{x}(\alpha,\sigma), \lambda(\alpha,\sigma), \mathbf{s}(\alpha,\sigma))$ stays in $\mathcal{N}_2(\theta)$, we need to find some $\bar{\alpha}$ such that, for $\forall\alpha \in (0, \bar{\alpha}]$, the following inequality holds:

$$\|\mathbf{x}(\alpha,\sigma)\circ\mathbf{s}(\alpha,\sigma) - \mu(\alpha,\sigma)\mathbf{e}\| \leq \theta\mu(\alpha,\sigma) = \theta\mu(1-\alpha(1-\sigma)). \tag{3.51}$$

Since

$$\begin{aligned} & \|\mathbf{x}(\alpha,\sigma)\circ\mathbf{s}(\alpha,\sigma) - \mu(\alpha,\sigma)\mathbf{e}\| \\ = & \|(\mathbf{x} - \alpha\dot{\mathbf{x}})\circ(\mathbf{s} - \alpha\dot{\mathbf{s}}) - \mu(1-\alpha(1-\sigma))\mathbf{e}\| \\ = & \|\mathbf{x}\circ\mathbf{s} - \alpha(\mathbf{x}\circ\dot{\mathbf{s}} + \dot{\mathbf{x}}\circ\mathbf{s}) + \alpha^2\dot{\mathbf{x}}\circ\dot{\mathbf{s}} - \mu(1-\alpha(1-\sigma))\mathbf{e}\| \end{aligned}$$

$$
\begin{aligned}
&= \|\mathbf{x}\circ\mathbf{s}-\alpha(\mathbf{x}\circ\mathbf{s}-\sigma\mu\mathbf{e})+\alpha^2\dot{\mathbf{x}}\circ\dot{\mathbf{s}}-\mu(1-\alpha(1-\sigma))\mathbf{e}\| \\
&= \|(1-\alpha)(\mathbf{x}\circ\mathbf{s}-\mu\mathbf{e})+\alpha^2(\mathbf{p}-\sigma\mathbf{q}+\sigma^2\mathbf{r})\| \\
&\leq (1-\alpha)\|\mathbf{x}\circ\mathbf{s}-\mu\mathbf{e}\|+\alpha^2\|\mathbf{p}-\sigma\mathbf{q}+\sigma^2\mathbf{r}\|,
\end{aligned}
$$

and $\|\mathbf{x}\circ\mathbf{s}-\mu\mathbf{e}\| \leq \theta\mu$, we conclude that equation (3.51) holds if

$$h(\sigma) := \|\mathbf{p}-\sigma\mathbf{q}+\sigma^2\mathbf{r}\|^2 \leq \frac{\theta^2\sigma^2\mu^2}{\alpha^2}. \tag{3.52}$$

Equation (3.52) is a quartic polynomial (in terms of σ) inequality constraint which can be written as

$$f(\sigma,\alpha) := h(\sigma) - \frac{\theta^2\mu^2\sigma^2}{\alpha^2} = a_4\sigma^4 - a_3\sigma^3 + \left(a_2 - \frac{\theta^2\mu^2}{\alpha^2}\right)\sigma^2 - a_1\sigma + a_0 \leq 0, \tag{3.53}$$

with $h(\sigma) = a_4\sigma^4 - a_3\sigma^3 + a_2\sigma^2 - a_1\sigma + a_0 \geq 0$, and

$$
\begin{aligned}
a_0 &= \mathbf{p}^{\mathrm{T}}\mathbf{p} \geq 0, \\
a_1 &= \mathbf{q}^{\mathrm{T}}\mathbf{p}+\mathbf{p}^{\mathrm{T}}\mathbf{q}, \\
a_2 &= \mathbf{p}^{\mathrm{T}}\mathbf{r}+\mathbf{r}^{\mathrm{T}}\mathbf{p}+\mathbf{q}^{\mathrm{T}}\mathbf{q}, \\
a_3 &= \mathbf{q}^{\mathrm{T}}\mathbf{r}+\mathbf{r}^{\mathrm{T}}\mathbf{q}, \\
a_4 &= \mathbf{r}^{\mathrm{T}}\mathbf{r} \geq 0.
\end{aligned} \tag{3.54}
$$

Here, a_i, $i = 0,1,2,3,4$ are all known constants since they are functions of $\mathbf{x}$ and $\mathbf{s}$ which are known at the beginning of every iteration.

It is important to note that $f(\sigma,\alpha)$ is a monotonically increasing function of α. Therefore, for any fixed $\sigma \in [0,1]$, if for some $\bar{\alpha}$, $f(\sigma,\bar{\alpha}) \leq 0$ holds, then $f(\sigma,\alpha) \leq 0$ holds for $\forall\alpha \in (0,\bar{\alpha}]$. Using the relation that $\|\mathbf{x}(\alpha,\sigma)\circ\mathbf{s}(\alpha,\sigma)-\mu(\alpha,\sigma)\mathbf{e}\| \leq \theta\mu(\alpha,\sigma)$, we have $x_i(\alpha,\sigma)s_i(\alpha,\sigma) \geq (1-\theta)\mu(1-\alpha(1-\sigma)) > 0$ for all $\forall\alpha \in (0,\bar{\alpha}]$. This means that $(\mathbf{x}(\alpha,\sigma),\mathbf{s}(\alpha,\sigma)) > \mathbf{0}$ for all $\forall\alpha \in (0,\bar{\alpha}]$. Therefore, in the remaining discussions, we simply use α instead of $\bar{\alpha}$.

Assuming that the initial point $(\mathbf{x}^0,\lambda^0,\mathbf{s}^0) \in \mathcal{N}_2(\theta)$, we want to minimize the duality measure $\mu(\alpha,\sigma)$ in each iteration under the constraint

$$(\mathbf{x}(\alpha,\sigma),\lambda(\alpha,\sigma),\mathbf{s}(\alpha,\sigma)) \in \mathcal{N}_2(\theta).$$

Because of Lemma 3.16, (3.51), (3.52), and (3.53), the selection of α and σ in each iteration is reduced to the following optimization problem.

$$
\begin{aligned}
\min_{\alpha,\sigma} \quad & \mu(1-\alpha(1-\sigma)) \\
s.t. \quad & 0 \leq \alpha \leq 1, \quad 0 \leq \sigma \leq 1, \quad f(\sigma,\alpha) \leq 0.
\end{aligned} \tag{3.55}
$$

Since $0 \leq \alpha \leq 1$ and $0 \leq \sigma \leq 1$, we have $0 \leq \alpha(1-\sigma) \leq 1$. i.e., $0 \leq (1-\alpha(1-\sigma)) \leq 1$. This means that $0 \leq \mu(\alpha,\sigma) = \mu(1-\alpha(1-\sigma)) \leq \mu$. Clearly, if $a_0 = 0$, then, the optimization problem has a solution of $\sigma = 0$ and $\alpha = 1$ with the objective function $\mu(\alpha,\sigma) = 0$. One iteration will find the solution of (1.4). Therefore, in the rest discussions, we do not consider this simple case. Instead, we assume that $a_0 > 0$ holds in all the iterations. Let the Lagrange function be defined as follows.

$$L = \mu(1-\alpha(1-\sigma)) - \nu_1\alpha - \nu_2(1-\alpha) - \nu_3\sigma - \nu_4(1-\sigma) + \nu_5 f(\sigma,\alpha),$$

where ν_i, $i = 1,2,3,4,5$, are Lagrange multipliers. The KKT conditions for Problem (3.55) are as follows.

$$\frac{\partial L}{\partial \alpha} = -(1-\sigma)\mu - \nu_1 + \nu_2 + 2\nu_5 \frac{\sigma^2\theta^2\mu^2}{\alpha^3} = 0, \tag{3.56a}$$

$$\frac{\partial L}{\partial \sigma} = \alpha\mu - \nu_3 + \nu_4 + \nu_5\left(4a_4\sigma^3 - 3a_3\sigma^2 + 2\left(a_2 - \frac{\theta^2\mu^2}{\alpha^2}\right)\sigma - a_1\right) = 0, \tag{3.56b}$$

$$\nu_1 \geq 0,\ \nu_2 \geq 0,\ \nu_3 \geq 0,\ \nu_4 \geq 0,\ \nu_5 \geq 0, \tag{3.56c}$$

$$\nu_1\alpha = 0,\ \nu_2(1-\alpha) = 0,\ \nu_3\sigma = 0,\ \nu_4(1-\sigma) = 0,\ \nu_5 f(\sigma,\alpha) = 0, \tag{3.56d}$$

$$0 \leq \alpha \leq 1, \quad 0 \leq \sigma \leq 1, \quad f(\sigma,\alpha) \leq 0. \tag{3.56e}$$

Relations in (3.56) can be simplified because of the following claims.

Claim 1 : $\alpha \neq 0$. Otherwise, $\mu(\alpha,\sigma) = \mu$ will be the maximum.

Claim 2 : $\nu_1 = 0$ because of (3.56d) and Claim 1.

Claim 3 : $\sigma \neq 1$. Otherwise, $\mu(\alpha,\sigma) = \mu$ will be the maximum.

Claim 4 : $\nu_4 = 0$ because of (3.56d) and Claim 3.

Claim 5 : $\sigma \neq 0$. Otherwise (3.53) does not hold since $a_0 = p^{\mathrm{T}}p > 0$ is assumed.

Claim 6 : $\nu_3 = 0$ because of (3.56d) and Claim 5.

Therefore, we can rewrite the KKT conditions as follows.

$$(\sigma-1)\mu + \nu_2 + 2\nu_5\frac{\sigma^2\theta^2\mu^2}{\alpha^3} = 0, \tag{3.57a}$$

$$\alpha\mu + \nu_5\left(4a_4\sigma^3 - 3a_3\sigma^2 + 2\left(a_2 - \frac{\theta^2\mu^2}{\alpha^2}\right)\sigma - a_1\right) = 0, \tag{3.57b}$$

$$\nu_2 \geq 0,\ \nu_5 \geq 0, \tag{3.57c}$$

$$\nu_2(1-\alpha) = 0,\ \nu_5 f(\sigma,\alpha) = 0, \tag{3.57d}$$

$$0 < \alpha \leq 1, \quad 0 < \sigma < 1, \quad f(\sigma,\alpha) \leq 0. \tag{3.57e}$$

Notice that $f(\sigma,1)<0$ cannot hold for all $\sigma \in (0,1)$, otherwise let $\sigma \to 0$, then $f(\sigma,1) \to p^{\mathrm{T}}p > 0$. Therefore, we divide our discussion into two cases.

Case 1: $f(\sigma,1)=0$ has solution(s) in $\sigma \in (0,1)$. First, in view of the fact that $f(0,1)=\mathbf{p}^{\mathrm{T}}\mathbf{p}>0$, it is straightforward to check that the smallest solution of $f(\sigma,1)=0$ in $\sigma \in (0,1)$ and $\alpha = 1$ is a feasible solution and a candidate of the optimal solution that minimizes $\mu(\alpha,\sigma)=\mu(1-\alpha(1-\sigma))$ under all the constraints. Then, let us consider other feasible solutions which meet KKT condition but $\alpha<1$. Since $\alpha \neq 1$, we conclude that $v_2=0$ from (3.57d). From (3.57a), we have

$$v_5 = \frac{(1-\sigma)\alpha^3}{2\mu\sigma^2\theta^2} \neq 0.$$

The last relation follows from the facts that $\alpha \neq 0$ and $\sigma \neq 1$. Substituting v_5 into (3.57b) yields

$$\mu + \frac{(1-\sigma)\alpha^2}{2\mu\sigma^2\theta^2}\left(4a_4\sigma^3 - 3a_3\sigma^2 + 2\left(a_2 - \frac{\theta^2\mu^2}{\alpha^2}\right)\sigma - a_1)\right) = 0. \tag{3.58}$$

Since $v_5 \neq 0$, from (3.57d), we have

$$f(\sigma,\alpha) = a_4\sigma^4 - a_3\sigma^3 + \left(a_2 - \frac{\theta^2\mu^2}{\alpha^2}\right)\sigma^2 - a_1\sigma + a_0 = 0,$$

which gives,

$$0 < \frac{\theta^2\mu^2}{\alpha^2}\sigma^2 = a_4\sigma^4 - a_3\sigma^3 + a_2\sigma^2 - a_1\sigma + a_0 := h(\sigma). \tag{3.59}$$

Substituting this relation into (3.58) and simplifying the result yield

$$g(\sigma) := (2a_4 - a_3)\sigma^4 + (2a_2 - a_3)\sigma^3 - 3a_1\sigma^2 + (4a_0 + a_1)\sigma - 2a_0 = 0. \tag{3.60}$$

For all $\sigma \in (0,1)$ such that $g(\sigma)=0$, we can calculate $h(\sigma) = a_4\sigma^4 - a_3\sigma^3 + a_2\sigma^2 - a_1\sigma + a_0$, and find

$$\alpha = \frac{\theta\mu\sigma}{\sqrt{h(\sigma)}}. \tag{3.61}$$

For all pairs $(\sigma,\alpha) \in (0,1)\times(0,1)$ obtained this way, they are candidates of the optimal solutions of (3.55).

Case 2: $f(\sigma,1)>0$ for all $\sigma \in (0,1)$. For any fixed σ, since $f(\sigma,\alpha)$ is a monotonic increasing function of α and $f(\sigma,0)=-\infty$, there exists an $\alpha \in (0,1)$ such that $f(\sigma,\alpha)=0$. It is easy to see that $\alpha \neq 1$ (otherwise the constraint $f(\sigma,\alpha)\leq 0$ will not hold). Therefore, all arguments for $\alpha \neq 1$ in Case 1 apply here. Furthermore, in this case, we have a stronger condition than (3.59), i.e.,

$$\frac{\theta^2\mu^2}{\alpha^2}\sigma^2 = a_4\sigma^4 - a_3\sigma^3 + a_2\sigma^2 - a_1\sigma + a_0 := h(\sigma) > f(\sigma,1) > 0, \quad \forall \sigma \in (0,1). \tag{3.62}$$

In view of the facts that $g(0) = -2a_0 < 0$ and $g(1) = 2(a_4 - a_3 + a_2 - a_1 + a_0) = 2h(1) > 0$, $g(\sigma) = 0$ has solution(s) in $\sigma \in (0,1)$.

For any candidate pair (σ, α) of the optimal solution obtained in Cases 1 and 2, we use (3.46) to calculate $\mu(\alpha, \sigma)$ for all candidate pairs. The smallest $\mu(\alpha, \sigma)$ among all candidate pairs (σ, α) is the solution of (3.55). Now we are ready to present the algorithm.

Algorithm 3.4

Data: $\mathbf{A}$, $\mathbf{b}$, $\mathbf{c}$, $\hat{\mathbf{A}}$. *Parameters:* $\theta \in (0,1)$. *Iinitial point:* $(\mathbf{x}^0, \mathbf{y}^0, \mathbf{s}^0) \in \mathcal{F}^0$, *and* $\mu_0 = \frac{\mathbf{x}^{0^{\mathrm{T}}}\mathbf{s}^0}{n}$.

for *iteration* $k = 0, 1, 2, \ldots$

Step 1: *Calculate* $\mathbf{p}_x$, $\mathbf{q}_x$, $\mathbf{p}_s$, $\mathbf{q}_s$, $\dot{\mathbf{x}}(\sigma)$, $\dot{\mathbf{y}}(\sigma)$, *and* $\dot{\mathbf{s}}(\sigma)$ *using (3.48);* $\mathbf{p}$, $\mathbf{q}$, *and* $\mathbf{r}$ *using (3.50);* a_0, a_1, a_2, a_3, *and* a_4 *using (3.54).*

Step 2: *Select* α *and* σ *as follows.*

1. *If* $a_0 = 0$
   set $\sigma = 0$ *and* $\alpha = 1$.
2. *else* $a_0 > 0$
   (a) *Solve* $f(\sigma, 1) = 0$. *If* $f(\sigma, 1)$ *has solution(s) in* $\sigma \in (0,1)$, *the smallest solution* $\sigma \in (0,1)$ *and* $\alpha = 1$ *is a candidate of optimal solution.*
   (b) *Solve* $g(\sigma) = 0$. *If* $g(\sigma)$ *has solutions in* $\sigma \in (0,1)$, *calculate* $h(\sigma)$ *and* α *using (3.59) and (3.61); for each pair of* (σ, α), *if the pair meets* $0 < \sigma < 1$ *and* $0 < \alpha < 1$, *the pair is a candidate of solution.*
   (c) *Calculate* $\mu(\alpha, \sigma)$ *using (3.46) for all candidate pairs; select* σ *and* α *that generate the smallest* $\mu(\alpha, \sigma)$.

Step 3: *Set*

$$(\mathbf{x}(k+1), \mathbf{y}(k+1), \mathbf{s}(k+1)) = (\mathbf{x} - \alpha\dot{\mathbf{x}}(\sigma), \mathbf{y} - \alpha\dot{\mathbf{y}}(\sigma), \mathbf{s} - \alpha\dot{\mathbf{s}}(\sigma)).$$

end (for)

Based on the analysis in this section, it is easy to see the following:

Theorem 3.4
Algorithm 3.4 finds the optimal solution of problem (3.55).

Remark 3.5 The most expensive computation is in Step 1, which involves matrix inverse and products of matrices and vectors. It is worthwhile to note that the update of λ is not necessary but it is included. The computations in Step 2 involve the quartic polynomial solutions of $f(\sigma,1)$ and $g(\sigma)$ which are negligible [125]. The computational details for quartic solution are described in [162]. ∎

Remark 3.6 Comparing the the short-step path-following algorithm 3.1 and the modified short-step path-following algorithm 3.4, the former uses the fixed parameters $\theta = 0.4$, $\sigma = 1 - 0.4/\sqrt{n}$ and $\alpha = 1$, while the latter uses optimized parameters that minimize the duality measure in every iteration. Notice that the polynomial bound for algorithm 3.1 is $\mathcal{O}(\sqrt{n}\log(\frac{1}{\varepsilon}))$, therefore, the polynomial bound of Algorithm 3.4 is at least the same as or better than $\mathcal{O}(\sqrt{n}\log(\frac{1}{\varepsilon}))$. ∎

We summarize the discussion in this section in the following theorem.

Theorem 3.5
Algorithm 3.4 is convergent with the polynomial bound at least the same as or better than $\mathcal{O}(\sqrt{n}\log(\frac{1}{\varepsilon}))$.

3.4.2 Implementation and numerical test

Algorithm 3.4 is implemented in MATLAB® and test is conducted for Netlib test problems. We provide the implementation details and discuss the test result in this section.

3.4.2.1 Implementation

Algorithm 3.4 is presented in a simple form which is convenient for analysis. Some implementation details are provided here.

First, to have a large step size, we need to have a large central-path neighborhood, therefore, parameter $\theta = 0.99$ is used. Second, the program needs a stopping criterion to avoid an infinity loop, the code stops if

$$\frac{\mu}{\max\{1, \|\mathbf{c}^{\mathrm{T}}\mathbf{x}\|, \|\mathbf{b}^{\mathrm{T}}\boldsymbol{\lambda}\|\}} < 10^{-8}$$

holds, which is similar to the stopping criterion of *linprog* [172].

Our experience shows that, when iterations approach an optimal point, some x_i and/or s_j approach to zero, which introduces large numerical error in the matrix inverses of (3.48). Therefore, the following alternative formulas are used to replace (3.48). Using the QR decomposition, we can write

$$\mathbf{X}^{-0.5}\mathbf{S}^{0.5}\hat{\mathbf{A}} = \mathbf{Q}_1\mathbf{R}_1,$$

where $\mathbf{Q}_1$ is an orthonormal matrix in $\mathbb{R}^{n\times(n-m)}$, and $\mathbf{R}_1$ is an invertible triangle matrix in $\mathbf{R}^{(n-m)\times(n-m)}$. Then, we have

$$\hat{\mathbf{A}}\left(\hat{\mathbf{A}}^{\mathrm{T}}\mathbf{S}\mathbf{X}^{-1}\hat{\mathbf{A}}\right)^{-1}\hat{\mathbf{A}}^{\mathrm{T}} \tag{3.63}$$

$$= \mathbf{X}^{0.5}\mathbf{S}^{-0.5}\left(\mathbf{X}^{-0.5}\mathbf{S}^{0.5}\hat{\mathbf{A}}\left(\hat{\mathbf{A}}^{\mathrm{T}}\mathbf{S}\mathbf{X}^{-1}\hat{\mathbf{A}}\right)^{-1}\hat{\mathbf{A}}^{\mathrm{T}}\mathbf{X}^{-0.5}\mathbf{S}^{0.5}\right)\mathbf{X}^{0.5}\mathbf{S}^{-0.5}$$
$$= \mathbf{X}^{0.5}\mathbf{S}^{-0.5}\mathbf{Q}_1\mathbf{Q}_1^{\mathrm{T}}\mathbf{X}^{0.5}\mathbf{S}^{-0.5}.$$

Therefore,

$$\mathbf{p}_x = \mathbf{X}^{0.5}\mathbf{S}^{-0.5}\mathbf{Q}_1\mathbf{Q}_1^{\mathrm{T}}\mathbf{X}^{0.5}\mathbf{S}^{0.5}\mathbf{e}, \quad \mathbf{q}_x = \mu\mathbf{X}^{0.5}\mathbf{S}^{-0.5}\mathbf{Q}_1\mathbf{Q}_1^{\mathrm{T}}\mathbf{X}^{-0.5}\mathbf{S}^{-0.5}\mathbf{e}, \tag{3.64}$$

Similarly, we can write

$$\mathbf{X}^{0.5}\mathbf{S}^{-0.5}\mathbf{A}^{\mathrm{T}} = \mathbf{Q}_2\mathbf{R}_2,$$

where $\mathbf{Q}_2$ is an orthonormal matrix in $\mathbb{R}^{n\times m}$, and $\mathbf{R}_2$ is an invertible triangle matrix in $\mathbb{R}^{m\times m}$,

$$\mathbf{A}^{\mathrm{T}}\left(\mathbf{A}\mathbf{S}^{-1}\mathbf{X}\mathbf{A}^{\mathrm{T}}\right)^{-1}\mathbf{A} = \mathbf{X}^{-0.5}\mathbf{S}^{0.5}\mathbf{Q}_2\mathbf{Q}_2^{\mathrm{T}}\mathbf{X}^{-0.5}\mathbf{S}^{0.5}, \tag{3.65}$$

and

$$\mathbf{p}_s = \mathbf{X}^{-0.5}\mathbf{S}^{0.5}\mathbf{Q}_2\mathbf{Q}_2^{\mathrm{T}}\mathbf{X}^{0.5}\mathbf{S}^{0.5}\mathbf{e}, \quad \mathbf{q}_s = \mu\mathbf{X}^{-0.5}\mathbf{S}^{0.5}\mathbf{Q}_2\mathbf{Q}_2^{\mathrm{T}}\mathbf{X}^{-0.5}\mathbf{S}^{-0.5}\mathbf{e}, \tag{3.66}$$

Remark 3.7 It is observed that formulas (3.64) and (3.66) produce a much more accurate result than (3.48) when iterations approach to the optimal solution. For sparse matrix **A**, we can use sparse QR decomposition [26]. ∎

3.4.2.2 Test on Netlib problems

Numerical tests have been performed for linear programming problems in Netlib library. For Netlib problems, [20] has classified these problems into two categories: Problems with strict interior-point and problems without strict interior-point. Though the MATLAB code and other existing codes can solve problems without strict interior-point, we are most interested in the problems with strict interior-point that is assumed by all feasible interior-point methods. Among these problems, we only choose problems which are presented in standard form and whose **A** matrices are full rank. The selected problems are solved using our MATLAB function *optimalAlphaSigma* and function *linprog* in MATLAB optimization toolbox.

For several reasons, it is impossible to be completely fair in the comparison of the test results obtained by *optimalAlphaSigma* and *linprog*. First, there is no detail about the initial point selection in *linprog*. Second, *linprog* does not allow to start from user selected initial point other than the one provided by *linprog*. Third, there is no information on what pre-process is actually used before *linprog* starts to run MPC, we only know from [82] that pre-process "generally increases computational efficiency, often substantial".

We compare the two codes simply by using the iteration numbers for the tested problem which are listed in Table 3.1. Only two Netlib problems that are classified as problems with strict interior-point and are presented in standard form are not included in the table because the PC computer used in the testing does not have enough memory to handle problems of this size.

Table 3.1: Iteration counts for test problems in Netlib and MATLAB.

Problem	iterations used by different algorithms optimalAlphaSigma	linprog	source
AFIRO	4	7	Netlib
blend	13	12	Netlib
SCAGR25	5	16	Netlib
SCAGR7	7	12	Netlib
SCSD1	18	10	Netlib
SCSD6	26	12	Netlib
SCSD8	19	11	Netlib
SCTAP1	17	17	Netlib
SCTAP2	17	18	Netlib
SCTAP3	18	18	Netlib
SHARE1B	11	22	Netlib

For all problems, *optimalAlphaSigma* starts with $\mathbf{x} = \mathbf{s} = \mathbf{e}$. A pre-process described in [20] is used to find an initial point before Algorithm 3.4 runs. The initial point used in *linprog* is said to be similar to the one used in [82], with some minor modifications (see [172]).

This result is very impressive because *optimalAlphaSigma* does not have a "corrector step" which is used by MPC and many other algorithms. Although corrector step is not as expensive as "predictor step", it still requires some substantial numerical operations.

3.5 A Higher-Order Feasible Interior-Point Algorithm

Monteiro, Adler, and Resende [97] proposed a higher-order path-following algorithm using the ∞-norm neighborhood

$$\max_{i=1,\ldots,n} |x_i s_i - \mu| \leq \theta\mu. \tag{3.67}$$

The main idea is to use higher-order derivatives to approximate the central path and search for a new iterate along this approximated path. They proved that, for r-order path-following algorithm, this algorithm has the polynomial bound of

$$\mathcal{O}\left(n^{\frac{1+1/r}{2}}[\max(\log n, \log(1/\varepsilon), \log\mu_0)]^{(1+1/r)}\right).$$

When $r \to \infty$, this bound approaches the best bound of the short step path-following algorithm. However, for second-order ($r = 2$) path-following algorithm, this bound is approximately $\mathcal{O}(n^{\frac{3}{4}}[\max(\log n, \log(1/\varepsilon), \log\mu_0)]^{(3/2)}\big)$, which is worse than the bound of the short step path-following algorithm but comparable to the long step path-following algorithm.

Let

$$f_i(\mathbf{x},\mathbf{s}) = x_i s_i, \quad f_{\min} = \min_{i=1,\ldots,n} f_i, \quad f_{\max} = \max_{i=1,\ldots,n} f_i,$$

and

$$\mu_0 = \frac{f_{\max}(\mathbf{x}^0,\mathbf{s}^0) + f_{\min}(\mathbf{x}^0,\mathbf{s}^0)}{2}, \quad \theta_0 = \frac{f_{\max}(\mathbf{x}^0,\mathbf{s}^0) - f_{\min}(\mathbf{x}^0,\mathbf{s}^0)}{f_{\max}(\mathbf{x}^0,\mathbf{s}^0) + f_{\min}(\mathbf{x}^0,\mathbf{s}^0)}.$$

Denote $\gamma = 1/(1-\theta_0)$ and $q(r) = \sum_{k=r+1}^{2r} p(k)$ with the sequence $p(k)$ defined recursively as follows:

$$p(1) = 1, \quad p(k) = \sum_{j=1}^{k-1} p(j)p(k-j), \quad k \geq 2.$$

Further, let $\mathbf{v}^r(\mathbf{x},\lambda,\mathbf{s},\alpha)$ be the rth-order Taylor polynomial that approximates the central path, $r \geq 1$, at $\alpha = 0$. The algorithm is presented below.

Algorithm 3.5

Data: **A**, **b**, **c**. *Parameters:* $\varepsilon \in (0,1)$. *Initial point:* $(\mathbf{x}^0,\lambda^0,\mathbf{s}^0)$ *meets (3.67), and* $\mu_0 = \frac{\mathbf{x}^{0^{\mathrm{T}}}\mathbf{s}^0}{n}$.
for *iteration* $k = 0,1,2,\ldots$

Step 1: *Calculate r-order derivatives as described in [97].*

Step 2: *Select step size* α *as follows:*

$$\alpha = \left(\left\lceil 2^{\frac{r+3}{2r}} \gamma^{\frac{r+1}{2r}} q(r)^{1/r} n^{\frac{r+1}{2r}} \lceil \log(2n\varepsilon^{-1}\mu_0) \rceil^{1/r} \right\rceil\right)^{-1}$$

Step 3: *Set* $(\mathbf{x}^{k+1},\lambda^{k+1},\mathbf{s}^{k+1}) = (\mathbf{x}^k,\lambda^k,\mathbf{s}^k) + \alpha\mathbf{v}^r(\mathbf{x},\lambda,\mathbf{s},\alpha)$.

Step 4: *Set k:=k+1.*

end (for)

For a more detailed convergence analysis, the readers are referred to [97].

3.6 Concluding Remarks

In this chapter, we have presented five feasible interior-point algorithms. We also discussed three methods that are aimed at improving the efficiency of the path-following algorithms: (a) enlarge the neighborhood, (b) optimize both α and σ at the same time, (c) use high-order derivative in the algorithms. All these methods are proved to improve the computational efficiency. However, except for method (b), these methods appear to increase the polynomial bounds of the algorithms, which makes these methods less attractive in theory. We will resolve this dilemma in Chapters 5–8, when we introduce arc-search methods where method (b) plays an important role.

Chapter 4

Infeasible Interior-Point Method Algorithms for LP

All algorithms introduced so far are feasible interior-point algorithms that require the initial point to meet the equality constraints in the KKT conditions (1.8) and strictly feasible (i.e., inside $\mathcal{F}^o$). This requirement, however, turns out to be computationally difficult for two reasons. First, it may take substantial effort to find a feasible interior-point to start with; second, many practical problems (or benchmark problems like Netlib problems) do not even have an interior-point [20]. A brilliant idea is to start with an infeasible interior-point that meets only strict positive condition and then reduce residual of the equality constraints constantly during the iteration process.

This idea was one of the reasons that Mehrotra's method had a huge success. At first, people thought that it may not be possible to have a rigorous analysis leading to a global and polynomial convergence result for infeasible interior-point algorithms. Surprisingly, Kojima, Megiddo, and Mizuno [68] and Zhang [171] proved polynomial convergence for some long step infeasible interior-point algorithms. Miao [90] gave a proof for short step infeasible interior-point algorithms. This chapter provides the derivations of the polynomial bounds for infeasible interior-point algorithms obtained in the 1990s. These polynomial bounds, however, are worse than the ones established in Chapter 3 for the feasible interior-point algorithms, which contradicts our expectation because better polynomial bounds should correspond to more efficient algorithms. We will show in Chapters 6 and 8 that, by using arc-search method, the polynomial bounds for infeasible interior-point algorithms can be as good as the ones for feasible interior-point algorithms.

4.1 A Short Step Infeasible Interior-Point Algorithm

This algorithm is proposed by Miao [90]. Although infeasible interior-point algorithms may start from a point that does not meet the equality constraints of the KKT conditions, these constraints have to be met when the algorithms converge. To characterize how close the iterate is to meeting the equality requirements, the residuals of primal programming and dual programming are defined as:

$$\begin{aligned} \mathbf{r}_b^k &= \mathbf{A}\mathbf{x}^k - \mathbf{b}, \\ \mathbf{r}_c^k &= \mathbf{A}^{\mathrm{T}}\lambda^k + \mathbf{s}^k - \mathbf{c}. \end{aligned} \tag{4.1}$$

To simplify the notation, we oftentimes omit the superscript k if it will not cause confusion. The narrow infeasible neighborhood under consideration is given by

$$\mathcal{N}_2^I(\theta) = \{(\mathbf{x},\mathbf{s}) \mid (\mathbf{x},\mathbf{s}) > \mathbf{0}, \quad \|\mathbf{X}\mathbf{s} - \mu\mathbf{e}\| \leq \theta\mu\}, \tag{4.2}$$

where $\theta \in (0,1)$. The initial point of the short step (or MTY-type) infeasible interior-point algorithm is in $\mathcal{N}_2^I(0.25)$. We also define the ε-solution set as

$$\mathcal{S}_\varepsilon = \{(\mathbf{x},\lambda,\mathbf{s}) \mid (\mathbf{x},\mathbf{s}) > \mathbf{0}, \quad \mathbf{x}^{\mathrm{T}}\mathbf{s} \leq \varepsilon, \quad \|\mathbf{r}_b\| \leq \varepsilon, \quad \|\mathbf{r}_c\| \leq \varepsilon\}. \tag{4.3}$$

A linear system of equations to be solved in the predictor step of the short step infeasible interior-point algorithm is

$$\begin{bmatrix} \mathbf{A} & \mathbf{0} & \mathbf{0} \\ \mathbf{0} & \mathbf{A}^{\mathrm{T}} & \mathbf{I} \\ \mathbf{S}^k & \mathbf{0} & \mathbf{X}^k \end{bmatrix} \begin{bmatrix} \Delta\mathbf{x}^p \\ \Delta\lambda^p \\ \Delta\mathbf{s}^p \end{bmatrix} = \begin{bmatrix} -\mathbf{r}_b^k \\ -\mathbf{r}_c^k \\ -\mathbf{x}^k \circ \mathbf{s}^k \end{bmatrix}, \tag{4.4}$$

and the iterate at predictor step is obtained by searching for an α such that the duality measure is reduced and

$$(\mathbf{x}(\alpha),\lambda(\alpha),\mathbf{s}(\alpha)) = (\mathbf{x}^k,\lambda^k,\mathbf{s}^k) + \alpha(\Delta\mathbf{x}^p,\Delta\lambda^p,\Delta\mathbf{s}^p) \in \mathcal{N}_2^I(0.5). \tag{4.5}$$

The linear system of equations to be solved in corrector step of the short step infeasible interior-point algorithm is

$$\begin{bmatrix} \mathbf{A} & \mathbf{0} & \mathbf{0} \\ \mathbf{0} & \mathbf{A}^{\mathrm{T}} & \mathbf{I} \\ \mathbf{S}(\alpha) & \mathbf{0} & \mathbf{X}(\alpha) \end{bmatrix} \begin{bmatrix} \Delta\mathbf{x}^c \\ \Delta\lambda^c \\ \Delta\mathbf{s}^c \end{bmatrix} = \begin{bmatrix} \mathbf{0} \\ \mathbf{0} \\ (1-\alpha_k)\mu_k\mathbf{e} - \mathbf{x}(\alpha) \circ \mathbf{s}(\alpha) \end{bmatrix}, \tag{4.6}$$

where α_k is the step size taken in the predictor step.

The short step infeasible interior-point algorithm discussed in this section can be formally stated below:

Algorithm 4.1

Data: $\mathbf{A}$, $\mathbf{b}$, $\mathbf{c}$.

Parameters: $\varepsilon > 0$.

Initial point $(\mathbf{x}^0,\lambda^0,\mathbf{s}^0) \in \mathcal{N}_2^I(0.25)$, *and* $\mu_0 = \frac{\mathbf{x}^{0^{\mathrm{T}}}\mathbf{s}^0}{n}$.

for *iteration* $k = 0,1,2,\ldots$

If $(\mathbf{x}^k, \mathbf{s}^k) \in \mathcal{S}_\varepsilon$, *stop. Otherwise continue.*

If k is even (predictor):

Solve the linear systems of equations (4.4) to get $(\Delta\mathbf{x}^p, \Delta\lambda^p, \Delta\mathbf{s}^p)$.

Choose the largest $\alpha_k \in [0,1]$ *such that*

$$(\mathbf{x}(\alpha), \lambda(\alpha), \mathbf{s}(\alpha)) \in \mathcal{N}_2^I(0.5), \quad \alpha \leq \alpha_k.$$

Set $\mu(\alpha_k) = (1-\alpha_k)\mu_k$.

Else (corrector)

Solve the linear systems of equations (4.6) to get $(\Delta\mathbf{x}^c, \Delta\lambda^c, \Delta\mathbf{s}^c)$.

Set

$$(\mathbf{x}^{k+1}, \lambda^{k+1}, \mathbf{s}^{k+1}) = (\mathbf{x}(\alpha_k), \lambda(\alpha_k), \mathbf{s}(\alpha_k)) + (\Delta\mathbf{x}^c, \Delta\lambda^c, \Delta\mathbf{s}^c),$$

and $\mu_{k+1} = \frac{\mathbf{x}^{k+1^{\mathrm{T}}}\mathbf{s}^{k+1}}{n}$.

Set $k+1 \rightarrow k$.

end (for)

Remark 4.1 It must be emphasized that, unlike most algorithms, where $\mu(\alpha_k) := \mathbf{x}(\alpha_k)^{\mathrm{T}}\mathbf{s}(\alpha_k)/n$ is the duality measure, for this algorithm, we define $\mu(\alpha_k) = (1-\alpha_k)\mu_k$. ▮

The rest of the section is to establish the polynomial bound for this algorithm. Let parameter ν_k be defined as

$$\nu_0 = 1, \quad \nu_k = \prod_{i=0}^{k-1}(1-\alpha_i) \tag{4.7}$$

In view of (4.6), it is easy to see

$$\Delta\mathbf{x}^{c^{\mathrm{T}}}\Delta\mathbf{s}^c = 0, \quad \mathbf{x}(\alpha_k)^{\mathrm{T}}\Delta\mathbf{s}^c + \mathbf{s}(\alpha_k)^{\mathrm{T}}\Delta\mathbf{x}^c = (1-\alpha_k)n\mu_k - \mathbf{x}(\alpha_k)^{\mathrm{T}}\mathbf{s}(\alpha_k). \tag{4.8}$$

The first lemma shows that the primal and dual residuals and duality gap decrease in every iteration.

Lemma 4.1
Let $(\mathbf{x}^k, \lambda^k, \mathbf{s}^k)$ *be generated by Algorithm 4.1. Then,*

$$\mathbf{r}_b^{k+1} = (1-\alpha_k)\mathbf{r}_b^k = \nu_{k+1}\mathbf{r}_b^0, \quad \mathbf{r}_c^{k+1} = (1-\alpha_k)\mathbf{r}_c^k = \nu_{k+1}\mathbf{r}_c^0. \tag{4.9}$$

Moreover,

$$\mathbf{x}^{k+1^{\mathrm{T}}}\mathbf{s}^{k+1} = (1-\alpha_k)\mathbf{x}^{k^{\mathrm{T}}}\mathbf{s}^k = \nu_{k+1}\mathbf{x}^{0^{\mathrm{T}}}\mathbf{s}^0. \tag{4.10}$$

Proof 4.1 Using the definition of $\mathbf{r}_b^{k+1}$, (4.6), and (4.4), we have

$$\begin{aligned}
\mathbf{r}_b^{k+1} &= \mathbf{A}\mathbf{x}^{k+1} - \mathbf{b} = \mathbf{A}(\mathbf{x}(\alpha_k) + \Delta\mathbf{x}^c) - \mathbf{b} \\
&= \mathbf{A}\mathbf{x}(\alpha_k) - \mathbf{b} = \mathbf{A}(\mathbf{x}^k + \alpha_k\Delta\mathbf{x}^p) - \mathbf{b} \\
&= \mathbf{A}\mathbf{x}^k - \mathbf{b} - \alpha_k\mathbf{r}_b^k = (1-\alpha_k)\mathbf{r}_b^k.
\end{aligned}$$

Similarly, using the definition of $\mathbf{r}_c^{k+1}$, (4.6), and (4.4), we can prove the second equality of (4.9).

To show (4.10), using (4.8), we have

$$\begin{aligned}
\mathbf{x}^{k+1^{\mathrm{T}}}\mathbf{s}^{k+1} &= (\mathbf{x}(\alpha_k) + \Delta\mathbf{x}^c)^{\mathrm{T}}(\mathbf{s}(\alpha_k) + \Delta\mathbf{s}^c) \\
&= \mathbf{x}(\alpha_k)^{\mathrm{T}}\mathbf{s}(\alpha_k) + (1-\alpha_k)n\mu_k - \mathbf{x}(\alpha_k)^{\mathrm{T}}\mathbf{s}(\alpha_k) \\
&= (1-\alpha_k)\mathbf{x}^{k^{\mathrm{T}}}\mathbf{s}^k.
\end{aligned}$$

This completes the proof. ■

The next lemma confirms that the predictor step is well-defined:

Lemma 4.2
If $(\mathbf{x}^k, \mathbf{s}^k) > \mathbf{0}$, then, $\mathbf{x}(\alpha_k)$ and $\mathbf{s}(\alpha_k)$ satisfies

$$\|\mathbf{x}(\alpha)\circ\mathbf{s}(\alpha) - \mu(\alpha)\mathbf{e}\| \le \theta\mu(\alpha) = 0.5(1-\alpha)\mu_k = 0.5\mu(\alpha). \tag{4.11}$$

Moreover, $(\mathbf{x}(\alpha_k), \mathbf{s}(\alpha_k)) > \mathbf{0}$.

Proof 4.2 The first claim follows directly from predictor step of the algorithm. From (4.11), the following inequality

$$x_i(\alpha)s_i(\alpha) \ge (1-\theta)(1-\alpha)\mu_k > 0$$

holds for all α with $\alpha_k \ge \alpha > 0$. There is no chance for $x_i(\alpha) = 0$ or $s_i(\alpha) = 0$. Therefore, $(\mathbf{x}^k, \mathbf{s}^k) > \mathbf{0}$ implies that $(\mathbf{x}(\alpha_k), \mathbf{s}(\alpha_k)) > \mathbf{0}$. ■

Remark 4.2 Lemma 4.2 indicates that $(\mathbf{x}(\alpha_k), \lambda(\alpha_k), \mathbf{s}(\alpha_k)) \in \mathcal{N}_2^I(0.5)$. ■

The following lemma is given in [91, 90].

Lemma 4.3
Let $(\Delta\mathbf{x}^c, \Delta\mathbf{s}^c)$ be given by solving (4.6). Then,

$$\|\Delta\mathbf{x}^c \circ \Delta\mathbf{s}^c\| \le \frac{\sqrt{2}}{4}\|(\mathbf{X}(\alpha_k)\mathbf{S}(\alpha_k))^{-\frac{1}{2}}(\mathbf{x}(\alpha_k)\circ\mathbf{s}(\alpha_k) - \mu(\alpha_k)\mathbf{e})\|^2. \tag{4.12}$$

Proof 4.3 Applying Lemma 3.3 to the last row of (4.6) proves this lemma. ∎

Another basic property of the algorithm is given below.

Lemma 4.4
Assume that $(\mathbf{x}^0, \lambda^0, \mathbf{s}^0) \in \mathcal{N}_2^I(0.25)$. *Let* $(\mathbf{x}^k, \lambda^k, \mathbf{s}^k)$ *be generated by Algorithm 4.1. Then, for all* $k \geq 0$

$$(\mathbf{x}^{k+1}, \lambda^{k+1}, \mathbf{s}^{k+1}) \in \mathcal{N}_2^I(0.25). \tag{4.13}$$

Proof 4.4 In view of Lemma 4.2, we have $(\mathbf{x}(\alpha), \lambda(\alpha), \mathbf{s}(\alpha)) \in \mathcal{N}_2^I(0.5)$. Therefore, $\|\mathbf{x}(\alpha) \circ \mathbf{s}(\alpha) - \mu(\alpha)\mathbf{e}\| \leq 0.5\mu(\alpha)$ holds. This gives $0.5\mu(\alpha) \leq x_i(\alpha)s_i(\alpha)$, which means that

$$\left\|(\mathbf{X}(\alpha_k)\mathbf{S}(\alpha_k))^{-\frac{1}{2}}\right\|^2 = \left\|\text{diag}\left(\frac{1}{\sqrt{x_i(\alpha)s_i(\alpha)}}\right)\right\|^2 = \frac{1}{\min_i x_i(\alpha)s_i(\alpha)} \leq \frac{1}{0.5\mu(\alpha)}.$$

Using the last row of (4.6), we have

$$\begin{aligned} \mathbf{x}^{k+1} \circ \mathbf{s}^{k+1} - \mu_{k+1}\mathbf{e} &= (\mathbf{x}(\alpha_k) + \Delta\mathbf{x}^c) \circ (\mathbf{s}(\alpha_k) + \Delta\mathbf{s}^c) - \mu(\alpha_k)\mathbf{e} \\ &= \Delta\mathbf{x}^c \circ \Delta\mathbf{s}^c. \end{aligned}$$

Then, from these two relations and Lemma 4.3, we have

$$\begin{aligned} \|\mathbf{x}^{k+1} \circ \mathbf{s}^{k+1} - \mu_{k+1}\mathbf{e}\| &\leq \|\Delta\mathbf{x}^c \circ \Delta\mathbf{s}^c\| \\ &\leq \tfrac{\sqrt{2}}{4}\|[\mathbf{X}(\alpha_k)\mathbf{S}(\alpha_k)]^{-\frac{1}{2}}[\mathbf{x}(\alpha_k) \circ \mathbf{s}(\alpha_k) - \mu(\alpha_k)\mathbf{e}]\|^2 \\ &\leq \tfrac{\sqrt{2}}{4}\|(\mathbf{X}(\alpha_k)\mathbf{S}(\alpha_k))^{-\frac{1}{2}}\|^2\|\mathbf{x}(\alpha_k) \circ \mathbf{s}(\alpha_k) - \mu(\alpha_k)\mathbf{e}\|^2 \\ \text{[use Lemma 4.2]} \quad &\leq \tfrac{\sqrt{2}}{4}\frac{0.25\mu(\alpha_k)^2}{0.5\mu(\alpha_k)} \\ [\mu_{k+1} = \mu(\alpha_k)] \quad &= \tfrac{\sqrt{2}}{4}0.5\mu_{k+1} \\ &\leq 0.25\mu_{k+1}. \end{aligned}$$

Still, we need to show that $(\mathbf{x}^{k+1}, \mathbf{s}^{k+1}) > \mathbf{0}$. This proof is very similar to the argument used in Theorem 6.2, therefore, we omit it here. ∎

Note, that in all iterations, $(\mathbf{x}^k, \mathbf{s}^k) > \mathbf{0}$ is maintained. Lemma 4.1 indicates that if at some iteration k, $\alpha_k = 1$, then an ε-optimal solution is found. Therefore, in the rest of the analysis, we assume that $\alpha_k < 1$ for all iterations. Similar to Lemma 3.12 which was used for the MTY algorithm, the next lemma is used for short step infeasible interior-point algorithm.

Lemma 4.5
Suppose that $(\mathbf{x}^k, \lambda^k, \mathbf{s}^k) \in \mathcal{N}_2^I(0.25)$ *be generated by Algorithm 4.1. Then,* $(\mathbf{x}(\alpha), \lambda(\alpha), \mathbf{s}(\alpha)) \in \mathcal{N}_2^I(0.5)$ *for all* $\alpha \in [0, \bar{\alpha}]$, *where*

$$\bar{\alpha} = \min\left(\frac{1}{2}, \left(\frac{\mu_k}{8\|\Delta \mathbf{x}^p \circ \Delta \mathbf{s}^p\|}\right)^{\frac{1}{2}}\right), \tag{4.14}$$

i.e., $\alpha_k \geq \bar{\alpha}$.

Proof 4.5 Note that $\mu(\alpha)$ is defined as $\mu(\alpha) = (1-\alpha)\mu_k$ and

$$\begin{aligned}
\mathbf{x}(\alpha) \circ \mathbf{s}(\alpha) &= (\mathbf{x} + \alpha\Delta\mathbf{x}^p) \circ (\mathbf{s} + \alpha\Delta\mathbf{s}^p) \\
&= \mathbf{x} \circ \mathbf{s} + \alpha(\mathbf{x} \circ \Delta\mathbf{s}^p + \mathbf{s} \circ \Delta\mathbf{x}^p) + \alpha^2\Delta\mathbf{x}^p \circ \Delta\mathbf{s}^p \\
&= (1-\alpha)\mathbf{x} \circ \mathbf{s} + \alpha^2\Delta\mathbf{x}^p \circ \Delta\mathbf{s}^p.
\end{aligned}$$

By using these two relations, $\bar{\alpha} \leq 0.5$, and $\sigma = 0$, we have

$$\begin{aligned}
\|\mathbf{x}(\alpha) \circ \mathbf{s}(\alpha) - \mu(\alpha)\mathbf{e}\| &= \|\mathbf{x} \circ \mathbf{s}(1-\alpha) + \alpha^2\Delta\mathbf{x}^p \circ \Delta\mathbf{s}^p - (1-\alpha)\mu_k\mathbf{e}\| \\
&\leq (1-\alpha)\|\mathbf{x} \circ \mathbf{s} - \mu_k\mathbf{e}\| + \alpha^2\|\Delta\mathbf{x}^p \circ \Delta\mathbf{s}^p\| \\
[\text{use (4.14)}] \quad &\leq (1-\alpha)\|\mathbf{x} \circ \mathbf{s} - \mu_k\mathbf{e}\| + \tfrac{\mu_k\|\Delta\mathbf{x}^p \circ \Delta\mathbf{s}^p\|}{8\|\Delta\mathbf{x}^p \circ \Delta\mathbf{s}^p\|} \\
[(\mathbf{x}, \lambda, \mathbf{s}) \in \mathcal{N}_2^I(0.25)] \quad &\leq (1-\alpha)\tfrac{\mu_k}{4} + \tfrac{\mu_k}{8} \\
[\alpha \leq 0.5] \quad &\leq (1-\alpha)\tfrac{\mu_k}{4} + (1-\alpha)\tfrac{\mu_k}{4} \\
&\leq 0.5\mu(\alpha).
\end{aligned}$$

This completes the proof. ∎

Lemma 4.6
Suppose that the initial point of Algorithm 4.1 meets the conditions:

$$(\mathbf{x}^0, \mathbf{s}^0) \in \mathcal{N}_2^I(0.25), \quad \mathbf{x}^* \leq \rho\mathbf{x}^0, \quad \mathbf{s}^* \leq \rho\mathbf{s}^0, \tag{4.15}$$

for some sufficiently large $\rho \geq 1$. *Suppose further* $(\Delta\mathbf{x}^p, \Delta\lambda^p, \Delta\mathbf{s}^p)$ *be generated by solving (4.4). Then, there exists a positive constant* $C_1 > 1$, *independent of n such that*

$$\|\Delta\mathbf{x}^p \circ \Delta\mathbf{s}^p\| \leq C_1 n^2 \mu_k. \tag{4.16}$$

Proof 4.6 It is easy to see that (4.4) and (6.3) are identical, except that the notations are slightly different, i.e., $(\Delta\mathbf{x}^p, \Delta\lambda^p, \Delta\mathbf{s}^p)$ is used in (4.4) and $(\dot{\mathbf{x}}, \dot{\lambda}, \dot{\mathbf{s}})$ is used in (6.3). Therefore, the inequality (4.16) should be the same as (6.56), including the argument of the proof. To avoid repetition, the proof is, therefore, omitted here. ∎

Using Lemma 4.6, we have a simpler and useful expression for step size of $\bar{\alpha}$.

Theorem 4.1
Suppose that $(\mathbf{x}^k, \lambda^k, \mathbf{s}^k) \in \mathcal{N}_2^I(0.25)$ *be generated by Algorithm 4.1. Then,* $(\mathbf{x}(\alpha), \lambda(\alpha), \mathbf{s}(\alpha)) \in \mathcal{N}_2^I(0.5)$ *for all* $\alpha \in [0, \bar{\alpha}]$, *where*

$$\bar{\alpha} \geq \frac{1}{\sqrt{8}n}. \tag{4.17}$$

Therefore, from Lemma 4.1, we have

$$\mu_{k+1} \leq \left(1 - \frac{1}{\sqrt{8n}}\right)\mu_k, \tag{4.18}$$

$$\mathbf{r}_b^{k+1} \leq \left(1 - \frac{1}{\sqrt{8n}}\right)\mathbf{r}_b^k, \tag{4.19}$$

$$\mathbf{r}_c^{k+1} \leq \left(1 - \frac{1}{\sqrt{8n}}\right)\mathbf{r}_c^k. \tag{4.20}$$

Proof 4.7 Substituting (4.16) into (4.14), and noticing $C_1 > 1$ and $n > 1$ yield

$$\begin{aligned} \bar{\alpha} &\geq \min\left(\frac{1}{2}, \left(\frac{\mu_k}{8\|\Delta\mathbf{x}^p \circ \Delta\mathbf{s}^p\|}\right)^{\frac{1}{2}}\right) \\ &\geq \min\left(\frac{1}{2}, \left(\frac{\mu_k}{8C_1 n^2 \mu_k}\right)^{\frac{1}{2}}\right) \\ &\geq \min\left(\frac{1}{2}, \left(\frac{1}{8n^2}\right)^{\frac{1}{2}}\right) \\ &= \left(\frac{1}{\sqrt{8n}}\right). \end{aligned} \tag{4.21}$$

Substituting $\alpha_k \geq \bar{\alpha}$ in Lemma 4.1 proves the last claim. ■

Combining Theorems 1.4 and 4.1 gives the convergence result and polynomial bound of Algorithm 4.1.

Theorem 4.2
Let $\{(\mathbf{x}^k, \lambda^k, \mathbf{s}^k)\}$ be generated by Algorithm 4.1 with $\{(\mathbf{x}^0, \lambda^0, \mathbf{s}^0)\}$ satisfying conditions (4.15). For any given $\varepsilon \in (0,1)$, the algorithm will terminate with $\{(\mathbf{x}^k, \lambda^k, \mathbf{s}^k)\} \in \mathcal{S}_\varepsilon$ in at most $\mathcal{O}(nL)$ iterations where

$$L = \max\{\log[(\mathbf{x}^{0^{\mathrm{T}}}\mathbf{s}^0)/\varepsilon], \log(\|\mathbf{A}\mathbf{x}^0 - \mathbf{b}\|/\varepsilon), \log(\|\mathbf{A}^{\mathrm{T}}\lambda^0 + \mathbf{s}^0 - \mathbf{c}\|/\varepsilon).\}$$

4.2 A Long Step Infeasible Interior-Point Algorithm

As pointed out earlier, enlarging the search neighborhood will likely find longer step, thereby enhancing the computational performance of the algorithm. Therefore, this section introduces a long step infeasible interior-point algorithm which is based on the neighborhood of $\mathcal{N}_{-\infty}(\gamma)$, with some modifications. Similar to the short step infeasible interior-point algorithm, it is also assumed that the initial point is chosen to be a special form

$$(\mathbf{x}^0, \lambda^0, \mathbf{s}^0) = (\rho\mathbf{e}, \mathbf{0}, \rho\mathbf{e}) \tag{4.22}$$

for ρ sufficiently large. This condition is imposed for getting a convergence result. For the same purpose, another condition

$$\|(\mathbf{r}_b,\mathbf{r}_c)\| \le \left(\|(\mathbf{r}_b^0,\mathbf{r}_c^0)\|/\mu_0\right)\beta\mu,$$

is imposed, where $\beta \ge 1$ is a parameter that makes sure the initial point $(\mathbf{x}^0,\lambda^0,\mathbf{s}^0)$ is also in the modified neighborhood defined as follows:

$$\begin{aligned}\mathcal{N}^I_{-\infty}(\gamma,\beta) &= \{(\mathbf{x},\lambda,\mathbf{s}) \mid \|(\mathbf{r}_b,\mathbf{r}_c)\| \le \left(\|(\mathbf{r}_b^0,\mathbf{r}_c^0)\|/\mu_0\right)\beta\mu, \\ &\quad (\mathbf{x},\mathbf{s}) > \mathbf{0},\ x_is_i \ge \gamma\mu,\ i=1,\dots,n\}, \end{aligned} \tag{4.23}$$

where $\gamma \in (0,1)$ is a given parameter. For all points in the infeasible neighborhood $\mathcal{N}^I_{-\infty}(\gamma,\beta)$, the infeasibility is uniformly bounded by a parameter times the duality measure μ. The condition $(\mathbf{x},\mathbf{s}) > \mathbf{0}$ is required by the interior-point method; the condition $x_is_i \ge \gamma\mu$ is to prevent the iterate from going to zero before the iterate approach to an optimal solution. By forcing μ_k to zero and restricting all iterates $(\mathbf{x}^k,\lambda^k,\mathbf{s}^k)$ in the neighborhood $\mathcal{N}^I_{-\infty}(\gamma,\beta)$, it is expected that as $k\to\infty$, $(\mathbf{r}_b^k,\mathbf{r}_c^k)\to\mathbf{0}$ and an optimal solution of (1.4) is found.

The long step infeasible interior-point algorithm is formally stated as follows:

Algorithm 4.2

Data: **A**, **b**, **c**.

Parameters: $\varepsilon \in (0,1)$, $\gamma \in (0,1)$, $\beta \ge 1$, *and* $0 < \sigma_{\min} < \sigma_{\max} \le 0.5$.

Initial point $(\mathbf{x}^0,\lambda^0,\mathbf{s}^0) \in \mathcal{N}^I_{-\infty}(\gamma,\beta)$, *and* $\mu_0 = \frac{\mathbf{x}^{0\mathrm{T}}\mathbf{s}^0}{n}$.

for *iteration* $k = 0,1,2,\dots$

Step 1. If $(\mathbf{x}^k,\mathbf{s}^k) \in \mathcal{S}_\varepsilon$, *stop. Otherwise continue.*

Step 2. Solve the linear systems of equations

$$\begin{bmatrix} \mathbf{A} & \mathbf{0} & \mathbf{0} \\ \mathbf{0} & \mathbf{A}^{\mathrm{T}} & \mathbf{I} \\ \mathbf{S}^k & \mathbf{0} & \mathbf{X}^k \end{bmatrix} \begin{bmatrix} \Delta\mathbf{x}^k \\ \Delta\lambda^k \\ \Delta\mathbf{s}^k \end{bmatrix} = \begin{bmatrix} -\mathbf{r}_b^k \\ -\mathbf{r}_c^k \\ -\mathbf{x}^k\circ\mathbf{s}^k + \sigma_k\mu_k\mathbf{e} \end{bmatrix}, \tag{4.24}$$

to get $(\Delta\mathbf{x}^k,\Delta\lambda^k,\Delta\mathbf{s}^k)$.

Step 3. Choose the largest $\alpha_k \in [0,1]$ *such that*

$$(\mathbf{x}(\alpha),\lambda(\alpha),\mathbf{s}(\alpha)) = (\mathbf{x}^k,\lambda^k,\mathbf{s}^k) + \alpha(\Delta\mathbf{x}^k,\Delta\lambda^k,\Delta\mathbf{s}^k) \in \mathcal{N}^I_{-\infty}(\gamma,\beta). \tag{4.25}$$

and

$$\mu(\alpha) = \frac{\mathbf{x}(\alpha)^{\mathrm{T}}\mathbf{s}(\alpha)}{n} \le (1-0.01\alpha)\mu_k \tag{4.26}$$

hold.

Step 4. Set

$$(\mathbf{x}^{k+1}, \lambda^{k+1}, \mathbf{s}^{k+1}) = (\mathbf{x}(\alpha_k), \lambda(\alpha_k), \mathbf{s}(\alpha_k)).$$

and

$$\mu_{k+1} = \mu(\alpha_k).$$

Step 5. Set $k+1 \rightarrow k$.

end (for)

To prove the convergence of Algorithm 4.2, we need to make the following assumption:

$$\|(\mathbf{x}^*, \mathbf{s}^*)\|_\infty \le \rho. \tag{4.27}$$

The following simple relations are useful and listed as a lemma.

Lemma 4.7
Let $\mathbf{x}$ and $\mathbf{s}$ be two vectors. Then, we have

$$\max\{\|\mathbf{x}\|_\infty, \|\mathbf{s}\|_\infty\} = \|(\mathbf{x}, \mathbf{s})\|_\infty, \quad \|\mathbf{x}\|_1 + \|\mathbf{s}\|_1 = \|(\mathbf{x}, \mathbf{s})\|_1, \tag{4.28}$$

$$\|\mathbf{x}\|_\infty \le \|\mathbf{x}\|_2 \le \|\mathbf{x}\|_1 \le \sqrt{n}\|\mathbf{x}\|_2 \le n\|\mathbf{x}\|_\infty. \tag{4.29}$$

Similar to the proof of Lemma 4.1, it is easy to show the following lemma.

Lemma 4.8
Let $(\mathbf{x}^k, \lambda^k, \mathbf{s}^k)$ be generated by Algorithm 4.2. Then,

$$\mathbf{r}_b^{k+1} = (1-\alpha_k)\mathbf{r}_b^k = \nu_{k+1}\mathbf{r}_b^0, \quad \mathbf{r}_c^{k+1} = (1-\alpha_k)\mathbf{r}_c^k = \nu_{k+1}\mathbf{r}_c^0. \tag{4.30}$$

Moreover,

$$\mathbf{x}^{k+1^{\mathrm{T}}}\mathbf{s}^{k+1} = (1-\alpha_k)\mathbf{x}^{k^{\mathrm{T}}}\mathbf{s}^k = \nu_{k+1}\mathbf{x}^{0^{\mathrm{T}}}\mathbf{s}^0. \tag{4.31}$$

Proof 4.8 The proof is similar to the one of Lemma 4.1, therefore, it is omitted. ■

Assume $(\mathbf{x}^k, \lambda^k, \mathbf{s}^k) \in \mathcal{N}_{-\infty}^I(\gamma, \beta)$, Lemma 4.8 indicates that

$$\nu_k\|(\mathbf{r}_b^0, \mathbf{r}_c^0)\|/\mu_k = \|(\mathbf{r}_b^k, \mathbf{r}_c^k)\|/\mu_k \le \beta\|(\mathbf{r}_b^0, \mathbf{r}_c^0)\|/\mu_0.$$

Since this chapter considers infeasible interior-point algorithms, $(\mathbf{r}_b^0, \mathbf{r}_c^0) \neq \mathbf{0}$, which means that

$$\nu_k \le \beta\mu_k/\mu_0. \tag{4.32}$$

Let $(\bar{\mathbf{x}}, \bar{\lambda}, \bar{\mathbf{s}})$ satisfy the following conditions:

$$\mathbf{A}\bar{\mathbf{x}} = \mathbf{0}, \quad \mathbf{A}^{\mathrm{T}}\bar{\lambda} + \bar{\mathbf{s}} = \mathbf{0}, \tag{4.33}$$

then,

$$\bar{\mathbf{x}}^{\mathrm{T}}\bar{\mathbf{s}} = -\bar{\mathbf{x}}^{\mathrm{T}}\mathbf{A}^{\mathrm{T}}\bar{\lambda} = 0, \tag{4.34}$$

The following lemma is based on this simple relation.

Lemma 4.9

There exists a positive constant C_1 (depending on n, initial point, and the optimal solution) such that

$$\nu_k \|(\mathbf{x}^k, \mathbf{s}^k)\|_1 \le C_1 \mu_k, \quad \textit{for all } k. \tag{4.35}$$

Proof 4.9 Let $(\mathbf{x}^*, \lambda^*, \mathbf{s}^*)$ be any primal dual solution for (1.4) and (1.5) and $(\bar{\mathbf{x}}, \bar{\lambda}, \bar{\mathbf{s}})$ be defined as

$$(\bar{\mathbf{x}}, \bar{\lambda}, \bar{\mathbf{s}}) = \nu_k(\mathbf{x}^0, \lambda^0, \mathbf{s}^0) + (1-\nu_k)(\mathbf{x}^*, \lambda^*, \mathbf{s}^*) - (\mathbf{x}^k, \lambda^k, \mathbf{s}^k). \tag{4.36}$$

From $\mathbf{A}\mathbf{x}^k - \mathbf{b} = \mathbf{r}_b^k$ and $\nu_k \mathbf{A}\mathbf{x}^0 = \nu_k(\mathbf{r}_b^0 + \mathbf{b})$, we have

$$\begin{aligned}
\mathbf{A}\bar{\mathbf{x}} &= \mathbf{A}(\nu_k \mathbf{x}^0 + (1-\nu_k)\mathbf{x}^* - \mathbf{x}^k) \\
&= \nu_k \mathbf{A}\mathbf{x}^0 + (1-\nu_k)\mathbf{A}\mathbf{x}^* - \mathbf{A}\mathbf{x}^k \\
&= \mathbf{r}_b^k + \nu_k \mathbf{b} + (1-\nu_k)\mathbf{b} - \mathbf{r}_b^k - \mathbf{b} \\
&= \mathbf{0},
\end{aligned}$$

and

$$\begin{aligned}
\mathbf{A}^{\mathrm{T}}\bar{\lambda} + \bar{\mathbf{s}} &= \mathbf{A}^{\mathrm{T}}(\nu_k \lambda^0 + (1-\nu_k)\lambda^* - \lambda^k) + (\nu_k \mathbf{s}^0 + (1-\nu_k)\mathbf{s}^* - \mathbf{s}^k) \\
&= \nu_k \mathbf{A}^{\mathrm{T}}\lambda^0 - \mathbf{A}^{\mathrm{T}}\lambda^k + \nu_k \mathbf{s}^0 - \mathbf{s}^k + (1-\nu_k)(\mathbf{A}^{\mathrm{T}}\lambda^* + \mathbf{s}^*) \\
&= \nu_k(\mathbf{A}^{\mathrm{T}}\lambda^0 + \mathbf{s}^0) - (\mathbf{A}^{\mathrm{T}}\lambda^k + \mathbf{s}^k) + (1-\nu_k)\mathbf{c} \\
&= \nu_k(\mathbf{A}^{\mathrm{T}}\lambda^0 + \mathbf{s}^0 - \mathbf{c}) - (\mathbf{A}^{\mathrm{T}}\lambda^k + \mathbf{s}^k - \mathbf{c}) \\
&= \nu_k \mathbf{r}_c^0 - \mathbf{r}_c^k \\
&= \mathbf{0}.
\end{aligned}$$

This verifies that condition (4.33) holds, therefore, (4.34) holds. This gives

$$\begin{aligned}
0 &= \bar{\mathbf{x}}^{\mathrm{T}}\bar{\mathbf{s}} \\
&= \left(\nu_k \mathbf{x}^0 + (1-\nu_k)\mathbf{x}^* - \mathbf{x}^k\right)^{\mathrm{T}} \left(\nu_k \mathbf{s}^0 + (1-\nu_k)\mathbf{s}^* - \mathbf{s}^k\right) \\
&= \nu_k^2 \mathbf{x}^{0\mathrm{T}}\mathbf{s}^0 + (1-\nu_k)^2 \mathbf{x}^{*\mathrm{T}}\mathbf{s}^* + \nu_k(1-\nu_k)\left(\mathbf{x}^{0\mathrm{T}}\mathbf{s}^* + \mathbf{s}^{0\mathrm{T}}\mathbf{x}^*\right) \\
&\quad + \mathbf{x}^{k\mathrm{T}}\mathbf{s}^k - \nu_k\left(\mathbf{x}^{k\mathrm{T}}\mathbf{s}^0 + \mathbf{s}^{k\mathrm{T}}\mathbf{x}^0\right) - (1-\nu_k)\left(\mathbf{s}^{k\mathrm{T}}\mathbf{x}^* + \mathbf{x}^{k\mathrm{T}}\mathbf{s}^*\right).
\end{aligned} \tag{4.37}$$

Since $(\mathbf{x}^k, \mathbf{s}^k) \ge \mathbf{0}$ and $(\mathbf{x}^*, \mathbf{s}^*) \ge \mathbf{0}$, we have $\mathbf{s}^{k\mathrm{T}}\mathbf{x}^* + \mathbf{x}^{k\mathrm{T}}\mathbf{s}^* \ge 0$. Since $(\mathbf{x}^*, \lambda^*, \mathbf{s}^*)$ is an optimal solution, we have $\mathbf{x}^{*\mathrm{T}}\mathbf{s}^* = 0$. Simple rearranging of (4.37) yields

$$\nu_k\left(\mathbf{x}^{k\mathrm{T}}\mathbf{s}^0 + \mathbf{s}^{k\mathrm{T}}\mathbf{x}^0\right)$$

$$\leq \nu_k^2 \mathbf{x}^{0^{\mathrm{T}}} \mathbf{s}^0 + \nu_k(1-\nu_k)\left(\mathbf{x}^{0^{\mathrm{T}}} \mathbf{s}^* + \mathbf{s}^{0^{\mathrm{T}}} \mathbf{x}^*\right) + \mathbf{x}^{k^{\mathrm{T}}} \mathbf{s}^k. \tag{4.38}$$

Since $(\mathbf{x}^0, \mathbf{s}^0) > \mathbf{0}$, we have

$$\xi = \min_{i=1,\ldots,n} \min\{x_i^0, s_i^0\} > 0. \tag{4.39}$$

Using Lemma 4.7 gives

$$\begin{aligned} \xi \|(\mathbf{x}^k, \mathbf{s}^k)\|_1 &\leq \min_{i=1,\ldots,n} \{x_i^0\} \|\mathbf{s}^k\|_1 + \min_{i=1,\ldots,n} \{s_i^0\} \|\mathbf{x}^k\|_1 \\ &\leq \left(\mathbf{x}^{k^{\mathrm{T}}} \mathbf{s}^0 + \mathbf{s}^{k^{\mathrm{T}}} \mathbf{x}^0\right). \end{aligned} \tag{4.40}$$

Combining (4.40) and (4.38) yields

$$\begin{aligned} \xi \nu_k \|(\mathbf{x}^k, \mathbf{s}^k)\|_1 &\leq \nu_k \left(\mathbf{x}^{k^{\mathrm{T}}} \mathbf{s}^0 + \mathbf{s}^{k^{\mathrm{T}}} \mathbf{x}^0\right) \\ &\leq \nu_k^2 \mathbf{x}^{0^{\mathrm{T}}} \mathbf{s}^0 + \mathbf{x}^{k^{\mathrm{T}}} \mathbf{s}^k + \nu_k(1-\nu_k)\left(\mathbf{x}^{0^{\mathrm{T}}} \mathbf{s}^* + \mathbf{s}^{0^{\mathrm{T}}} \mathbf{x}^*\right) \\ \text{[use (4.28)]} \quad &\leq \nu_k^2 n\mu_0 + n\mu_k + \nu_k(1-\nu_k)\left(\|\mathbf{x}^0\|_\infty \|\mathbf{s}^*\|_1 + \|\mathbf{s}^0\|_\infty \|\mathbf{x}^*\|_1\right) \\ \text{[use } \nu_k \in (0,1)] \quad &\leq \nu_k n\mu_0 + n\mu_k + \nu_k\left(\|(\mathbf{x}^0, \mathbf{s}^0)\|_\infty \|(\mathbf{s}^*, \mathbf{x}^*)\|_1\right) \\ \text{[use (4.32)]} \quad &\leq \beta n\mu_k + n\mu_k + \beta\mu_k\left(\|(\mathbf{x}^0, \mathbf{s}^0)\|_\infty \|(\mathbf{s}^*, \mathbf{x}^*)\|_1/\mu_0\right). \end{aligned} \tag{4.41}$$

Setting

$$C_1 = \xi^{-1}\left(\beta n + n + \beta\|(\mathbf{x}^0, \mathbf{s}^0)\|_\infty \|(\mathbf{s}^*, \mathbf{x}^*)\|_1/\mu_0\right)$$

proves the lemma. ■

Applying the assumptions (4.22) and (4.27) into Lemma 4.9, we have

Lemma 4.10
Suppose that the initial point is selected to satisfy the assumptions (4.22) and (4.27). Then, we have

$$\rho \nu_k \|(\mathbf{x}^k, \mathbf{s}^k)\|_1 \leq 4\beta n\mu_k, \quad \textit{for all } k. \tag{4.42}$$

Proof 4.10 It is easy to see from (4.22) and (4.39) that $\xi = \rho$. Due to the choice of ξ and $(\mathbf{x}^0, \lambda^0, \mathbf{s}^0)$, we have

$$\begin{aligned} \|(\mathbf{x}^0, \mathbf{s}^0)\|_\infty &\leq \rho, \\ \|(\mathbf{s}^*, \mathbf{x}^*)\|_1 \leq 2n\|(\mathbf{s}^*, \mathbf{x}^*)\|_\infty &\leq 2n\rho, \\ \mu_0 = \mathbf{x}^{0^{\mathrm{T}}} \mathbf{s}^0/n &= \rho^2. \end{aligned}$$

Substituting these relations into (4.41) gives

$$\begin{aligned}
\rho\nu_k\|(\mathbf{x}^k,\mathbf{s}^k)\|_1 &\leq \beta n\mu_k + n\mu_k + \beta\mu_k\Big(\|(\mathbf{x}^0,\mathbf{s}^0)\|_\infty\|(\mathbf{s}^*,\mathbf{x}^*)\|_1/\mu_0\Big) \\
&\leq \beta n\mu_k + n\mu_k + \beta(\mu_k/\mu_0)2n\rho^2 \\
\text{[use } \beta\geq 1] \quad &\leq 4\beta n\mu_k.
\end{aligned}$$

This proves the lemma. ■

In the following analysis, we will use inequalities for induced norms $\|\mathbf{A}\mathbf{x}\|_p \leq \|\mathbf{A}\|_p\|\mathbf{x}\|_p$ which is given in (1.3).

Lemma 4.11
Suppose that the initial point is chosen as in (4.22) and assumption (4.27) holds. Then, there exists a constant C_2 independent of n such that

$$\|\mathbf{D}^{-1}\Delta\mathbf{x}\| \leq C_2 n\sqrt{\mu_k}, \quad \|\mathbf{D}\Delta\mathbf{s}\| \leq C_2 n\sqrt{\mu_k}. \tag{4.43}$$

Proof 4.11 Let

$$(\bar{\mathbf{x}},\bar{\lambda},\bar{\mathbf{s}}) = (\Delta\mathbf{x},\Delta\lambda,\Delta\mathbf{s}) + \nu_k(\mathbf{x}^0,\lambda^0,\mathbf{s}^0) - \nu_k(\mathbf{x}^*,\lambda^*,\mathbf{s}^*). \tag{4.44}$$

Then, using the first row of (4.24), we have

$$\begin{aligned}
\mathbf{A}\bar{\mathbf{x}} &= \mathbf{A}\Delta\mathbf{x} + \nu_k\mathbf{A}\mathbf{x}^0 - \nu_k\mathbf{A}\mathbf{x}^* \\
&= -\mathbf{r}_b^k + \nu_k(\mathbf{r}_b^0 + \mathbf{b}) - \nu_k\mathbf{b} \\
&= \mathbf{0}
\end{aligned} \tag{4.45}$$

and using the second row of (4.24), we have

$$\begin{aligned}
\mathbf{A}^{\mathrm{T}}\bar{\lambda} + \bar{\mathbf{s}} &= \mathbf{A}^{\mathrm{T}}\Delta\lambda + \nu_k\mathbf{A}^{\mathrm{T}}\lambda^0 - \nu_k\mathbf{A}^{\mathrm{T}}\lambda^* + \Delta\mathbf{s} + \nu_k\mathbf{s}^0 - \nu_k\mathbf{s}^* \\
&= (\mathbf{A}^{\mathrm{T}}\Delta\lambda + \Delta\mathbf{s}) + \nu_k(\mathbf{A}^{\mathrm{T}}\lambda^0 + \mathbf{s}^0 - \mathbf{c}) - \nu_k(\mathbf{A}^{\mathrm{T}}\lambda^* + \mathbf{s}^* - \mathbf{c}) \\
&= -\mathbf{r}_c^k + \nu_k\mathbf{r}_c^0 = \mathbf{0}.
\end{aligned} \tag{4.46}$$

This shows that $(\bar{\mathbf{x}},\bar{\lambda},\bar{\mathbf{s}})$ satisfies (4.33). From (4.34), we have

$$0 = \bar{\mathbf{x}}^{\mathrm{T}}\bar{\mathbf{s}} = (\Delta\mathbf{x} + \nu_k\mathbf{x}^0 - \nu_k\mathbf{x}^*)^{\mathrm{T}}(\Delta\mathbf{s} + \nu_k\mathbf{s}^0 - \nu_k\mathbf{s}^*). \tag{4.47}$$

Using the last row of (4.24), we have

$$\begin{aligned}
&\mathbf{S}(\Delta\mathbf{x} + \nu_k(\mathbf{x}^0 - \mathbf{x}^*)) + \mathbf{X}(\Delta\mathbf{s} + \nu_k(\mathbf{s}^0 - \mathbf{s}^*)) \\
= \; &-\mathbf{X}\mathbf{S}\mathbf{e} + \sigma_k\mu_k\mathbf{e} + \nu_k[\mathbf{S}(\mathbf{x}^0 - \mathbf{x}^*) + \mathbf{X}(\mathbf{s}^0 - \mathbf{s}^*)].
\end{aligned}$$

Pre-multiplying $(\mathbf{X}\mathbf{S})^{-1/2}$ on both sides gives

$$\begin{aligned}
&\mathbf{D}^{-1}(\Delta\mathbf{x} + \nu_k(\mathbf{x}^0 - \mathbf{x}^*)) + \mathbf{D}(\Delta\mathbf{s} + \nu_k(\mathbf{s}^0 - \mathbf{s}^*)) \\
= \; &-(\mathbf{X}\mathbf{S})^{-1/2}(\mathbf{X}\mathbf{S}\mathbf{e} - \sigma_k\mu_k\mathbf{e}) + \nu_k[\mathbf{D}^{-1}(\mathbf{x}^0 - \mathbf{x}^*) + \mathbf{D}(\mathbf{s}^0 - \mathbf{s}^*)].
\end{aligned} \tag{4.48}$$

Using (4.47), we have

$$\begin{aligned} & \|\mathbf{D}^{-1}(\Delta\mathbf{x}+\nu_k(\mathbf{x}^0-\mathbf{x}^*))+\mathbf{D}(\Delta\mathbf{s}+\nu_k(\mathbf{s}^0-\mathbf{s}^*))\|^2 \\ = & \|\mathbf{D}^{-1}(\Delta\mathbf{x}+\nu_k(\mathbf{x}^0-\mathbf{x}^*))\|^2+\|\mathbf{D}(\Delta\mathbf{s}+\nu_k(\mathbf{s}^0-\mathbf{s}^*))\|^2. \end{aligned} \tag{4.49}$$

Combining (4.48) and (4.49) gives

$$\begin{aligned} & \|\mathbf{D}^{-1}(\Delta\mathbf{x}+\nu_k(\mathbf{x}^0-\mathbf{x}^*))\|^2+\|\mathbf{D}(\Delta\mathbf{s}+\nu_k(\mathbf{s}^0-\mathbf{s}^*))\|^2 \\ \le & \left(\|(\mathbf{XS})^{-1/2}\|\|(\mathbf{XSe}-\sigma_k\mu_k\mathbf{e})\|+\nu_k\|\mathbf{D}^{-1}(\mathbf{x}^0-\mathbf{x}^*)\|+\nu_k\|\mathbf{D}(\mathbf{s}^0-\mathbf{s}^*)\|\right)^2. \end{aligned}$$

This shows

$$\begin{aligned} \|\mathbf{D}^{-1}(\Delta\mathbf{x}+\nu_k(\mathbf{x}^0-\mathbf{x}^*))\| \le & \|(\mathbf{XS})^{-1/2}\|\|(\mathbf{XSe}-\sigma_k\mu_k\mathbf{e})\| \\ & +\nu_k\left(\|\mathbf{D}^{-1}(\mathbf{x}^0-\mathbf{x}^*)\|+\|\mathbf{D}(\mathbf{s}^0-\mathbf{s}^*)\|\right). \end{aligned} \tag{4.50}$$

Using triangle inequality and (4.50) yields

$$\begin{aligned} \|\mathbf{D}^{-1}\Delta\mathbf{x}\| \le & \|\mathbf{D}^{-1}(\Delta\mathbf{x}+\nu_k(\mathbf{x}^0-\mathbf{x}^*))-\nu_k\mathbf{D}^{-1}(\mathbf{x}^0-\mathbf{x}^*)\| \\ \le & \|\mathbf{D}^{-1}(\Delta\mathbf{x}+\nu_k(\mathbf{x}^0-\mathbf{x}^*))\|+\|\nu_k\mathbf{D}^{-1}(\mathbf{x}^0-\mathbf{x}^*)\| \\ \le & \|(\mathbf{XS})^{-1/2}\|\|(\mathbf{XSe}-\sigma_k\mu_k\mathbf{e})\| \\ & +2\nu_k\|\mathbf{D}^{-1}(\mathbf{x}^0-\mathbf{x}^*)\|+2\nu_k\|\mathbf{D}(\mathbf{s}^0-\mathbf{s}^*)\|. \end{aligned} \tag{4.51}$$

Noticing that

$$\begin{aligned} \|(\mathbf{XSe}-\sigma_k\mu_k\mathbf{e})\|^2 & = \|\mathbf{XSe}\|_2^2-2\sigma_k\mu_k\mathbf{x}^{\mathrm{T}}\mathbf{s}+\sigma_k^2\mu_k^2 n \\ \text{[use (4.29)]} & \le \|\mathbf{XSe}\|_1^2-2\sigma_k n\mu_k^2+\sigma_k^2 n\mu_k^2 \\ & \le (n\mu_k)^2-2\sigma_k n\mu_k^2+\sigma_k^2 n\mu_k^2 \\ \text{[use } \sigma_k\le 0.5\text{]} & \le n^2\mu_k^2, \end{aligned}$$

and

$$\|(\mathbf{XS})^{-1/2}\| = \max_i \frac{1}{\sqrt{x_i s_i}} \le \frac{1}{\sqrt{\gamma\mu_k}}, \tag{4.52}$$

combining these two inequality gives

$$\|(\mathbf{XS})^{-1/2}\|\|(\mathbf{XSe}-\sigma_k\mu_k\mathbf{e})\| \le \frac{n}{\sqrt{\gamma}}\sqrt{\mu_k}. \tag{4.53}$$

To have an estimate for the last two terms in (4.51), we will use the following relations.

$$\|\mathbf{D}^{-1}\mathbf{e}\|_2 = \|(\mathbf{XS})^{-1/2}\mathbf{Se}\|_2 \le \|(\mathbf{XS})^{-1/2}\|\|\mathbf{s}\|_1,$$
$$\|\mathbf{De}\| \le \|(\mathbf{XS})^{-1/2}\|\|\mathbf{x}\|_1.$$

Since the initial point is chosen as in (4.22) and (4.27), we have

$$\rho\mathbf{e} \ge \mathbf{x}^0-\mathbf{x}^* \ge \mathbf{0}, \quad \rho\mathbf{e} \ge \mathbf{s}^0-\mathbf{s}^* \ge \mathbf{0},$$

therefore,

$$\|\mathbf{D}^{-1}(\mathbf{x}^0-\mathbf{x}^*)\|+\|\mathbf{D}(\mathbf{s}^0-\mathbf{s}^*)\| \le \rho(\|\mathbf{D}^{-1}\mathbf{e}\|+\|\mathbf{D}\mathbf{e}\|).$$

Using these relations, (4.52), and Lemma 4.10, we can estimate the last two terms in (4.51);

$$\begin{aligned}
& 2\nu_k\left(\|\mathbf{D}^{-1}(\mathbf{x}^0-\mathbf{x}^*)\|+\|\mathbf{D}(\mathbf{s}^0-\mathbf{s}^*)\|\right) \\
\le\ & 2\nu_k\rho\left(\|\mathbf{D}^{-1}\mathbf{e}\|+\|\mathbf{D}\mathbf{e}\|\right) \\
\le\ & 2\nu_k\rho\left(\|(\mathbf{XS})^{-1/2}\mathbf{s}\|+\|(\mathbf{XS})^{-1/2}\mathbf{x}\|\right) \\
\le\ & 2\nu_k\rho\|(\mathbf{x},\mathbf{s})\|_1\|(\mathbf{XS})^{-1/2}\| \\
\le\ & 8\beta n\mu_k\frac{1}{\sqrt{\gamma\mu_k}} \\
\le\ & \frac{8\beta n}{\sqrt{\gamma}}\sqrt{\mu_k} \qquad (4.54)
\end{aligned}$$

Let

$$C_2=\frac{9\beta}{\sqrt{\gamma}}.$$

Substituting (4.53) and (4.54) into (4.51) proves the first claim of the lemma. A similar argument can prove the second claim. ■

The last important lemma in this section is given as follows.

Lemma 4.12
Suppose that the initial point is chosen as in (4.22) and assumption (4.27) holds. Then, there exists a constant $\bar{\alpha}\in(0,1)$ such that the following three conditions are satisfied for all $\alpha\in[0,\bar{\alpha}]$ and all $k\ge 0$:

$$(\mathbf{x}^k+\alpha\Delta\mathbf{x})^{\mathrm{T}}(\mathbf{s}^k+\alpha\Delta\mathbf{s})\ge(1-\alpha)(\mathbf{x}^k)^{\mathrm{T}}\mathbf{s}^k, \qquad (4.55)$$

$$(x_i^k+\alpha\Delta x_i)(s_i^k+\alpha\Delta s_i)\ge\frac{\gamma}{n}(\mathbf{x}^k+\alpha\Delta\mathbf{x})^{\mathrm{T}}(\mathbf{s}^k+\alpha\Delta\mathbf{s}), \qquad (4.56)$$

$$(\mathbf{x}^k+\alpha\Delta\mathbf{x})^{\mathrm{T}}(\mathbf{s}^k+\alpha\Delta\mathbf{s})\le(1-0.01\alpha)(\mathbf{x}^k)^{\mathrm{T}}\mathbf{s}^k. \qquad (4.57)$$

More specifically, there exists some $C_3>0$, independent of n such that

$$\bar{\alpha}\ge\frac{C_3}{n^2}. \qquad (4.58)$$

In addition, the conditions (4.25) and (4.26) are satisfied for all $\alpha\in[0,\bar{\alpha}]$ and all $k\ge 0$

Proof 4.12 In view of (4.43), we have

$$\Delta\mathbf{x}^{\mathrm{T}}\Delta\mathbf{s} = (\mathbf{D}^{-1}\Delta\mathbf{x})^{\mathrm{T}}(\mathbf{D}\Delta\mathbf{s}) \leq \|\mathbf{D}^{-1}\Delta\mathbf{x}\|\|\mathbf{D}\Delta\mathbf{s}\| \leq C_2^2 n^2 \mu_k \tag{4.59}$$

$$|\Delta x_i \Delta s_i| = |\mathbf{D}_{ii}^{-1}\Delta x_i||\mathbf{D}_{ii}\Delta s_i| \leq \|\mathbf{D}^{-1}\Delta\mathbf{x}\|\|\mathbf{D}\Delta\mathbf{s}\| \leq C_2^2 n^2 \mu_k \tag{4.60}$$

From the last row of (4.24), we have

$$s_i\Delta x_i + x_i\Delta s_i = -x_i s_i + \sigma_k \mu_k. \tag{4.61}$$

Taking summation over $i = 1,\ldots,n$ gives

$$\mathbf{s}^{\mathrm{T}}\Delta\mathbf{x} + \mathbf{x}^{\mathrm{T}}\Delta\mathbf{s} = (\sigma_k - 1)\mathbf{x}^{\mathrm{T}}\mathbf{s}. \tag{4.62}$$

Now we find the condition for α such that (4.55) holds. Using (4.62), we have

$$\begin{array}{lcl}
(\mathbf{x}+\alpha\Delta\mathbf{x})^{\mathrm{T}}(\mathbf{s}+\alpha\Delta\mathbf{s}) & = & \mathbf{x}^{\mathrm{T}}\mathbf{s} + \alpha(\sigma_k - 1)\mathbf{x}^{\mathrm{T}}\mathbf{s} + \alpha^2\Delta\mathbf{x}^{\mathrm{T}}\Delta\mathbf{s} \\
\text{[use (4.59)]} & \geq & (1-\alpha)\mathbf{x}^{\mathrm{T}}\mathbf{s} + \alpha\sigma_k\mathbf{x}^{\mathrm{T}}\mathbf{s} - \alpha^2 C_2^2 n^2 \mu_k \\
 & \geq & (1-\alpha)\mathbf{x}^{\mathrm{T}}\mathbf{s} + (\alpha\sigma_{\min} - \alpha^2 C_2^2 n)\mathbf{x}^{\mathrm{T}}\mathbf{s}.
\end{array}$$

If the last term of the above inequality is greater than zero, i.e.,

$$\alpha \leq \frac{\sigma_{\min}}{n C_2^2}, \tag{4.63}$$

then, inequality (4.55) holds. Next, we find the condition for α such that (4.56) holds. Using (4.60) and (4.61) and $x_i^k s_i^k \geq \gamma\mu_k$, we have

$$\begin{array}{lcl}
(x_i^k + \alpha\Delta x_i)(s_i^k + \alpha\Delta s_i) & \geq & x_i^k s_i^k (1-\alpha) + \alpha\sigma_k\mu_k - \alpha^2 C_2^2 n^2 \mu_k \\
 & \geq & \gamma(1-\alpha)\mu_k + \alpha\sigma_k\mu_k - \alpha^2 C_2^2 n^2 \mu_k.
\end{array}$$

In view of (4.24) and (4.59), we have

$$\begin{array}{lcl}
\frac{\gamma}{n}(\mathbf{x}+\alpha\Delta\mathbf{x})^{\mathrm{T}}(\mathbf{s}+\alpha\Delta\mathbf{s}) & \leq & \frac{\gamma}{n}[(1-\alpha)\mathbf{x}^{\mathrm{T}}\mathbf{s} + (\alpha\sigma_k + \alpha^2 C_2^2 n)\mathbf{x}^{\mathrm{T}}\mathbf{s}] \\
 & \leq & \gamma[(1-\alpha)\mu_k + (\alpha\sigma_k + \alpha^2 C_2^2 n)\mu_k]
\end{array} \tag{4.64}$$

Combining the above two inequalities gives

$$\begin{array}{l}
\quad (x_i^k + \alpha\Delta x_i)^{\mathrm{T}}(s_i^k + \alpha\Delta s_i) - \frac{\gamma}{n}(\mathbf{x}+\alpha\Delta\mathbf{x})^{\mathrm{T}}(\mathbf{s}^{\mathrm{T}}+\alpha\Delta\mathbf{s}) \\
\geq \alpha\sigma_k\mu_k - \alpha^2 C_2^2 n^2 \mu_k - \gamma(\alpha\sigma_k + \alpha^2 C_2^2 n)\mu_k \\
\geq (1-\gamma)\alpha\sigma_{\min}\mu_k - 2\alpha^2 C_2^2 n^2 \mu_k.
\end{array}$$

Therefore, inequality (4.56) holds if the right hand side of above inequality is greater than zero, i.e., if

$$\alpha \leq \frac{(1-\gamma)\sigma_{\min}}{2n^2 C_2^2}. \tag{4.65}$$

Finally, we find the condition for α such that (4.57) holds. Using the inequality of (4.64) and $\sigma \le 0.5$, we have

$$\begin{aligned}
& \frac{1}{n}(\mathbf{x}+\alpha\Delta\mathbf{x})^{\mathrm{T}}(\mathbf{s}+\alpha\Delta\mathbf{s}) - (1-0.01\alpha)\mu_k \\
\le\ & [(1-\alpha)\mu_k + (\alpha\sigma_k + \alpha^2 C_2^2 n^2)\mu_k] - (1-0.01\alpha)\mu_k \\
\le\ & -0.99\alpha\mu_k + (0.5\alpha + \alpha^2 C_2^2 n^2)\mu_k \\
\le\ & -0.49\alpha\mu_k + \alpha^2 C_2^2 n^2 \mu_k
\end{aligned}$$

Therefore, inequality (4.57) holds if the right hand side of above inequality is smaller than zero, i.e., if

$$\alpha \le \frac{0.49}{n^2 C_2^2}. \tag{4.66}$$

Setting

$$C_3 \le \min\left\{ \frac{\sigma_{\min}}{C_2^2}, \frac{(1-\gamma)\sigma_{\min}}{2C_2^2}, \frac{0.49}{C_2^2} \right\} \tag{4.67}$$

proves (4.58). To show the last claim of the lemma, in view of (4.32), we have $\beta/\mu_0 \ge \nu_k/\mu_k$. Since

$$\mathbf{r}_b(\alpha) = \mathbf{A}\mathbf{x}(\alpha) - \mathbf{b} = (1-\alpha)\mathbf{r}_b^k, \quad \mathbf{r}_c(\alpha) = \mathbf{A}^{\mathrm{T}}\lambda(\alpha) + \mathbf{s}(\alpha) - \mathbf{c} = (1-\alpha)\mathbf{r}_c^k,$$

we have

$$\frac{\|(\mathbf{r}_b(\alpha), \mathbf{r}_c(\alpha))\|}{\mu(\alpha)} \le \frac{(1-\alpha)\|(\mathbf{r}_b^k, \mathbf{r}_c^k)\|}{(1-\alpha)\mu_k} \le \frac{\|\nu_k(\mathbf{r}_b^0, \mathbf{r}_c^0)\|}{\mu_k} \le \beta\frac{\|(\mathbf{r}_b^0, \mathbf{r}_c^0)\|}{\mu_0}.$$

This shows that the first inequality of (4.23) holds. The positivity is guaranteed because $(\mathbf{x}^0, \mathbf{s}^0) > \mathbf{0}$ and (4.55) holds for all $\alpha \le \bar{\alpha}$. Finally, $x_i^k s_i^k \ge \gamma\mu_k$ is guaranteed by (4.56) and the initial point selection. ∎

The convergence theorem is obtained by combining Lemma 4.12 and Theorem 1.4

Theorem 4.3
Let $\varepsilon \in (0,1)$ be given. Suppose that the initial point is chosen as in (4.22) and assumption (4.27) holds. Then, there exists an index K with $K = \mathcal{O}(n^2\log(1/\varepsilon))$ such that all iterates $\{(\mathbf{x}^k, \lambda^k, \mathbf{s}^k)\}$ generated by Algorithm 4.2 satisfy

$$\mu_k \le \varepsilon, \quad \textit{all } k \ge K. \tag{4.68}$$

Remark 4.3 The polynomial bound for short step infeasible interior-point algorithm, which searches for the optimizer in a narrow neighborhood, is $K = \mathcal{O}(n\log(1/\varepsilon))$, while the polynomial bound for long step infeasible interior-point algorithm, which searches for the optimizer in a wider neighborhood, is $K = \mathcal{O}(n^2\log(1/\varepsilon))$. However, computational experience shows that searching over a wider neighborhood is more efficient than searching over a narrow neighborhood. This dilemma will be resolved in Chapters 6 and 8 when arc-search is introduced. ∎

4.3 Mehrotra's Infeasible Interior-Point Algorithm

Since the 1990s, most interior-point software packages have been based on Mehrotra's algorithm because it was demonstrated to be very efficient in computational practice, which is one of the main reasons that interior-point methods have attracted so many researchers. As a matter of fact, Mehrotra's algorithm is the most efficient one among all algorithms introduced up to now in this book. Mehrotra's algorithm is based on several brilliant intuitions: (a) start from an infeasible point which avoids the computational effort for finding a feasible starting point and makes the overall algorithm much more efficient, (b) search for the optimizer in a wide neighborhood to achieve a long step size and reduce the iteration numbers, (c) scale back a little from the calculated step size to prevent the iterate from coming too close to the boundary, which may cause difficulties in the following searches, (d) make full use of Cholesky factorization so that the corrector direction can be calculated in an efficient way, (e) use a good heuristic to select the centering parameter σ, and (f) use different step sizes for primal and dual variables. The original Mehrotra's algorithm [89] is slightly different from the one discussed in this section. The original one suggested searching for the optimizer along an arc similar to [97], which was briefly discussed in Chapter 3. Because the Taylor approximation for the central-path is good only in a small area, the later version described in [144] changed the search over a corrector direction (a straight line).

The algorithm is formally provided below:

Algorithm 4.3

Data: $\mathbf{A}$, $\mathbf{b}$, $\mathbf{c}$.

Parameters: $\varepsilon > 0$,

Initial point $(\mathbf{x}^0, \lambda^0, \mathbf{s}^0)$ *with* $(\mathbf{x}^0, \mathbf{s}^0) > 0$, *and* $\mu_0 = \frac{\mathbf{x}^{0^{\mathrm{T}}}\mathbf{s}^0}{n}$.

for *iteration* $k = 0, 1, 2, \ldots$

Step 1. If $(\mathbf{x}^k, \mathbf{s}^k) \in \mathcal{S}_\varepsilon$, *stop. Otherwise continue.*

Step 2. Solve the linear systems of equations (4.4) to get $(\Delta\mathbf{x}^p, \Delta\lambda^p, \Delta\mathbf{s}^p)$.

Step 3. Calculate

$$\alpha_p = \arg\max\{\alpha \in [0,1] \mid \mathbf{x}^k + \alpha\Delta\mathbf{x}^p \geq \mathbf{0}\},$$
$$\alpha_d = \arg\max\{\alpha \in [0,1] \mid \mathbf{s}^k + \alpha\Delta\mathbf{s}^p \geq \mathbf{0}\} \tag{4.69}$$

and calculate

$$\mu_p = (\mathbf{x} + \alpha_p\Delta\mathbf{x}^p)^{\mathrm{T}}(\mathbf{s} + \alpha_d\Delta\mathbf{s}^p)/n. \tag{4.70}$$

Step 4. Set

$$\sigma_k = \left(\frac{\mu_p}{\mu_k}\right)^3. \tag{4.71}$$

Step 5. Solve the linear systems of equations

$$\begin{bmatrix} \mathbf{A} & \mathbf{0} & \mathbf{0} \\ \mathbf{0} & \mathbf{A}^{\mathrm{T}} & \mathbf{I} \\ \mathbf{S}^k & \mathbf{0} & \mathbf{X}^k \end{bmatrix} \begin{bmatrix} \Delta\mathbf{x}^c \\ \Delta\lambda^c \\ \Delta\mathbf{s}^c \end{bmatrix} = \begin{bmatrix} \mathbf{0} \\ \mathbf{0} \\ \sigma_k\mu_k\mathbf{e} - \Delta\mathbf{x}^p \circ \Delta\mathbf{s}^p \end{bmatrix} \tag{4.72}$$

to get $(\Delta\mathbf{x}^c, \Delta\lambda^c, \Delta\mathbf{s}^c)$.

Step 6. Calculate the search direction

$$(\Delta\mathbf{x}^k, \Delta\lambda^k, \Delta\mathbf{s}^k) = (\Delta\mathbf{x}^p, \Delta\lambda^p, \Delta\mathbf{s}^p) + (\Delta\mathbf{x}^c, \Delta\lambda^c, \Delta\mathbf{s}^c)$$

Step 7. Calculate the step size

$$\begin{aligned} \alpha_k^p &= \arg\max\{\alpha \in [0,1] \mid \mathbf{x}^k + \alpha\Delta\mathbf{x}^k \geq \mathbf{0}\}, \\ \alpha_k^d &= \arg\max\{\alpha \in [0,1] \mid \mathbf{s}^k + \alpha\Delta\mathbf{s}^k \geq \mathbf{0}\} \end{aligned} \tag{4.73}$$

Set $\alpha_k^p = \min(0.99\alpha_k^p, 1)$ *and* $\alpha_k^d = \min(0.99\alpha_k^d, 1)$

Step 8. Calculate the next iterate

$$\begin{aligned} \mathbf{x}^{k+1} &= \mathbf{x}^k + \alpha_k^p \Delta\mathbf{x}^k, \\ (\lambda^{k+1}, \mathbf{s}^{k+1}) &= (\lambda^k, \mathbf{s}^k) + \alpha_k^d (\Delta\lambda^k, \Delta\mathbf{s}^k). \end{aligned} \tag{4.74}$$

and $\mu_{k+1} = \frac{\mathbf{x}^{k+1^{\mathrm{T}}}\mathbf{s}^{k+1}}{n}$.

Step 9. Set $k+1 \rightarrow k$.

end (for)

We identify the intuitions from (a) to (f) applied in the algorithm one by one: (a) it is clear that initial point is infeasible because it meets only one condition $(\mathbf{x}^0, \mathbf{s}^0) > \mathbf{0}$, (b) the algorithm takes the longest step in the widest neighborhood with the only restriction $(\mathbf{x}^k, \mathbf{s}^k) \geq \mathbf{0}$ in Steps 3 and 7 when step sizes α_p, α_d, α_k^p, and α_k^d are selected. This neighborhood is wider than both wide feasible neighborhood and wide infeasible neighborhood, (c) to prevent the iterate from being too close to the boundary, a scaling back factor 0.99 is used in Step 7, (d) Cholosky factorization is performed one time but used twice in Steps 2 and 5 because the matrices in (4.4) and (4.72) are the same, which reduces the computational cost, (e) the centering parameter σ is selected using a heuristic method in Step 4, and (f) the step sizes for primal and dual variables are different, this can be seen from Steps 3 and 7.

Although Mehrotra's algorithm is computationally very efficient, much more efficient than all algorithms presented so far in this book, it has not been proved to be convergent. As a matter of fact, there exists analysis that supports a guess that Mehrotra's algorithm does not converge [19].

Remark 4.4 With extensive computational experience, this author believes that Intuition (f) may have least impact to enhance the efficiency of the algorithm if there is any. Instead of using some heuristics to select the the centering parameter σ (Intuition (e)), a better way is to optimally select the step size α and the centering parameter σ at the same time, as suggested for the modified short step path-following algorithm that was discussed in Chapter 3. Intuition (d) should always be used when higher-order methods are used because it significantly reduces the computational cost, in fact, higher-order methods should always be considered, as they find better approximation for the central path, which leads to a longer step size; in addition, arc-search should be used instead of line search as proposed in Mehrotra's method. Using intuitions (c) and (b) together is more efficient than the narrow and wide neighborhood considered so far in this book even though an analytic neighborhood may be helpful in convergence analysis. In summary, starting with infeasible starting point, using higher-order derivative and widest neighborhood, and optimally selecting σ and α at the same time, are the most important strategies for interior-point algorithms. ∎

4.4 Concluding Remarks

At this point, we have discussed the most important interior-point algorithms. We know three useful strategies: (a) higher-order method to give better search direction, (b) wide neighborhood to increase the step size, and (c) infeasible initial point to avoid the cost of finding a feasible initial point, that enhances the efficiency of the algorithms. We also know which strategies are used for what algorithm(s). In addition, we know which polynomial bounds are associated with what algorithm(s). In this section, we summarize these results and hope that this information will give us some useful insights that will direct our research.

We list these algorithms in increasing order of complexity bounds, starting from the lowest polynomial bound, which is supposed to be the most efficient algorithm(s).

(1a) Algorithm 3.1: **Feasible** short step interior-point algorithm; it uses **first-order derivatives** to get the search direction, searches for the next iterate in a **narrow neighborhood**, and its polynomial bound is $\mathcal{O}(\sqrt{n}\log L)$, where $\log L := \log(1/\varepsilon)$.

(1b) Algorithm 3.3: **Feasible** predictor-corrector interior-point algorithm; it uses **first-order derivatives** twice to first predict the iterate in a narrow neighborhood with a large θ then correct the iterate in a **narrow neighborhood** with a small θ, and its polynomial bound is $\mathcal{O}(\sqrt{n}\log L)$, where $\log L := \log(1/\varepsilon)$.

(2a) Algorithm 3.5: **Feasible** short step interior-point algorithm; it uses **higher-order derivatives** to get the search direction, searches for the next iterate in a **narrow neighborhood**, its polynomial bound is $\mathcal{O}(n^{3/4}\log L^{3/2})$ when the second-order derivatives is used, where $\log L := [\max(\log n, \log(1/\varepsilon), \log \mu_0)]$.

(2b) Algorithm 3.2: **Feasible** long step interior-point algorithm; it uses **first-order derivatives** to get the search direction, searches for the next iterate in a **wide neighborhood**, and its polynomial bound is $\mathcal{O}(n\log L)$, where $\log L := \log(1/\varepsilon)$.

(2c) Algorithm 4.1: **Infeasible** short step interior-point algorithm; it uses **second-order derivatives** to get the search direction, searches for the next iterate in a **narrow neighborhood**, and its polynomial bound is $\mathcal{O}(n\log L)$, where $\log L := \max\left(\log(\frac{\mathbf{x}^{0\mathrm{T}}\mathbf{s}^0}{\varepsilon}), \log(\frac{\|\mathbf{A}\mathbf{x}^0-\mathbf{b}\|}{\varepsilon}), \log(\frac{\|\mathbf{A}^{\mathrm{T}}\lambda^0+\mathbf{s}^0-\mathbf{c}\|}{\varepsilon})\right)$.

(3) Algorithm 4.2: **Infeasible** long step interior-point algorithm; it uses **first-order derivatives** to get the search direction, searches for the next iterate in a **wide neighborhood**, and its polynomial bound is $\mathcal{O}(n^2\log L)$, where $\log L := \log(1/\varepsilon)$.

(4) Algorithm 4.3: **Infeasible** long step interior-point algorithm; it uses **second-order derivatives** to get the search direction, searches for the next iterate in the **widest neighborhood**, and its convergence and polynomial bound are unknown.

While researchers have known for a long time that incorporating infeasible starting point, wide neighborhood, and higher-order derivatives will significantly enhance the computational efficiency, this list shows that incorporating more strategies leads to worse polynomial bounds. Given that the polynomial bound is supposed to be used to measure the efficiency of the algorithms, this is a dilemma that perplexed researchers for a long time, i.e., there is a big gap between the theory and practical computational experience.

Closing this gap is the main focus of the next part of the book. In addition to using the three known strategies, we will introduce two new key ideas: (a) use the arc-search technique, and (b) select step size α and centering parameter σ at the same time using a systematic and optimal method. Both of these ideas were proposed by this author in [151, 154].

II

ARC-SEARCH INTERIOR-POINT METHODS FOR LINEAR PROGRAMMING

Chapter 5

A Feasible Arc-Search Algorithm for LP

Since Karmarkar's algorithm [57] was proved to be polynomial and demonstrated the potential to bc computationally competitive in comparison to the simplex method, many interior-point polynomial algorithms, for example [42, 70, 71, 95, 97, 93, 114], have been developed. Since then, several software packages that demonstrated computational efficiency of the interior-point methods have been released. The most efficient ones, such as `OB1, linprog, PCx, and LOGO,`[1] are based on MPC (Mehrotra's predictor-corrector algorithm), proposed by Mehrotra [89] and refined by other researchers.

It is now believed that the state-of-the-art interior-point algorithms can be more efficient than simplex algorithm for some large problems and may not be as efficient as simplex algorithm for some other problems [133]. The popular interior-point algorithms use three strategies to improve efficiency: (a) start from infeasible initial interior-point, (b) use higher-order derivatives, and (c) search in a larger neighborhood. These strategies were discussed in previous chapters and resulted in poorer polynomial bounds.

This chapter discusses a higher-order interior-point algorithm using an arc-search idea proposed in [151, 153] for linear programming. The algorithm searches for optimizers along an ellipse that approximates the central path. This strategy is different from other higher-order methods, such as [97, 89, 41, 5], which search for optimizers either along an arc of power series approximation or along a straight line related to the first and higher-order derivatives of the central path. This chapter will prove that the proposed algorithm has polynomial complexity bound $O(n^{\frac{1}{2}}\log(1/\varepsilon))$ which is equal to the best known complexity bound established in Chapter 3 for the short

[1] `HOPDM` is based on some other higher-order algorithms in [41, 5].

step interior-point algorithm and better than the complexity bound for higher-order polynomial algorithms derived in [97]. Some numerical test will be given to show its advantage over algorithms that search for the optimizer along straight lines.

5.1 A Polynomial Arc-Search Algorithm for LP

We consider the linear programming in the standard form, described as (1.4), and its dual programming also in standard form, described as (1.5).

Throughout this chapter, we make the following assumptions.

Assumptions:

1. $\mathbf{A}$ is a full rank matrix.
2. $\mathcal{F}^o$ is not empty.

Assumption 1 is trivial, as $\mathbf{A}$ can always be reduced to meet this condition with operation counts bounded by polynomial complexity. Assumption 2 implies the existence of a central path. When Assumption 2 does not hold, it may be a problem for the algorithm discussed in this chapter. This case will be discussed in Chapters 7 and 8, where infeasible interior-point algorithms are the main topic. It is well known that $\mathbf{x} \in \mathbb{R}^n$ is an optimal solution of (1.4) if and only if $\mathbf{x}$, λ, and $\mathbf{s}$ satisfy the KKT conditions (1.8), which are restated below:

$$\mathbf{A}\mathbf{x} = \mathbf{b}, \tag{5.1a}$$

$$\mathbf{A}^{\mathrm{T}}\mathbf{y} + \mathbf{s} = \mathbf{c}, \tag{5.1b}$$

$$(\mathbf{x}, \mathbf{s}) \geq \mathbf{0}, \tag{5.1c}$$

$$x_i s_i = 0, \quad i = 1, \ldots, n. \tag{5.1d}$$

The first three conditions imply that $\mathbf{x}$ is a feasible solution of the primal problem and $(\lambda, \mathbf{s})$ is a feasible solution of the dual problem. The last condition implies that the duality gap is zero. The central path $\mathcal{C}$ is parametrized by a scalar $\tau > 0$ as follows. For each interior point $(\mathbf{x}, \lambda, \mathbf{s}) \in \mathcal{C}$ on the central path, there is a $\tau > 0$ such that

$$\mathbf{A}\mathbf{x} = \mathbf{b} \tag{5.2a}$$

$$\mathbf{A}^{\mathrm{T}}\lambda + \mathbf{s} = \mathbf{c} \tag{5.2b}$$

$$(\mathbf{x}, \mathbf{s}) > \mathbf{0} \tag{5.2c}$$

$$x_i s_i = \tau, \quad i = 1, \ldots, n. \tag{5.2d}$$

Therefore, the central path is an arc in $\mathbb{R}^{2n+m}$ parametrized as a function of τ and is denoted as

$$\mathcal{C} = \{(\mathbf{x}(\tau), \lambda(\tau), \mathbf{s}(\tau)) : \tau > 0\}. \tag{5.3}$$

As $\tau \rightarrow 0$, the central path $(\mathbf{x}(\tau), \lambda(\tau), \mathbf{s}(\tau))$ represented by (5.2) approaches to a solution of linear programming represented by (1.4). Theoretical analyses demonstrated [99] that searching along the central path is an ideal way to find optimizers.

However, there is no practical way to calculate the entire arc of the central path. All path-following algorithms discussed in Chapter 3 try (a) to search, from the current point $(\mathbf{x}, \lambda, \mathbf{s})$ along certain directions related to the tangent of the central path, to a new point that reduces the value of $\mathbf{x}^{\mathrm{T}}\mathbf{s}$ (the duality gap) and simultaneously satisfies (5.2a), (5.2b), and (5.2c), thereby moving the current point towards the solution, and (b) to stay close to the central path, thereby being able to make good progress in the next search. We will consider a central path-following algorithm that searches for the optimizers (located at the boundary of $\mathcal{F}$) along an arc that approximates the central path $\mathcal{C} \in \mathcal{F}^o \subset \mathcal{F}$.

5.1.1 Ellipsoidal approximation of the central path

First, the ellipse in three dimensional space [18] is extended to $2n+m$ dimensional space and denoted by $\mathcal{E}$. This ellipse is used to approximate the central path $\mathcal{C}$ described by (5.2), where

$$\mathcal{E} = \{(\mathbf{x}(\alpha), \lambda(\alpha), \mathbf{s}(\alpha)) : (\mathbf{x}(\alpha), \lambda(\alpha), \mathbf{s}(\alpha)) = \vec{\mathbf{a}}\cos(\alpha) + \vec{\mathbf{b}}\sin(\alpha) + \vec{\mathbf{c}}\}, \tag{5.4}$$

$\vec{\mathbf{a}} \in \mathbb{R}^{2n+m}$ and $\vec{\mathbf{b}} \in \mathbb{R}^{2n+m}$ are the axes of the ellipse, and $\vec{\mathbf{c}} \in \mathbb{R}^{2n+m}$ is the center of the ellipse. Given a point $\mathbf{y} = (\mathbf{x}, \lambda, \mathbf{s}) = (\mathbf{x}(\alpha_0), \lambda(\alpha_0), \mathbf{s}(\alpha_0)) \in \mathcal{E}$, which is close to or on the central path, we will determine $\vec{\mathbf{a}}$, $\vec{\mathbf{b}}$, $\vec{\mathbf{c}}$ and α_0 such that the first and second order derivatives of $(\mathbf{x}(\alpha_0), \lambda(\alpha_0), \mathbf{s}(\alpha_0))$ have the form as if they were on the central path (though they may not be on the central path). Therefore, we want the first and second order derivatives at $(\mathbf{x}(\alpha_0), \lambda(\alpha_0), \mathbf{s}(\alpha_0)) \in \mathcal{E}$ to satisfy

$$\begin{bmatrix} \mathbf{A} & \mathbf{0} & \mathbf{0} \\ \mathbf{0} & \mathbf{A}^{\mathrm{T}} & \mathbf{I} \\ \mathbf{S} & \mathbf{0} & \mathbf{X} \end{bmatrix} \begin{bmatrix} \dot{\mathbf{x}} \\ \dot{\lambda} \\ \dot{\mathbf{s}} \end{bmatrix} = \begin{bmatrix} \mathbf{0} \\ \mathbf{0} \\ \mathbf{x} \circ \mathbf{s} \end{bmatrix}, \tag{5.5}$$

$$\begin{bmatrix} \mathbf{A} & \mathbf{0} & \mathbf{0} \\ \mathbf{0} & \mathbf{A}^{\mathrm{T}} & \mathbf{I} \\ \mathbf{S} & \mathbf{0} & \mathbf{X} \end{bmatrix} \begin{bmatrix} \ddot{\mathbf{x}} \\ \ddot{\lambda} \\ \ddot{\mathbf{s}} \end{bmatrix} = \begin{bmatrix} \mathbf{0} \\ \mathbf{0} \\ -2\dot{\mathbf{x}} \circ \dot{\mathbf{s}} \end{bmatrix}. \tag{5.6}$$

Intuitively, this ellipse approximates the central path well when $(\mathbf{x}(\alpha_0), \lambda(\alpha_0), \mathbf{s}(\alpha_0))$ is close to the central path and $\alpha_0 \pm \varepsilon \to \alpha_0$. To simplify the notation, let

$$\mathbf{y}(\alpha) = (\mathbf{x}(\alpha), \lambda(\alpha), \mathbf{s}(\alpha)) = \vec{\mathbf{a}}\cos(\alpha) + \vec{\mathbf{b}}\sin(\alpha) + \vec{\mathbf{c}}. \tag{5.7}$$

Then,

$$\dot{\mathbf{y}}(\alpha) = (\dot{\mathbf{x}}(\alpha), \dot{\lambda}(\alpha), \dot{\mathbf{s}}(\alpha)) = -\vec{\mathbf{a}}\sin(\alpha) + \vec{\mathbf{b}}\cos(\alpha), \tag{5.8}$$

$$\ddot{\mathbf{y}}(\alpha) = (\ddot{\mathbf{x}}(\alpha), \ddot{\lambda}(\alpha), \ddot{\mathbf{s}}(\alpha)) = -\vec{\mathbf{a}}\cos(\alpha) - \vec{\mathbf{b}}\sin(\alpha). \tag{5.9}$$

It is straightforward to verify from (5.7), (5.8), and (5.9) that

$$\vec{\mathbf{a}} = -\dot{\mathbf{y}}\sin(\alpha) - \ddot{\mathbf{y}}\cos(\alpha), \quad \vec{\mathbf{b}} = \dot{\mathbf{y}}\cos(\alpha) - \ddot{\mathbf{y}}\sin(\alpha), \quad \vec{\mathbf{c}} = \mathbf{y} + \ddot{\mathbf{y}}. \tag{5.10}$$

The search along the ellipse will be carried out on the interval $\alpha \in [0, \frac{\pi}{2}]$. In the next subsection, we will show that the calculation of α_0 can be avoided.

5.1.2 Search along the approximate central path

Although one can search for a better feasible point with reduced duality gap along the ellipse defined by (5.7) which needs to compute $\vec{\mathbf{a}}$, $\vec{\mathbf{b}}$, and $\vec{\mathbf{c}}$, we will use a simplified formula that reduces the operation counts slightly and is more convenient for convergence analysis. Denote

$$\vec{\mathbf{a}} = \begin{bmatrix} \mathbf{a}_x \\ \mathbf{a}_\lambda \\ \mathbf{a}_s \end{bmatrix} = -\dot{\mathbf{y}}\sin(\alpha) - \ddot{\mathbf{y}}\cos(\alpha) = \begin{bmatrix} -\dot{\mathbf{x}}\sin(\alpha) - \ddot{\mathbf{x}}\cos(\alpha) \\ -\dot{\lambda}\sin(\alpha) - \ddot{\lambda}\cos(\alpha) \\ -\dot{\mathbf{s}}\sin(\alpha) - \ddot{\mathbf{s}}\cos(\alpha) \end{bmatrix},$$

$$\vec{\mathbf{b}} = \begin{bmatrix} \mathbf{b}_x \\ \mathbf{b}_\lambda \\ \mathbf{b}_s \end{bmatrix} = \dot{\mathbf{y}}\cos(\alpha) - \ddot{\mathbf{y}}\sin(\alpha) = \begin{bmatrix} \dot{\mathbf{x}}\cos(\alpha) - \ddot{\mathbf{x}}\sin(\alpha) \\ \dot{\lambda}\cos(\alpha) - \ddot{\lambda}\sin(\alpha) \\ \dot{\mathbf{s}}\cos(\alpha) - \ddot{\mathbf{s}}\sin(\alpha) \end{bmatrix},$$

and

$$\vec{\mathbf{c}} = \begin{bmatrix} \mathbf{c}_x \\ \mathbf{c}_\lambda \\ \mathbf{c}_s \end{bmatrix} = \mathbf{y} + \ddot{\mathbf{y}} = \begin{bmatrix} \mathbf{x} + \ddot{\mathbf{x}} \\ \lambda + \ddot{\lambda} \\ \mathbf{s} + \ddot{\mathbf{s}} \end{bmatrix}.$$

Let $\mathbf{x}(\alpha)$ and $\mathbf{s}(\alpha)$ be the updated $\mathbf{x}$ and $\mathbf{s}$ after the search, this means, $\mathbf{x} = \mathbf{x}(\alpha_0) = \mathbf{a}_x\cos(\alpha_0) + \mathbf{b}_x\sin(\alpha_0) + \mathbf{c}_x$. Using this relation, we have

$$\begin{aligned} \mathbf{x}(\alpha) &= \mathbf{a}_x\cos(\alpha_0 - \alpha) + \mathbf{b}_x\sin(\alpha_0 - \alpha) + \mathbf{c}_x \qquad (5.11) \\ &= \mathbf{a}_x(\cos(\alpha_0)\cos(\alpha) + \sin(\alpha_0)\sin(\alpha)) + \mathbf{b}_x(\sin(\alpha_0)\cos(\alpha) - \cos(\alpha_0)\sin(\alpha)) \\ &\quad + \mathbf{c}_x - \mathbf{c}_x\cos(\alpha) + \mathbf{c}_x\cos(\alpha) \\ &= \mathbf{x}\cos(\alpha) + \mathbf{a}_x\sin(\alpha_0)\sin(\alpha) - \mathbf{b}_x\cos(\alpha_0)\sin(\alpha) + \mathbf{c}_x(1 - \cos(\alpha)) \\ &= \mathbf{x}\cos(\alpha) - (\dot{\mathbf{x}}\sin(\alpha_0) + \ddot{\mathbf{x}}\cos(\alpha_0))\sin(\alpha_0)\sin(\alpha) \\ &\quad - (\dot{\mathbf{x}}\cos(\alpha_0) - \ddot{\mathbf{x}}\sin(\alpha_0))\cos(\alpha_0)\sin(\alpha) + (\mathbf{x} + \ddot{\mathbf{x}})(1 - \cos(\alpha)) \\ &= \mathbf{x} - \dot{\mathbf{x}}(\sin^2(\alpha_0)\sin(\alpha) + \cos^2(\alpha_0)\sin(\alpha)) \\ &\quad + \ddot{\mathbf{x}}(-\sin(\alpha_0)\cos(\alpha_0)\sin(\alpha) + \sin(\alpha_0)\cos(\alpha_0)\sin(\alpha) + (1 - \cos(\alpha))) \\ &= \mathbf{x} - \dot{\mathbf{x}}\sin(\alpha) + \ddot{\mathbf{x}}(1 - \cos(\alpha)). \qquad (5.12) \end{aligned}$$

Similarly,

$$\mathbf{s}(\alpha) = \mathbf{s} - \dot{\mathbf{s}}\sin(\alpha) + \ddot{\mathbf{s}}(1 - \cos(\alpha)), \quad \lambda(\alpha) = \lambda - \dot{\lambda}\sin(\alpha) + \ddot{\lambda}(1 - \cos(\alpha)). \quad (5.13)$$

As pointed out above, (5.12) and (5.13) do not explicitly depend on α_0. We summarize the above discussion as the following

Theorem 5.1
Let $(\mathbf{x}(\alpha), \lambda(\alpha), \mathbf{s}(\alpha))$ be an arc defined by (5.4) passing through a point $(\mathbf{x}, \lambda, \mathbf{s}) \in \mathcal{E}$, and its first and second derivatives at $(\mathbf{x}, \lambda, \mathbf{s})$ be $(\dot{\mathbf{x}}, \dot{\lambda}, \dot{\mathbf{s}})$ and $(\ddot{\mathbf{x}}, \ddot{\lambda}, \ddot{\mathbf{s}})$, which are defined by (5.5) and (5.6). Then, an ellipsoidal approximation of the central path is given by

$$\mathbf{x}(\alpha) = \mathbf{x} - \dot{\mathbf{x}}\sin(\alpha) + \ddot{\mathbf{x}}(1 - \cos(\alpha)), \quad (5.14)$$

$$\lambda(\alpha) = \lambda - \dot{\lambda}\sin(\alpha) + \ddot{\lambda}(1-\cos(\alpha)) \tag{5.15}$$

$$\mathbf{s}(\alpha) = \mathbf{s} - \dot{\mathbf{s}}\sin(\alpha) + \ddot{\mathbf{s}}(1-\cos(\alpha)). \tag{5.16}$$

Assuming $(\mathbf{x},\mathbf{s}) > \mathbf{0}$, one can easily see that if $\frac{\dot{\mathbf{x}}}{\mathbf{x}}$, $\frac{\ddot{\mathbf{x}}}{\mathbf{x}}$, $\frac{\dot{\mathbf{s}}}{\mathbf{s}}$, and $\frac{\ddot{\mathbf{s}}}{\mathbf{s}}$ are bounded by some constants, and if α is small enough, then $\mathbf{x}(\alpha) > \mathbf{0}$ and $\mathbf{s}(\alpha) > \mathbf{0}$.

Lemma 5.1
Let $\dot{\mathbf{x}}$, $\dot{\mathbf{s}}$, $\ddot{\mathbf{x}}$, and $\ddot{\mathbf{s}}$ be the solution of (5.5) *and* (5.6). *Then,*

$$\dot{\mathbf{x}}^{\mathrm{T}}\dot{\mathbf{s}} = 0, \quad \ddot{\mathbf{x}}^{\mathrm{T}}\dot{\mathbf{s}} = 0, \quad \dot{\mathbf{x}}^{\mathrm{T}}\ddot{\mathbf{s}} = 0, \quad \ddot{\mathbf{x}}^{\mathrm{T}}\ddot{\mathbf{s}} = 0.$$

Proof 5.1 Pre-multiplying $\dot{\mathbf{x}}^{\mathrm{T}}$ or $\ddot{\mathbf{x}}^{\mathrm{T}}$ to the second rows of (5.5) and (5.6), and using the first rows of (5.5) and (5.6) gives the results. ■

Denote the duality measure by

$$\mu = \frac{\mathbf{x}^{\mathrm{T}}\mathbf{s}}{n}, \tag{5.17}$$

We will show that searching along the ellipse, at least for a small step, will reduce the duality measure, i.e., $\mu(\alpha) = \frac{\mathbf{x}(\alpha)^{\mathrm{T}}\mathbf{s}(\alpha)}{n} < \mu$. If $(\mathbf{x}(\alpha), \mathbf{s}(\alpha)) > \mathbf{0}$ holds in all iterations, reducing the duality measure to zero means approaching to the solution of the linear programming. Notice that

$$\begin{aligned}
\mu(\alpha) &= \frac{[\mathbf{x} - \dot{\mathbf{x}}\sin(\alpha) + \ddot{\mathbf{x}}(1-\cos(\alpha))]^{\mathrm{T}}[\mathbf{s} - \dot{\mathbf{s}}\sin(\alpha) + \ddot{\mathbf{s}}(1-\cos(\alpha))]}{n} \\
&= \frac{[\mathbf{x}^{\mathrm{T}}\mathbf{s} - \sin(\alpha)(\mathbf{x}^{\mathrm{T}}\dot{\mathbf{s}} + \dot{\mathbf{x}}^{\mathrm{T}}\mathbf{s})]}{n} \\
&= \mu(1-\sin(\alpha))
\end{aligned} \tag{5.18}$$

holds for any choice of $\alpha \in [0, \frac{\pi}{2}]$ due to Lemma 5.1 and equation (5.5), this means that the larger the α is, the more improvement the $\mu(\alpha)$ will be.

5.1.3 A polynomial arc-search algorithm

Let $\theta \in (0, 0.5)$, and

$$\mathcal{N}_2(\theta) = \{(\mathbf{x},\lambda,\mathbf{s}) : \mathbf{A}\mathbf{x} = \mathbf{b}, \mathbf{A}^{\mathrm{T}}\lambda + \mathbf{s} = \mathbf{c}, (\mathbf{x},\mathbf{s}) > \mathbf{0}, \|\mathbf{x}\circ\mathbf{s} - \mu\mathbf{e}\| \le \theta\mu\}. \tag{5.19}$$

Similar to a strategy used in Chapter 3 (which was developed in [93]), we present a predictor-corrector type polynomial algorithm which uses $\mathcal{N}_2(\theta)$ and $\mathcal{N}_2(2\theta)$. The algorithm starts the iterate inside $\mathcal{N}_2(\theta)$ and restricts the arc-search in $\mathcal{N}_2(2\theta)$. After the search finds an iterate with smaller duality measure, a corrector step brings the iterate from $\mathcal{N}_2(2\theta)$ back to $\mathcal{N}_2(\theta)$ without changing the duality measure.

The following lemma indicates that if the initial iterate satisfies the equality constraints in $\mathcal{N}_2(\theta)$, a search along the ellipse will satisfy the equality constraints in $\mathcal{N}_2(2\theta)$.

Lemma 5.2
Let $(\mathbf{x},\lambda,\mathbf{s})$ *be a strictly feasible point of* (1.4) *and* (1.5), $(\dot{\mathbf{x}},\dot{\lambda},\dot{\mathbf{s}})$ *and* $(\ddot{\mathbf{x}},\ddot{\lambda},\ddot{\mathbf{s}})$ *satisfy* (5.5) *and* (5.6), $(\mathbf{x}(\alpha),\lambda(\alpha),\mathbf{s}(\alpha))$ *be calculated using* (5.12) *and* (5.13), *then the following conditions hold.*

$$\mathbf{A}\mathbf{x}(\alpha)=\mathbf{b},\quad \mathbf{A}^{\mathrm{T}}\lambda(\alpha)+\mathbf{s}(\alpha)=\mathbf{c}.$$

Proof 5.2 Direct calculation verifies the result. ∎

Before we present the algorithm, we will introduce several simple lemmas that will be used repeatedly in the convergence analysis.

Lemma 5.3
Let $p>0$, $q>0$, *and* $r>0$. *If* $p+q\le r$, *then* $pq\le \frac{r^2}{4}$.

Lemma 5.4
For $\alpha\in[0,\frac{\pi}{2}]$,

$$\sin(\alpha)\ge\sin^2(\alpha)=1-\cos^2(\alpha)\ge 1-\cos(\alpha)\ge\frac{1}{2}\sin^2(\alpha). \tag{5.20}$$

The next Lemma shows that $\frac{\dot{\mathbf{x}}}{\mathbf{x}}$, $\frac{\ddot{\mathbf{x}}}{\mathbf{x}}$, $\frac{\dot{\mathbf{s}}}{\mathbf{s}}$, and $\frac{\ddot{\mathbf{s}}}{\mathbf{s}}$ are bounded by some constants. The lemma also gives some useful estimations and notations to be used later.

Lemma 5.5
Let $(\dot{\mathbf{x}},\dot{\lambda},\dot{\mathbf{s}})$ *be calculated by* (5.5) *and* $(\ddot{\mathbf{x}},\ddot{\lambda},\ddot{\mathbf{s}})$ *be calculated by* (5.6). *Assuming that* $(\mathbf{x},\lambda,\mathbf{s})\in\mathcal{N}_2(\theta)$, $\mu=\frac{\mathbf{x}^{\mathrm{T}}\mathbf{s}}{n}$, *then*

$$\left\|\frac{\dot{\mathbf{x}}}{\mathbf{x}}\right\|^2\le\frac{(1+\theta)}{(1-\theta)}n,\quad \left\|\frac{\dot{\mathbf{s}}}{\mathbf{s}}\right\|^2\le\frac{(1+\theta)}{(1-\theta)}n, \tag{5.21a}$$

$$\left\|\frac{\ddot{\mathbf{x}}}{\mathbf{x}}\right\|^2\le 4\left(\frac{1+\theta}{1-\theta}\right)^3 n^2,\quad \left\|\frac{\ddot{\mathbf{s}}}{\mathbf{s}}\right\|^2\le 4\left(\frac{1+\theta}{1-\theta}\right)^3 n^2, \tag{5.21b}$$

$$\left\|\frac{\dot{\mathbf{x}}\circ\dot{\mathbf{s}}}{\mu}\right\|\le\frac{(1+\theta)^2}{(1-\theta)}n,\quad \left\|\frac{\ddot{\mathbf{x}}\circ\ddot{\mathbf{s}}}{\mu}\right\|\le\frac{4(1+\theta)^4}{(1-\theta)^3}n^2, \tag{5.21c}$$

$$\left\|\frac{\ddot{\mathbf{x}}\circ\dot{\mathbf{s}}}{\mu}\right\|\le\frac{2(1+\theta)^3}{(1-\theta)^2}n^{\frac{3}{2}},\quad \left\|\frac{\dot{\mathbf{x}}\circ\ddot{\mathbf{s}}}{\mu}\right\|\le\frac{2(1+\theta)^3}{(1-\theta)^2}n^{\frac{3}{2}}. \tag{5.21d}$$

Proof 5.3 Let $\hat{\mathbf{A}}$ denote the orthonormal basis of the null space of $\mathbf{A}$, i.e., $\mathbf{A}\hat{\mathbf{A}} = \mathbf{0}$. It is easy to see from the first row of (5.5) that there exists a vector $\mathbf{v}$ such that

$$\frac{\dot{\mathbf{x}}}{\mathbf{x}} = \mathbf{X}^{-1}\hat{\mathbf{A}}\mathbf{v}. \tag{5.22}$$

From the second row and the third row of (5.5), we have

$$\mathbf{S}^{-1}\mathbf{A}^{\mathrm{T}}\dot{\lambda} + \frac{\dot{\mathbf{s}}}{\mathbf{s}} = \mathbf{0}, \quad \frac{\dot{\mathbf{s}}}{\mathbf{s}} + \frac{\dot{\mathbf{x}}}{\mathbf{x}} = \mathbf{e}. \tag{5.23}$$

Combining (5.22) and (5.23) gives

$$\left[\mathbf{X}^{-1}\hat{\mathbf{A}}, -\mathbf{S}^{-1}\mathbf{A}^{\mathrm{T}}\right] \begin{bmatrix} \mathbf{v} \\ \dot{\lambda} \end{bmatrix} = \mathbf{e}. \tag{5.24}$$

Since both $\mathbf{A}$ and $\hat{\mathbf{A}}$ are full rank, we have

$$\begin{bmatrix} (\hat{\mathbf{A}}^{\mathrm{T}}\mathbf{S}\mathbf{X}^{-1}\hat{\mathbf{A}})^{-1}\hat{\mathbf{A}}^{\mathrm{T}}\mathbf{S} \\ -(\mathbf{A}\mathbf{X}\mathbf{S}^{-1}\mathbf{A}^{\mathrm{T}})^{-1}\mathbf{A}\mathbf{X} \end{bmatrix} \left[\mathbf{X}^{-1}\hat{\mathbf{A}}, -\mathbf{S}^{-1}\mathbf{A}^{\mathrm{T}}\right] = \mathbf{I}.$$

Taking the inverse in (5.24) gives

$$\begin{bmatrix} \mathbf{v} \\ \dot{\lambda} \end{bmatrix} = \begin{bmatrix} (\hat{\mathbf{A}}^{\mathrm{T}}\mathbf{S}\mathbf{X}^{-1}\hat{\mathbf{A}})^{-1}\hat{\mathbf{A}}^{\mathrm{T}}\mathbf{S} \\ -(\mathbf{A}\mathbf{X}\mathbf{S}^{-1}\mathbf{A}^{\mathrm{T}})^{-1}\mathbf{A}\mathbf{X} \end{bmatrix} \mathbf{e}.$$

Substituting this relation to (5.22) gives

$$\frac{\dot{\mathbf{x}}}{\mathbf{x}} = \mathbf{X}^{-1}\hat{\mathbf{A}}(\hat{\mathbf{A}}^{\mathrm{T}}\mathbf{S}\mathbf{X}^{-1}\hat{\mathbf{A}})^{-1}\hat{\mathbf{A}}^{\mathrm{T}}\mathbf{S}\mathbf{e} = \left[\mathbf{I} - \mathbf{S}^{-1}\mathbf{A}^{\mathrm{T}}(\mathbf{A}\mathbf{X}\mathbf{S}^{-1}\mathbf{A}^{\mathrm{T}})^{-1}\mathbf{A}\mathbf{X}\right]\mathbf{e}, \tag{5.25}$$

and

$$\frac{\dot{\mathbf{s}}}{\mathbf{s}} = \mathbf{S}^{-1}\mathbf{A}^{\mathrm{T}}(\mathbf{A}\mathbf{X}\mathbf{S}^{-1}\mathbf{A}^{\mathrm{T}})^{-1}\mathbf{A}\mathbf{X}\mathbf{e} = \left[\mathbf{I} - \mathbf{X}^{-1}\hat{\mathbf{A}}(\hat{\mathbf{A}}^{\mathrm{T}}\mathbf{S}\mathbf{X}^{-1}\hat{\mathbf{A}})^{-1}\hat{\mathbf{A}}^{\mathrm{T}}\mathbf{S}\right]\mathbf{e}. \tag{5.26}$$

Since $(\mathbf{x}, \lambda, \mathbf{s}) \in \mathcal{N}_2(\theta)$, we have $(1-\theta)\mu\mathbf{I} \le \mathbf{X}\mathbf{S} \le (1+\theta)\mu\mathbf{I}$. Repeatedly using this estimation and (5.25) yields

$$\begin{aligned} \left\|\frac{\dot{\mathbf{x}}}{\mathbf{x}}\right\|^2 &= \mathbf{e}^{\mathrm{T}}\mathbf{S}^{\mathrm{T}}\hat{\mathbf{A}}(\hat{\mathbf{A}}^{\mathrm{T}}\mathbf{S}\mathbf{X}^{-1}\hat{\mathbf{A}})^{-1}\hat{\mathbf{A}}^{\mathrm{T}}\mathbf{X}^{-1}\mathbf{X}^{-1}\hat{\mathbf{A}}(\hat{\mathbf{A}}^{\mathrm{T}}\mathbf{S}\mathbf{X}^{-1}\hat{\mathbf{A}})^{-1}\hat{\mathbf{A}}^{\mathrm{T}}\mathbf{S}\mathbf{e} \\ &\le ((1-\theta)\mu)^{-1}\mathbf{e}^{\mathrm{T}}\mathbf{S}^{\mathrm{T}}\hat{\mathbf{A}}(\hat{\mathbf{A}}^{\mathrm{T}}\mathbf{S}\mathbf{X}^{-1}\hat{\mathbf{A}})^{-1}\hat{\mathbf{A}}^{\mathrm{T}}\mathbf{S}\mathbf{X}^{-1}\hat{\mathbf{A}}(\hat{\mathbf{A}}^{\mathrm{T}}\mathbf{S}\mathbf{X}^{-1}\hat{\mathbf{A}})^{-1}\hat{\mathbf{A}}^{\mathrm{T}}\mathbf{S}\mathbf{e} \\ &= ((1-\theta)\mu)^{-1}\mathbf{e}^{\mathrm{T}}\mathbf{S}\hat{\mathbf{A}}(\hat{\mathbf{A}}^{\mathrm{T}}\mathbf{S}\mathbf{X}^{-1}\hat{\mathbf{A}})^{-1}\hat{\mathbf{A}}^{\mathrm{T}}\mathbf{S}\mathbf{e} \\ &\le \frac{(1+\theta)}{(1-\theta)}\mathbf{e}^{\mathrm{T}}\mathbf{S}\hat{\mathbf{A}}(\hat{\mathbf{A}}^{\mathrm{T}}\mathbf{S}^2\hat{\mathbf{A}})^{-1}\hat{\mathbf{A}}^{\mathrm{T}}\mathbf{S}\mathbf{e}. \end{aligned}$$

Using QR decomposition

$$\mathbf{S}\hat{\mathbf{A}} = \mathbf{Q}\begin{bmatrix} \mathbf{R}_1 \\ \mathbf{0} \end{bmatrix} = [\mathbf{Q}_1, \mathbf{Q}_2]\begin{bmatrix} \mathbf{R}_1 \\ \mathbf{0} \end{bmatrix} = \mathbf{Q}_1\mathbf{R}_1,$$

where $\mathbf{Q}_1$ and $\mathbf{Q}_2$ are orthonormal matrices and orthogonal to each other, we have

$$\left\|\frac{\dot{\mathbf{x}}}{\mathbf{x}}\right\|^2 \le \frac{(1+\theta)}{(1-\theta)}\mathbf{e}^{\mathrm{T}}\mathbf{Q}_1\mathbf{Q}_1^{\mathrm{T}}\mathbf{e} \le \frac{(1+\theta)}{(1-\theta)}\|\mathbf{e}\|^2 = \frac{(1+\theta)}{(1-\theta)}n. \tag{5.27}$$

Similarly,

$$\left\|\frac{\dot{\mathbf{s}}}{\mathbf{s}}\right\|^2 \le \frac{(1+\theta)}{(1-\theta)}\mathbf{e}^{\mathrm{T}}\mathbf{X}\mathbf{A}^{\mathrm{T}}(\mathbf{A}\mathbf{X}^2\mathbf{A}^{\mathrm{T}})^{-1}\mathbf{A}\mathbf{X}\mathbf{e} \le \frac{(1+\theta)}{(1-\theta)}n. \tag{5.28}$$

From these two inequalities, we have

$$\left\|\frac{\dot{\mathbf{x}}}{\mathbf{x}} \circ \frac{\dot{\mathbf{s}}}{\mathbf{s}}\right\| \le \left\|\frac{\dot{\mathbf{x}}}{\mathbf{x}}\right\|\left\|\frac{\dot{\mathbf{s}}}{\mathbf{s}}\right\| \le \frac{(1+\theta)}{(1-\theta)}n. \tag{5.29}$$

Since $\|\mathbf{x}\circ\mathbf{s}-\mu\mathbf{e}\| \le \theta\mu$, for any i, it follows that $(1-\theta)\mu \le x_i s_i \le (1+\theta)\mu$, or equivalently,

$$\frac{x_i s_i}{1+\theta} \le \frac{\max_i x_i s_i}{1+\theta} \le \mu \le \frac{\min_i x_i s_i}{1-\theta} \le \frac{x_i s_i}{1-\theta}. \tag{5.30}$$

Using (5.29), we have

$$\frac{\left\|\dot{\mathbf{x}}\circ\dot{\mathbf{s}}\right\|}{\mu} \le (1+\theta)\frac{\left\|\dot{\mathbf{x}}\circ\dot{\mathbf{s}}\right\|}{\max_i(x_i s_i)} \le (1+\theta)\left\|\frac{\dot{\mathbf{x}}}{\mathbf{x}}\circ\frac{\dot{\mathbf{s}}}{\mathbf{s}}\right\| \le \frac{(1+\theta)^2}{1-\theta}n.$$

Let $\phi = -2\left(\frac{\dot{\mathbf{x}}}{\mathbf{x}}\circ\frac{\dot{\mathbf{s}}}{\mathbf{s}}\right)$. From (5.6), there exists a vector $\mathbf{v}$ such that $\frac{\ddot{\mathbf{x}}}{\mathbf{x}} = \mathbf{X}^{-1}\hat{\mathbf{A}}\mathbf{v}$, $\mathbf{S}^{-1}\mathbf{A}^{\mathrm{T}}\ddot{\lambda} + \frac{\ddot{\mathbf{s}}}{\mathbf{s}} = \mathbf{0}$, and $\frac{\ddot{\mathbf{x}}}{\mathbf{x}} + \frac{\ddot{\mathbf{s}}}{\mathbf{s}} = \phi$. Following a proof similar to the one used above, it is easy to get

$$\begin{aligned}
\left\|\frac{\ddot{\mathbf{x}}}{\mathbf{x}}\right\|^2 &\le \frac{1+\theta}{1-\theta}\|\phi\|^2 \\
&\le \frac{1+\theta}{1-\theta}\left\|-2\left(\frac{\dot{\mathbf{x}}}{\mathbf{x}}\circ\frac{\dot{\mathbf{s}}}{\mathbf{s}}\right)\right\|^2 \\
&\le 4\left(\frac{1+\theta}{1-\theta}\right)^3 n^2,
\end{aligned} \tag{5.31a}$$

$$\begin{aligned}
\left\|\frac{\ddot{\mathbf{s}}}{\mathbf{s}}\right\|^2 &\le \frac{1+\theta}{1-\theta}\|\phi\|^2 \\
&\le 4\left(\frac{1+\theta}{1-\theta}\right)^3 n^2.
\end{aligned} \tag{5.31b}$$

From (5.30), (5.27), (5.28), and (5.31), we get

$$\frac{\left\|\dot{\mathbf{x}}\circ\ddot{\mathbf{s}}\right\|}{\mu} \le (1+\theta)\left\|\frac{\dot{\mathbf{x}}}{\mathbf{x}}\circ\frac{\ddot{\mathbf{s}}}{\mathbf{s}}\right\| \le (1+\theta)\left\|\frac{\dot{\mathbf{x}}}{\mathbf{x}}\right\|\left\|\frac{\ddot{\mathbf{s}}}{\mathbf{s}}\right\| \le 2\frac{(1+\theta)^3}{(1-\theta)^2}n^{\frac{3}{2}}.$$

$$\frac{\left\|\ddot{\mathbf{x}}\circ\dot{\mathbf{s}}\right\|}{\mu} \le (1+\theta)\left\|\frac{\ddot{\mathbf{x}}}{\mathbf{x}}\circ\frac{\dot{\mathbf{s}}}{\mathbf{s}}\right\| \le (1+\theta)\left\|\frac{\ddot{\mathbf{x}}}{\mathbf{x}}\right\|\left\|\frac{\dot{\mathbf{s}}}{\mathbf{s}}\right\| \le 2\frac{(1+\theta)^3}{(1-\theta)^2}n^{\frac{3}{2}}.$$

$$\frac{\left\|\ddot{\mathbf{x}} \circ \ddot{\mathbf{s}}\right\|}{\mu} \leq (1+\theta)\left\|\frac{\ddot{\mathbf{x}}}{\mathbf{x}} \circ \frac{\ddot{\mathbf{s}}}{\mathbf{s}}\right\| \leq (1+\theta)\left\|\frac{\ddot{\mathbf{x}}}{\mathbf{x}}\right\|\left\|\frac{\ddot{\mathbf{s}}}{\mathbf{s}}\right\| \leq 4\frac{(1+\theta)^4}{(1-\theta)^3}n^2.$$

This finishes the proof. ■

For the sake of simplicity in the analysis, we assume that an initial point inside $\mathcal{N}_2(\theta)$ is available (this can be achieved by calling Algorithm 1 of [20] prior to calling the algorithm below). The proposed arc-search algorithm is given as follows:

Algorithm 5.1

Data: $\mathbf{A}$, $\mathbf{b}$, $\mathbf{c}$, $\theta = 0.292$, $\varepsilon > 0$, *initial point* $(\mathbf{x}^0, \lambda^0, \mathbf{s}^0) \in \mathcal{N}_2(\theta)$, *and* $\mu_0 = \frac{\mathbf{x}^{0^{\mathrm{T}}}\mathbf{s}^0}{n}$.
for *iteration* $k = 0, 1, 2, \ldots$

Step 1: If $\mu_k < \varepsilon$, *and*

$$\|\mathbf{r}_b\| = \|\mathbf{A}\mathbf{x}^k - \mathbf{b}\| \leq \varepsilon,\ \|\mathbf{r}_c\| = \|\mathbf{A}^{\mathrm{T}}\lambda^k + \mathbf{s}^k - \mathbf{c}\| \leq \varepsilon,\ \|\mathbf{r}_t\| = \|\mathbf{x}^k\mathbf{s}^k\mathbf{e} - \mu_k\mathbf{e}\| \leq \varepsilon. \quad (5.32)$$

hold, stop. Otherwise continue.

Step 2: Solve the linear systems of equations (5.5) *and* (5.6) *to get* $(\dot{\mathbf{x}}, \dot{\lambda}, \dot{\mathbf{s}})$ *and* $(\ddot{\mathbf{x}}, \ddot{\lambda}, \ddot{\mathbf{s}})$.

Step 3: Find the smallest positive $\sin(\alpha)$ *that satisfies quartic polynomial in terms of* $\sin(\alpha)$

$$q(\alpha) = \left(\left\|\ddot{\mathbf{x}} \circ \ddot{\mathbf{s}}\right\| + \left\|\dot{\mathbf{x}} \circ \dot{\mathbf{s}}\right\|\right)\sin^4(\alpha) + \left(\left\|\dot{\mathbf{x}} \circ \ddot{\mathbf{s}}\right\| + \left\|\ddot{\mathbf{x}} \circ \dot{\mathbf{s}}\right\|\right)\sin^3(\alpha) + \theta\mu_k\sin(\alpha) - \theta\mu_k = 0. \quad (5.33)$$

Set

$$(\mathbf{x}(\alpha), \lambda(\alpha), \mathbf{s}(\alpha)) = (\mathbf{x}^k, \lambda^k, \mathbf{s}^k) - (\dot{\mathbf{x}}, \dot{\lambda}, \dot{\mathbf{s}})\sin(\alpha) + (\ddot{\mathbf{x}}, \ddot{\lambda}, \ddot{\mathbf{s}})(1 - \cos(\alpha)), \quad (5.34)$$

and

$$\mu(\alpha) = \mu_k(1 - \sin(\alpha)). \quad (5.35)$$

Step 4: Calculate $(\Delta\mathbf{x}, \Delta\lambda, \Delta\mathbf{s})$ *by solving*

$$\begin{bmatrix} \mathbf{A} & \mathbf{0} & \mathbf{0} \\ \mathbf{0} & \mathbf{A}^{\mathrm{T}} & \mathbf{I} \\ \mathbf{S}(\alpha) & \mathbf{0} & \mathbf{X}(\alpha) \end{bmatrix} \begin{bmatrix} \Delta\mathbf{x} \\ \Delta\lambda \\ \Delta\mathbf{s} \end{bmatrix} = \begin{bmatrix} \mathbf{0} \\ \mathbf{0} \\ \mu(\alpha)\mathbf{e} - \mathbf{x}(\alpha) \circ \mathbf{s}(\alpha) \end{bmatrix}. \quad (5.36)$$

Set

$$(\mathbf{x}^{k+1}, \lambda^{k+1}, \mathbf{s}^{k+1}) = (\mathbf{x}(\alpha), \lambda(\alpha), \mathbf{s}(\alpha)) + (\Delta\mathbf{x}, \Delta\lambda, \Delta\mathbf{s}) \quad (5.37)$$

and $\mu_{k+1} = \frac{\mathbf{x}^{k+1^{\mathrm{T}}}\mathbf{s}^{k+1}}{n}$. *Set* $k+1 \to k$. *Go back to Step 1.*

end (for)

Using (5.5), (5.6), (5.34), and Lemma 5.1, it is easy to check that (5.35) satisfies the definition of duality measure.

$$\begin{aligned}
\mathbf{x}(\alpha)^{\mathrm{T}}\mathbf{s}(\alpha) &= (\mathbf{x}^k - \dot{\mathbf{x}}\sin(\alpha) + \ddot{\mathbf{x}}(1-\cos(\alpha))^{\mathrm{T}}(\mathbf{s}^k - \dot{\mathbf{s}}\sin(\alpha) + \ddot{\mathbf{s}}(1-\cos(\alpha)) \\
&= \mathbf{x}^{k^{\mathrm{T}}}\mathbf{s}^k - (\dot{\mathbf{x}}^{\mathrm{T}}\mathbf{s}^k + \dot{\mathbf{s}}^{\mathrm{T}}\mathbf{x}^k)\sin(\alpha) + (\ddot{\mathbf{x}}^{\mathrm{T}}\mathbf{s}^k + \ddot{\mathbf{s}}^{\mathrm{T}}\mathbf{x}^k)(1-\cos(\alpha)) \\
&= \mathbf{x}^{k^{\mathrm{T}}}\mathbf{s}^k - \mathbf{x}^{k^{\mathrm{T}}}\mathbf{s}^k\sin(\alpha) - 2\dot{\mathbf{x}}^{\mathrm{T}}\dot{\mathbf{s}}(1-\cos(\alpha)) \\
&= \mathbf{x}^{k^{\mathrm{T}}}\mathbf{s}^k(1-\sin(\alpha)).
\end{aligned}$$

Dividing both sides by n shows the claim.

Remark 5.1 For any feasible point, if $\alpha = \frac{\pi}{2}$, then $\mu(\alpha) = 0$ means an optimal solution is obtained. In the remaining discussion in this chapter, we exclude this simple case and assume that $\alpha < \frac{\pi}{2}$. ■

Step 3 of the algorithm finds a step size $\sin(\alpha)$ such that $(\mathbf{x}(\alpha), \lambda(\alpha), \mathbf{s}(\alpha)) \in \mathcal{N}_2(2\theta)$. Lemma 5.6 will show that the step size $\sin(\alpha)$ can be determined by the smallest positive solution of (5.33). Since (5.33) is a quartic polynomial which accepts analytic solutions [113], the computational cost of solving (5.33) is negligible. It is easy to see that the quartic polynomial (5.33) is a monotonic increasing function of $\alpha \in [0, \frac{\pi}{2}]$, for $(\mathbf{x}^k, \mathbf{s}^k) > \mathbf{0}$ (which will be shown in Lemma 5.7), $q(0) < 0$, and $q(\frac{\pi}{2}) \geq 0$. Therefore, $q(\alpha)$ has only one real solution in $[0, \frac{\pi}{2}]$, and $\alpha = \sin^{-1}(\alpha)$ is well-defined. In the rest of this section, we will show that if $(\mathbf{x}^k, \lambda^k, \mathbf{s}^k) \in \mathcal{N}_2(0.292)$, then $(\mathbf{x}(\alpha), \lambda(\alpha), \mathbf{s}(\alpha)) \in \mathcal{N}_2(0.584)$ and $(\mathbf{x}^{k+1}, \lambda^{k+1}, \mathbf{s}^{k+1}) \in \mathcal{N}_2(0.292)$. We will also estimate the size of α, which will be used to prove the polynomiality at the end of this section.

Lemma 5.6
Let $(\mathbf{x}^k, \lambda^k, \mathbf{s}^k) \in \mathcal{N}_2(\theta)$, $(\dot{\mathbf{x}}, \dot{\lambda}, \dot{\mathbf{s}})$ *and* $(\ddot{\mathbf{x}}, \ddot{\lambda}, \ddot{\mathbf{s}})$ *be defined by* (5.5) *and* (5.6), $\sin(\alpha) > 0$ *be the solution of* (5.33) *and* $(\mathbf{x}(\alpha), \lambda(\alpha), \mathbf{s}(\alpha))$ *be updated using* (5.34). *Then,* $(\mathbf{x}(\alpha), \lambda(\alpha), \mathbf{s}(\alpha)) \in \mathcal{N}_2(2\theta)$.

Proof 5.4 First, it is worthwhile to note that Lemmas 5.1 and 5.2 hold in this case. From (5.33), we have

$$\begin{aligned}
&\left(\left\|\ddot{\mathbf{x}} \circ \ddot{\mathbf{s}}\right\| + \left\|\dot{\mathbf{x}} \circ \dot{\mathbf{s}}\right\|\right)\sin^4(\alpha) + \left(\left\|\dot{\mathbf{x}} \circ \ddot{\mathbf{s}}\right\| + \left\|\ddot{\mathbf{x}} \circ \dot{\mathbf{s}}\right\|\right)\sin^3(\alpha) \\
= \;& \theta\mu_k(1-\sin(\alpha)) = \theta\mu(\alpha).
\end{aligned} \tag{5.38}$$

Also, the following identity is useful.

$$\begin{aligned}
& -2\dot{\mathbf{x}} \circ \dot{\mathbf{s}}(1-\cos(\alpha)) + \dot{\mathbf{x}} \circ \dot{\mathbf{s}}\sin^2(\alpha) \\
=\; & \dot{\mathbf{x}} \circ \dot{\mathbf{s}}(-2 + 2\cos(\alpha) + \sin^2(\alpha)) \\
=\; & \dot{\mathbf{x}} \circ \dot{\mathbf{s}}(-1 + 2\cos(\alpha) - \cos^2(\alpha)) \\
=\; & -\dot{\mathbf{x}} \circ \dot{\mathbf{s}}(1-\cos(\alpha))^2.
\end{aligned}$$

Using this identity, relations (5.5), (5.6), (5.34), (5.35), and Lemma 5.4, we have

$$\begin{aligned}
& \|\mathbf{x}(\alpha)\circ\mathbf{s}(\alpha)-\mu(\alpha)\mathbf{e}\| \\
= & \|[\mathbf{x}^k-\dot{\mathbf{x}}\sin(\alpha)+\ddot{\mathbf{x}}(1-\cos(\alpha))]\circ[\mathbf{s}^k-\dot{\mathbf{s}}\sin(\alpha)+\ddot{\mathbf{s}}(1-\cos(\alpha))]-\mu_k(1-\sin(\alpha))\mathbf{e}\| \\
= & \|(\mathbf{x}^k\circ\mathbf{s}^k-\mu_k\mathbf{e})(1-\sin(\alpha))+(\mathbf{x}^k\circ\ddot{\mathbf{s}}+\mathbf{s}^k\circ\ddot{\mathbf{x}})(1-\cos(\alpha))+\dot{\mathbf{x}}\circ\dot{\mathbf{s}}\sin^2(\alpha) \\
& -(\dot{\mathbf{x}}\circ\ddot{\mathbf{s}}+\dot{\mathbf{s}}\circ\ddot{\mathbf{x}})\sin(\alpha)(1-\cos(\alpha))+\ddot{\mathbf{x}}\circ\ddot{\mathbf{s}}(1-\cos(\alpha))^2\| \qquad (5.39) \\
\leq & (1-\sin(\alpha))\|\mathbf{x}^k\circ\mathbf{s}^k-\mu_k\mathbf{e}\| \\
& +\|(\ddot{\mathbf{x}}\circ\ddot{\mathbf{s}}-\dot{\mathbf{x}}\circ\dot{\mathbf{s}})(1-\cos(\alpha))^2-(\dot{\mathbf{x}}\circ\ddot{\mathbf{s}}+\dot{\mathbf{s}}\circ\ddot{\mathbf{x}})\sin(\alpha)(1-\cos(\alpha))\| \\
\leq & (1-\sin(\alpha))\theta\mu_k+\left(\left\|\ddot{\mathbf{x}}\circ\ddot{\mathbf{s}}\right\|+\left\|\dot{\mathbf{x}}\circ\dot{\mathbf{s}}\right\|\right)\sin^4(\alpha)+\left(\left\|\dot{\mathbf{x}}\circ\ddot{\mathbf{s}}\right\|+\left\|\ddot{\mathbf{x}}\circ\dot{\mathbf{s}}\right\|\right)\sin^3(\alpha) \\
= & \theta\mu(\alpha)+\theta\mu(\alpha)=2\theta\mu(\alpha). \qquad (5.40)
\end{aligned}$$

Since

$$x_i(\alpha)s_i(\alpha)\geq(1-2\theta)\mu(\alpha)=\mu_k(1-2\theta)(1-\sin(\alpha))>0,$$

$\mathbf{x}(0)=\mathbf{x}>\mathbf{0}$ and $\mathbf{s}(0)=\mathbf{s}>\mathbf{0}$ means $\mathbf{x}(\alpha)>\mathbf{0}$ and $\mathbf{s}(\alpha)>\mathbf{0}$. This finishes the proof. ∎

The following lemma is similar to Lemma 3.14.

Lemma 5.7
Let $(\mathbf{x}(\alpha),\lambda(\alpha),\mathbf{s}(\alpha))\in\mathcal{N}_2(2\theta)$. Then, for $0<\theta\leq 0.292$, we have $(\mathbf{x}^{k+1},\lambda^{k+1},\mathbf{s}^{k+1})\in\mathcal{N}_2(\theta)$ and $\mu_{k+1}=\mu(\alpha)$.

Proof 5.5 Denote $(\mathbf{x}^{k+1}(t),\mathbf{s}^{k+1}(t))=(\mathbf{x}(\alpha)+t\Delta\mathbf{x},\mathbf{s}(\alpha)+t\Delta\mathbf{s}))$, and $(\mathbf{x}^{k+1},\mathbf{s}^{k+1})=(\mathbf{x}^{k+1}(1),\mathbf{s}^{k+1}(1))$. Using the third row of (5.36), we have $\frac{\mathbf{x}(\alpha)^{\mathrm{T}}\Delta\mathbf{s}+\mathbf{s}(\alpha)^{\mathrm{T}}\Delta\mathbf{x}}{n}=0$. Using the second row of (5.36), we have $\Delta\mathbf{x}^{\mathrm{T}}\Delta\mathbf{s}=0$. Therefore, we have

$$\mu_{k+1}(t)=\frac{(\mathbf{x}(\alpha)+t\Delta\mathbf{x})^{\mathrm{T}}(\mathbf{s}(\alpha)+t\Delta\mathbf{s})}{n}=\frac{\mathbf{x}(\alpha)^{\mathrm{T}}\mathbf{s}(\alpha)}{n}=\mu(\alpha).$$

Let $\mathbf{D}=\mathbf{X}(\alpha)^{\frac{1}{2}}\mathbf{S}(\alpha)^{-\frac{1}{2}}$. Pre-multiplying $(\mathbf{X}(\alpha)\mathbf{S}(\alpha))^{-\frac{1}{2}}$ in the last row of (5.36) yields

$$\mathbf{D}\Delta\mathbf{s}+\mathbf{D}^{-1}\Delta\mathbf{x}=(\mathbf{X}(\alpha)\mathbf{S}(\alpha))^{-\frac{1}{2}}(\mu(\alpha)\mathbf{e}-\mathbf{X}(\alpha)\mathbf{S}(\alpha)\mathbf{e}).$$

Denote $\mathbf{u}=\mathbf{D}\Delta\mathbf{s}$ and $\mathbf{v}=\mathbf{D}^{-1}\Delta\mathbf{x}$. Using Lemma 3.2 and the assumption that $(\mathbf{x}(\alpha),\lambda(\alpha),\mathbf{s}(\alpha))\in\mathcal{N}_2(2\theta)$, we have

$$\begin{aligned}
\|\Delta\mathbf{x}\circ\Delta\mathbf{s}\| &= \|\mathbf{u}\circ\mathbf{v}\|\leq 2^{-\frac{3}{2}}\|(\mathbf{X}(\alpha)\mathbf{S}(\alpha))^{-\frac{1}{2}}(\mu(\alpha)\mathbf{e}-\mathbf{X}(\alpha)\mathbf{S}(\alpha)\mathbf{e})\|^2 \\
&= 2^{-\frac{3}{2}}\sum_{i=1}^{n}\frac{(\mu(\alpha)-x_i(\alpha)s_i(\alpha))^2}{x_i(\alpha)s_i(\alpha)}\leq 2^{-\frac{3}{2}}\frac{\|\mu(\alpha)\mathbf{e}-\mathbf{x}(\alpha)\circ\mathbf{s}(\alpha)\|^2}{\min_i x_i(\alpha)s_i(\alpha)} \\
&\leq 2^{-\frac{3}{2}}\frac{(2\theta)^2\mu(\alpha)^2}{(1-2\theta)\mu(\alpha)}=2^{\frac{1}{2}}\frac{\theta^2\mu(\alpha)}{(1-2\theta)}=2^{\frac{1}{2}}\frac{\theta^2}{(1-2\theta)}\mu_{k+1}. \qquad (5.41)
\end{aligned}$$

Using this result, (5.37), the last row of (5.36), we have

$$\begin{aligned} & \|\mathbf{x}^{k+1}(t)\circ\mathbf{s}^{k+1}(t)-\mu_{k+1}(t)\mathbf{e}\| = \left\|\left(\mathbf{x}(\alpha)+t\Delta\mathbf{x}\right)\circ\left(\mathbf{s}(\alpha)+t\Delta\mathbf{s}\right)-\mu(\alpha)\mathbf{e}\right\| \\ = & \left\|(1-t)\left(\mathbf{x}(\alpha)\circ\mathbf{s}(\alpha)-\mu(\alpha)\mathbf{e}\right)+t^2\Delta\mathbf{x}\circ\Delta\mathbf{s}\right\| \le (1-t)2\theta\mu(\alpha)+t^2\|\Delta\mathbf{x}\circ\Delta\mathbf{s}\| \\ \le & \left((1-t)2\theta+2^{\frac{1}{2}}t^2\frac{\theta^2}{(1-2\theta)}\right)\mu_{k+1} := h(t,\theta)\mu_{k+1}. \end{aligned} \tag{5.42}$$

Taking $t=1$ gives $\|\mathbf{x}^{k+1}\circ\mathbf{s}^{k+1}-\mu_{k+1}\mathbf{e}\| \le 2^{\frac{1}{2}}\frac{\theta^2}{(1-2\theta)}$. It is easy to verify that for $0<\theta\le 0.292$,

$$2^{\frac{1}{2}}\frac{\theta^2}{(1-2\theta)} \le \theta.$$

Since, for $0<\theta\le 0.292$ and $t\in[0,1]$, $0<h(t,\theta)\le h(t,0.292)<1$, we have, for $\alpha\in[0,\frac{\pi}{2})$,

$$\begin{aligned} \mathbf{x}_i^{k+1}(t)\mathbf{s}_i^{k+1}(t) & \ge (1-h(t,\theta))\mu_{k+1} \\ & = (1-h(t,\theta))\mu(\alpha) \\ & = (1-h(t,\theta))(1-\sin(\alpha))\mu_k > 0. \end{aligned}$$

Therefore, $(\mathbf{x}^{k+1},\mathbf{s}^{k+1})>\mathbf{0}$. This finishes the proof. ■

The choice of $\theta=0.25$ is used in Chapter 3. Taking a larger θ will allow a longer step size in arc-search, which may reduce the number of iterations to converge to the optimal solution.

Lemma 5.8
Let $0<\theta\le 0.292$, *and* $\sin(\alpha)$, $\alpha\in[0,\pi/2]$, *be the positive real solution of* (5.33). *Then,*

$$\sin(\alpha)\ge\theta^2 n^{-\frac{1}{2}}.$$

Proof 5.6 Since $q(\sin(\alpha))$ is a monotonic increasing function of $\sin(\alpha)\in[0,1]$ with $q(\sin(0))<0$ and $q(\sin(\frac{\pi}{2}))\ge 0$, we need only to show that $q(\theta^2 n^{-\frac{1}{2}})<0$. Using Lemma 5.5, for $\sin(\alpha)\le\theta^2 n^{-\frac{1}{2}}$, we have

$$\begin{aligned} q(\alpha) & = \left(\left\|\ddot{\mathbf{x}}\circ\ddot{\mathbf{s}}\right\|+\left\|\dot{\mathbf{x}}\circ\dot{\mathbf{s}}\right\|\right)\sin^4(\alpha)+\left(\left\|\dot{\mathbf{x}}\circ\ddot{\mathbf{s}}\right\|+\left\|\ddot{\mathbf{x}}\circ\dot{\mathbf{s}}\right\|\right)\sin^3(\alpha) \\ & \quad +\theta\mu_k\sin(\alpha)-\theta\mu_k \\ & \le \left(\frac{4(1+\theta)^4}{(1-\theta)^3}n^2+\frac{(1+\theta)^2}{(1-\theta)}n\right)\left(\theta^8 n^{-2}\right)\mu_k \\ & \quad +\frac{4(1+\theta)^3}{(1-\theta)^2}n^{\frac{3}{2}}\left(\theta^6 n^{-\frac{3}{2}}\right)\mu_k+\theta\mu_k\left(\theta^2 n^{-\frac{1}{2}}\right)-\theta\mu_k \\ & = \theta\mu_k\left(\frac{4\theta^7(1+\theta)^4}{(1-\theta)^3}+\frac{\theta^7(1+\theta)^2}{(1-\theta)n}+\frac{4\theta^5(1+\theta)^3}{(1-\theta)^2}+\frac{\theta^2}{\sqrt{n}}-1\right) \\ & := \theta\mu_k f(\theta,n). \end{aligned} \tag{5.43}$$

Clearly, $f(\theta,n)$ is a monotonic increasing function of $\theta \in (0,1)$, and a monotonic decreasing function of $n \in \{1,2,\ldots\}$. For $\theta \leq 0.292$, and $n \geq 1$, $f(\theta,n) < 0$, hence, for $\sin(\alpha) \leq \theta^2 n^{-\frac{1}{2}}$, $q(\sin(\alpha)) < 0$. This proves $\sin(\alpha) \geq \theta^2 n^{-\frac{1}{2}}$. ■

Remark 5.2 Clearly, if $\left\|\frac{\dot{\mathbf{x}}}{\mathbf{x}}\right\|$, $\left\|\frac{\ddot{\mathbf{x}}}{\mathbf{x}}\right\|$, $\left\|\frac{\dot{\mathbf{s}}}{\mathbf{s}}\right\|$, $\left\|\frac{\ddot{\mathbf{s}}}{\mathbf{s}}\right\|$ are smaller than some constant independent of m and n, then the arc-search algorithm proposed above would reduce the duality measure at a constant rate independent of m and n, a nice feature (polynomial bound would be independent of n) that appears to hold according to our numerical test, but we cannot prove it. ■

Let $(\mathbf{x}^*,\lambda^*,\mathbf{s}^*)$ be any solution of (5.1). Let index sets $\mathcal{B}$, $\mathcal{N}$ be defined as

$$\mathcal{B} = \{j \in \{1,\ldots,n\} \mid x_j^* \neq 0\}. \tag{5.44}$$

$$\mathcal{N} = \{j \in \{1,\ldots,n\} \mid s_j^* \neq 0\}. \tag{5.45}$$

Let $\mathbf{x}^*$ be a solution of the primal linear programming and $(\lambda^*,\mathbf{s}^*)$ be a solution of the dual linear programming, such that, $\mathcal{B} \cap \mathcal{N} = \emptyset$ and $\mathcal{B} \cup \mathcal{N} = \{1,\ldots,n\}$, i.e., $\mathbf{x}^* \circ \mathbf{s}^* = \mathbf{0}$, and $\mathbf{x}^* + \mathbf{s}^* > \mathbf{0}$. An optimal solution with this property is called strictly complementary. Now we are ready to state our main result.

Theorem 5.2
The sequence $(\mathbf{x}^k,\lambda^k,\mathbf{s}^k)$ generated by Algorithm 5.1 globally converges to a set of limit points $(\mathbf{x}^*,\lambda^*,\mathbf{s}^*)$. Moreover, Algorithm 5.1 is a polynomial algorithm with polynomial complexity bound of $O(n^{\frac{1}{2}}\log(1/\varepsilon))$. For every limit point $(\mathbf{x}^*,\lambda^*,\mathbf{s}^*)$, $\mathbf{x}^*$ is the optimal solution of the primal problem, $(\lambda^*,\mathbf{s}^*)$ is the optimal solution of the dual problem, and $(\mathbf{x}^*,\mathbf{s}^*)$ is strictly complementary.

Proof 5.7 In view of Lemma 5.8, (5.18), and (1.4), Algorithms 5.1 converges in polynomial time with the complexity bound of $O(n^{\frac{1}{2}}\log(1/\varepsilon))$. The rest of the proof is to show that the solution is complementary. In view of Assumption 2 in this chapter, $\mathcal{F}^0$ is not empty. Theorem 1.1 shows that the optimal solution set is bounded, i.e., there is a constant M such that

$$\|(\mathbf{x}^*,\mathbf{s}^*)\| \leq M. \tag{5.46}$$

Let P^* be the set of all primal optimizers $\mathbf{x}^*$, D^* be the set of all dual optimizers $(\lambda^*,\mathbf{s}^*)$, and

$$\varepsilon(\mathbf{A},\mathbf{b},\mathbf{c}) = \min\left(\min_{i\in\mathcal{B}} \sup_{\mathbf{x}^*\in P^*} x_i^*, \min_{i\in\mathcal{N}} \sup_{(\lambda^*,\mathbf{s}^*)\in D^*} s_i^* \right).$$

For any feasible $(\mathbf{x},\lambda,\mathbf{s}) \in \mathcal{F}^o$ with $x_i s_i > \gamma\mu$, we have $\mathbf{A}\mathbf{x} = \mathbf{A}\mathbf{x}^* = \mathbf{b}$ and $\mathbf{A}^{\mathrm{T}}\lambda + \mathbf{s} = \mathbf{A}^{\mathrm{T}}\lambda^* + \mathbf{s}^* = \mathbf{c}$. Therefore,

$$(\mathbf{x}-\mathbf{x}^*)^{\mathrm{T}}(\mathbf{s}-\mathbf{s}^*) = -(\mathbf{x}-\mathbf{x}^*)^{\mathrm{T}}\mathbf{A}^{\mathrm{T}}(\lambda-\lambda^*) = 0.$$

Since Theorem 1.3 implies $x_i^* = 0$ for $i \in \mathcal{N}$ and $s_i^* = 0$ for $i \in \mathcal{B}$, we can rearrange this expression to obtain

$$n\mu = \mathbf{x}^{\mathrm{T}}\mathbf{s}^* + \mathbf{s}^{\mathrm{T}}\mathbf{x}^* = \sum_{i\in\mathcal{N}} x_i s_i^* + \sum_{i\in\mathcal{B}} s_i x_i^*$$

Since each of these summations is nonnegative, each term is bounded by $n\mu$. Hence, for any $i \in \mathcal{N}$ with $s_i^* > 0$, it must have

$$0 < x_i s_i^* \le n\mu, \rightarrow x_i \le \frac{n\mu}{s_i^*}.$$

This relation holds for any $(\mathbf{x}^*, \lambda^*, \mathbf{s}^*)$ with $s_i^* > 0$, which implies

$$0 < x_i \le \frac{n\mu}{\sup_{(\lambda^*,\mathbf{s}^*)\in D^*} s_i^*}.$$

This relation holds for any x_i with $i \in \mathcal{N}$, which implies

$$0 < \max_{i\in\mathcal{N}} x_i \le \frac{n\mu}{\min_{i\in\mathcal{N}} \sup_{(\lambda^*,\mathbf{s}^*)\in D^*} s_i^*}.$$

Similarly,

$$0 < \max_{i\in\mathcal{B}} s_i \le \frac{n\mu}{\min_{i\in\mathcal{B}} \sup_{\mathbf{x}^*\in P^*} x_i^*}.$$

Adjoining this two inequalities gives

$$\max\left(\max_{i\in\mathcal{N}} x_i, \max_{i\in\mathcal{B}} s_i\right) \le \frac{n\mu}{\varepsilon(\mathbf{A},\mathbf{b},\mathbf{c})}.$$

Setting $C_1 = \frac{\varepsilon(\mathbf{A},\mathbf{b},\mathbf{c})}{n}$ given

$$0 < x_i \le \mu/C_1 \quad (i \in \mathcal{N}), \qquad 0 < s_i \le \mu/C_1 \quad (i \in \mathcal{B}). \tag{5.47}$$

Theorem 1.3 asserts the existence of a strictly complementary solution, which indicates that $\varepsilon(\mathbf{A},\mathbf{b},\mathbf{c}) > 0$ and so is C_1. Since $x_i s_i \ge \gamma\mu$ for all $i = 1,2,\ldots,n$, from (5.47), we have

$$s_i \ge \frac{\gamma\mu}{x_i} \ge \frac{\gamma\mu}{\mu/C_1} \ge C_1\gamma \quad (i \in \mathcal{N}), \tag{5.48a}$$

$$x_i \ge \frac{\gamma\mu}{s_i} \ge \frac{\gamma\mu}{\mu/C_1} \ge C_1\gamma \quad (i \in \mathcal{B}). \tag{5.48b}$$

Therefore, we conclude that every limit point is a strictly complementary primary-dual solution of the linear programming, i.e.,

$$s_i^* \ge C_2\gamma \quad (i \in \mathcal{N}), \qquad x_i^* \ge C_2\gamma \quad (i \in \mathcal{B}). \tag{5.49}$$

■

5.2 Numerical Examples

In this section, a simple problem is used as an example to show the central path, the ellipse approximation, and the arc-search in every iteration by plots. From these plots, one can intuitively see that searching along an ellipse is more attractive than searching along a straight line. Some larger scale test problems are also provided in order to show the efficiency of the proposed algorithm.

5.2.1 A simple illustrative example

Let us consider

$$\min x_1, \quad s.t. \;\; x_1 + x_2 = 5, \;\; x_1 \geq 0, \;\; x_2 \geq 0.$$

The central path $(\mathbf{x}, \lambda, \mathbf{s})$ satisfies the following conditions.

$$x_1 + x_2 = 5,$$

$$\begin{bmatrix} 1 \\ 1 \end{bmatrix} \lambda + \begin{bmatrix} s_1 \\ s_2 \end{bmatrix} = \begin{bmatrix} 1 \\ 0 \end{bmatrix},$$

$$x_1 s_1 = \mu, \quad x_2 s_2 = \mu.$$

The optimizer is given by $x_1 = 0$, $x_2 = 5$, $\lambda = 0$, $s_1 = 1$, and $s_2 = 0$. The central path for this problem is given analytically as

$$\lambda = \frac{5 - 2\mu - \sqrt{(5-2\mu)^2 + 20\mu}}{10},$$

$$s_1 = 1 - \lambda, \quad s_2 = -\lambda, \quad x_1 s_1 = \mu, \quad x_2 s_2 = \mu.$$

The central path is an arc in 5-dimensional space $(\lambda, x_1, s_1, x_2, s_2)$. If we project the central path to 2-dimensional space spanned by (s_1, x_1), it is an arc in 2-dimensional space. Similarly, we can project the ellipse in 5-dimensional space to the same 2-dimensional space spanned by (s_1, x_1). Figure 5.1 shows all iterations of Algorithm 5.1 in the two dimensional subspace spanned by (s_1, x_1). The first iteration moves the iterate very close to the solution, and the remaining iterations have to be rescaled to show the details. In Figure 5.1, $(\dot{\mathbf{x}}, \dot{\mathbf{s}}, \dot{\lambda})$ is calculated by using (5.5), $(\ddot{\mathbf{x}}, \ddot{\mathbf{s}}, \ddot{\lambda})$ is calculated by using (5.6), the projected central path is the continuous line in black, the projected ellipse approximations are the dotted lines in blue in every iteration (they may sometimes look like a continuous line because many dots are used), the initial point (s_1^0, x_1^0) is marked by 'x' in red, after moving 'x' towards the central path, (s_1^k, x_1^k) is marked by 'o' in red, after arc-search, the point (s_1^k, x_1^k) on the ellipse is marked by '+' in green, the optimal solution $(\mathbf{s}^*, \mathbf{x}^*)$ is marked by '*' in red. More detailed information in the second iteration, the third iteration, and the final result are presented in Figures 5.2, 5.3, and 5.4 which are amplified parts of Figure 5.1. It is clear that after the first iteration, the central path is close to a straight line, and the ellipses approximate the central path very well, regardless of whether the central path is close to a straight line or not. We expect that the developed algorithm is

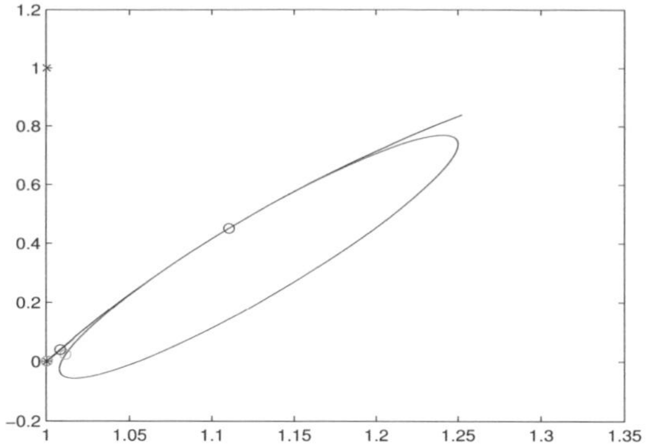

Figure 5.1: Arc-search for the simple example.

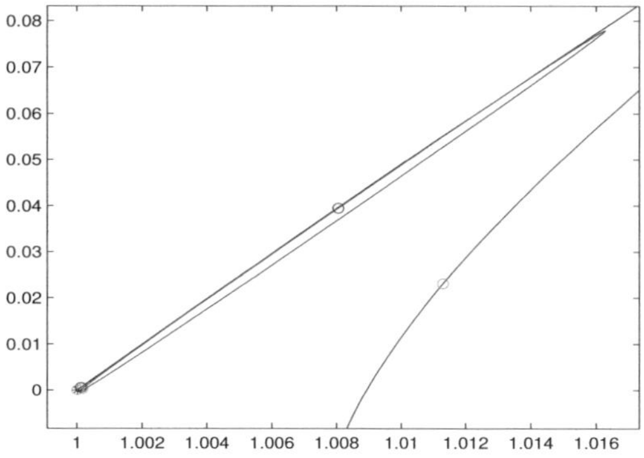

Figure 5.2: Arc-search of the second iteration for the simple example.

more efficient in the early stage than MPC and other interior-point algorithms based on line search when the central path is not close to the straight line, as shown in Figure 5.1.

5.2.2 Some Netlib test examples

The algorithm developed in this chapter is implemented in MATLAB®. function `arc2`. Numerical tests have been performed for linear programming problems in Netlib LP library. Cartis and Gould [20] have classified Netlib LP problems into two categories: Problems with strict interior-point and problems without strict interior-point. The test is performed for problems with strict interior-point and expressed in standard form with **A** being full rank matrices. The selected problems are solved using the implemented MATLAB function and the `linprog` function in the MATLAB optimization toolBox. The iterations used to solve these problems are compared and

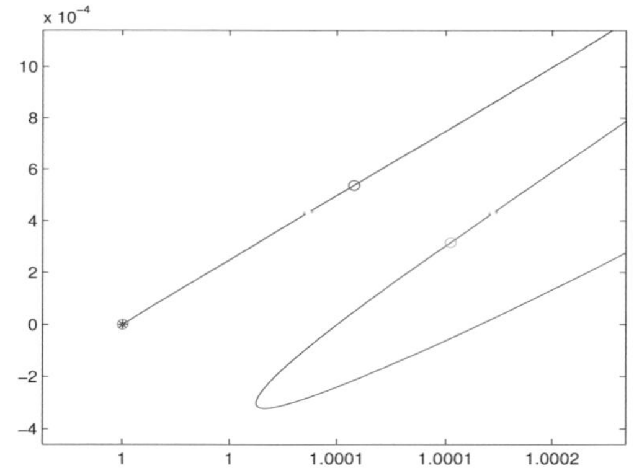

Figure 5.3: Arc-search of the third iteration for the simple example.

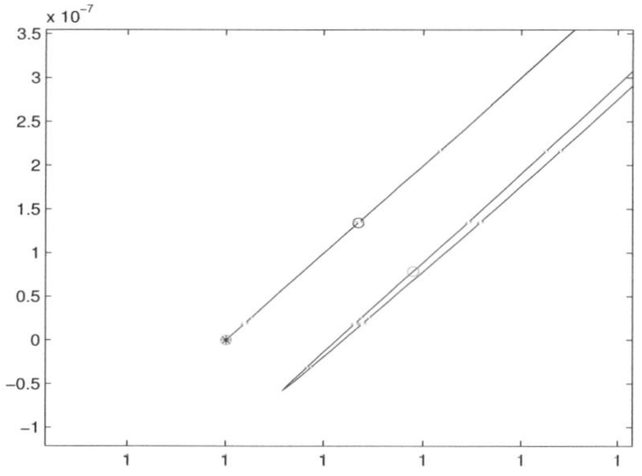

Figure 5.4: Arc-search of the final result for the simple example.

the iteration numbers are listed in Table 5.1. Only two Netlib problems that are classified as problems with strict interior-point and are presented in standard form are not included in the table because the PC computer used in the test did not have enough memory to handle problems of big size.

In the implementation, the stopping criterion used in Algorithm 5.1 is the same as `linprog` [172]

$$\frac{\|\mathbf{r}_b\|}{\max\{1,\|\mathbf{b}\|\}}+\frac{\|\mathbf{r}_c\|}{\max\{1,\|\mathbf{c}\|\}}+\frac{\mu}{\max\{1,\|\mathbf{c}^{\mathrm{T}}\mathbf{x}\|,\|\mathbf{b}^{\mathrm{T}}\lambda\|\}}<10^{-8}.$$

For all problems, the initial point is set to $\mathbf{x}=\mathbf{s}=\mathbf{e}$. The initial point used in `linprog` is essentially the same as was used in [83], with some minor modifications (see [172]).

Table 5.1: Iteration counts for test problems in Netlib and MATLAB.

	iterations used by different algorithms		
Problem	arc2	linprog	source
AFIRO	4	7	Netlib
blend	10	8	Netlib
SCAGR25	5	16	Netlib
SCAGR7	6	12	Netlib
SCSD1	13	10	Netlib
SCSD6	17	12	Netlib
SCSD8	13	11	Netlib
SCTAP1	14	17	Netlib
SCTAP2	13	18	Netlib
SCTAP3	14	18	Netlib
SHARE1B	9	22	Netlib

5.3 Concluding Remarks

This chapter proposes an arc-search interior-point path-following algorithm that searches for the optimizers along the ellipses that approximate a central path. The main purpose of this chapter is to show that higher-order interior-point path-following algorithms can achieve the best polynomial bound by using the arc-search method, with the potential to be very efficient. The convergence analysis proved the first claim. Some preliminary numerical result is presented to demonstrate that the proposed algorithm has numerical merit in comparison to some well-implemented algorithm based on Mehrotra's method, which is known to be very efficient but has no convergence result.

Chapter 6

A MTY-Type Infeasible Arc-Search Algorithm for LP

From Chapters 2, 3, and 4, we know that there are three strategies that can be used to improve the efficiency of the interior-point algorithms, namely, (a) search for optimizers in a larger neighborhood, (b) use infeasible initial point, and (c) approximate the central path using higher-order derivatives. However, when line search method is used with the interior-point methods, these strategies adversely affect the polynomial bounds, i.e., the theoretical analysis contradicts the computational experience [133]. In Chapter 5, we introduced an arc-search method and showed that the higher-order method does not undermine the polynomial bound if the arc-search is used because the lowest bound $\mathcal{O}(\sqrt{n}L)$ can be achieved for the algorithm that uses the arc-search (higher-order derivatives). Preliminary computational experience also demonstrated promising result.

In this chapter, we will incorporate the arc-search to another important strategy, i.e., starting from infeasible initial point. We show that, for a higher-order infeasible-interior-point method using arc-search, the polynomial bound is $\mathcal{O}(nL)$, which is as good as the infeasible first order method using line search [144] and the best bound for infeasible-interior-point algorithms that was obtained by [90]. This method also improves the polynomial bound obtained by Yang et al. [149]. In Chapter 8, we will further improve this bound to $\mathcal{O}(\sqrt{n}L)$. The material of this chapter is based on [161, 67].

6.1 An Infeasible Predictor-Corrector Algorithm

The standard form of linear programming considered in this chapter is the same as (1.4). The dual problem of (1.4) is given in (1.5). To simplify the analysis, we assume that the rank of matrix $\mathbf{A}$ equals to m. However, this requirement is not essential because $\mathbf{A}$ can always be reduced to a full rank matrix in polynomial time. We use $\mathcal{S}$ to denote the set of all the optimal solutions $(\mathbf{x}^*, \lambda^*, \mathbf{s}^*)$ of (1.4) and (1.5). It is well known that the primal-dual vector $(\mathbf{x}, \lambda, \mathbf{s})$ is an optimal solution of (1.4) and (1.5) if and only if it satisfies the KKT conditions (5.1).

We will consider the residuals of primal programming and dual programming defined in Chapter 4:

$$\begin{aligned} \mathbf{r}_b^k &= \mathbf{A}\mathbf{x}^k - \mathbf{b}, \\ \mathbf{r}_c^k &= \mathbf{A}^{\mathrm{T}}\lambda^k + \mathbf{s}^k - \mathbf{c}. \end{aligned} \tag{6.1}$$

Given a strictly positive current point $(\mathbf{x}^k, \mathbf{s}^k) > \mathbf{0}$, the infeasible-predictor-corrector algorithm is to find the solution of (1.4) approximately along a curve $\mathcal{C}(t)$ defined by the following system

$$\begin{gathered} \mathbf{A}\mathbf{x}(t) - \mathbf{b} = t\mathbf{r}_b^k \\ \mathbf{A}^{\mathrm{T}}\lambda(t) + \mathbf{s}(t) - \mathbf{c} = t\mathbf{r}_c^k \\ \mathbf{x}(t) \circ \mathbf{s}(t) = t\mathbf{x}^k \circ \mathbf{s}^k \\ (\mathbf{x}(t), \mathbf{s}(t)) > \mathbf{0}, \end{gathered} \tag{6.2}$$

where $t \in (0, 1]$. As $t \to 0$, $(\mathbf{x}(t), \lambda(t), \mathbf{s}(t))$ approaches the solution of (1.4). Since it is not easy to obtain $\mathcal{C}(t)$, an ellipse $\mathcal{E}$ (defined in (5.4)) in the $2n+m$ dimensional space will be used to approximate the curve defined by (6.2). Taking the derivatives for t in (6.2) gives (see [153])

$$\begin{bmatrix} \mathbf{A} & \mathbf{0} & \mathbf{0} \\ \mathbf{0} & \mathbf{A}^{\mathrm{T}} & \mathbf{I} \\ \mathbf{S}^k & \mathbf{0} & \mathbf{X}^k \end{bmatrix} \begin{bmatrix} \dot{\mathbf{x}} \\ \dot{\lambda} \\ \dot{\mathbf{s}} \end{bmatrix} = \begin{bmatrix} \mathbf{r}_b^k \\ \mathbf{r}_c^k \\ \mathbf{x}^k \circ \mathbf{s}^k \end{bmatrix}, \tag{6.3}$$

$$\begin{bmatrix} \mathbf{A} & \mathbf{0} & \mathbf{0} \\ \mathbf{0} & \mathbf{A}^{\mathrm{T}} & \mathbf{I} \\ \mathbf{S}^k & \mathbf{0} & \mathbf{X}^k \end{bmatrix} \begin{bmatrix} \ddot{\mathbf{x}} \\ \ddot{\lambda} \\ \ddot{\mathbf{s}} \end{bmatrix} = \begin{bmatrix} \mathbf{0} \\ \mathbf{0} \\ -2\dot{\mathbf{x}} \circ \dot{\mathbf{s}} \end{bmatrix}. \tag{6.4}$$

Here, $\mathbf{X}^k$ and $\mathbf{S}^k$ are diagonal matrices whose diagonal elements are given by $\mathbf{x}^k$ and $\mathbf{s}^k$, respectively.

We require the ellipse to pass the same point $(\mathbf{x}^k, \lambda^k, \mathbf{s}^k)$ on $\mathcal{C}(t)$ and to have the same derivatives given by (6.3) and (6.4). The ellipse can be derived similar to the derivation of Theorem 5.1 and is given as below.

Theorem 6.1

Let $(\mathbf{x}(\alpha), \lambda(\alpha), \mathbf{s}(\alpha))$ *be an arc defined by (5.4) passing through a point* $(\mathbf{x}, \lambda, \mathbf{s}) \in \mathcal{E} \cap \mathcal{C}(t)$*, and its first and second derivatives at* $(\mathbf{x}, \lambda, \mathbf{s})$ *be* $(\dot{\mathbf{x}}, \dot{\lambda}, \dot{\mathbf{s}})$ *and* $(\ddot{\mathbf{x}}, \ddot{\lambda}, \ddot{\mathbf{s}})$ *which are defined by (6.3) and (6.4). Then, the ellipse approximation of* (6.2) *is given by*

$$\mathbf{x}(\alpha) = \mathbf{x} - \dot{\mathbf{x}}\sin(\alpha) + \ddot{\mathbf{x}}(1 - \cos(\alpha)), \tag{6.5}$$

$$\lambda(\alpha) = \lambda - \dot{\lambda}\sin(\alpha) + \ddot{\lambda}(1 - \cos(\alpha)), \tag{6.6}$$

$$\mathbf{s}(\alpha) = \mathbf{s} - \dot{\mathbf{s}}\sin(\alpha) + \ddot{\mathbf{s}}(1 - \cos(\alpha)). \tag{6.7}$$

■

Let the duality measure be defined as (5.17), and define the set of neighborhood by

$$\mathcal{N}_2^I(\theta) := \{(\mathbf{x}, \mathbf{s}) \mid (\mathbf{x}, \mathbf{s}) > \mathbf{0}, \ \ \|\mathbf{x} \circ \mathbf{s} - \mu \mathbf{e}\| \leq \theta\mu\}. \tag{6.8}$$

The proposed algorithm searches for an optimizer along the ellipse while staying inside $\mathcal{N}_2^I(\theta)$.

Algorithm 6.1

Data: **A**, **b**, **c**, $\theta \in (0, \frac{1}{2+\sqrt{2}}]$, $\varepsilon > 0$, *initial point* $(\mathbf{x}^0, \lambda^0, \mathbf{s}^0) \in \mathcal{N}_2^I(\theta)$.
for *iteration* $k = 0, 1, 2, \ldots$

Step 1: If

$$\mu_k \leq \varepsilon, \tag{6.9a}$$

$$\|\mathbf{r}_b^k\| = \|\mathbf{A}\mathbf{x}^k - \mathbf{b}\| \leq \varepsilon, \tag{6.9b}$$

$$\|\mathbf{r}_c^k\| = \|\mathbf{A}^{\mathrm{T}}\lambda^k + \mathbf{s}^k - \mathbf{c}\| \leq \varepsilon, \tag{6.9c}$$

$$(\mathbf{x}^k, \mathbf{s}^k) > \mathbf{0}. \tag{6.9d}$$

stop. Otherwise continue.

Step 2: Solve the linear systems of equations (6.3) and (6.4) to get $(\dot{\mathbf{x}}, \dot{\lambda}, \dot{\mathbf{s}})$ *and* $(\ddot{\mathbf{x}}, \ddot{\lambda}, \ddot{\mathbf{s}})$.

Step 3: Find the largest positive $\alpha_k \in (0, \pi/2]$ *such that for all* $\alpha \in (0, \alpha_k]$, $(\mathbf{x}(\alpha), \mathbf{s}(\alpha)) > \mathbf{0}$ *and*

$$\|(\mathbf{x}(\alpha) \circ \mathbf{s}(\alpha)) - (1 - \sin(\alpha))\mu_k \mathbf{e}\| \leq 2\theta(1 - \sin(\alpha))\mu_k. \tag{6.10}$$

Set

$$(\mathbf{x}(\alpha_k), \lambda(\alpha_k), \mathbf{s}(\alpha_k)) = (\mathbf{x}^k, \lambda^k, \mathbf{s}^k) - (\dot{\mathbf{x}}, \dot{\lambda}, \dot{\mathbf{s}})\sin(\alpha_k) + (\ddot{\mathbf{x}}, \ddot{\lambda}, \ddot{\mathbf{s}})(1 - \cos(\alpha_k)). \tag{6.11}$$

Step 4: Calculate $(\Delta\mathbf{x},\Delta\lambda,\Delta\mathbf{s})$ *by solving*

$$\begin{bmatrix} \mathbf{A} & \mathbf{0} & \mathbf{0} \\ \mathbf{0} & \mathbf{A}^{\mathrm{T}} & \mathbf{I} \\ \mathbf{S}(\alpha_k) & \mathbf{0} & \mathbf{X}(\alpha_k) \end{bmatrix} \begin{bmatrix} \Delta\mathbf{x} \\ \Delta\lambda \\ \Delta\mathbf{s} \end{bmatrix} = \begin{bmatrix} \mathbf{0} \\ \mathbf{0} \\ (1-\sin(\alpha_k))\mu_k\mathbf{e} - \mathbf{x}(\alpha_k)\circ\mathbf{s}(\alpha_k) \end{bmatrix}. \tag{6.12}$$

Update

$$(\mathbf{x}^{k+1},\lambda^{k+1},\mathbf{s}^{k+1}) = (\mathbf{x}(\alpha_k),\lambda(\alpha_k),\mathbf{s}(\alpha_k)) + (\Delta\mathbf{x},\Delta\lambda,\Delta\mathbf{s}) \tag{6.13}$$

and

$$\mu_{k+1} = \frac{\mathbf{x}^{k+1^{\mathrm{T}}}\mathbf{s}^{k+1}}{n}. \tag{6.14}$$

Step 5: Set $k+1 \to k$*. Go back to Step 1.*

end (for) ■

In the rest of this chapter, we will show (1) $\mathbf{r}_b^k \to \mathbf{0}$, $\mathbf{r}_c^k \to \mathbf{0}$, and $\mu_k \to 0$; (2) there exist $\alpha_k \in (0,\pi/2]$ such that $(\mathbf{x}(\alpha),\mathbf{s}(\alpha)) > \mathbf{0}$ and (6.10) hold for any $\alpha \in (0,\alpha_k]$; (3) $(\mathbf{x}^k,\mathbf{s}^k) \in \mathcal{N}_2^I(\theta)$. It is easy to show that $\mathbf{r}_b^k$, $\mathbf{r}_c^k$, and μ_k decrease at the same rate in every iteration.

Lemma 6.1
Consider the sequence of points generated by Algorithm 3.1. Then,

$$\begin{aligned} \mathbf{r}_b^{k+1} &= \mathbf{r}_b^k(1-\sin(\alpha_k)), \\ \mathbf{r}_c^{k+1} &= \mathbf{r}_c^k(1-\sin(\alpha_k)), \\ \mu_{k+1} &= \mu_k(1-\sin(\alpha_k)). \end{aligned} \tag{6.15}$$

Proof 6.1 Using (6.1), (6.13), (6.12), (6.11), (6.4), and (6.3), we have

$$\begin{aligned} & \mathbf{r}_b^{k+1} - \mathbf{r}_b^k = \mathbf{A}(\mathbf{x}^{k+1}-\mathbf{x}^k) = \mathbf{A}(\mathbf{x}(\alpha_k)+\Delta\mathbf{x}-\mathbf{x}^k) \\ = \; & \mathbf{A}(\mathbf{x}^k - \dot{\mathbf{x}}\sin(\alpha_k) - \mathbf{x}^k) = -\mathbf{A}\dot{\mathbf{x}}\sin(\alpha_k) = -\mathbf{r}_b^k\sin(\alpha_k). \end{aligned}$$

This shows the first relation. The second relation follows a similar derivation. From (6.12), it holds that $(\Delta\mathbf{x})^{\mathrm{T}}\Delta\mathbf{s} = (\Delta\mathbf{x})^{\mathrm{T}}(-\mathbf{A}^{\mathrm{T}}\Delta\lambda) = -(\mathbf{A}\Delta\mathbf{x})^{\mathrm{T}}\Delta\lambda = 0$. Using (6.13) and (6.12), we have

$$\begin{aligned} & \mathbf{x}^{k+1^{\mathrm{T}}}\mathbf{s}^{k+1} \\ = \; & (\mathbf{x}(\alpha_k)+\Delta\mathbf{x})^{\mathrm{T}}(\mathbf{s}(\alpha_k)+\Delta\mathbf{s}) \\ = \; & \mathbf{x}(\alpha_k)^{\mathrm{T}}\mathbf{s}(\alpha_k) + \mathbf{x}(\alpha_k)^{\mathrm{T}}\Delta\mathbf{s} + \mathbf{s}(\alpha_k)^{\mathrm{T}}\Delta\mathbf{x} \\ = \; & \mathbf{x}(\alpha_k)^{\mathrm{T}}\mathbf{s}(\alpha_k) + (1-\sin(\alpha_k))\mu_k n - \mathbf{x}(\alpha_k)^{\mathrm{T}}\mathbf{s}(\alpha_k) \\ = \; & (1-\sin(\alpha_k))\mu_k n. \end{aligned}$$

Dividing both sides by n proves the last relation. ■

Clearly, if we can take $\sin(\alpha_k) = 1$ ($\alpha_k = \frac{\pi}{2}$) at some kth iteration, we will exactly reach the optimal solution (allowing some $x_i = 0$ and/or $s_j = 0$) at this iteration, which is rarely the case. Therefore, from now on, we assume $\alpha_k \in (0, \frac{\pi}{2})$ through the execution of the algorithm. We will use the following lemma of [91, 90].

Lemma 6.2
Let $(\Delta\mathbf{x}, \Delta\mathbf{s})$ *be given by (6.12). Then,*

$$\|\Delta\mathbf{x} \circ \Delta\mathbf{s}\| \leq \frac{\sqrt{2}}{4}\|(\mathbf{X}(\alpha_k)\mathbf{S}(\alpha_k))^{-\frac{1}{2}}(\mathbf{x}(\alpha_k) \circ \mathbf{s}(\alpha_k) - \mu_{k+1}\mathbf{e})\|^2. \tag{6.16}$$

Proof 6.2 Noticing that $\mu_{k+1} = (1 - \sin(\alpha_k))\mu_k$ from Lemma 6.1, and applying Lemma 3.2 to the last row of (6.12) proves this lemma. ∎

Theorem 6.2
Assume $(\mathbf{x}^k, \mathbf{s}^k) \in \mathcal{N}_2^I(\theta)$. *Then,*

(i) there is an $\alpha_k \in (0, \frac{\pi}{2})$, *such that* $(\mathbf{x}(\alpha), \mathbf{s}(\alpha)) > \mathbf{0}$ *and (6.10) hold for any* $\alpha \in (0, \alpha_k]$.

(ii) if $\theta \leq \frac{1}{2+\sqrt{2}}$, *then* $(\mathbf{x}^{k+1}, \mathbf{s}^{k+1}) \in \mathcal{N}_2^I(\theta)$.

Proof 6.3 First, note that the last rows of (6.3) and (6.4) are equivalent to

$$\mathbf{s}^k \circ \dot{\mathbf{x}} + \mathbf{x}^k \circ \dot{\mathbf{s}} = \mathbf{x}^k \circ \mathbf{s}^k, \qquad \mathbf{s}^k \circ \ddot{\mathbf{x}} + \mathbf{x}^k \circ \ddot{\mathbf{s}} = -2\dot{\mathbf{x}} \circ \dot{\mathbf{s}}. \tag{6.17}$$

Also, we have the following simple identity:

$$\sin^2(\alpha) - 2(1 - \cos(\alpha)) = -(1 - \cos(\alpha))^2. \tag{6.18}$$

Using (6.17) and (6.18), we have

$$\begin{aligned}
& \mathbf{x}(\alpha) \circ \mathbf{s}(\alpha) \\
= \; & (\mathbf{x}^k - \dot{\mathbf{x}}\sin(\alpha) + \ddot{\mathbf{x}}(1 - \cos(\alpha)) \circ (\mathbf{s}^k - \dot{\mathbf{s}}\sin(\alpha) + \ddot{\mathbf{s}}(1 - \cos(\alpha)) \\
= \; & \mathbf{x}^k \circ \mathbf{s}^k - (\mathbf{x}^k \circ \dot{\mathbf{s}} + \mathbf{s}^k \circ \dot{\mathbf{x}})\sin(\alpha) + (\mathbf{x}^k \circ \ddot{\mathbf{s}} + \mathbf{s}^k \circ \ddot{\mathbf{x}})(1 - \cos(\alpha)) \\
& + \dot{\mathbf{x}} \circ \dot{\mathbf{s}}\sin^2(\alpha) - (\ddot{\mathbf{x}} \circ \dot{\mathbf{s}} + \dot{\mathbf{x}} \circ \ddot{\mathbf{s}})\sin(\alpha)(1 - \cos(\alpha)) \\
& + \ddot{\mathbf{x}} \circ \ddot{\mathbf{s}}(1 - \cos(\alpha))^2 \\
= \; & \mathbf{x}^k \circ \mathbf{s}^k - \mathbf{x}^k \circ \mathbf{s}^k\sin(\alpha) - 2\dot{\mathbf{x}} \circ \dot{\mathbf{s}}(1 - \cos(\alpha)) \\
& + \dot{\mathbf{x}} \circ \dot{\mathbf{s}}\sin^2(\alpha) - (\ddot{\mathbf{x}} \circ \dot{\mathbf{s}} + \dot{\mathbf{x}} \circ \ddot{\mathbf{s}})\sin(\alpha)(1 - \cos(\alpha)) \\
& + \ddot{\mathbf{x}} \circ \ddot{\mathbf{s}}(1 - \cos(\alpha))^2 \\
= \; & \mathbf{x}^k \circ \mathbf{s}^k(1 - \sin(\alpha)) + \dot{\mathbf{x}} \circ \dot{\mathbf{s}}(\sin^2(\alpha) - 2(1 - \cos(\alpha))) \\
& - (\ddot{\mathbf{x}} \circ \dot{\mathbf{s}} + \dot{\mathbf{x}} \circ \ddot{\mathbf{s}})\sin(\alpha)(1 - \cos(\alpha)) + \ddot{\mathbf{x}} \circ \ddot{\mathbf{s}}(1 - \cos(\alpha))^2 \\
= \; & \mathbf{x}^k \circ \mathbf{s}^k(1 - \sin(\alpha)) + (\ddot{\mathbf{x}} \circ \ddot{\mathbf{s}} - \dot{\mathbf{x}} \circ \dot{\mathbf{s}})(1 - \cos(\alpha))^2 \\
& - (\ddot{\mathbf{x}} \circ \dot{\mathbf{s}} + \dot{\mathbf{x}} \circ \ddot{\mathbf{s}})\sin(\alpha)(1 - \cos(\alpha)).
\end{aligned} \tag{6.19}$$

Furthermore, for $\alpha \in (0, \frac{\pi}{2})$, it holds that $0 \le 1-\cos(\alpha) \le 1-\cos^2(\alpha) = \sin^2(\alpha)$. Using (6.19), $0 \le 1-\cos(\alpha) \le \sin^2(\alpha)$ and $(\mathbf{x}^k, \mathbf{s}^k) \in \mathcal{N}_2^I(\theta)$, we have

$$\begin{aligned}
& \|\mathbf{x}(\alpha) \circ \mathbf{s}(\alpha) - (1-\sin(\alpha))\mu_k \mathbf{e}\| \\
= & \|\mathbf{x}^k \circ \mathbf{s}^k (1-\sin(\alpha)) + (\ddot{\mathbf{x}} \circ \ddot{\mathbf{s}} - \dot{\mathbf{x}} \circ \dot{\mathbf{s}})(1-\cos(\alpha))^2 \\
& -(\ddot{\mathbf{x}} \circ \dot{\mathbf{s}} + \dot{\mathbf{x}} \circ \ddot{\mathbf{s}}) \sin(\alpha)(1-\cos(\alpha)) - (1-\sin(\alpha))\mu_k \mathbf{e}\| \\
= & \|(\mathbf{x}^k \circ \mathbf{s}^k - \mu_k \mathbf{e})(1-\sin(\alpha)) + (\ddot{\mathbf{x}} \circ \ddot{\mathbf{s}} - \dot{\mathbf{x}} \circ \dot{\mathbf{s}})(1-\cos(\alpha))^2 \\
& -(\ddot{\mathbf{x}} \circ \dot{\mathbf{s}} + \dot{\mathbf{x}} \circ \ddot{\mathbf{s}}) \sin(\alpha)(1-\cos(\alpha))\| \\
\le & \theta\mu_k(1-\sin(\alpha)) + (\|\ddot{\mathbf{x}} \circ \ddot{\mathbf{s}}\| + \|\dot{\mathbf{x}} \circ \dot{\mathbf{s}}\|) \sin^4(\alpha) \\
& +(\|\ddot{\mathbf{x}} \circ \dot{\mathbf{s}}\| + \|\dot{\mathbf{x}} \circ \ddot{\mathbf{s}}\|) \sin^3(\alpha).
\end{aligned} \tag{6.20}$$

Clearly, if

$$\begin{aligned}
q(\alpha) \quad := \quad & \left(\left\|\ddot{\mathbf{x}} \circ \ddot{\mathbf{s}}\right\| + \left\|\dot{\mathbf{x}} \circ \dot{\mathbf{s}}\right\|\right) \sin^4(\alpha) + \left(\left\|\dot{\mathbf{x}} \circ \ddot{\mathbf{s}}\right\| + \left\|\ddot{\mathbf{x}} \circ \dot{\mathbf{s}}\right\|\right) \sin^3(\alpha) \\
& +\theta\mu_k \sin(\alpha) - \theta\mu_k \le 0,
\end{aligned} \tag{6.21}$$

then, (6.10) holds. Indeed, since $q(0) = -\theta\mu_k < 0$ and $q(\alpha)$ is a monotonically increasing function of α, by continuity, there exists an $\alpha_k \in (0, \frac{\pi}{2})$ such that (6.21) holds for all $\alpha \in (0, \alpha_k]$. This shows that (6.10) holds. From (6.10), we have

$$x_i(\alpha)s_i(\alpha) \ge (1-2\theta)(1-\sin(\alpha))\mu_k > 0, \quad \forall \theta \in [0, 0.5) \quad \text{and} \quad \forall \alpha \in (0, \alpha_k].$$

This shows $(\mathbf{x}(\alpha), \mathbf{s}(\alpha)) > 0$, therefore, this completes the proof of part (i).

From Lemma 6.1, inequality (6.10) is equivalent to $\|\mathbf{x}(\alpha) \circ \mathbf{s}(\alpha) - \mu_{k+1}\mathbf{e}\| \le 2\theta\mu_{k+1}$. Using (6.13), (6.12), Lemmas 6.1 and 6.2, and part (i) of this theorem, we have

$$\begin{aligned}
& \|\mathbf{x}^{k+1} \circ \mathbf{s}^{k+1} - \mu_{k+1}\mathbf{e}\| \\
= & \|(\mathbf{x}(\alpha_k) + \Delta\mathbf{x}) \circ (\mathbf{s}(\alpha_k) + \Delta\mathbf{s}) - \mu_{k+1}\mathbf{e}\| \\
= & \|\Delta\mathbf{x} \circ \Delta\mathbf{s}\| \le \frac{\sqrt{2}}{4} \|(\mathbf{X}(\alpha_k)\mathbf{S}(\alpha_k))^{-\frac{1}{2}} (\mathbf{x}(\alpha_k) \circ \mathbf{s}(\alpha_k) - \mu_{k+1}\mathbf{e})\|^2 \\
\le & \frac{\sqrt{2}}{4} \frac{\|\mathbf{x}(\alpha_k) \circ \mathbf{s}(\alpha_k) - \mu_{k+1}\mathbf{e}\|^2}{\min_i x_i(\alpha_k)s_i(\alpha_k)} \\
\le & \frac{\sqrt{2}(2\theta)^2\mu_{k+1}^2}{4(1-2\theta)\mu_{k+1}} \\
\le & \frac{\sqrt{2}\theta^2}{(1-2\theta)}\mu_{k+1}.
\end{aligned} \tag{6.22}$$

It is easy to check that for $\theta \le \frac{1}{2+\sqrt{2}} \approx 0.29289$, $\frac{\sqrt{2}\theta^2}{(1-2\theta)} \le \theta$ holds, therefore, for $\theta \le \frac{1}{2+\sqrt{2}}$, we have

$$\|\mathbf{x}^{k+1} \circ \mathbf{s}^{k+1} - \mu_{k+1}\mathbf{e}\| \le \theta\mu_{k+1}.$$

We now show that $(\mathbf{x}^{k+1}, \mathbf{s}^{k+1}) > 0$. Let $\mathbf{x}^{k+1}(t) = \mathbf{x}(\alpha_k) + t\Delta\mathbf{x}$ and $\mathbf{s}^{k+1}(t) = \mathbf{s}(\alpha_k) + t\Delta\mathbf{s}$. Then, $\mathbf{x}^{k+1}(0) = \mathbf{x}(\alpha_k)$ and $\mathbf{x}^{k+1}(1) = \mathbf{x}^{k+1}$. Since

$$\begin{aligned} & \mathbf{x}^{k+1}(t) \circ \mathbf{s}^{k+1}(t) = (\mathbf{x}(\alpha_k) + t\Delta\mathbf{x}) \circ (\mathbf{s}(\alpha_k) + t\Delta\mathbf{s}) \\ = \ & \mathbf{x}(\alpha_k) \circ \mathbf{s}(\alpha_k) + t(\mathbf{x}(\alpha_k) \circ \Delta\mathbf{s} + \mathbf{s}(\alpha_k) \circ \Delta\mathbf{x}) + t^2 \Delta\mathbf{x} \circ \Delta\mathbf{s}, \end{aligned}$$

using (6.12), Lemma 6.1, (6.10), (6.22), and the assumption that $\theta \le \frac{1}{2+\sqrt{2}}$, we have

$$\begin{aligned} & \|\mathbf{x}^{k+1}(t) \circ \mathbf{s}^{k+1}(t) - \mu_{k+1}\mathbf{e}\| \\ = \ & \|\mathbf{x}(\alpha_k) \circ \mathbf{s}(\alpha_k) + t(\mathbf{x}(\alpha_k) \circ \Delta\mathbf{s} + \mathbf{s}(\alpha_k) \circ \Delta\mathbf{x}) + t^2 \Delta\mathbf{x} \circ \Delta\mathbf{s} - \mu_{k+1}\mathbf{e}\| \\ = \ & \|\mathbf{x}(\alpha_k) \circ \mathbf{s}(\alpha_k) + t[(1 - \sin(\alpha_k))\mu_k\mathbf{e} - \mathbf{x}(\alpha_k) \circ \mathbf{s}(\alpha_k)] + t^2 \Delta\mathbf{x} \circ \Delta\mathbf{s} - \mu_{k+1}\mathbf{e}\| \\ = \ & \|(1-t)(\mathbf{x}(\alpha_k) \circ \mathbf{s}(\alpha_k) - \mu_{k+1}\mathbf{e}) + t^2 \Delta\mathbf{x} \circ \Delta\mathbf{s}\| \\ \le \ & 2(1-t)\theta\mu_{k+1} + t^2 \frac{\sqrt{2}\theta^2}{1-2\theta}\mu_{k+1} \\ \le \ & (2(1-t) + t^2)\theta\mu_{k+1} := f(t)\theta\mu_{k+1}. \end{aligned} \tag{6.23}$$

The function $f(t)$ is a monotonically decreasing function of $t \in [0,1]$, and $f(0) = 2$. This proves $\|\mathbf{x}^{k+1}(t) \circ \mathbf{s}^{k+1}(t) - \mu_{k+1}\mathbf{e}\| \le 2\theta\mu_{k+1}$. Therefore, $x_i^{k+1}(t)s_i^{k+1}(t) \ge (1 - 2\theta)\mu_{k+1} > 0$ for all $t \in [0,1]$, which means $(\mathbf{x}^{k+1}, \mathbf{s}^{k+1}) > 0$. This completes the proof of part (ii). ■

This theorem indicates that the proposed algorithm is well-defined.

6.2 Polynomiality

Let the initial point be selected to satisfy

$$(\mathbf{x}^0, \mathbf{s}^0) \in \mathcal{N}_2^I(\theta), \quad \mathbf{x}^* \le \rho\mathbf{x}^0, \quad \mathbf{s}^* \le \rho\mathbf{s}^0, \tag{6.24}$$

where $\rho \ge 1$ and $(\mathbf{x}^*, \lambda^*, \mathbf{s}^*) \in \mathcal{S}$. Let $(\bar{\mathbf{x}}, \bar{\lambda}, \bar{\mathbf{s}})$ be a member of the set:

$$\bar{\mathcal{S}} = \{(\bar{\mathbf{x}}, \bar{\lambda}, \bar{\mathbf{s}}) \mid \mathbf{A}\bar{\mathbf{x}} = \mathbf{b}, \quad \mathbf{A}^{\mathrm{T}}\bar{\lambda} + \bar{\mathbf{s}} = \mathbf{c}\}. \tag{6.25}$$

Let ω^f and ω^o be the qualities of the initial point which are the "distances" from feasibility and optimality given by

$$\omega^f = \min_{(\bar{\mathbf{x}}, \bar{\lambda}, \bar{\mathbf{s}}) \in \bar{\mathcal{S}}} \left\{ \max\left\{ \|(\mathbf{X}^0)^{-1}(\bar{\mathbf{x}} - \mathbf{x}^0)\|_\infty, \|(\mathbf{S}^0)^{-1}(\bar{\mathbf{s}} - \mathbf{s}^0)\|_\infty \right\} \right\}. \tag{6.26}$$

and

$$\omega^0 = \min_{\mathbf{x}^*, \lambda^*, \mathbf{s}^*} \left\{ \max\left\{ \frac{\mathbf{x}^{*\mathrm{T}}\mathbf{s}^0}{\mathbf{x}^{0\mathrm{T}}\mathbf{s}^0}, \frac{\mathbf{s}^{*\mathrm{T}}\mathbf{x}^0}{\mathbf{x}^{0\mathrm{T}}\mathbf{s}^0}, 1 \right\} \mid (\mathbf{x}^*, \lambda^*, \mathbf{s}^*) \in \mathcal{S} \right\}. \tag{6.27}$$

From the definition of ω^f, there are $(\bar{\mathbf{x}}, \bar{\lambda}, \bar{\mathbf{s}}) \in \bar{\mathcal{S}}$ such that

$$|x_i^0 - \bar{x}_i| \le \omega^f x_i^0,$$

$$|s_i^0 - \bar{s}_i| \leq \omega^f s_i^0. \tag{6.28}$$

From the definition of ω^o, there is the optimal solution $(\mathbf{x}^*, \lambda^*, \mathbf{s}^*) \in \mathcal{S}$ such that

$$\mathbf{x}^{0^{\mathrm{T}}} \mathbf{s}^*, \mathbf{s}^{0^{\mathrm{T}}} \mathbf{x}^* \leq \omega^0 \mathbf{x}^{0^{\mathrm{T}}} \mathbf{s}^0. \tag{6.29}$$

Let $\nu_k = \prod_{i=0}^{k-1} (1 - \sin(\alpha_k))$, it has

$$\nu_k \omega^0 \mathbf{x}^{0^{\mathrm{T}}} \mathbf{s}^0 \leq \omega^0 \mathbf{x}^{k^{\mathrm{T}}} \mathbf{s}^k. \tag{6.30}$$

Let ω_p^r and ω_d^r be the "ratios" of the feasibility and the total complementarity defined by

$$\begin{aligned} \omega_p^r &= \frac{\|\mathbf{A}\mathbf{x}^0 - \mathbf{b}\|}{\mathbf{x}^{0^{\mathrm{T}}} \mathbf{s}^0}, \\ \omega_d^r &= \frac{\|\mathbf{A}^{\mathrm{T}} \lambda^0 + \mathbf{s}^0 - \mathbf{c}\|}{\mathbf{x}^{0^{\mathrm{T}}} \mathbf{s}^0}. \end{aligned} \tag{6.31}$$

In view of Lemma 6.1, we have that

$$\begin{aligned} \|\mathbf{A}\mathbf{x}^k - \mathbf{b}\| &= \omega_p^r \mathbf{x}^{k^{\mathrm{T}}} \mathbf{s}^k, \\ \|\mathbf{A}^{\mathrm{T}} \lambda^k + \mathbf{s}^k - \mathbf{c}\| &= \omega_d^r \mathbf{x}^{k^{\mathrm{T}}} \mathbf{s}^k. \end{aligned} \tag{6.32}$$

Several lemmas, which will be used to prove the polynomiality, are provided below.

Lemma 6.3
Let $(\mathbf{x}^0, \mathbf{s}^0)$ be defined by (6.24). Then,

$$\omega^f \leq \rho, \quad \omega^o \leq \rho. \tag{6.33}$$

Proof 6.4 First, from (6.24), (6.27), and $\rho \geq 1$, it is easy to see $\omega^o \leq \rho$. Now, let

$$\rho^f = \max \left\{ \min\{\|\mathbf{x}\| \mid \mathbf{A}\mathbf{x} = \mathbf{b}\}, \quad \min\{\|\mathbf{s}\| \mid \mathbf{A}^{\mathrm{T}}\lambda + \mathbf{s} = \mathbf{c}\} \right\}$$

there exists a constant ω^* such that

$$\rho^f (\mathbf{e}, \mathbf{e}) \leq \omega^* (\mathbf{x}^0, \mathbf{s}^0).$$

and by the definition of ρ^f, we see that

$$\begin{aligned} |\bar{x}_i| \leq \rho^f \leq \omega^* x_i^0, \\ |\bar{s}_i| \leq \rho^f \leq \omega^* s_i^0. \end{aligned}$$

Therefore,

$$\begin{aligned} \omega^f &= \min_{(\bar{\mathbf{x}}, \bar{\lambda}, \bar{\mathbf{s}}) \in \bar{\mathcal{S}}} \left\{ \max\{\|(\mathbf{X}^0)^{-1}(\bar{\mathbf{x}} - \mathbf{x}^0)\|_\infty, \|(\mathbf{S}^0)^{-1}(\bar{\mathbf{s}} - \mathbf{s}^0)\|_\infty\} \right\} \\ &\leq \max \left\{ \frac{|\bar{x}_i - x_i^0|}{x_i^0}, \frac{|\bar{s}_i - s_i^0|}{s_i^0} \right\} \end{aligned}$$

$$
\begin{aligned}
&\leq \max\left\{\frac{|\bar{x}_i|+x_i^0}{x_i^0}, \frac{|\bar{s}_i|+s_i^0}{s_i^0}\right\} \\
&\leq \max\left\{\frac{\omega^* x_i^0+x_i^0}{x_i^0}, \frac{\omega^* s_i^0+s_i^0}{s_i^0}\right\} \\
&= 1+\omega^*. \qquad (6.34)
\end{aligned}
$$

Resetting $\rho = \max\{\rho, 1+\omega^*\}$ finishes the proof. ■

The next lemma is also useful.

Lemma 6.4
Let $(\mathbf{x}^0, \mathbf{s}^0)$ be defined by (6.24). Then,

$$
\nu_k(\mathbf{x}^{k^{\mathrm{T}}}\mathbf{s}^0 + \mathbf{s}^{k^{\mathrm{T}}}\mathbf{x}^0) \leq (1+2\omega^0)\mathbf{x}^{k^{\mathrm{T}}}\mathbf{s}^k \qquad (6.35)
$$

Proof 6.5 If $\omega^0 = \infty$, then the lemma holds. Therefore, it is assumed that $\omega^0 < \infty$. Let $(\mathbf{x}^*, \lambda^*, \mathbf{s}^*) \in \mathcal{S}$ be the optimal solution that achieves the minimum of (6.27). From Lemma 6.1, it follows that

$$
\mathbf{A}\mathbf{x}^k - \mathbf{b} = \nu_k(\mathbf{A}\mathbf{x}^0 - \mathbf{b}), \quad \mathbf{A}^{\mathrm{T}}\lambda^k + \mathbf{s}^k - \mathbf{c} = \nu_k(\mathbf{A}^{\mathrm{T}}\lambda^0 + \mathbf{s}^0 - \mathbf{c}), \quad \mathbf{x}^{k^{\mathrm{T}}}\mathbf{s}^k = \nu_k \mathbf{x}^{0^{\mathrm{T}}}\mathbf{s}^0.
$$

Let

$$
\begin{aligned}
\hat{\mathbf{x}} &= \nu_k \mathbf{x}^0 + (1-\nu_k)\mathbf{x}^*, \\
(\hat{\lambda}, \hat{\mathbf{s}}) &= \nu_k(\lambda^0, \mathbf{s}^0) + (1-\nu_k)(\lambda^*, \mathbf{s}^*),
\end{aligned}
$$

then

$$
\begin{aligned}
\mathbf{A}\hat{\mathbf{x}} - \mathbf{b} &= \nu_k(\mathbf{A}\mathbf{x}^0 - \mathbf{b}), \\
\mathbf{A}^{\mathrm{T}}\hat{\lambda} + \hat{\mathbf{s}} - \mathbf{c} &= \nu_k(\mathbf{A}^{\mathrm{T}}\lambda^0 + \mathbf{s}^0 - \mathbf{c}).
\end{aligned}
$$

Since

$$
\hat{\mathbf{s}} = \mathbf{c} - \mathbf{A}^{\mathrm{T}}\hat{\lambda} + \nu_k(\mathbf{A}^{\mathrm{T}}\lambda^0 + \mathbf{s}^0 - \mathbf{c})
$$

and

$$
\mathbf{s}^k = \mathbf{c} - \mathbf{A}^{\mathrm{T}}\lambda^k + \nu_k(\mathbf{A}^{\mathrm{T}}\lambda^0 + \mathbf{s}^0 - \mathbf{c}),
$$

it follows that

$$
\hat{\mathbf{s}} - \mathbf{s}^k = \mathbf{c} - \mathbf{A}^{\mathrm{T}}\hat{\lambda} - \mathbf{c} + \mathbf{A}^{\mathrm{T}}\lambda^k = -\mathbf{A}^{\mathrm{T}}(\hat{\lambda} - \lambda^k).
$$

Noticing that $\mathbf{A}(\hat{\mathbf{x}} - \mathbf{x}^k) = \mathbf{0}$, we have

$$
\begin{aligned}
0 &= (\hat{\mathbf{x}} - \mathbf{x}^k)^{\mathrm{T}}(\hat{\mathbf{s}} - \mathbf{s}^k) \\
&= (\nu_k \mathbf{x}^0 + (1-\nu_k)\mathbf{x}^* - \mathbf{x}^k)^{\mathrm{T}}(\nu_k \mathbf{s}^0 + (1-\nu_k)\mathbf{s}^* - \mathbf{s}^k). \qquad (6.36)
\end{aligned}
$$

The last equation is equivalent to

$$\begin{aligned} & (\nu_k \mathbf{x}^0 + (1-\nu_k)\mathbf{x}^*)^{\mathrm{T}} \mathbf{s}^k + (\nu_k \mathbf{s}^0 + (1-\nu_k)\mathbf{s}^*)^{\mathrm{T}} \mathbf{x}^k \\ = \; & (\nu_k \mathbf{x}^0 + (1-\nu_k)\mathbf{x}^*)^{\mathrm{T}} (\nu_k \mathbf{s}^0 + (1-\nu_k)\mathbf{s}^*) + \mathbf{x}^{k^{\mathrm{T}}} \mathbf{s}^k \end{aligned} \tag{6.37}$$

Using this relation, ${\mathbf{x}^*}^{\mathrm{T}} \mathbf{s}^* = 0$, (6.29), and (6.30), and noticing that $\nu_k \in [0,1]$ and $(\mathbf{x}^*, \mathbf{s}^*) \geq \mathbf{0}$, it follows that

$$\begin{array}{lcl} \nu_k(\mathbf{x}^{0^{\mathrm{T}}} \mathbf{s}^k + \mathbf{s}^{0^{\mathrm{T}}} \mathbf{x}^k) & \leq & (\nu_k \mathbf{x}^{0^{\mathrm{T}}} + (1-\nu_k)\mathbf{x}^*)^{\mathrm{T}} \mathbf{s}^k + (\nu_k \mathbf{s}^{0^{\mathrm{T}}} + (1-\nu_k)\mathbf{s}^*)^{\mathrm{T}} \mathbf{x}^k \\ \text{[use (6.37)]} & = & (\nu_k \mathbf{x}^0 + (1-\nu_k)\mathbf{x}^*)^{\mathrm{T}} (\nu_k \mathbf{s}^0 + (1-\nu_k)\mathbf{s}^*) + \mathbf{x}^{k^{\mathrm{T}}} \mathbf{s}^k \\ \text{[use } {\mathbf{x}^*}^{\mathrm{T}} \mathbf{s}^* = 0\text{]} & = & \nu_k^2 \mathbf{x}^{0^{\mathrm{T}}} \mathbf{s}^0 + \nu_k(1-\nu_k)(\mathbf{x}^{0^{\mathrm{T}}} \mathbf{s}^* + \mathbf{s}^{0^{\mathrm{T}}} \mathbf{x}^*) + \mathbf{x}^{k^{\mathrm{T}}} \mathbf{s}^k \\ \text{[use (6.29)]} & \leq & \nu_k^2 \mathbf{x}^{0^{\mathrm{T}}} \mathbf{s}^0 + 2\nu_k(1-\nu_k)\omega^o \mathbf{x}^{0^{\mathrm{T}}} \mathbf{s}^0 + \mathbf{x}^{k^{\mathrm{T}}} \mathbf{s}^k \\ \text{[use } \omega^o \geq 1\text{]} & \leq & 2\nu_k \omega^o \mathbf{x}^{0^{\mathrm{T}}} \mathbf{s}^0 + \mathbf{x}^{k^{\mathrm{T}}} \mathbf{s}^k \\ \text{[use (6.30)]} & = & (1+2\omega^o)\mathbf{x}^{k^{\mathrm{T}}} \mathbf{s}^k. \end{array}$$

This finishes the proof. ■

The next Lemma will be used a few times.

Lemma 6.5
Let $\mathbf{D}^k = (\mathbf{X}^k)^{\frac{1}{2}} (\mathbf{S}^k)^{-\frac{1}{2}}$. *For* $i = 1,2,3$, *let* $(\delta \mathbf{x}^i, \delta \lambda^i, \delta \mathbf{s}^i)$ *be the solution of*

$$\begin{bmatrix} \mathbf{A} & \mathbf{0} & \mathbf{0} \\ \mathbf{0} & \mathbf{A}^{\mathrm{T}} & \mathbf{I} \\ \mathbf{S} & \mathbf{0} & \mathbf{X} \end{bmatrix} \begin{bmatrix} \delta \mathbf{x}^i \\ \delta \lambda^i \\ \delta \mathbf{s}^i \end{bmatrix} = \begin{bmatrix} \mathbf{0} \\ \mathbf{0} \\ \mathbf{r}^i \end{bmatrix} \tag{6.38}$$

and $(\delta \mathbf{x}, \delta \lambda, \delta \mathbf{s})$ *be the solution of*

$$\begin{bmatrix} \mathbf{A} & \mathbf{0} & \mathbf{0} \\ \mathbf{0} & \mathbf{A}^{\mathrm{T}} & \mathbf{I} \\ \mathbf{S} & \mathbf{0} & \mathbf{X} \end{bmatrix} \begin{bmatrix} \delta \mathbf{x} \\ \delta \lambda \\ \delta \mathbf{s} \end{bmatrix} = \begin{bmatrix} \mathbf{0} \\ \mathbf{0} \\ \mathbf{r}^1 + \mathbf{r}^2 + \mathbf{r}^3 \end{bmatrix}, \tag{6.39}$$

where $\mathbf{r}^1 = \mathbf{x} \circ \mathbf{s}$, $\mathbf{r}^2 = -\nu_k \mathbf{s} \circ (\mathbf{x}^0 - \bar{\mathbf{x}})$, *and* $\mathbf{r}^3 = -\nu_k \mathbf{x} \circ (\mathbf{s}^0 - \bar{\mathbf{s}})$. *Then,*

$$\|\mathbf{D}^{-1}\delta \mathbf{x}^1\|, \|\mathbf{D}\delta \mathbf{s}^1\| \leq \|\mathbf{D}^{-1}\delta \mathbf{x}^1 + \mathbf{D}\delta \mathbf{s}^1\| = \|(\mathbf{X}\mathbf{s})^{\frac{1}{2}}\| = \sqrt{\mathbf{x}^{\mathrm{T}}\mathbf{s}} = \sqrt{n\mu}, \tag{6.40a}$$

$$\|\mathbf{D}^{-1}\delta \mathbf{x}^2\|, \|\mathbf{D}\delta \mathbf{s}^2\| \leq \|\mathbf{D}^{-1}\delta \mathbf{x}^2 + \mathbf{D}\delta \mathbf{s}^2\| = \nu_k \|\mathbf{D}^{-1}(\mathbf{x}^0 - \bar{\mathbf{x}})\|, \tag{6.40b}$$

$$\|\mathbf{D}^{-1}\delta \mathbf{x}^3\|, \|\mathbf{D}\delta \mathbf{s}^3\| \leq \|\mathbf{D}^{-1}\delta \mathbf{x}^3 + \mathbf{D}\delta \mathbf{s}^3\| = \nu_k \|\mathbf{D}(\mathbf{s}^0 - \bar{\mathbf{s}})\|. \tag{6.40c}$$

Proof 6.6 Clearly, we have

$$\begin{aligned} \delta \mathbf{x} &= \delta \mathbf{x}^1 + \delta \mathbf{x}^2 + \delta \mathbf{x}^3 \\ \delta \lambda &= \delta \lambda^1 + \delta \lambda^2 + \delta \lambda^3 \\ \delta \mathbf{s} &= \delta \mathbf{s}^1 + \delta \mathbf{s}^2 + \delta \mathbf{s}^3. \end{aligned}$$

From the second row of (6.38), we have $(\mathbf{D}^{-1}\delta\mathbf{x}^i)^{\mathrm{T}}(\mathbf{D}\delta\mathbf{s}^i) = 0$, for $i = 1,2,3$, therefore,

$$\|\mathbf{D}^{-1}\delta\mathbf{x}^i\|^2, \|\mathbf{D}\delta\mathbf{s}^i\|^2 \le \|\mathbf{D}^{-1}\delta\mathbf{x}^i\|^2 + \|\mathbf{D}\delta\mathbf{s}^i\|^2 = \|\mathbf{D}^{-1}\delta\mathbf{x}^i + \mathbf{D}\delta\mathbf{s}^i\|^2. \tag{6.41}$$

Applying $(\mathbf{XS})^{-1/2}(\mathbf{S}\delta\mathbf{x}^i + \mathbf{X}\delta\mathbf{s}^i) = (\mathbf{XS})^{-1/2}\mathbf{r}^i$ to (6.41) for $i = 1,2,3$, respectively, we obtain (6.40). This finishes the proof. ∎

Lemma 6.6
Let $(\dot{\mathbf{x}}, \dot{\mathbf{s}})$ *be defined by (6.3), and* $\mathbf{D}^k = (\mathbf{X}^k)^{\frac{1}{2}}(\mathbf{S}^k)^{-\frac{1}{2}}$. *Then,*

$$\max\{\|(\mathbf{D}^k)^{-1}\dot{\mathbf{x}}\|, \|(\mathbf{D}^k)\dot{\mathbf{s}}\|\} \le \|(\mathbf{x}^k \circ \mathbf{s}^k)^{\frac{1}{2}}\| + \omega^f(1+2\omega^o)\frac{(\mathbf{x}^k)^{\mathrm{T}}\mathbf{s}^k}{\min_i (x_i^k s_i^k)^{\frac{1}{2}}}. \tag{6.42}$$

Proof 6.7 Let $\bar{\mathbf{x}}$ be a feasible solution of (1.4) and $(\bar{\lambda}, \bar{\mathbf{s}})$ be a feasible solution of (1.5) such that the minimum of (6.26) is achieved. Since

$$\mathbf{A}\dot{\mathbf{x}} = \mathbf{r}_b^k = \nu_k \mathbf{r}_b^0 = \nu_k(\mathbf{A}\mathbf{x}^0 - \mathbf{b}) = \nu_k \mathbf{A}(\mathbf{x}^0 - \bar{\mathbf{x}}),$$

we have

$$\mathbf{A}(\dot{\mathbf{x}} - \nu_k(\mathbf{x}^0 - \bar{\mathbf{x}})) = \mathbf{0}.$$

Similarly, since

$$\mathbf{A}^{\mathrm{T}}\dot{\lambda} + \dot{\mathbf{s}} = \mathbf{r}_c^k = \nu_k \mathbf{r}_c^0 = \nu_k(\mathbf{A}^{\mathrm{T}}\lambda^0 + \mathbf{s}^0 - \mathbf{c}) = \nu_k(\mathbf{A}^{\mathrm{T}}(\lambda^0 - \bar{\lambda}) + (\mathbf{s}^0 - \bar{\mathbf{s}})),$$

we have

$$\mathbf{A}^{\mathrm{T}}(\dot{\lambda} - \nu_k(\lambda^0 - \bar{\lambda})) + (\dot{\mathbf{s}} - \nu_k(\mathbf{s}^0 - \bar{\mathbf{s}})) = \mathbf{0}.$$

Using the last row of (6.3), we have

$$\mathbf{s} \circ (\dot{\mathbf{x}} - \nu_k(\mathbf{x}^0 - \bar{\mathbf{x}})) + \mathbf{x} \circ (\dot{\mathbf{s}} - \nu_k(\mathbf{s}^0 - \bar{\mathbf{s}})) = \mathbf{x} \circ \mathbf{s} - \nu_k \mathbf{s} \circ (\mathbf{x}^0 - \bar{\mathbf{x}}) - \nu_k \mathbf{x} \circ (\mathbf{s}^0 - \bar{\mathbf{s}}).$$

Thus, in matrix form, we have

$$\begin{bmatrix} \mathbf{A} & \mathbf{0} & \mathbf{0} \\ \mathbf{0} & \mathbf{A}^{\mathrm{T}} & \mathbf{I} \\ \mathbf{S} & \mathbf{0} & \mathbf{X} \end{bmatrix} \begin{bmatrix} \dot{\mathbf{x}} - \nu_k(\mathbf{x}^0 - \bar{\mathbf{x}}) \\ \dot{\lambda} - \nu_k(\lambda^0 - \bar{\lambda}) \\ \dot{\mathbf{s}} - \nu_k(\mathbf{s}^0 - \bar{\mathbf{s}}) \end{bmatrix} = \begin{bmatrix} \mathbf{0} \\ \mathbf{0} \\ \mathbf{x} \circ \mathbf{s} - \nu_k \mathbf{s} \circ (\mathbf{x}^0 - \bar{\mathbf{x}}) - \nu_k \mathbf{x} \circ (\mathbf{s}^0 - \bar{\mathbf{s}}) \end{bmatrix} \tag{6.43}$$

Denote

$$(\delta\mathbf{x}, \delta\lambda, \delta\mathbf{s}) = (\dot{\mathbf{x}} - \nu_k(\mathbf{x}^0 - \bar{\mathbf{x}}), \dot{\lambda} - \nu_k(\lambda^0 - \bar{\lambda}), \dot{\mathbf{s}} - \nu_k(\mathbf{s}^0 - \bar{\mathbf{s}}))$$

and

$$\mathbf{r} = \mathbf{r}^1 + \mathbf{r}^2 + \mathbf{r}^3$$

with

$$(\mathbf{r}^1, \mathbf{r}^2, \mathbf{r}^3) = (\mathbf{x} \circ \mathbf{s}, -\nu_k \mathbf{s} \circ (\mathbf{x}^0 - \bar{\mathbf{x}}), -\nu_k \mathbf{x} \circ (\mathbf{s}^0 - \bar{\mathbf{s}})).$$

Applying Lemma 6.5, it follows that equations (6.40) holds. Considering (6.38) with $i = 2$, we have

$$\mathbf{S}\delta\mathbf{x}^2 + \mathbf{X}\delta\mathbf{s}^2 = \mathbf{r}^2 = -\nu_k \mathbf{S}(\mathbf{x}^0 - \bar{\mathbf{x}}),$$

which is equivalent to

$$\delta\mathbf{x}^2 = -\nu_k(\mathbf{x}^0 - \bar{\mathbf{x}}) - \mathbf{D}^2\delta\mathbf{s}^2. \tag{6.44}$$

Thus, from (6.44) and (6.40), we have

$$\begin{aligned}
\|\mathbf{D}^{-1}\dot{\mathbf{x}}\| &= \|\mathbf{D}^{-1}[\delta\mathbf{x}^1 + \delta\mathbf{x}^2 + \delta\mathbf{x}^3 + \nu_k(\mathbf{x}^0 - \bar{\mathbf{x}})]\| \\
&= \|\mathbf{D}^{-1}\delta\mathbf{x}^1 - \mathbf{D}\delta\mathbf{s}^2 + \mathbf{D}^{-1}\delta\mathbf{x}^3\| \\
&\leq \|\mathbf{D}^{-1}\delta\mathbf{x}^1\| + \|\mathbf{D}\delta\mathbf{s}^2\| + \|\mathbf{D}^{-1}\delta\mathbf{x}^3\|.
\end{aligned} \tag{6.45}$$

Considering (6.38) with $i = 3$, we have

$$\mathbf{S}\delta\mathbf{x}^3 + \mathbf{X}\delta\mathbf{s}^3 = \mathbf{r}^3 = -\nu_k \mathbf{X}(\mathbf{s}^0 - \bar{\mathbf{s}}),$$

which is equivalent to

$$\delta\mathbf{s}^3 = -\nu_k(\mathbf{s}^0 - \bar{\mathbf{s}}) - \mathbf{D}^{-2}\delta\mathbf{x}^3. \tag{6.46}$$

Thus, from (6.46) and (6.40), we have

$$\begin{aligned}
\|\mathbf{D}\dot{\mathbf{s}}\| &= \|\mathbf{D}[\delta\mathbf{s}^1 + \delta\mathbf{s}^2 + \delta\mathbf{s}^3 + \nu_k(\mathbf{s}^0 - \bar{\mathbf{s}})]\| \\
&= \|\mathbf{D}\delta\mathbf{s}^1 + \mathbf{D}\delta\mathbf{s}^2 - \mathbf{D}^{-1}\delta\mathbf{x}^3\| \\
&\leq \|\mathbf{D}\delta\mathbf{s}^1\| + \|\mathbf{D}\delta\mathbf{s}^2\| + \|\mathbf{D}^{-1}\delta\mathbf{x}^3\|.
\end{aligned} \tag{6.47}$$

In view of (6.40a), it follows that

$$\|\mathbf{D}\dot{\mathbf{s}}\|, \|\mathbf{D}^{-1}\dot{\mathbf{x}}\| \leq \sqrt{n\mu} + \|\mathbf{D}\delta\mathbf{s}^2\| + \|\mathbf{D}^{-1}\delta\mathbf{x}^3\|. \tag{6.48}$$

Using (6.40b), we have

$$\begin{aligned}
\|\mathbf{D}\delta\mathbf{s}^2\| &\leq \nu_k\|\mathbf{D}^{-1}(\mathbf{x}^0 - \bar{\mathbf{x}})\| \\
&\leq \frac{\nu_k}{\min\sqrt{x_i s_i}}\|\mathbf{S}(\mathbf{x}^0 - \bar{\mathbf{x}})\| \\
[\text{use (6.28)}] &\leq \frac{\nu_k\omega^f}{\min\sqrt{x_i s_i}}\|\mathbf{S}\mathbf{x}^0\| \\
[\text{use } \|\cdot\|_2 \leq \|\cdot\|_1] &\leq \frac{\nu_k\omega^f}{\min\sqrt{x_i s_i}}\mathbf{s}^{\mathrm{T}}\mathbf{x}^0
\end{aligned}$$

Using (6.40c), we have

$$\begin{aligned}
\|\mathbf{D}^{-1}\delta\mathbf{x}^3\| &\leq \nu_k\|\mathbf{D}(\mathbf{s}^0 - \bar{\mathbf{s}})\| \\
&\leq \frac{\nu_k}{\min\sqrt{x_i s_i}}\|\mathbf{X}(\mathbf{s}^0 - \bar{\mathbf{s}})\| \\
[\text{use (6.28)}] &\leq \frac{\nu_k\omega^f}{\min\sqrt{x_i s_i}}\|\mathbf{X}\mathbf{s}^0\| \\
[\text{use } \|\cdot\|_2 \leq \|\cdot\|_1] &\leq \frac{\nu_k\omega^f}{\min\sqrt{x_i s_i}}\mathbf{x}^{\mathrm{T}}\mathbf{s}^0.
\end{aligned}$$

Using Lemme 6.4, this shows that

$$\begin{aligned} \|\mathbf{D}\delta\mathbf{s}^2\| + \|\mathbf{D}^{-1}\delta\mathbf{x}^3\| &\leq \frac{\nu_k \omega^f}{\min\sqrt{x_i s_i}}(\mathbf{s}^{\mathrm{T}}\mathbf{x}^0 + \mathbf{x}^{\mathrm{T}}\mathbf{s}^0) \\ &\leq \omega^f(1+2\omega^o)\frac{(\mathbf{x}^k)^{\mathrm{T}}\mathbf{s}^k}{\min_i(x_i^k s_i^k)^{\frac{1}{2}}}. \end{aligned} \tag{6.49}$$

This complete the proof. ■

This leads to the following lemma.

Lemma 6.7
Let $(\dot{\mathbf{x}}, \dot{\mathbf{s}})$ be defined by (6.3). Then, there exists a positive constant C_0, independent of n, such that

$$\max\{\|(\mathbf{D}^k)^{-1}\dot{\mathbf{x}}\|, \|(\mathbf{D}^k)\dot{\mathbf{s}}\|\} \leq C_0\sqrt{n(\mathbf{x}^k)^{\mathrm{T}}\mathbf{s}^k}. \tag{6.50}$$

Proof 6.8 First, it is easy to see

$$\|(\mathbf{x}^k \circ \mathbf{s}^k)^{\frac{1}{2}}\| = \sqrt{\sum_i x_i^k s_i^k} = \sqrt{(\mathbf{x}^k)^{\mathrm{T}}\mathbf{s}^k}. \tag{6.51}$$

Since $(\mathbf{x}^k, \mathbf{s}^k) \in \mathcal{N}_2^I(\theta)$, we have $\min_i(x_i^k s_i^k) \geq (1-\theta)\mu_k - (1-\theta)\frac{(\mathbf{x}^k)^{\mathrm{T}}\mathbf{s}^k}{n}$. Therefore,

$$\frac{(\mathbf{x}^k)^{\mathrm{T}}\mathbf{s}^k}{\min_i(x_i^k s_i^k)^{\frac{1}{2}}} \leq \sqrt{\frac{n(\mathbf{x}^k)^{\mathrm{T}}\mathbf{s}^k}{(1-\theta)}}. \tag{6.52}$$

Substituting (6.51) and (6.52) into (6.42) and using Lemma 6.3 prove (6.50) with $C_0 \geq 1 + \frac{\omega^f(1+2\omega^o)}{\sqrt{(1-\theta)}}$. ■

From Lemma 6.7, we can establish several useful inequalities. The following simple facts will be used several times. Let $\mathbf{u}$ and $\mathbf{v}$ be two vectors with the same dimension, then

$$\|\mathbf{u} \circ \mathbf{v}\|^2 = \sum_i (u_i v_i)^2 \leq \left(\sum_i u_i^2\right)\left(\sum_i v_i^2\right) = \|\mathbf{u}\|^2\|\mathbf{v}\|^2. \tag{6.53}$$

If $\mathbf{u}$ and $\mathbf{v}$ satisfy $\mathbf{u}^{\mathrm{T}}\mathbf{v} = 0$, then,

$$\max\{\|\mathbf{u}\|^2, \|\mathbf{v}\|^2\} \leq \|\mathbf{u}\|^2 + \|\mathbf{v}\|^2 = \|\mathbf{u}+\mathbf{v}\|^2, \tag{6.54}$$

and (see Lemma 3.2)

$$\|\mathbf{u} \circ \mathbf{v}\| \leq 2^{-\frac{3}{2}}\|\mathbf{u}+\mathbf{v}\|^2. \tag{6.55}$$

Lemma 6.8
Let $(\dot{\mathbf{x}},\dot{\mathbf{s}})$ *and* $(\ddot{\mathbf{x}},\ddot{\mathbf{s}})$ *be defined by (6.3) and (6.4), respectively. Then, there exist positive constants* C_1, C_2, C_3, *and* C_4, *independent of n, such that*

$$\|\dot{\mathbf{x}}\circ\dot{\mathbf{s}}\| \le C_1 n^2 \mu_k, \tag{6.56}$$

$$\|\ddot{\mathbf{x}}\circ\ddot{\mathbf{s}}\| \le C_2 n^4 \mu_k, \tag{6.57}$$

$$\max\{\|(\mathbf{D}^k)^{-1}\ddot{\mathbf{x}}\|, \|(\mathbf{D}^k)\ddot{\mathbf{s}}\|\} \le C_3 n^2 \sqrt{\mu_k}, \tag{6.58}$$

$$\max\{\|\ddot{\mathbf{x}}\circ\dot{\mathbf{s}}\|, \|\dot{\mathbf{x}}\circ\ddot{\mathbf{s}}\|\} \le C_4 n^3 \mu_k \tag{6.59}$$

Proof 6.9 First, using (6.53) and Lemma 6.7, we have

$$\begin{aligned}\|\dot{\mathbf{x}}\circ\dot{\mathbf{s}}\| &= \|(\mathbf{D}^k)^{-1}\dot{\mathbf{x}}\circ(\mathbf{D}^k)\dot{\mathbf{s}}\| \\ &\le \|(\mathbf{D}^k)^{-1}\dot{\mathbf{x}}\|\|(\mathbf{D}^k)\dot{\mathbf{s}}\| \le C_0^2 n(\mathbf{x}^k)^{\mathrm{T}}\mathbf{s}^k := C_1 n^2 \mu_k.\end{aligned} \tag{6.60}$$

Second, using (6.55), (6.4), (6.56), and (6.51), we have

$$\begin{aligned}\|\ddot{\mathbf{x}}\circ\ddot{\mathbf{s}}\| &= \|(\mathbf{D}^k)^{-1}\ddot{\mathbf{x}}\circ(\mathbf{D}^k)\ddot{\mathbf{s}}\| \\ &\le 2^{-\frac{3}{2}}\|(\mathbf{D}^k)^{-1}\ddot{\mathbf{x}}+(\mathbf{D}^k)\ddot{\mathbf{s}}\|^2 \\ &\le 2^{-\frac{3}{2}}\left\|-2(\mathbf{X}^k\mathbf{S}^k)^{-\frac{1}{2}}(\dot{\mathbf{x}}\circ\dot{\mathbf{s}})\right\|^2 \\ &= 2^{\frac{1}{2}}\sum_{i=1}^{n}\left(\frac{\dot{x}_i\dot{s}_i}{\sqrt{x_i^k}\sqrt{s_i^k}}\right)^2 = 2^{\frac{1}{2}}\sum_{i=1}^{n}\frac{(\dot{x}_i\dot{s}_i)^2}{x_i^k s_i^k} \\ &\le 2^{\frac{1}{2}}\frac{\sum_{i=1}^{n}(\dot{x}_i\dot{s}_i)^2}{\min_{i=1,\ldots,n}x_i^k s_i^k} \\ &\le 2^{\frac{1}{2}}\frac{\|\dot{\mathbf{x}}\circ\dot{\mathbf{s}}\|^2}{(1-\theta)\mu_k} \le 2^{\frac{1}{2}}\frac{C_1^2 n^4 \mu_k^2}{(1-\theta)\mu_k} \\ &= 2^{\frac{1}{2}}\frac{C_1^2 n^4 \mu_k}{1-\theta} := C_2 n^4 \mu_k.\end{aligned} \tag{6.61}$$

Third, using (6.54), (6.4), and (6.56), we have

$$\begin{aligned}&\max\{\|(\mathbf{D}^k)^{-1}\ddot{\mathbf{x}}\|^2, \|(\mathbf{D}^k)\ddot{\mathbf{s}}\|^2\} \le \|(\mathbf{D}^k)^{-1}\ddot{\mathbf{x}}+(\mathbf{D}^k)\ddot{\mathbf{s}}\|^2 \\ &= \left\|-2(\mathbf{X}^k\mathbf{S}^k)^{-\frac{1}{2}}(\dot{\mathbf{x}}\circ\dot{\mathbf{s}})\right\|^2 \le \frac{4C_1^2 n^4 \mu_k}{1-\theta} := C_3^2 n^4 \mu_k.\end{aligned} \tag{6.62}$$

Taking the square root on both sides proves (6.58). Finally, using (6.53), (6.58), and Lemma 6.7, we have

$$\begin{aligned}\|\ddot{\mathbf{x}}\circ\dot{\mathbf{s}}\| &= \|(\mathbf{D}^k)^{-1}\ddot{\mathbf{x}}\circ(\mathbf{D}^k)\dot{\mathbf{s}}\| \\ &\le \|(\mathbf{D}^k)^{-1}\ddot{\mathbf{x}}\|\|(\mathbf{D}^k)\dot{\mathbf{s}}\| \\ &\le (C_3 n^2\sqrt{\mu_k})(C_0 n\sqrt{\mu_k}) := C_4 n^3 \mu_k.\end{aligned} \tag{6.63}$$

Similarly, we can show

$$\|\dot{\mathbf{x}} \circ \ddot{\mathbf{s}}\| \le C_4 n^3 \mu_k. \tag{6.64}$$

This completes the proof. ■

Now we are ready to estimate a conservative bound for $\sin(\alpha_k)$.

Lemma 6.9
Let $(\mathbf{x}^k, \lambda^k, \mathbf{s}^k)$ be generated by Algorithm 6.1. Then, α_k obtained in Step 3 satisfies the following inequality.

$$\sin(\alpha_k) \ge \frac{\theta}{2Cn}, \tag{6.65}$$

where $C = \max\{1, C_4^{\frac{1}{3}}, (C_1 + C_2)^{\frac{1}{4}}\}$.

Proof 6.10 Let $\alpha \in (0, \frac{\pi}{2}]$ satisfy $\sin(\alpha) = \frac{\theta}{2Cn}$. In view of (6.21) and Lemma 6.8, we have

$$\begin{aligned} q(\alpha) &\le \mu_k((C_1+C_2)n^4 \sin^4(\alpha) + 2C_4 n^3 \sin^3(\alpha) + \theta \sin(\alpha) - \theta) := \mu_k p(\alpha) \\ &= \mu_k \left(\frac{(C_1+C_2)\theta^4}{16C^4} + \frac{2C_4\theta^3}{8C^3} + \frac{\theta^2}{2Cn} - \theta \right) \\ &\le \mu_k \left(\frac{\theta^4}{16} + \frac{\theta^3}{4} + \frac{\theta^2}{2} - 0 \right) \le 0. \end{aligned}$$

Since $p(\alpha)$ is a monotonic function of $\sin(\alpha)$, for all $\sin(\alpha) \le \frac{\theta}{2Cn}$, the above inequalities hold (the last inequality holds because of $\theta \le 1$). Therefore, for all $\sin(\alpha) \le \frac{\theta}{2Cn}$, the inequality (6.10) holds. This completes the proof. ■

Remark 6.1 It is worthwhile to point out that the constant C depends on C_0 which depends on ρ, but ρ is an unknown before we find the solution. Also, we can always find a better step-length $\sin(\alpha)$ by solving the quartic $q(\alpha) = 0$ and the calculation of the roots for a quartic polynomial is deterministic, negligible, and independent to n [125, 153]. ■

Using the standard argument developed in Theorem 1.4, we have the following theorem.

Theorem 6.3
Let $(\mathbf{x}^k, \lambda^k, \mathbf{s}^k)$ be generated by Algorithm 6.1 with an initial point given by (6.24). For any $\varepsilon > 0$, the algorithm will terminate with $(\mathbf{x}^k, \lambda^k, \mathbf{s}^k)$ satisfying (6.9) in at most $\mathcal{O}(nL)$ iterations, where

$$L = \max\{\ln((\mathbf{x}^0)^{\mathrm{T}}\mathbf{s}^0/\varepsilon), \ln(\|\mathbf{r}_b^0\|/\varepsilon), \ln(\|\mathbf{r}_c^0\|/\varepsilon)\}.$$

Proof 6.11 In view of Lemma 6.1, $\mathbf{r}_b^k$, $\mathbf{r}_c^k$, and μ_k all decrease at the same rate $(1-\sin(\alpha))$ in every iteration. In view of Lemma 6.9, the rate is at least $(1-\frac{\theta}{2Cn})$. Invoking Theorem 1.4 proves the claim. ∎

6.3 Concluding Remarks

An infeasible arc-search interior-point algorithm was proposed in this Chapter. The algorithm searches for the optimizer along an ellipse that approximates the central-path. It is shown that the arc-search algorithm attained a polynomial bound $\mathcal{O}(nL)$, which is at least as good as the best existing bound for infeasible-interior-point algorithms for linear programming using line search. Numerical test result reported in [161] is promising. In the next two chapters, the result will be improved further in terms of both complexity bound and computational efficiency.

Chapter 7

A Mehrotra-Type Infeasible Arc-Search Algorithm for LP

Interior-point method was regarded as a mature technique of linear programming [144, page 2], following many important developments between 1980-1990, such as a proposal of path-following method [88], the establishment of polynomial bounds for path-following algorithms [70, 71], the development of Mehrotra's predictor-corrector (MPC) algorithm [89] and independent implementation and verification [82, 83], and the proof of the polynomiality of some related infeasible interior-point algorithms [92, 171]. Although many more algorithms were proposed between 1990-2010 (see, for example, [93, 90, 116, 120, 63]), there was no significant improvement in the best polynomial bound for interior-point algorithms, and there was no report of a better algorithm than an MPC variation, described in [144, page 198], for general linear programming problems[1]. In fact, most popular interior-point method software packages implemented the MPC variation described in [144, page 198], for example, OB1 [82], LOQO [139], PCx [23], and LIPSOL [172].

However, as pointed out in previous chapters, researchers knew that there were still some dilemmas between theoretical polynomial bounds and computational experience. Namely, strategies implemented in the popular software packages, such as (a) search optimizers in large neighborhood, (b) use higher-order derivatives to approximate central path, and (c) start from infeasible interior-point were proved to be efficient, but theoretical analysis gave worse polynomial bounds when these strategies were considered. These dilemmas were discussed in Chapters 2, 3, and 4. In

[1] There was some noticeable progress focused on problems with special structures, for example [142].

Chapter 5, we introduced arc-search to replace the traditional line search and showed that this higher-order method works for feasible short step algorithm, i.e., higher-order algorithm with arc-search achieves the best polynomial bound and has promising numerical performance. In Chapter 6, we showed that arc-search also improves polynomial bound for algorithm starting from infeasible initial point.

In this chapter, our goal is to demonstrate that arc-search incorporating all three strategies mentioned above will be more efficient than the best line search algorithm, Mehrotra's method. Instead of theoretical analysis, we perform extensive numerical testing in order to demonstrate the claim. We discuss an implementation of an infeasible arc-search interior-point algorithm, which allows us to test a lot more Netlib problems, because very few problems in benchmark Netlib test set have an interior-point, as noted in [20]. The implemented algorithm is almost identical to Mehrotra's, except the former uses arc-search discussed in Chapters 5 and 6 while the latter uses line search. We show that the implemented arc-search infeasible interior-point algorithm is very competitive in computation by testing all Netlib problems in standard form and comparing the results to those obtained by the MPC described in [144, page 198]. To have a fair comparison, both algorithms are implemented in MATLAB® (`curvelp.m` for arc-search algorithm and `mehrotra.m` for Mehrotra's algorithm); for all test problems, the two MATLAB codes use the same pre-processor and post-processor, start from the same initial point, use the same parameters, and terminate with the same stopping criterion. Since the main cost in computation for both algorithms is to solve linear systems of equations, which is exactly the same for both algorithms, and the arc-search infeasible interior-point algorithm uses less iterations in most tested problems than the MPC described in [144, page 198], we believe that the proposed arc-search algorithm is more attractive than the MPC algorithm. The material in this Chapter was first presented in [156] (which was the first arc-search infeasible interior-point algorithm for LP) and was published in [158] a few years later.

7.1 Installation and Basic Structure of CurveLP

CurveLP is available from Netlib (http://www.netlib.org/numeralgo/) as the na43 package. It is distributed as a single file CurveLP.zip. To install the library, download and unpack the zip archive, start MATLAB, change to the root folder CurveLP. The readers will see four m-files: curvelp.m is the implementation of the proposed algorithm, mehrotra.m is the implementation of Mehrotra's algorithm which is mainly used for the purpose of comparison for the performance of CurveLP and Mehrotra's algorithms, extract.m is used to extract $(\mathbf{A}, \mathbf{b}, \mathbf{c})$ from Netlib linear programming test problems which are in the same folder in MATLAB format, and infeasibleexample.m is the MATLAB code used to generate Figure 7.2. README describes the simple procedures for testing and running the code.

All four m-files are developed in MATLAB version 2006a and tested in MATLAB versions of 2006a and 2012b on Windows.

7.2 An Infeasible Long-Step Arc-Search Algorithm for Linear Programming

Consider the linear programming in the standard form described in (1.4) and its dual programming that is also presented in the standard form described in (1.5). As pointed out before, if a problem is not represented in the standard form, it is easy to transform it to the standard form by using the method described in [144, pages 226-230].

Similar to the aforementioned discussions, we will denote the residuals of the equality constraints (the deviation from the feasibility) by

$$\mathbf{r}_b = \mathbf{A}\mathbf{x} - \mathbf{b}, \quad \mathbf{r}_c = \mathbf{A}^{\mathrm{T}}\boldsymbol{\lambda} + \mathbf{s} - \mathbf{c}, \tag{7.1}$$

the duality measure by

$$\mu = \frac{\mathbf{x}^{\mathrm{T}}\mathbf{s}}{n}. \tag{7.2}$$

The central path $\mathcal{C}(t)$ of the primal-dual linear programming problem is parametrized by a scalar $t > 0$ as follows. For each interior point $(\mathbf{x}, \boldsymbol{\lambda}, \mathbf{s}) \in \mathcal{C}(t)$ on the central path, there is a $t > 0$ such that

$$\mathbf{A}\mathbf{x} = \mathbf{b} \tag{7.3a}$$

$$\mathbf{A}^{\mathrm{T}}\boldsymbol{\lambda} + \mathbf{s} = \mathbf{c} \tag{7.3b}$$

$$(\mathbf{x}, \mathbf{s}) \geq \mathbf{0} \tag{7.3c}$$

$$x_i s_i = t, \quad i = 1, \ldots, n. \tag{7.3d}$$

Therefore, the central path is an arc in $\mathbf{R}^{2n+m}$ parametrized as a function of t and is denoted as

$$\mathcal{C}(t) = \{(\mathbf{x}(t), \boldsymbol{\lambda}(t), \mathbf{s}(t)) : t > 0\}. \tag{7.4}$$

As $t \to 0$, the central path $(\mathbf{x}(t), \boldsymbol{\lambda}(t), \mathbf{s}(t))$ represented by (7.3) approaches to a solution of LP represented by (1.4) because (7.3) reduces to the KKT condition as $t \to 0$.

Because of the high cost of finding an initial feasible point and the central path described in (7.3), we consider a modified problem which allows infeasible initial point:

$$\mathbf{A}\mathbf{x} - \mathbf{b} = \mathbf{r}_b \tag{7.5a}$$

$$\mathbf{A}^{\mathrm{T}}\boldsymbol{\lambda} + \mathbf{s} - \mathbf{c} = \mathbf{r}_c \tag{7.5b}$$

$$(\mathbf{x}, \mathbf{s}) \geq \mathbf{0} \tag{7.5c}$$

$$x_i s_i = t, \quad i = 1, \ldots, n. \tag{7.5d}$$

We search for the optimizer along an infeasible central path neighborhood. The infeasible central path neighborhood $\mathcal{F}(\gamma)$ considered in this chapter is defined as a collection of points that satisfy the following conditions,

$$\mathcal{F}(\gamma(t)) = \{(\mathbf{x}, \boldsymbol{\lambda}, \mathbf{s}) : \|(\mathbf{r}_b(t), \mathbf{r}_c(t))\| \leq \gamma(t)\|(\mathbf{r}_b^0, \mathbf{r}_c^0)\|, (\mathbf{x}, \mathbf{s}) > \mathbf{0}\}, \tag{7.6}$$

where $\mathbf{r}_b(1) = \mathbf{r}_b^0$, $\mathbf{r}_c(1) = \mathbf{r}_c^0$, $\gamma(t) \in [0,1]$ is a monotonic function of t such that $\gamma(1) = 1$ and $\gamma(t) \to 0$ as $t \to 0$. In the next section, we will show (Lemma 7.2) that $\|(\mathbf{r}_b(t), \mathbf{r}_c(t))\| \le \gamma(t)\|(\mathbf{r}_b^0, \mathbf{r}_c^0)\|$ always holds with $\gamma(0) = \prod_{j=0}^{\infty}(1-\sin(\alpha_j^x)) \to 0$ because $\frac{\pi}{2} \ge \alpha_j^x > 0$. Therefore, the only restriction of this neighborhood is $(\mathbf{x},\mathbf{s}) > \mathbf{0}$. This shows that the central path neighborhood is the widest in comparison to any neighborhood we have considered.

Starting from any initial point $(\mathbf{x}^0, \lambda^0, \mathbf{s}^0)$ in a central path neighborhood that satisfies $(\mathbf{x}^0, \mathbf{s}^0) > \mathbf{0}$, for $k \ge 0$, we consider a special arc parametrized by t and defined by the current iterate as follows:

$$\mathbf{A}\mathbf{x}(t) - \mathbf{b} = t\mathbf{r}_b^k, \tag{7.7a}$$

$$\mathbf{A}^{\mathrm{T}}\lambda(t) + \mathbf{s}(t) - \mathbf{c} = t\mathbf{r}_c^k, \tag{7.7b}$$

$$(\mathbf{x}(t), \mathbf{s}(t)) > \mathbf{0}, \tag{7.7c}$$

$$\mathbf{x}(t) \circ \mathbf{s}(t) = t\mathbf{x}^k \circ \mathbf{s}^k. \tag{7.7d}$$

Clearly, each iteration starts at $t = 1$; and $(\mathbf{x}(1), \lambda(1), \mathbf{s}(1)) = (\mathbf{x}^k, \lambda^k, \mathbf{s}^k)$. We want the iterate to stay inside $\mathcal{F}(\gamma)$ as t decreases. We denote the infeasible central path defined by (7.7) as

$$\mathcal{H}(t) = \{(\mathbf{x}(t), \lambda(t), \mathbf{s}(t)) : t \ge \tau \ge 0\}. \tag{7.8}$$

If this arc is inside $\mathcal{F}(\gamma)$ for $\tau = 0$, then as $t \to 0$, $(\mathbf{r}_b(t), \mathbf{r}_c(t)) := t(\mathbf{r}_b^k, \mathbf{r}_c^k) \to \mathbf{0}$; and equation (7.7d) implies that $\mu(t) \to 0$; hence, the arc will approach to an optimal solution of (1.4) because (7.7) reduces to KKT condition as $t \to 0$. To avoid computing the entire infeasible central path $\mathcal{H}(t)$, we will search along an approximation of $\mathcal{H}(t)$ and keep the iterate in $\mathcal{F}(\gamma)$. Therefore, we will use an ellipse $\mathcal{E}(\alpha)$ in $2n+m$ dimensional space to approximate the infeasible central path $\mathcal{H}(t)$, where $\mathcal{E}(\alpha)$ is given by

$$\mathcal{E}(\alpha) = \{(\mathbf{x}(\alpha), \lambda(\alpha), \mathbf{s}(\alpha)) : (\mathbf{x}(\alpha), \lambda(\alpha), \mathbf{s}(\alpha)) = \vec{\mathbf{a}}\cos(\alpha) + \vec{\mathbf{b}}\sin(\alpha) + \vec{\mathbf{c}}\}, \tag{7.9}$$

$\vec{\mathbf{a}} \in \mathbb{R}^{2n+m}$ and $\vec{\mathbf{b}} \in \mathbb{R}^{2n+m}$ are the axes of the ellipse, and $\vec{\mathbf{c}} \in \mathbb{R}^{2n+m}$ is the center of the ellipse. Given the current iterate $\mathbf{y} = (\mathbf{x}^k, \lambda^k, \mathbf{s}^k) = (\mathbf{x}(\alpha_0), \lambda(\alpha_0), \mathbf{s}(\alpha_0)) \in \mathcal{E}(\alpha)$ which is also on $\mathcal{H}(t)$, we will determine $\vec{\mathbf{a}}$, $\vec{\mathbf{b}}$, $\vec{\mathbf{c}}$ and α_0 such that the first and second derivatives of $\mathcal{E}(\alpha)$ at $(\mathbf{x}(\alpha_0), \lambda(\alpha_0), \mathbf{s}(\alpha_0))$ are the same as those of $\mathcal{H}(t)$ at $(\mathbf{x}(\alpha_0), \lambda(\alpha_0), \mathbf{s}(\alpha_0))$. Therefore, by taking the first order derivative for (7.7) at $(\mathbf{x}(\alpha_0), \lambda(\alpha_0), \mathbf{s}(\alpha_0)) = (\mathbf{x}^k, \lambda^k, \mathbf{s}^k) \in \mathcal{E}$, we have

$$\begin{bmatrix} \mathbf{A} & \mathbf{0} & \mathbf{0} \\ \mathbf{0} & \mathbf{A}^{\mathrm{T}} & \mathbf{I} \\ \mathbf{S}^k & \mathbf{0} & \mathbf{X}^k \end{bmatrix} \begin{bmatrix} \dot{\mathbf{x}} \\ \dot{\lambda} \\ \dot{\mathbf{s}} \end{bmatrix} = \begin{bmatrix} \mathbf{r}_b^k \\ \mathbf{r}_c^k \\ \mathbf{x}^k \circ \mathbf{s}^k \end{bmatrix}, \tag{7.10}$$

These linear systems of equations are very similar to those used in Chapter 5 except that equality constraints in (7.3) are not assumed to be satisfied. By taking the second derivative, we have

$$\begin{bmatrix} \mathbf{A} & \mathbf{0} & \mathbf{0} \\ \mathbf{0} & \mathbf{A}^{\mathrm{T}} & \mathbf{I} \\ \mathbf{S}^k & \mathbf{0} & \mathbf{X}^k \end{bmatrix} \begin{bmatrix} \ddot{\mathbf{x}} \\ \ddot{\lambda} \\ \ddot{\mathbf{s}} \end{bmatrix} = \begin{bmatrix} \mathbf{0} \\ \mathbf{0} \\ -2\dot{\mathbf{x}} \circ \dot{\mathbf{s}} \end{bmatrix}. \tag{7.11}$$

One of the brilliant contributions of [89] is that, instead of using (7.11), the second order term ('corrector') of the primal-dual trajectory is computed with the centering direction. Following the idea of [89], we modify (7.11) slightly to make sure that a substantial segment of the ellipse stays in $\mathcal{F}(t)$, thereby making sure that the step size along the ellipse is significantly greater than zero,

$$\begin{bmatrix} \mathbf{A} & \mathbf{0} & \mathbf{0} \\ \mathbf{0} & \mathbf{A}^{\mathrm{T}} & \mathbf{I} \\ \mathbf{S}^k & \mathbf{0} & \mathbf{X}^k \end{bmatrix} \begin{bmatrix} \ddot{\mathbf{x}}(\sigma_k) \\ \ddot{\lambda}(\sigma_k) \\ \ddot{\mathbf{s}}(\sigma_k) \end{bmatrix} = \begin{bmatrix} \mathbf{0} \\ \mathbf{0} \\ \sigma_k \mu_k \mathbf{e} - 2\dot{\mathbf{x}} \circ \dot{\mathbf{s}} \end{bmatrix}, \tag{7.12}$$

where the duality measure μ_k is evaluated at $(\mathbf{x}^k, \lambda^k, \mathbf{s}^k)$, and we set the centering parameter σ_k satisfying $0 < \sigma_k < \sigma_{\max} \leq 0.5$. Several slightly different heuristics have been proposed in literature. Mehrotra's original idea in [89] was a subroutine CENPAR which was fine tuned by Lustig et al. in [82, page 440] in order to improve computational performance, which, in turn, was modified again and the modified centering method was still credited to Mehrotra in [144, page 196] (please note that Mehrotra is one of the authors of PCx [23]). The formula of [144] is widely believed to be the best method and is now implemented in most state-of-the-art software packages and published papers, such as [120, 139, 23, 172]. We follow the common practice and use the formula of [144] in our implementation. We emphasize that the second derivatives are functions of σ_k, which is the idea discussed in Section 3.4 and is used to speed up the convergence of Algorithm 3.1. Several relations follow immediately from (7.10) and (7.12).

Lemma 7.1
Let $(\dot{\mathbf{x}}, \dot{\lambda}, \dot{\mathbf{s}})$ and $(\ddot{\mathbf{x}}, \ddot{\lambda}, \ddot{\mathbf{s}})$ be defined in (7.10) and (7.12). Then, the following relations hold.

$$\mathbf{s}^{\mathrm{T}}\dot{\mathbf{x}} + \mathbf{x}^{\mathrm{T}}\dot{\mathbf{s}} = \mathbf{x}^{\mathrm{T}}\mathbf{s} = n\mu, \quad \mathbf{s}^{\mathrm{T}}\ddot{\mathbf{x}} + \mathbf{x}^{\mathrm{T}}\ddot{\mathbf{s}} = \sigma\mu n - 2\dot{\mathbf{x}}^{\mathrm{T}}\dot{\mathbf{s}}, \quad \ddot{\mathbf{x}}^{\mathrm{T}}\ddot{\mathbf{s}} = 0. \tag{7.13}$$

Equations (7.10) and (7.12) can be solved in either unreduced form, or augmented system form, or normal equation form, as suggested in [144]. We solve the normal equations for $(\dot{\mathbf{x}}, \dot{\lambda}, \dot{\mathbf{s}})$ and $(\ddot{\mathbf{x}}, \ddot{\lambda}, \ddot{\mathbf{s}})$ as follows:

$$(\mathbf{AXS}^{-1}\mathbf{A}^{\mathrm{T}})\dot{\lambda} = \mathbf{AXS}^{-1}\mathbf{r}_c - \mathbf{b}, \tag{7.14a}$$

$$\dot{\mathbf{s}} = \mathbf{r}_c - \mathbf{A}^{\mathrm{T}}\dot{\lambda}, \tag{7.14b}$$

$$\dot{\mathbf{x}} = \mathbf{x} - \mathbf{XS}^{-1}\dot{\mathbf{s}}, \tag{7.14c}$$

and

$$(\mathbf{AXS}^{-1}\mathbf{A}^{\mathrm{T}})\ddot{\lambda} = -\mathbf{AS}^{-1}(\sigma\mu\mathbf{e} - 2\dot{\mathbf{x}} \circ \dot{\mathbf{s}}), \tag{7.15a}$$

$$\ddot{\mathbf{s}} = -\mathbf{A}^{\mathrm{T}}\ddot{\lambda}. \tag{7.15b}$$

$$\ddot{\mathbf{x}} = \mathbf{S}^{-1}(\sigma\mu\mathbf{e} - \mathbf{X}\ddot{\mathbf{s}} - 2\dot{\mathbf{x}} \circ \dot{\mathbf{s}}). \tag{7.15c}$$

Given the first and second derivatives defined by (7.10) and (7.12), an analytic expression of the ellipse that is used to approximate the central path is derived in Chapter 5.

Theorem 7.1
Let $(\mathbf{x}(\alpha),\lambda(\alpha),\mathbf{s}(\alpha))$ *be an arc defined by (7.9) passing through a point* $(\mathbf{x},\lambda,\mathbf{s}) \in \mathcal{E} \cap \mathcal{H}$*, and its first and second derivatives at* $(\mathbf{x},\lambda,\mathbf{s})$ *be* $(\dot{\mathbf{x}},\dot{\lambda},\dot{\mathbf{s}})$ *and* $(\ddot{\mathbf{x}},\ddot{\lambda},\ddot{\mathbf{s}})$*, which are defined by (7.10) and (7.12). Then, the ellipse approximation of* $\mathcal{H}(t)$ *is given by*

$$\mathbf{x}(\alpha,\sigma) = \mathbf{x} - \dot{\mathbf{x}}\sin(\alpha) + \ddot{\mathbf{x}}(\sigma)(1-\cos(\alpha)). \tag{7.16}$$

$$\lambda(\alpha,\sigma) = \lambda - \dot{\lambda}\sin(\alpha) + \ddot{\lambda}(\sigma)(1-\cos(\alpha)). \tag{7.17}$$

$$\mathbf{s}(\alpha,\sigma) = \mathbf{s} - \dot{\mathbf{s}}\sin(\alpha) + \ddot{\mathbf{s}}(\sigma)(1-\cos(\alpha)). \tag{7.18}$$

In the algorithm proposed below, we suggest taking step size $\alpha_k^s = \alpha_k^\lambda$ which may not be equal to the step size of α_k^x.

Algorithm 7.1

Data: $\mathbf{A}$, $\mathbf{b}$, $\mathbf{c}$, *and step size scaling factor* $\beta \in (0,1)$.
Initial point: $\lambda^0 = \mathbf{0}$, $\mathbf{x}^0 > \mathbf{0}$, $\mathbf{s}^0 > \mathbf{0}$, *and* $\mu_0 = \frac{\mathbf{x}^{0\mathrm{T}}\mathbf{s}^0}{n}$.
for *iteration* $k = 0,1,2,\ldots$

Step 1: Calculate $(\dot{\mathbf{x}},\dot{\lambda},\dot{\mathbf{s}})$ *using (7.14) and set*

$$\alpha_x^a := \arg\max\{\alpha \in [0,1] \mid \mathbf{x}^k - \alpha\dot{\mathbf{x}} \geq \mathbf{0}\}, \tag{7.19a}$$

$$\alpha_s^a := \arg\max\{\alpha \in [0,1] \mid \mathbf{s}^k - \alpha\dot{\mathbf{s}} \geq \mathbf{0}\}. \tag{7.19b}$$

Step 2: Calculate $\mu^a = \frac{(\mathbf{x}^k+\alpha_x^a\dot{\mathbf{x}})^{\mathrm{T}}(\mathbf{s}^k+\alpha_s^a\dot{\mathbf{s}})}{n}$ *and compute the centering parameter*

$$\sigma = \left(\frac{\mu^a}{\mu_k}\right)^3. \tag{7.20}$$

Step 3: Compute $(\ddot{\mathbf{x}},\ddot{\lambda},\ddot{\mathbf{s}})$ *using (7.15) with calculated* σ *from (7.20).*

Step 4: Set

$$\alpha^x = \arg\max\{\alpha \in [0,\frac{\pi}{2}] \mid \mathbf{x}^k - \dot{\mathbf{x}}\sin(\alpha) + \ddot{\mathbf{x}}(1-\cos(\alpha)) \geq \mathbf{0}\}, \tag{7.21a}$$

$$\alpha^s = \arg\max\{\alpha \in [0,\frac{\pi}{2}] \mid \mathbf{s}^k - \dot{\mathbf{s}}\sin(\alpha) + \ddot{\mathbf{s}}(1-\cos(\alpha)) \geq \mathbf{0}\}. \tag{7.21b}$$

Step 5: Scale the step size by $\alpha_k^x = \beta\alpha^x$ and $\alpha_k^s = \beta\alpha^s$ such that the update

$$\mathbf{x}^{k+1} = \mathbf{x}^k - \dot{\mathbf{x}}\sin(\alpha_k^x) + \ddot{\mathbf{x}}(1-\cos(\alpha_k^x)) > \mathbf{0}, \tag{7.22a}$$

$$\lambda^{k+1} = \lambda^k - \dot{\lambda}\sin(\alpha_k^s) + \ddot{\lambda}(1-\cos(\alpha_k^s)), \tag{7.22b}$$

$$\mathbf{s}^{k+1} = \mathbf{s}^k - \dot{\mathbf{s}}\sin(\alpha_k^s) + \ddot{\mathbf{s}}(1-\cos(\alpha_k^s)) > \mathbf{0}. \tag{7.22c}$$

Step 6: Set $k \leftarrow k+1$. Go back to Step 1.

end (for)

Remark 7.1 The main difference between the proposed algorithm and Mehrotra's MPC algorithm described in [144, 198] is in Steps 4 and 5 where the iterate moves along the ellipse instead of a straight line. More specifically, instead of using (7.21) and (7.22), Mehrotra's method uses

$$\alpha^x = \arg\max\{\alpha \in [0,1] \mid \mathbf{x}^k - \alpha(\dot{\mathbf{x}} - \ddot{\mathbf{x}}) \geq \mathbf{0}\}, \tag{7.23a}$$

$$\alpha^s = \arg\max\{\alpha \in [0,1] \mid \mathbf{s}^k - \alpha(\dot{\mathbf{s}} - \ddot{\mathbf{s}}) \geq \mathbf{0}\}. \tag{7.23b}$$

and

$$\mathbf{x}^{k+1} = \mathbf{x}^k - \alpha_k^x(\dot{\mathbf{x}} - \ddot{\mathbf{x}}) > \mathbf{0}, \tag{7.24a}$$

$$\lambda^{k+1} = \lambda^k - \alpha_k^s(\dot{\lambda} - \ddot{\lambda}), \tag{7.24b}$$

$$\mathbf{s}^{k+1} = \mathbf{s}^k - \alpha_k^s(\dot{\mathbf{s}} - \ddot{\mathbf{s}}) > \mathbf{0}. \tag{7.24c}$$

Note that the end points of arc-search algorithm $(\alpha_k^x, \alpha_k^s) = (0,0)$ and $(\alpha_k^x, \alpha_k^s) = (\frac{\pi}{2}, \frac{\pi}{2})$ in (7.21) and (7.22) are equal to the end points of Mehrotra's formulas in (7.23) and (7.24); for any (α_k^x, α_k^s) between $(0, \frac{\pi}{2})$, the ellipse is a better approximation of the infeasible central path. Therefore, the proposed algorithm should have a larger step size than Mehrotra's method and be more efficient. This intuitive has been verified by the numerical test result reported in Section 7.4. ▮

The following lemma shows that searching along the ellipse in iterations will reduce the residuals of the equality constraints to zero as $k \to \infty$, provided that α_k^x and α_k^s are bounded below and away from zero.

Lemma 7.2
Let $\mathbf{r}_b^k = \mathbf{A}\mathbf{x}^k - \mathbf{b}$, $\mathbf{r}_c^k = \mathbf{A}^{\mathrm{T}}\lambda^k + \mathbf{s}^k - \mathbf{c}$, $\rho_k = \prod_{j=0}^{k-1}(1-\sin(\alpha_j^x))$, and $\nu_k = \prod_{j=0}^{k-1}(1-\sin(\alpha_j^s))$. Then, the following relations hold.

$$\mathbf{r}_b^k = \mathbf{r}_b^{k-1}(1-\sin(\alpha_{k-1}^x)) = \cdots = \mathbf{r}_b^0\prod_{j=0}^{k-1}(1-\sin(\alpha_j^x)) = \mathbf{r}_b^0\rho_k, \tag{7.25a}$$

$$\mathbf{r}_c^k = \mathbf{r}_c^{k-1}(1-\sin(\alpha_{k-1}^s)) = \cdots = \mathbf{r}_c^0\prod_{j=0}^{k-1}(1-\sin(\alpha_j^s)) = \mathbf{r}_c^0\nu_k. \tag{7.25b}$$

Proof 7.1 From Theorem 7.1, searching along ellipse generates iterate as follows.

$$\mathbf{x}^{k+1} - \mathbf{x}^k = -\dot{\mathbf{x}}\sin(\alpha_k^x) + \ddot{\mathbf{x}}(1-\cos(\alpha_k^x)),$$
$$\lambda^{k+1} - \lambda^k = -\dot{\lambda}\sin(\alpha_k^s) + \ddot{\lambda}(1-\cos(\alpha_k^s)),$$
$$\mathbf{s}^{k+1} - \mathbf{s}^k = -\dot{\mathbf{s}}\sin(\alpha_k^s) + \ddot{\mathbf{s}}(1-\cos(\alpha_k^s)).$$

In view of (7.10) and (7.12), we have

$$\begin{aligned} \mathbf{r}_b^{k+1} - \mathbf{r}_b^k &= \mathbf{A}(\mathbf{x}^{k+1} - \mathbf{x}^k) = \mathbf{A}(-\dot{\mathbf{x}}\sin(\alpha_k^x) + \ddot{\mathbf{x}}(1-\cos(\alpha_k^x)) \\ &= -\mathbf{A}\dot{\mathbf{x}}\sin(\alpha_k^x) = -\mathbf{r}_b^k\sin(\alpha_k^x), \end{aligned} \tag{7.26}$$

therefore, $\mathbf{r}_b^{k+1} = \mathbf{r}_b^k(1-\sin(\alpha_k^x))$; this proves (7.25a). Similarly,

$$\begin{aligned} \mathbf{r}_c^{k+1} - \mathbf{r}_c^k &= \mathbf{A}^{\mathrm{T}}(\lambda^{k+1} - \lambda^k) + (\mathbf{s}^{k+1} - \mathbf{s}^k) \\ &= \mathbf{A}^{\mathrm{T}}(-\dot{\lambda}\sin(\alpha_k^s) + \ddot{\lambda}(1-\cos(\alpha_k^s)) - \dot{\mathbf{s}}\sin(\alpha_k^s) + \ddot{\mathbf{s}}(1-\cos(\alpha_k^s)) \\ &= -(\mathbf{A}^{\mathrm{T}}\dot{\lambda} + \dot{\mathbf{s}})\sin(\alpha_k^s) + (\mathbf{A}^{\mathrm{T}}\ddot{\lambda} + \ddot{\mathbf{s}})(1-\cos(\alpha_k^s)) \\ &= -\mathbf{r}_c^k\sin(\alpha_k^s), \end{aligned} \tag{7.27}$$

therefore, $\mathbf{r}_c^{k+1} = \mathbf{r}_c^k(1-\sin(\alpha_k^s))$; this proves (7.25b). ∎

To show that the duality measure decreases with iterations, we present the following lemma.

Lemma 7.3
Let α_x be the step length for $\mathbf{x}(\sigma,\alpha)$ and α_s be the step length for $\mathbf{s}(\sigma,\alpha)$ and $\lambda(\sigma,\alpha)$ defined in Theorem 7.1. Assume that $\alpha_x = \alpha_s := \alpha$, then, the updated duality measure can be expressed as

$$\begin{aligned} \mu(\alpha) &= \mu[1-\sin(\alpha) + \sigma(1-\cos(\alpha))] \\ &- \frac{1}{n}\left[(\ddot{\mathbf{x}}^{\mathrm{T}}\mathbf{r}_c - \ddot{\lambda}^{\mathrm{T}}\mathbf{r}_b)\sin(\alpha)(1-\cos(\alpha)) + (\dot{\mathbf{x}}^{\mathrm{T}}\mathbf{r}_c - \dot{\lambda}^{\mathrm{T}}\mathbf{r}_b)(1-\cos(\alpha))^2\right] \end{aligned} \tag{7.28}$$

Proof 7.2 First, from (7.10) and (7.12), we have

$$\dot{\mathbf{x}}^{\mathrm{T}}\mathbf{A}^{\mathrm{T}}\dot{\lambda} + \dot{\mathbf{x}}^{\mathrm{T}}\dot{\mathbf{s}} = \dot{\mathbf{x}}^{\mathrm{T}}\mathbf{r}_c,$$

this gives

$$\dot{\mathbf{x}}^{\mathrm{T}}\dot{\mathbf{s}} = -\dot{\lambda}^{\mathrm{T}}\mathbf{r}_b + \dot{\mathbf{x}}^{\mathrm{T}}\mathbf{r}_c.$$

Similarly,

$$\dot{\mathbf{x}}^{\mathrm{T}}\ddot{\mathbf{s}} = -\dot{\mathbf{x}}^{\mathrm{T}}\mathbf{A}^{\mathrm{T}}\ddot{\lambda} = -\ddot{\lambda}^{\mathrm{T}}\mathbf{r}_b, \qquad \ddot{\mathbf{x}}^{\mathrm{T}}\dot{\mathbf{s}} = \ddot{\mathbf{x}}^{\mathrm{T}}\dot{\mathbf{s}} + \ddot{\mathbf{x}}^{\mathrm{T}}\mathbf{A}^{\mathrm{T}}\dot{\lambda} = \ddot{\mathbf{x}}^{\mathrm{T}}\mathbf{r}_c.$$

Using these relations with (7.2) and Lemmas 7.1, we have

$$\mu(\alpha) = (\mathbf{x} - \dot{\mathbf{x}}\sin(\alpha_x) + \ddot{\mathbf{x}}(1-\cos(\alpha_x)))^{\mathrm{T}}(\mathbf{s} - \dot{\mathbf{s}}\sin(\alpha_s) + \ddot{\mathbf{s}}(1-\cos(\alpha_s)))/n$$

$$
\begin{aligned}
&= \frac{\mathbf{x}^{\mathrm{T}}\mathbf{s}}{n} - \frac{\mathbf{x}^{\mathrm{T}}\dot{\mathbf{s}}\sin(\alpha_s)+\mathbf{s}^{\mathrm{T}}\dot{\mathbf{x}}\sin(\alpha_x)}{n} \\
&\quad + \frac{\mathbf{x}^{\mathrm{T}}\ddot{\mathbf{s}}(1-\cos(\alpha_s))+\mathbf{s}^{\mathrm{T}}\ddot{\mathbf{x}}(1-\cos(\alpha_x))}{n} + \frac{\dot{\mathbf{x}}^{\mathrm{T}}\dot{\mathbf{s}}\sin(\alpha_s)\sin(\alpha_x)}{n} \\
&\quad - \frac{\dot{\mathbf{x}}^{\mathrm{T}}\ddot{\mathbf{s}}\sin(\alpha_x)(1-\cos(\alpha_s))+\dot{\mathbf{s}}^{\mathrm{T}}\ddot{\mathbf{x}}\sin(\alpha_s)(1-\cos(\alpha_x))}{n} \\
&= \mu[1-\sin(\alpha)+\sigma(1-\cos(\alpha))] + \frac{\dot{\mathbf{x}}^{\mathrm{T}}\dot{\mathbf{s}}\sin^2(\alpha)-2\dot{\mathbf{x}}^{\mathrm{T}}\dot{\mathbf{s}}(1-\cos(\alpha))}{n} \\
&\quad - \frac{\dot{\mathbf{x}}^{\mathrm{T}}\ddot{\mathbf{s}}\sin(\alpha)(1-\cos(\alpha))+\dot{\mathbf{s}}^{\mathrm{T}}\ddot{\mathbf{x}}\sin(\alpha)(1-\cos(\alpha))}{n} \\
&= \mu[1-\sin(\alpha)+\sigma(1-\cos(\alpha))] \\
&\quad - \frac{1}{n}\left[(\ddot{\mathbf{x}}^{\mathrm{T}}\mathbf{r}_c-\ddot{\lambda}^{\mathrm{T}}\mathbf{r}_b)\sin(\alpha)(1-\cos(\alpha))+(\dot{\mathbf{x}}^{\mathrm{T}}\mathbf{r}_c-\dot{\lambda}^{\mathrm{T}}\mathbf{r}_b)(1-\cos(\alpha))^2\right].
\end{aligned}
\tag{7.29}
$$

This finishes the proof. ■

Remark 7.2 In view of Lemma 7.2, if $\sin(\alpha)$ is bounded below and away from zero, then $\mathbf{r}_b \to \mathbf{0}$ and $\mathbf{r}_c \to \mathbf{0}$ as $k \to \infty$. Therefore, in view of Lemmas 7.3 and 5.4, we have, as $k \to \infty$,

$$\mu(\alpha) \approx \mu[1-\sin(\alpha)+\sigma(1-\cos(\alpha))] \leq \mu[1-\sin(\alpha)+\sigma\sin^2(\alpha)] < \mu$$

provided that $\dot{\lambda}$, $\dot{\mathbf{x}}$, $\ddot{\lambda}$, and $\ddot{\mathbf{x}}$ are bounded. This means that equation $\mu(\alpha) < \mu$ for any $\alpha \in (0, \frac{\pi}{2})$ as $k \to \infty$. As a matter of fact, in all numerical tests, the decrease of the duality measure was observed in every iteration, even for $\alpha_x \neq \alpha_s$. ■

Positivity of $\mathbf{x}(\sigma, \alpha_x)$ and $\mathbf{s}(\sigma, \alpha_s)$ is guaranteed if $(\mathbf{x}, \mathbf{s}) > \mathbf{0}$ holds and α_x and α_s are small enough. Assuming that $\dot{\mathbf{x}}$, $\dot{\mathbf{s}}$, $\ddot{\mathbf{x}}$, and $\ddot{\mathbf{s}}$ are bounded, the claim can easily be seen from the following relations

$$\mathbf{x}(\sigma, \alpha_x) = \mathbf{x} - \dot{\mathbf{x}}\sin(\alpha_x) + \ddot{\mathbf{x}}(1-\cos(\alpha_x)) > \mathbf{0}, \tag{7.30}$$

$$\mathbf{s}(\sigma, \alpha_s) = \mathbf{s} - \dot{\mathbf{s}}\sin(\alpha_s) + \ddot{\mathbf{x}}(1-\cos(\alpha_s)) > \mathbf{0}. \tag{7.31}$$

7.3 Implementation Details

In this section, we discuss factors that are normally not discussed in the main body of algorithms but affect noticeably, if not significantly, the effectiveness and efficiency of the infeasible interior-point algorithms. Most of these factors have been discussed in widely spread literatures, and they are likely implemented differently from code to code. We will address all of these implementation topics and provide the detailed information of the implementation in this chapter. As we will compare arc-search method and Mehrotra's method, to make a meaningful and fair comparison, we will implement everything discussed in this section the same way for both methods, so that the only differences of the two algorithms in our implementations are in Steps 4 and 5, where the arc-search method uses formulas (7.21) and (7.22) and Mehrotra's method uses (7.23) and (7.24). However, the difference of the computational cost is very small because these computations are all analytic and negligible compared to the cost of solving linear systems.

7.3.1 Initial point selection

Initial point selection has been known as an important factor in the computational efficiency for most infeasible interior-point algorithms. However, many commercial software packages do not provide sufficient details, for example, [23, 172]. We will use the methods proposed in [89, 83]. In [89, page 589], the initial point is selected as follows:

$$\hat{\lambda} = (\mathbf{A}\mathbf{A}^{\mathrm{T}})^{-1}\mathbf{A}\mathbf{c}; \quad \hat{\mathbf{s}} = \mathbf{c} - \mathbf{A}^{\mathrm{T}}\hat{\lambda}; \quad \hat{\mathbf{x}} = \mathbf{A}^{\mathrm{T}}(\mathbf{A}\mathbf{A}^{\mathrm{T}})^{-1}\mathbf{b}. \tag{7.32}$$

Let $\delta_x = \max(-1.1\min\{\hat{x}_i\}, 0)$ and $\delta_s = \max(-1.1\min\{\hat{s}_i\}, 0)$. We then calculate

$$\hat{\delta}_x = \delta_x + 0.5\frac{(\hat{\mathbf{x}} + \delta_x\mathbf{e})^{\mathrm{T}}(\hat{\mathbf{s}} + \delta_s\mathbf{e})}{\sum_{i=1}^{n}(\hat{s}_i + \delta_s)} \tag{7.33}$$

$$\hat{\delta}_s = \delta_s + 0.5\frac{(\hat{\mathbf{s}} + \delta_s\mathbf{e})^{\mathrm{T}}(\hat{\mathbf{x}} + \delta_x\mathbf{e})}{\sum_{i=1}^{n}(\hat{x}_i + \delta_x)} \tag{7.34}$$

and generate $\lambda^0 = \hat{\lambda}$ and $s_i^0 = \hat{s}_i + \hat{\delta}_s$, $i = 1, \ldots, n$, and $x_i^0 = \hat{x}_i + \hat{\delta}_x$, $i = 1, \ldots, n$ as a possible initial point.

In [83, page 445], the initial point is selected as follows:

$$\hat{\mathbf{x}} = \mathbf{A}^{\mathrm{T}}(\mathbf{A}\mathbf{A}^{\mathrm{T}})^{-1}\mathbf{b}. \tag{7.35}$$

Let $\xi_1 = \max(-100\min\{\hat{x}_i\}, 100, \|\mathbf{b}\|_1/100)$ and $\xi_2 = 1 + \|\mathbf{c}\|_1$, where $\|\cdot\|_1$ is the l_1 norm. We then calculate, for $i = 1, \ldots, n$, $x_i^0 = \max\{\xi_1, \hat{x}_i\}$ and

$$s_i^0 = \begin{cases} c_i + \xi_2, & \text{if } c_i > \xi_2 \\ -c_i, & \text{if } c_i < -\xi_2 \\ c_i + \xi_2, & \text{if } c_i \geq 0,\ c_i < \xi_2 \\ \xi_2, & \text{if } c_i \geq -\xi_2,\ c_i < 0, \end{cases} \tag{7.36}$$

and set $\lambda^0 = 0$.

For these two sets of initial points, we then calculate

$$\max\{\|\mathbf{A}\mathbf{x}^0 - \mathbf{b}\|, \|\mathbf{A}^{\mathrm{T}}\lambda^0 + \mathbf{s}^0 - \mathbf{c}\|, \mu_0\} \tag{7.37}$$

and select the initial point that yields a smaller value because we have observed from testing Netlib problems that this selection generally reduces the number of iterations.

7.3.2 Pre-process

Pre-process or pre-solver is a major factor that can significantly affect the numerical stability and computational efficiency. Many literatures have been focused on this topic, for example [144, 83, 12, 7, 84]. As we will test all linear programming problems in standard form in Netlib, we focus on the strategies only for the standard linear programming problems in the form of (1.4), which are solved in normal equations[2]. We will use $\mathbf{A}_{i,\cdot}$ for the ith row of $\mathbf{A}$, $\mathbf{A}_{\cdot,j}$ for the jth column of $\mathbf{A}$, and $A_{i,j}$ for the element at (i,j) position of $\mathbf{A}$. While reducing the problem, we will express the objective function in two parts, $\mathbf{c}^{\mathrm{T}}\mathbf{x} = f_{obj} + \sum_k c_k x_k$. The first part f_{obj} at the beginning is zero and is updated all the time as we reduce the problem (remove some c_k from $\mathbf{c}$); the terms in the summation in the second part are continuously reduced and c_k are updated as necessary when we reduce the problem.

When we select pre-process methods, besides considering the numerical stability, efficiency and effectiveness in solving the linear programming problem, we will also assess their impact on the seamless implementation in post-process, which is the process to restore the expression of the solution in the original coordinate system. To have a seamless implementation in the post-process, we store several vectors for every pre-process: $\mathbf{c}_{orig} = \mathbf{c}$ for the original coefficients of the objective function; $\mathbf{x}_{final}$, which is set to zero at the beginning, will be used to store the optimal solution in the original unreduced coordinate; at the beginning of the pre-process, $\mathbf{x}_{idx} = [1,2,\ldots,n]^{\mathrm{T}}$, $\mathbf{x}_{idx}$ will be reduced to keep a mapping between the orders of $\mathbf{x}$ in the reduced coordinate system and $\mathbf{x}_{final}$ in the unreduced coordinate system. For pre-process 9, we will store a sparse matrix and a few more vectors: $\mathbf{A}^{post}$, which is empty at the beginning of the pre-process, is used to store equations which are eliminated in the pre-process 9 but need to be resolved in the final stage to recover the variables in the original coordinate system; $\mathbf{b}^{post}$ is a vector associated with $\mathbf{A}^{post}$ to recover the variables in the original coordinate system; and $\mathbf{x}_{tobe}$ is a vector of coordinate information of the variables to be resolved in the final stage. These matrix and vectors need to be updated in the pre-process so that we can recover the solution expressed in the original coordinate system. We describe these updates only for the pre-processes that are chosen to be implemented.

The first 6 pre-process methods presented below were reported in various literatures, such as [144, 23, 12, 7, 84]; the rest of them, to the best of our knowledge, are not reported anywhere.

1. Empty row
If $\mathbf{A}_{i,\cdot} = \mathbf{0}$ and $b_i = 0$, this row can be removed. If $\mathbf{A}_{i,\cdot} = 0$ but $b_i \neq 0$, the problem is infeasible.

2. Duplicate rows
If there is a constant k such that $\mathbf{A}_{i,\cdot} = k\mathbf{A}_{j,\cdot}$ and $b_i = kb_j$, a duplicate row can be removed. If $\mathbf{A}_{i,\cdot} = k\mathbf{A}_{j,\cdot}$ but $b_i \neq kb_j$, the problem is infeasible.

[2] Some strategies are specifically designed for solving augmented system form, for example, the ones discussed in [43], which we will not discuss in this chapter.

3. Empty column

If $\mathbf{A}_{\cdot,i} = \mathbf{0}$ and $c_i \geq 0$, $x_i = 0$ is the right choice for the minimization. Remove the ith column $\mathbf{A}_{\cdot,i}$ and c_i. Also remove $x_{idx}(i)$. If $\mathbf{A}_{\cdot,i} = \mathbf{0}$ but $c_i < 0$, the problem is unbounded as $x_i \to \infty$.

4. Duplicate columns

If $\mathbf{A}_{\cdot,i} = \mathbf{A}_{\cdot,j}$, then $\mathbf{Ax} = \mathbf{b}$ can be expressed as $\mathbf{A}_{\cdot,i}(x_i + x_j) + \sum_{k \neq i,j} \mathbf{A}_{\cdot,k} x_k = \mathbf{b}$, Moreover, if $c_i = c_j$, $\mathbf{c}^{\mathrm{T}}\mathbf{x}$ can be expressed as $c_i(x_i + x_j) + \sum_{k \neq i,j} c_k x_k$. Since $x_i \geq 0$ and $x_j \geq 0$, we have $(x_i + x_j) \geq 0$. Hence, a duplicate column can be removed.

5. Row singleton

If $\mathbf{A}_{i,\cdot}$ has exact one nonzero element, i.e., $A_{i,k} \neq 0$ for some k, and for $\forall j \neq k$, $A_{i,j} = 0$; then $x_k = b_i / A_{i,k}$ and $\mathbf{c}^{\mathrm{T}}\mathbf{x} = c_k b_i / A_{i,k} + \sum_{j \neq k} c_j x_j$. For $\ell \neq i$, $\mathbf{A}_{\ell,\cdot}\mathbf{x} = b_\ell$ can be rewritten as $\sum_{j \neq k} A_{\ell,j} x_j = b_\ell - A_{\ell,k} b_i / A_{i,k}$. This suggests the following update: (i) if $x_k < 0$, the problem is infeasible, otherwise, continue, (ii) $f_{opt} + c_k b_i / A_{i,k} \to f_{opt}$, (iii) remove c_k from c, (iv) $b_\ell - A_{\ell,k} b_i / A_{i,k} \to b_\ell$, (v) insert $x_k = b_i / A_{i,k}$ into $\mathbf{x}_{final}(x_{idx}(k))$, and (vi) remove $x_{idx}(k)$. With these changes, we can remove the ith row and the kth column.

6. Free variable

If $\mathbf{A}_{\cdot,i} = -\mathbf{A}_{\cdot,j}$ and $c_i = -c_j$, then we can rewrite $\mathbf{Ax} = \mathbf{b}$ as $\mathbf{A}_{\cdot,i}(x_i - x_j) + \sum_{k \neq i,j} \mathbf{A}_{\cdot,k} x_k$, and $\mathbf{c}^{\mathrm{T}}\mathbf{x} = c_i(x_i - x_j) + \sum_{k \neq i,j} c_k x_k$. The new variable $x_i - x_j$ is a free variable which can be solved if $A_{\alpha,i} \neq 0$ for some row α (otherwise, it is an empty column, which has been discussed). This gives

$$x_i - x_j = \frac{1}{A_{\alpha,i}} \left(b_\alpha - \sum_{k \neq i,j} A_{\alpha,k} x_k \right).$$

For any $A_{\beta,i} \neq 0$, $\beta \neq \alpha$, $\mathbf{A}_{\beta,\cdot}\mathbf{x} = b_\beta$ can be expressed as

$$A_{\beta,i}(x_i - x_j) + \sum_{k \neq i,j} A_{\beta,k} x_k = b_\beta,$$

or

$$\frac{A_{\beta,i}}{A_{\alpha,i}} \left(b_\alpha - \sum_{k \neq i,j} A_{\alpha,k} x_k \right) + \sum_{k \neq i,j} A_{\beta,k} x_k = b_\beta,$$

or

$$\sum_{k \neq i,j} \left(A_{\beta,k} - \frac{A_{\beta,i} A_{\alpha,k}}{A_{\alpha,i}} \right) x_k = b_\beta - \frac{A_{\beta,i} b_\alpha}{A_{\alpha,i}}.$$

Also, $\mathbf{c}^{\mathrm{T}}\mathbf{x}$ can be rewritten as

$$\frac{c_i}{A_{\alpha,i}} \left(b_\alpha - \sum_{k \neq i,j} A_{\alpha,k} x_k \right) + \sum_{k \neq i,j} c_k x_k,$$

or

$$\frac{c_i b_\alpha}{A_{\alpha,i}} + \sum_{k \neq i,j} \left(c_k - \frac{c_i A_{\alpha,k}}{A_{\alpha,i}} \right) x_k.$$

This suggests the following update: (i) $f_{obj} + \frac{c_i b_\alpha}{A_{\alpha,i}} \to f_{obj}$, (ii) $c_k - \frac{c_i A_{\alpha,k}}{A_{\alpha,i}} \to c_k$, (iii) $A_{\beta,k} - \frac{A_{\beta,i} A_{\alpha,k}}{A_{\alpha,i}} \to A_{\beta,k}$, (iv) $b_\beta - \frac{A_{\beta,i} b_\alpha}{A_{\alpha,i}} \to b_\beta$, (v) delete $\mathbf{A}_{\alpha,\cdot}$, b_α, $m-1 \to m$, delete $\mathbf{A}_{\cdot,i}$, $\mathbf{A}_{\cdot,j}$, c_i, c_j, and $n-2 \to n$.

7. Fixed variable defined by a single row

If $b_i < 0$ and $\mathbf{A}_{i,\cdot} \geq \mathbf{0}$ with at least one j such that $A_{i,j} > 0$, then the problem is infeasible. Similarly, If $b_i > 0$ and $\mathbf{A}_{i,\cdot} \leq \mathbf{0}$ with at least one j such that $A_{i,j} < 0$, then the problem is infeasible. If $b_i = 0$, but either $\max(\mathbf{A}_{i,\cdot}) \leq 0$ or $\min(\mathbf{A}_{i,\cdot}) \geq 0$, then, for any j such that $A_{i,j} \neq 0$, $x_j = 0$ has to hold. Therefore, we can remove all such rows in $\mathbf{A}$ and $\mathbf{b}$, and such columns in $\mathbf{A}$ and $\mathbf{c}$. Also remove all corresponding elements in $\mathbf{x}_{idx}$.

8. Fixed variable defined by multiple rows

If $b_i = b_j$, but either $\max(\mathbf{A}_{i,\cdot} - \mathbf{A}_{j,\cdot}) \leq 0$ or $\min(\mathbf{A}_{i,\cdot} - \mathbf{A}_{j,\cdot}) \geq 0$, then for any k such that $A_{i,k} - A_{j,k} \neq 0$, $x_k = 0$ has to hold. This suggests the following update: (i) remove kth columns of $\mathbf{A}$ and $\mathbf{c}$ if $A_{i,k} - A_{j,k} \neq 0$, and (ii) remove either ith or jth row depending on which has more nonzeros. The same idea can be used for the case when $b_i + b_j = 0$.

9. Positive variable defined by signs of $\mathbf{A}_{i,\cdot}$ and b_i

Since

$$x_i = \frac{1}{A_{\alpha,i}} \left(b_\alpha - \sum_{k \neq i} A_{\alpha,k} x_k \right),$$

if the sign of $A_{\alpha,i}$ is the same as b_α and opposite to all $A_{\alpha,k}$ for $k \neq i$, then $x_i \geq 0$ is guaranteed. We can solve x_i, and substitute back into $\mathbf{Ax} = \mathbf{b}$ and $\mathbf{c}^{\mathrm{T}}\mathbf{x}$. This suggests taking the following actions: (i) if $A_{\beta,i} \neq 0$, $b_\beta - \frac{A_{\beta,i} b_\alpha}{A_{\alpha,i}} \to b_\beta$, (ii) moreover, if $A_{\alpha,k} \neq 0$, then $A_{\beta,k} - \frac{A_{\beta,i} A_{\alpha,k}}{A_{\alpha,i}} \to A_{\beta,k}$, (iii) $f_{obj} + \frac{c_i b_\alpha}{A_{\alpha,i}} \to f_{obj}$, (iv) $c_k - \frac{c_i A_{\alpha,k}}{A_{\alpha,i}} \to c_k$, (v) add a row to the end of $\mathbf{A}^{post}$ such that $A^{post}_{\alpha,i} = 1$ and $A_{\alpha,k}/A_{\alpha,i} \to A^{post}_{\alpha,k}$ if $A_{\alpha,k} \neq 0$, (vi) add $b(\alpha)/A_{\alpha,i}$ to the end of $\mathbf{b}^{post}$ and add $x_{idx}(i)$ to the end of $\mathbf{b}_{tobe}$, (vii) remove the αth row and ith column from $\mathbf{A}$ and remove $x_{idx}(i)$.

10. A singleton variable defined by two rows

If $\mathbf{A}_{i,\cdot} - \mathbf{A}_{j,\cdot}$ is a singleton and $A_{i,k} - A_{j,k} \neq 0$ for one and only one k, then $x_k = \frac{b_i - b_j}{A_{i,k} - A_{j,k}}$. This suggests the following update: (i) if $x_k \geq 0$ does not hold, the problem is infeasible, (ii) if $x_k \geq 0$ does hold, for $\forall \ell \neq i,j$ and $A_{\ell,k} \neq 0$, $b_\ell - A_{\ell,k} \frac{b_i - b_j}{A_{i,k} - A_{j,k}} \to b_\ell$, (iii) remove either the ith or the jth row, and remove the

kth column from $\mathbf{A}$, (iv) remove c_k from $\mathbf{c}$, and (v) update $f_{obj} + c_k \frac{b_i - b_j}{A_{i,k}} \rightarrow f_{obj}$.

We have tested all these ten pre-solvers, and they all work in terms of reducing the problem sizes and making the problems easier to solve, in most cases. However, pre-solvers 2, 4, 6, 8 and 10 are observed to be significantly more time consuming than pre-solvers 1, 3, 5, 7 and 9. Moreover, our experience shows that pre-solvers 1, 3, 5, 7 and 9 are more efficient in reducing the problem sizes than pre-solvers 2, 4, 6, 8 and 10. Therefore, in our implementation, we use only pre-solvers 1, 3, 5, 7, and 9 for all of our test problems.

The proposed pre-process set (1,3,5,7,9) is very efficient. Table 7.1 compares the reduced problem sizes obtained by the proposed pre-process and by the pre-process of PCx [23]. Among all the tested problems, the pre-process of [23] is slightly better only for 4 problems (`agg, agg2, agg3, fffff800`), while the proposed pre-process in this chapter is better or the same for all other problems. PCx [23] does not report the reduced sizes for some large problems (`Osa_07, Osa_14, Osa_30, Qap12, Qap15, Qap8`). For the only one reported large problem `Stocfor3`, the proposed pre-process generates a much smaller problem: $(m,n) = (5974, 12840)$ vs. $(m,n) = (15362, 22228)$. If a full pre-process set $(1-10)$, described in this section, is used, the reduced problem sizes are smaller for all tested problems than the ones obtained in [23].

Table 7.1: Pre-processes comparison for test problems in Netlib.

Problem	before prep		proposed pre-solve		pre-solve of [23]	
	m	n	m	n	m	n
Adlittle	56	138	54	136	55	137
Afiro	27	51	8	32	27	51
Agg	488	615	391	479	390	477
Agg2	516	758	514	755	514	750
Agg3	516	758	514	755	514	750
Bandm	305	472	192	347	240	395
Beaconfd	173	295	57	147	86	171
Blend	74	114	49	89	71	111
Bnl1	643	1586	429	1314	610	1491
Bnl2	2324	4486	1007	3066	1964	4009
Brandy	220	303	113	218	133	238
Degen2	444	757	440	753	444	757
Degen3	1503	2604	1490	2591	1503	2604
fffff800	525	1208	487	991	322	826
Israel	174	316	174	316	174	316
Lotfi	153	366	113	326	133	346
Maros_r7	3136	9408	2152	7440	2152	7440
Osa_07	1118	25067	1081	25030	-	-
Osa_14	2337	54797	2300	54760	-	-
Osa_30	4350	104374	4313	104337	-	-
Qap12	3192	8856	3048	8712	-	-
Qap15	6330	22275	6105	22050	-	-

Qap8	912	1632	848	1568	-	-
Sc105	105	163	44	102	104	162
Sc205	205	317	89	201	203	315
Sc50a	50	78	19	47	49	77
Sc50b	50	78	14	42	48	76
Scagr25	471	671	343	543	469	669
Scagr7	129	185	91	147	127	183
Scfxm1	330	600	238	500	305	568
Scfxm2	660	1200	479	1003	610	1136
Scfxm3	990	1800	720	1506	915	1704
Scrs8	490	1275	115	893	421	1199
Scsd1	77	760	77	760	77	760
Scsd6	147	1350	147	1350	147	1350
Scsd8	397	2750	397	2750	397	2750
Sctap1	300	660	284	644	284	644
Sctap2	1090	2500	1033	2443	1033	2443
Sctap3	1480	3340	1408	3268	1408	3268
Share1b	117	253	102	238	112	248
Share2b	96	162	87	153	96	162
Ship04l	402	2166	292	1905	292	1905
Ship04s	402	1506	216	1281	216	1281
Ship08l	778	4363	470	3121	470	3121
Ship08s	778	2467	274	1600	275	1604
Ship12l	1151	5533	610	4171	610	4171
Ship12s	1151	2869	340	1943	340	1943
Stocfor1	117	165	34	82	102	150
Stocfor2	2157	3045	766	1654	1980	2868
Stocfor3	16675	23541	5974	12840	15362	22228
Truss	1000	8806	1000	8806	1000	8806

An alternative way to see the superiority of the proposed pre-process is to use *performance profile* which, to our best knowledge, was first proposed in [132] and carefully analyzed in [29, 46]. Let $\mathcal{S}$ be the set of solvers and $\mathcal{P}$ be the set of test problems. Let n_s be the number of solvers, n_p be the number of test problems, $m_{p,s}$ be the merit function of using solver s for the problem p. The performance ratio is defined as [132]:

$$r_{p,s} = \frac{m_{p,s}}{\min\{m_{p,s} : s \in \mathcal{S}\}}. \tag{7.38}$$

The performance profile is defined as the distribution function of a performance (merit) metric [29]. For this problem, the merit function $m_{p,s}$ is the size of the reduced problem p using pre-process s. The performance profiles for the proposed pre-process and PCx's pre-process are given in Figure 7.1.

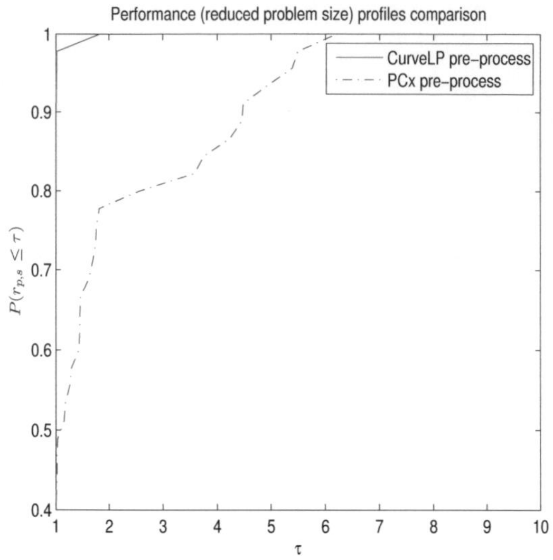

Figure 7.1: Performance profiles comparison for `CurveLP` and `PCx` pre-processes.

7.3.3 *Post-process*

With the implementation described in pre-process, the post-process is very simple and is given as the following MATLAB code.

for $i = 1 : length(\mathbf{x}_{idx})$

$x_{final}(x_{idx}(i)) = x(i);$

end

for ii=length($\mathbf{x}_{tobe}$):-1:1

idxp=find($\mathbf{A}^{post}$(ii,:));

$x_{final}(x_{tobe}$(ii))=b_{post}(ii);

for jj=1:length(idxp)

if idxp(jj)$\sim=$ x_{tobe}(ii)

$$x_{final}(x_{tobe}(ii)) = x_{final}(x_{tobe}(ii)) - A^{post}(ii, idxp(jj)) * x_{final}(idxp(jj));$$

end

end

end

Remark 7.3 After restoring the solution from the reduced $\mathbf{x}$ back to $\mathbf{x}_{final}$ in the original coordinate system, we can compare f_{opt} and $\mathbf{c}_{orig}\mathbf{x}_{final}$ and verify that our code is correctly implemented if $f_{opt} = \mathbf{c}_{orig}\mathbf{x}_{final}$. ∎

7.3.4 Matrix scaling

Scaling is believed (for example, [23]) to be a good practice for ill-conditioned matrix $\mathbf{A}$ where the ratio

$$\frac{\max |A_{i,j}|}{\min \{|A_{k,l}| \ : \ A_{k,l} \neq 0\}} \tag{7.39}$$

is big. PCx adopted a scaling strategy proposed in [22]. Let $\Phi = \text{diag}(\phi_1, \cdots, \phi_m)$ and $\Psi = \text{diag}(\psi_1, \cdots, \psi_n)$ be the diagonal scaling matrices of $\mathbf{A}$. The scaling for matrix $\mathbf{A}$ in [23, 22] is equivalent to minimize

$$\sum_{A_{ij} \neq 0} \log^2 \left| \frac{A_{ij}}{\phi_i \psi_j} \right|.$$

Various methods of solving this problem are proposed [23, 22]. We implemented these methods and some variations and tested against standard linear programming problems in Netlib. All of these methods have similar impact on the efficiency of infeasible interior-point algorithms. We ran the proposed interior-point algorithms and Mehrotra's algorithm with and without scaling for problems in Table 7.1 and compare the iteration counts. The following 9 problems use less iterations after the scaling is used: *Bnl1, Sc205, Scagr25, Scfxm1, Scfxm3, Sctap1, Sctap2, Ship04s, Stocfor1*. The following 20 problems use the same number of iterations after the scaling is used: *Agg, Bandm, Lotfi, Maros_r7, Qap8, Sc105, Scagr7, Scrs8, Scsd1, Scsd8, Sctap3, Share1b, Share2b, Ship04l, Ship08l, Ship08s, Ship12l, Stocfor2, Stocfor3, Truss*. The remaining 22 problems use more iterations after the scaling is used: *Adlittle, Afiro, Agg2, Agg3, Beaconfd, Blend, Bnl2, Brandy, Degen2, Degen3, fffff800, Israel, Osa_07, Osa_14, Osa_30, Qap_12, Qap_15, Sc50a, Sc50b, Scfxm2, Scsd6, Ship12s*.

These test results show that, although scaling can improve efficiency of infeasible interior-point algorithms for some problems, over all, it does not help. In addition, there are no clear criteria on what problems may benefit from scaling and what problems may be adversely affected by scaling. Therefore, we decide not to use scaling in all our test problems. However, the ratio defined in (7.39) is a good indicator which can be used to determine if pre-solver 9 should be applied or not.

7.3.5 Removing row dependency from A

Theoretically, convergence analyses in most existing literatures assume that the matrix $\mathbf{A}$ is full rank. Practically, row dependency causes some computational difficulties. However, many real world problems, including some problems in Netlib, have dependent rows. Though using standard Gaussian elimination method can reduce **A** into a full rank matrix, the sparse structure of **A** will be destroyed. In [6], Andersen reported an efficient method that removes row dependency of **A**. His paper also claimed that not only the numerical stability is improved by the method, but the cost of the effort can also be justified. One of the main ideas is to identify most independent rows of **A** in a cheap and easy way and separate these independent rows from those that may be dependent. A variation of Andersen's method can be summarized as follows.

First, it assumes that all empty rows have been removed by pre-solver. Second, matrix **A** often contains many column singletons (the column has only one nonzero), for example, slack variables are column singletons. Clearly, a row containing a column singleton cannot be dependent. If these rows are separated (temporarily removed) from the other rows of **A**, new column singletons may appear and more rows may be separated. This process may separate many rows from rest rows of **A** in practice. Permutation operations can be used to move the singletons to the diagonal elements of **A**. The dependent rows are among the rows left in the process. Then, Gaussian elimination method can be applied with pivot selection using Markowitz criterion [30, 28]. Some implementation details include (a) break ties by choosing element with the largest magnitude, and (b) use threshold pivoting.

We have tested and analyzed the impact of removing dependent rows using Andersen's method. Among 51 tested Netlib problems, this method is beneficial (in terms of finding accurate solution) to Mehrotra's algorithm for 9 problems: *Bnl2, Degen2, Degen3, Osa_07, Qap_15, Qap_8, Scfxm1, Scfxm3, Stocfor1*. It is beneficial (in terms of finding accurate solution) to the proposed arc-search algorithm for only one problem: *Stocfor1*. Considering the significant computational cost, we choose to not use this function unless we feel it is necessary when it is used as part of handling degenerate solutions discussed later. To have a fair comparison between the two algorithms, we will make it clear in the test report in Section 7.4 which algorithms and/or problems use this function and what algorithms and/or problems do not.

7.3.6 Linear algebra for sparse Cholesky matrix

Similar to Mehrotra's algorithm, the majority of the computational cost of our proposed algorithm is related to solving possibly ill-conditioned sparse Cholesky systems (7.14) and (7.15), which can be expressed as an abstract problem as follows.

$$\mathbf{A}\mathbf{D}^2\mathbf{A}^{\mathrm{T}}\mathbf{u} = \mathbf{v}, \tag{7.40}$$

where $\mathbf{D} = \mathbf{X}^{\frac{1}{2}}\mathbf{S}^{-\frac{1}{2}}$ is identical in (7.14) and (7.15), but $\mathbf{u}$ and $\mathbf{v}$ are different vectors. Many popular LP solvers [23, 172] call a modified software package[3] of [104] which uses some linear algebra specifically developed for the ill-conditioned sparse Cholesky decomposition [77]. Although MATLAB has a function `Chol`, it does not yet implement features for ill-conditioned matrices, which occur frequently in interior-point method in linear programming problems. This is the major difference between our implementation and other popular LP solvers, which is most likely the main reason that our test results are slightly different from test results reported in other literature.

7.3.7 Handling degenerate solutions

An important result in linear programming [40] is that there always exist strictly complementary optimal solutions that meet the conditions $\mathbf{x}^* \circ \mathbf{s}^* = \mathbf{0}$ and $\mathbf{x}^* + \mathbf{s}^* > \mathbf{0}$. Therefore, the columns of $\mathbf{A}$ can be partitioned as $B \subseteq \{1,2,\ldots,n\}$, the set of indices of the positive coordinates of $\mathbf{x}^*$, and $N \subseteq \{1,2,\ldots,n\}$, the set of indices of the positive coordinates of $\mathbf{s}^*$, such that $B \cup N = \{1,2,\ldots,n\}$ and $B \cap N = \emptyset$. Thus, we can partition $\mathbf{A} = (\mathbf{A}_B, \mathbf{A}_N)$, and define the primal and dual optimal faces by

$$\mathcal{P}_* = \{\mathbf{x} : \mathbf{A}_B\mathbf{x}_B = \mathbf{b}, \mathbf{x} \geq \mathbf{0}, \mathbf{x}_N = \mathbf{0}\},$$

and

$$\mathcal{D}_* = \{(\lambda, \mathbf{s}) : \mathbf{A}_N^{\mathrm{T}}\lambda + \mathbf{S}_N = \mathbf{c}_N, \mathbf{s}_B = \mathbf{0}, \mathbf{s} \geq \mathbf{0}\}.$$

However, not all optimal solutions in linear programming are strictly complementary. The following example shows this fact:

$$\min_{\mathbf{x} \in \mathbb{R}^3} x_1 \quad \text{subject to} \quad x_1 + x_2 + x_3 = 1, \quad (x_1, x_2, x_3) \geq \mathbf{0}.$$

whose dual is

$$\max_{\lambda \in \mathbb{R}, \mathbf{s} \in \mathbb{R}^3} \lambda \quad \text{subject to} \quad \begin{bmatrix} 1 \\ 1 \\ 1 \end{bmatrix} \lambda + \mathbf{s} = \begin{bmatrix} 1 \\ 0 \\ 0 \end{bmatrix}, \quad \mathbf{s} \geq \mathbf{0}.$$

The primal-dual solutions for this problem are

$$\mathbf{x}^* = (0, t, 1-t)^{\mathrm{T}}, \quad \lambda^* = 0, \quad \mathbf{s}^* = (1,0,0)^{\mathrm{T}}, \quad \text{for any} \quad t \in [0,1].$$

For $t \in (0,1)$, it is clear that $(\mathbf{x}^*, \lambda^*, \mathbf{s}^*)$ are strictly complementary solutions. However, for $t = 0$ or $t = 1$ the primal-dual solutions are no longer strictly complementary because there is an index i for which x_i^* and s_i^* are both zero.

Although many interior-point algorithms are proved to converge strictly to complementary solutions, this claim may not be true for Mehrotra's method and the arc-search method proposed in this chapter.

[3] See [144, Pages 216-219] for details.

Recall that the problem pair (1.4) and (1.5) is called to have a primal degenerate solution if a primal optimal solution $\mathbf{x}^*$ has less than m positive coordinates, and have a dual degenerate solution if a dual optimal solution $\mathbf{s}^*$ has less than $n-m$ positive coordinates. The pair $(\mathbf{x}^*, \mathbf{s}^*)$ is called degenerate if it is primal or dual degenerate. This means that as $\mathbf{x}^k \to \mathbf{x}^*$, equation (7.40) can be written as

$$(\mathbf{A}_B \mathbf{X}_B \mathbf{S}_B^{-1} \mathbf{A}_B^{\mathrm{T}})\mathbf{u} = \mathbf{v}, \tag{7.41}$$

If the problem converges to a primal degenerate solution, then some diagonal elements of $\mathbf{X}_B$ are zeros, hence, the rank of $(\mathbf{A}_B \mathbf{X}_B \mathbf{S}_B^{-1} \mathbf{A}_B^{\mathrm{T}})$ is less than m as $\mathbf{x}^k \to \mathbf{x}^*$. In this case, there is a difficulty in solving (7.41). Difficulty caused by degenerate solutions in interior-point methods for linear programming has been known for a long time [48]. We have observed this troublesome incidence in quite a few Netlib test problems. Similar observation was also reported in [44]. Though we don't see any special attention or report on this troublesome issue from some widely cited papers and LP solvers, such as [89, 82, 83, 23, 172], we noticed from [144, page 219] that some LP solvers [23, 172] twisted the sparse Cholesky decomposition code [104] in order to overcome the difficulty.

In our implementation, we use a different method to avoid the difficulty because we do not have access to the code of [104]. After each iteration, minimum $\mathbf{x}^k$ is examined. If $\min\{\mathbf{x}^k\} \le \varepsilon_x$, then, for all components of $\mathbf{x}$ satisfying $x_i \le \varepsilon_x$, we delete $\mathbf{A}_{.i}$, x_i, s_i, c_i, and the ith component of $\mathbf{r}_c$; use the method proposed in Subsection 7.4.5 to check if the updated $\mathbf{A}$ is full rank and make the updated $\mathbf{A}$ full rank if it is necessary.

The default ε_x is 10^{-6}. For problems that needs a different ε_x, we will make it clear in the report of the test results.

7.3.8 Analytic solution of α^x and α^s

We know that α^x and α^s in (7.23) can easily be calculated in analytic form. Similarly, α^x and α^s in (7.21) can also be calculated in analytic form as follows. For each $i \in \{1,\ldots,n\}$, we can select the largest α_{x_i} such that for any $\alpha \in [0, \alpha_{x_i}]$, the ith inequality of (7.21a) holds, and the largest α_{s_i} such that for any $\alpha \in [0, \alpha_{s_i}]$ the ith inequality of (7.21b) holds. We then define

$$\alpha^x = \min_{i \in \{1,\ldots,n\}} \{\alpha_{x_i}\}, \tag{7.42}$$

$$\alpha^s = \min_{i \in \{1,\ldots,n\}} \{\alpha_{s_i}\}. \tag{7.43}$$

α_{x_i} and α_{s_i} can be given in analytical forms according to the values of $\dot{x}_i$, $\ddot{x}_i$, $\dot{s}_i$, $\ddot{s}_i$. First, from (7.21), we have

$$x_i + \ddot{x}_i \ge \dot{x}_i \sin(\alpha) + \ddot{x}_i \cos(\alpha). \tag{7.44}$$

Clearly, let $\beta = \sin(\alpha)$, this is equivalent to finding $\beta \in (0, 1]$ such that

$$x_i - \dot{x}_i \beta + \ddot{x}_i (1 - \sqrt{1-\beta^2}) \ge 0. \tag{7.45}$$

However, we prefer to use (7.44) in the following analysis because of its geometric property.

Case 1 ($\dot{x}_i = 0$ *and* $\ddot{x}_i \neq 0$):

For $\ddot{x}_i \geq -x_i$, and for any $\alpha \in [0, \frac{\pi}{2}]$, $x_i(\alpha) \geq 0$ holds. For $\ddot{x}_i \leq -x_i$, to meet (7.44), we must have $\cos(\alpha) \geq \frac{x_i+\ddot{x}_i}{\ddot{x}_i}$, or, $\alpha \leq \cos^{-1}\left(\frac{x_i+\ddot{x}_i}{\ddot{x}_i}\right)$. Therefore,

$$\alpha_{x_i} = \begin{cases} \frac{\pi}{2} & \text{if } x_i + \ddot{x}_i \geq 0 \\ \cos^{-1}\left(\frac{x_i+\ddot{x}_i}{\ddot{x}_i}\right) & \text{if } x_i + \ddot{x}_i \leq 0. \end{cases} \tag{7.46}$$

Case 2 ($\ddot{x}_i = 0$ *and* $\dot{x}_i \neq 0$):

For $\dot{x}_i \leq x_i$, and for any $\alpha \in [0, \frac{\pi}{2}]$, $x_i(\alpha) \geq 0$ holds. For $\dot{x}_i \geq x_i$, to meet (7.44), we must have $\sin(\alpha) \leq \frac{x_i}{\dot{x}_i}$, or $\alpha \leq \sin^{-1}\left(\frac{x_i}{\dot{x}_i}\right)$. Therefore,

$$\alpha_{x_i} = \begin{cases} \frac{\pi}{2} & \text{if } \dot{x}_i \leq x_i \\ \sin^{-1}\left(\frac{x_i}{\dot{x}_i}\right) & \text{if } \dot{x}_i \geq x_i \end{cases} \tag{7.47}$$

Case 3 ($\dot{x}_i > 0$ *and* $\ddot{x}_i > 0$):

Let $\dot{x}_i = \sqrt{\dot{x}_i^2 + \ddot{x}_i^2}\cos(\beta)$, and $\ddot{x}_i = \sqrt{\dot{x}_i^2 + \ddot{x}_i^2}\sin(\beta)$, (7.44) can be rewritten as

$$x_i + \ddot{x}_i \geq \sqrt{\dot{x}_i^2 + \ddot{x}_i^2}\sin(\alpha + \beta), \tag{7.48}$$

where

$$\beta = \sin^{-1}\left(\frac{\ddot{x}_i}{\sqrt{\dot{x}_i^2 + \ddot{x}_i^2}}\right). \tag{7.49}$$

For $\ddot{x}_i + x_i \geq \sqrt{\dot{x}_i^2 + \ddot{x}_i^2}$, and for any $\alpha \in [0, \frac{\pi}{2}]$, $x_i(\alpha) \geq 0$ holds. For $\ddot{x}_i + x_i \leq \sqrt{\dot{x}_i^2 + \ddot{x}_i^2}$, to meet (7.48), we must have $\sin(\alpha + \beta) \leq \frac{x_i+\ddot{x}_i}{\sqrt{\dot{x}_i^2+\ddot{x}_i^2}}$, or $\alpha + \beta \leq \sin^{-1}\left(\frac{x_i+\ddot{x}_i}{\sqrt{\dot{x}_i^2+\ddot{x}_i^2}}\right)$. Therefore,

$$\alpha_{x_i} = \begin{cases} \frac{\pi}{2} & \text{if } x_i + \ddot{x}_i \geq \sqrt{\dot{x}_i^2 + \ddot{x}_i^2} \\ \sin^{-1}\left(\frac{x_i+\ddot{x}_i}{\sqrt{\dot{x}_i^2+\ddot{x}_i^2}}\right) - \sin^{-1}\left(\frac{\ddot{x}_i}{\sqrt{\dot{x}_i^2+\ddot{x}_i^2}}\right) & \text{if } x_i + \ddot{x}_i \leq \sqrt{\dot{x}_i^2 + \ddot{x}_i^2} \end{cases} \tag{7.50}$$

Case 4 ($\dot{x}_i > 0$ *and* $\ddot{x}_i < 0$):

Let $\dot{x}_i = \sqrt{\dot{x}_i^2 + \ddot{x}_i^2}\cos(\beta)$, and $\ddot{x}_i = -\sqrt{\dot{x}_i^2 + \ddot{x}_i^2}\sin(\beta)$, (7.44) can be rewritten as

$$x_i + \ddot{x}_i \geq \sqrt{\dot{x}_i^2 + \ddot{x}_i^2}\sin(\alpha - \beta), \tag{7.51}$$

where

$$\beta = \sin^{-1}\left(\frac{-\ddot{x}_i}{\sqrt{\dot{x}_i^2 + \ddot{x}_i^2}}\right). \tag{7.52}$$

For $\ddot{x}_i + x_i \geq \sqrt{\dot{x}_i^2 + \ddot{x}_i^2}$, and for any $\alpha \in [0, \frac{\pi}{2}]$, $x_i(\alpha) \geq 0$ holds. For $\ddot{x}_i + x_i \leq \sqrt{\dot{x}_i^2 + \ddot{x}_i^2}$, to meet (7.51), we must have $\sin(\alpha - \beta) \leq \frac{x_i + \ddot{x}_i}{\sqrt{\dot{x}_i^2 + \ddot{x}_i^2}}$, or $\alpha - \beta \leq \sin^{-1}\left(\frac{x_i + \ddot{x}_i}{\sqrt{\dot{x}_i^2 + \ddot{x}_i^2}}\right)$. Therefore,

$$\alpha_{x_i} = \begin{cases} \frac{\pi}{2} & \text{if } x_i + \ddot{x}_i \geq \sqrt{\dot{x}_i^2 + \ddot{x}_i^2} \\ \sin^{-1}\left(\frac{x_i + \ddot{x}_i}{\sqrt{\dot{x}_i^2 + \ddot{x}_i^2}}\right) + \sin^{-1}\left(\frac{-\ddot{x}_i}{\sqrt{\dot{x}_i^2 + \ddot{x}_i^2}}\right) & \text{if } x_i + \ddot{x}_i \leq \sqrt{\dot{x}_i^2 + \ddot{x}_i^2} \end{cases} \tag{7.53}$$

Case 5 ($\dot{x}_i < 0$ and $\ddot{x}_i < 0$):

Let $\dot{x}_i = -\sqrt{\dot{x}_i^2 + \ddot{x}_i^2}\cos(\beta)$, and $\ddot{x}_i = -\sqrt{\dot{x}_i^2 + \ddot{x}_i^2}\sin(\beta)$, (7.44) can be rewritten as

$$x_i + \ddot{x}_i \geq -\sqrt{\dot{x}_i^2 + \ddot{x}_i^2}\sin(\alpha + \beta), \tag{7.54}$$

where

$$\beta = \sin^{-1}\left(\frac{-\ddot{x}_i}{\sqrt{\dot{x}_i^2 + \ddot{x}_i^2}}\right). \tag{7.55}$$

For $\ddot{x}_i + x_i \geq 0$ and any $\alpha \in [0, \frac{\pi}{2}]$, $x_i(\alpha) \geq 0$ holds. For $\ddot{x}_i + x_i \leq 0$, to meet (7.54), we must have $\sin(\alpha + \beta) \geq \frac{-(x_i + \ddot{x}_i)}{\sqrt{\dot{x}_i^2 + \ddot{x}_i^2}}$, or $\alpha + \beta \leq \pi - \sin^{-1}\left(\frac{-(x_i + \ddot{x}_i)}{\sqrt{\dot{x}_i^2 + \ddot{x}_i^2}}\right)$. Therefore,

$$\alpha_{x_i} = \begin{cases} \frac{\pi}{2} & \text{if } x_i + \ddot{x}_i \geq 0 \\ \pi - \sin^{-1}\left(\frac{-(x_i + \ddot{x}_i)}{\sqrt{\dot{x}_i^2 + \ddot{x}_i^2}}\right) - \sin^{-1}\left(\frac{-\ddot{x}_i}{\sqrt{\dot{x}_i^2 + \ddot{x}_i^2}}\right) & \text{if } x_i + \ddot{x}_i \leq 0 \end{cases} \tag{7.56}$$

Case 6 ($\dot{x}_i < 0$ and $\ddot{x}_i > 0$):

Clearly (7.44) always holds for $\alpha \in [0, \frac{\pi}{2}]$. Therefore, we can take

$$\alpha_{x_i} = \frac{\pi}{2}. \tag{7.57}$$

Case 7 ($\dot{x}_i = 0$ and $\ddot{x}_i = 0$):

Clearly (7.44) always holds for $\alpha \in [0, \frac{\pi}{2}]$. Therefore, we can take

$$\alpha_{x_i} = \frac{\pi}{2}. \tag{7.58}$$

Similar analysis can be performed for α^s in (7.21) and similar results can be obtained for α_{s_i}. For completeness, we list the formulas without repeating the proofs.

Case 1a ($\dot{s}_i = 0$, $\ddot{s}_i \neq 0$):

$$\alpha_{s_i} = \begin{cases} \frac{\pi}{2} & \text{if } s_i + \ddot{s}_i \geq 0 \\ \cos^{-1}\left(\frac{s_i + \ddot{s}_i}{\ddot{s}_i}\right) & \text{if } s_i + \ddot{s}_i \leq 0. \end{cases} \tag{7.59}$$

Case 2a ($\ddot{s}_i = 0$ *and* $\dot{s}_i \neq 0$):

$$\alpha_{s_i} = \begin{cases} \frac{\pi}{2} & \text{if } \dot{s}_i \leq s_i \\ \sin^{-1}\left(\frac{s_i}{\dot{s}_i}\right) & \text{if } \dot{s}_i \geq s_i \end{cases} \tag{7.60}$$

Case 3a ($\dot{s}_i > 0$ *and* $\ddot{s}_i > 0$):

$$\alpha_{s_i} = \begin{cases} \frac{\pi}{2} & \text{if } s_i + \ddot{s}_i \geq \sqrt{\dot{s}_i^2 + \ddot{s}_i^2} \\ \sin^{-1}\left(\frac{s_i+\ddot{s}_i}{\sqrt{\dot{s}_i^2+\ddot{s}_i^2}}\right) - \sin^{-1}\left(\frac{\ddot{s}_i}{\sqrt{\dot{s}_i^2+\ddot{s}_i^2}}\right) & \text{if } s_i + \ddot{s}_i < \sqrt{\dot{s}_i^2 + \ddot{s}_i^2} \end{cases} \tag{7.61}$$

Case 4a ($\dot{s}_i > 0$ *and* $\ddot{s}_i < 0$):

$$\alpha_{s_i} = \begin{cases} \frac{\pi}{2} & \text{if } s_i + \ddot{s}_i \geq \sqrt{\dot{s}_i^2 + \ddot{s}_i^2} \\ \sin^{-1}\left(\frac{s_i+\ddot{s}_i}{\sqrt{\dot{s}_i^2+\ddot{s}_i^2}}\right) + \sin^{-1}\left(\frac{-\ddot{s}_i}{\sqrt{\dot{s}_i^2+\ddot{s}_i^2}}\right) & \text{if } s_i + \ddot{s}_i \leq \sqrt{\dot{s}_i^2 + \ddot{s}_i^2} \end{cases} \tag{7.62}$$

Case 5a ($\dot{s}_i < 0$ *and* $\ddot{s}_i < 0$):

$$\alpha_{s_i} = \begin{cases} \frac{\pi}{2} & \text{if } s_i + \ddot{s}_i \geq 0 \\ \pi - \sin^{-1}\left(\frac{-(s_i+\ddot{s}_i)}{\sqrt{\dot{s}_i^2+\ddot{s}_i^2}}\right) - \sin^{-1}\left(\frac{-\ddot{s}_i}{\sqrt{\dot{s}_i^2+\ddot{s}_i^2}}\right) & \text{if } s_i + \ddot{s}_i \leq 0 \end{cases} \tag{7.63}$$

Case 6a ($\dot{s}_i < 0$ *and* $\ddot{s}_i > 0$):

Clearly (7.44) always holds for $\alpha \in [0, \frac{\pi}{2}]$. Therefore, we can take

$$\alpha_{s_i} = \frac{\pi}{2}. \tag{7.64}$$

Case 7a ($\dot{s}_i = 0$ *and* $\ddot{s}_i = 0$):

Clearly (7.44) always holds for $\alpha \in [0, \frac{\pi}{2}]$. Therefore, we can take

$$\alpha_{s_i} = \frac{\pi}{2}. \tag{7.65}$$

7.3.9 Step scaling parameter

A fixed step scaling parameter is used in PCx [23]. A more sophisticated step scaling parameter is used in LIPSOL, according to [144, Pages 204-205]. In our implementation, we use an adaptive step scaling parameter which is given below

$$\beta = 1 - e^{-(k+2)}, \tag{7.66}$$

where k is the number of iterations. This parameter will approach to one as $k \to \infty$.

7.3.10 Terminate criteria

The main stopping criterion used in our implementations of the arc-search method and Mehrotra's method is similar to that of LIPSOL [172]

$$\frac{\|\mathbf{r}_b^k\|}{\max\{1,\|\mathbf{b}\|\}}+\frac{\|\mathbf{r}_c^k\|}{\max\{1,\|\mathbf{c}\|\}}+\frac{\mu_k}{\max\{1,\|\mathbf{c}^{\mathrm{T}}\mathbf{x}^k\|,\|\mathbf{b}^{\mathrm{T}}\lambda^k\|\}}<10^{-8}.$$

In case the algorithms fail to find a good search direction, the programs also stop if step sizes $\alpha_k^x < 10^{-8}$ and $\alpha_k^s < 10^{-8}$.

Finally, if due to the numerical problem, $\mathbf{r}_b^k$ or $\mathbf{r}_c^k$ does not decrease but $10\mathbf{r}_b^{k-1} < \mathbf{r}_b^k$ or $10\mathbf{r}_c^{k-1} < \mathbf{r}_c^k$, the programs stop.

7.4 Numerical Tests

In this section, we first examine a simple problem and show graphically what feasible central path and infeasible central path look like, why ellipsoidal approximation may be a better approximation to infeasible central path than a straight line, and how the arc-search is carried out for this simple problem. Using a figure, we can easily see that searching along the ellipse is more attractive than searching along a straight line. We then provide the numerical test results of larger scale Netlib test problems in order to validate our observation from this simple problem.

7.4.1 A simple illustrative example

Let us consider the following example:

$$\min x_1, \quad s.t. \;\; x_1+x_2=5, \;\; x_1\geq 0, \;\; x_2\geq 0.$$

The feasible central path $(\mathbf{x},\lambda,\mathbf{s})$ defined in (7.5) satisfies the following conditions:

$$\begin{gathered} x_1+x_2=5, \\ \begin{bmatrix} 1 \\ 1 \end{bmatrix}\lambda+\begin{bmatrix} s_1 \\ s_2 \end{bmatrix}=\begin{bmatrix} 1 \\ 0 \end{bmatrix}, \\ x_1s_1=\mu, \quad x_2s_2=\mu. \end{gathered}$$

The optimizer is given by $x_1=0$, $x_2=5$, $\lambda=0$, $s_1=1$, and $s_2=0$. The feasible central path of this problem is given analytically as

$$\lambda=\frac{5-2\mu-\sqrt{(5-2\mu)^2+20\mu}}{10}, \tag{7.67a}$$

$$s_1=1-\lambda, \quad s_2=-\lambda, \quad x_1=\mu/s_1, \quad x_2=\mu/s_2. \tag{7.67b}$$

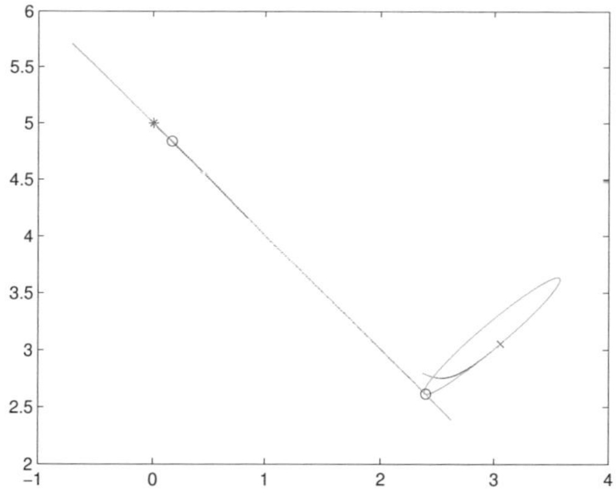

Figure 7.2: Arc-search for the simple example.

The feasible and infeasible central paths for this example are arcs in 5-dimensional space $(\lambda, x_1, s_1, x_2, s_2)$. If we project the central paths into 2-dimensional subspace spanned by (x_1, x_2), they are arcs in 2-dimensional subspace. Figure 7.2 shows the first two iterations of Algorithm 7.1 in the 2-dimensional subspace spanned by (x_1, x_2). In Figure 7.2, the initial point (x_1^0, x_2^0) is marked by 'x' in red; the optimal solution is marked by '*' in red; $(\dot{\mathbf{x}}, \dot{\mathbf{s}}, \dot{\lambda})$ is calculated by using (7.10); $(\ddot{\mathbf{x}}, \ddot{\mathbf{s}}, \ddot{\lambda})$ is calculated by using (7.12); the projected feasible central path $\mathcal{C}(t)$ near the optimal solution is calculated by using (7.67) and is plotted as a continuous line in black; the infeasible central path $\mathcal{H}(t)$ starting from current iterate is calculated by using (7.7) and plotted as the dotted lines in blue; and the projected ellipsoidal approximations $\mathcal{E}(\alpha)$ are the dotted lines in green (they may look like continuous lines sometimes because many dots are used). In the first iteration, the iterate 'x' moves along the ellipse (defined by in Theorem 7.1) to reach the next iterate marked as 'o' in red because the calculation of infeasible central path (the blue line) is very expensive and ellipse is cheap to calculate and a better approximation to the infeasible central path than a straight line. The remaining iterations are simply the repetition of the process until it reaches the optimal solution $(\mathbf{s}^*, \mathbf{x}^*)$. Only two iterations are plotted in Figure 7.2.

It is worthwhile to note that, in this simple problem, the infeasible central path has a sharp turn in the first iteration, which may happen a number of times for general problem, as discussed in [141]. The arc-search method is expected to perform better than Mehrotra's method in iterations that are close to the sharp turns. In this simple problem, after the first iteration, the feasible central path $\mathcal{C}(t)$, the infeasible central path $\mathcal{H}(t)$, and the ellipse $\mathcal{E}(\alpha)$ are all very close to each other and close to a straight line.

7.4.2 Netlib test problems

The algorithm developed in this chapter is implemented in a MATLAB function, we name it as `curvelp.m`. Mehrotra's algorithm is also implemented in a MATLAB function, we name it as `mehrotra.m`. They are almost identical. Both algorithms use exactly the same initial point, the same stopping criteria, the same pre-process and post-process, and the same parameters. The only difference of the two implementations is that the arc-search method searches for the optimizer along an ellipse and Mehrotra's method searches for the optimizer along a straight line. Netlib test problems represented in MATLAB format are available in [50]. Numerical tests for both algorithms have been performed for all Netlib LP problems that are presented in standard form, except `Osa_60` ($m = 10281$ and $n = 232966$) because the PC computer used for the testing does not have enough memory to handle this problem. The iteration numbers used to solve these problems are listed in Table 7.2.

Table 7.2: Numerical results for test problems in Netlib.

Problem	CurveLP.m			Mehrotra.m		
	iter	objective	infeas	iter	objective	infeas
Adlittle	15	2.2549e+05	1.0e-07	15	2.2549e+05	3.4e-08
Afiro	9	-464.7531	1.0e-11	9	-464.7531	8.0e-12
Agg	18	-3.5992e+07	5.0e-06	22	-3.5992e+07	5.2e-05
Agg2	18	-2.0239e+07	4.6e-07	20	-2.0239e+07	5.2e-07
Agg3	17	1.0312e+07	3.1e-08	18	1.0312e+07	8.8e-09
Bandm	19	-158.6280	3.2e-11	22	-158.6280	8.3e-10
Beaconfd	10	3.3592e+04	1.4e-12	11	3.3592e+04	1.4e-10
Blend	12	-30.8121	1.0e-09	14	-30.8122	4.9e-11
Bnl1	32	1.9776e+03	2.7e-09	35	1.9776e+03	3.4e-09
Bnl2+	31	1.8112e+03	5.4e-10	38	1.8112e+03	9.3e-07
Brandy	21	1.5185e+03	3.0e-06	19	1.5185e+03	6.2e-08
Degen2*	16	-1.4352e+03	1.9e-08	17	-1.4352e+03	2.0e-10
Degen3*	22	-9.8729e+02	7.0e-05	22	-9.8729e+02	1.2e-09
fffff800	26	5.5568e+005	4.3e-05	31	5.5568e+05	7.7e-04
Israel	23	-8.9664e+05	7.4e-08	29	-8.9665e+05	1.8e-08
Lotfi	14	-25.2647	3.5e-10	18	-25.2647	2.7e-07
Maros_r7	18	1.4972e+06	1.6e-08	21	1.4972e+06	6.4e-09
Osa_07*	37	5.3574e+05	4.2e-07	35	5.3578e+05	1.5e-07
Osa_14	35	1.1065e+06	2.0e-09	37	1.1065e+06	3.0e-08
Osa_30	32	2.1421e+06	1.0e-08	36	2.1421e+06	1.3e-08
Qap12	22	5.2289e+02	1.9e-08	24	5.2289e+02	6.2e-09
Qap15*	27	1.0411e+03	3.9e-07	44	1.0410e+03	1.5e-05
Qap8*	12	2.0350e+02	1.2e-12	13	2.0350e+02	7.1e-09
Sc105	10	-52.2021	3.8e-12	11	-52.2021	9.8e-11
Sc205	13	-52.2021	3.7e-10	12	-52.2021	8.8e-11
Sc50a	10	-64.5751	3.4e-12	9	-64.5751	8.3e-08
Sc50b	8	-70.0000	1.0e-10	8	-70.0000	9.1e-07
Scagr25	19	-1.4753e+07	5.0e-07	18	-1.4753e+07	4.6e-09

Scagr7	15	-2.3314e+06	2.7e-09	17	-2.3314e+06	1.1e-07
Scfxm1+	20	1.8417e+04	3.1e-07	22	1.8417e+04	1.6e-08
Scfxm2	23	3.6660e+04	2.3e-06	26	3.6660e+04	2.6e-08
Scfxm3+	24	5.4901e+04	1.9e-06	23	5.4901e+04	9.8e-08
Scrs8	23	9.0430e+02	1.2e-11	30	9.0430e+02	1.8e-10
Scsd1	12	8.6666	1.0e-10	13	8.6666	8.7e-14
Scsd6	14	50.5000	1.5e-13	16	50.5000	7.9e-13
Scsd8	13	9.0500e+02	6.7e-10	14	9.0500e+02	1.3e-10
Sctap1	20	1.4122e+03	2.6e-10	27	1.4123e+03	0.0031
Sctap2	21	1.7248e+03	2.1e-10	21	1.7248e+03	4.4e-07
Sctap3	20	1.4240e+03	5.7e-08	22	1.4240e+03	5.9e-07
Share1b	22	-7.6589e+04	6.5e-08	25	-7.6589e+04	1.5e-06
Share2b	13	-4.1573e+02	4.9e-11	15	-4.1573e+02	7.9e-10
Ship04l	17	1.7933e+06	5.2e-11	18	1.7933e+06	2.9e-11
Ship04s	17	1.7987e+06	2.2e-11	20	1.7987e+06	4.5e-09
Ship08l	19	1.9090e+06	1.6e-07	22	1.9091e+06	1.0e-10
Ship08s	17	1.9201e+06	3.7e-08	20	1.9201e+06	4.5e-12
Ship12l	21	1.4702e+06	4.7e-13	21	1.4702e+06	1.0e-08
Ship12s	17	1.4892e+06	1.0e-10	19	1.4892e+06	2.1e-13
Stocfor1**	20/14	-4.1132e+04	2.8e-10	14	-4.1132e+04	1.1e-10
Stocfor2	22	-3.9024e+04	2.1e-09	22	-3.9024e+04	1.6e-09
Stocfor3	34	-3.9976e+04	4.7e-08	38	-3.9976e+04	6.4e-08
Truss	25	4.5882e+05	1.7e-07	26	4.5882e+05	9.5e-06

Several problems have degenerate solutions which make them difficult to solve or need significantly more iterations. We choose to use the option described in Section 7.4.7 to solve these problems. For problems marked with '+', this option is called only for Mehrotra's algorithm because Mehrotra's algorithm cannot solve these problems without using this option. For problems marked with '*', both algorithms need to call this option for better results. For the problem with '**', only Mehrotra's algorithm needs to use the option described in Section 7.4.7; but if both algorithms use the option described in Section 7.4.7, the iteration counts are the same for both algorithms. We need to keep in mind that although using the option described in Section 7.4.7 reduces the iteration count significantly, these iterations are significantly more expensive. Therefore, simply comparing iteration counts for problem(s) marked with '+' will lead to a conclusion in favor of Mehrotra's method (which is what we will do in the following discussions).

Since the major cost in each iteration for both algorithms is solving linear systems of equations, which are identical in these two algorithms, we conclude that the iteration number is a good measure of efficiency. In view of Table 7.2, it is clear that Algorithm 7.1 uses less iterations than Mehrotra's algorithm to find the optimal solutions for majority tested problems. Among 51 tested problems, Mehrotra's method uses fewer iterations (7 iterations in total) than the arc-search method for only 6 problems (`brandy`, `osa_07`, `sc205`, `sc50a`, `scagr25`,

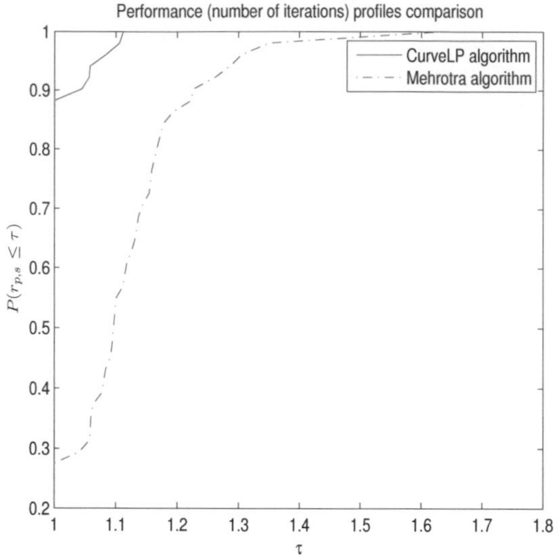

Figure 7.3: Performance profiles comparison for `curvelp.m` and `mehrotra.m`.

`scfxm3`[4]), while the arc-search method uses fewer iterations (115 iterations in total) than Mehrotra's method for 38 problems. For the rest 7 problems, both methods use the same number of iterations. The arc-search method is numerically more stable than Mehrotra's method because, for problems `bnl2`, `scfxm1`, `stocfer1`, the arc-search method does not need to use the option described in Section 7.4.7, but Mehrotra's method does need to use the option to solve the problems. For problem `scatp1`, Mehrotra's method terminated with a relatively large error.

We can also use performance profiles to explain superiority of the proposed algorithm. The performance function in this case is the number of iterations. Figure 7.3 gives the performance profiles for both `curvelp.m` and `mehrotra.m`.

7.5 Concluding Remarks

This chapter proposes and implements an arc-search interior-point path-following algorithm that searches for optimizers along the ellipse that approximates infeasible central path. The proposed algorithm is different from Mehrotra's method only in search path. Both the arc-search method and Mehrotra's

[4]For this problem, Mehrotra's method needs to use the option described in Section 7.4.7 but the arc-search method does not need to. As a result, Mehrotra's method uses noticeably more CPU time than the arc-search method.

method are implemented in MATLAB. Moreover, the two methods use exactly the same initial point, the same pre-process and post-process, the same parameters, and the same stopping criteria. By doing this, we can compare both algorithms in a fair and controlled way. Numerical test is conducted using Netlib problems for both methods. The results show that the proposed arc-search method is more efficient and reliable than the well-known Mehrotra's method. The MATLAB implementations for both algorithms are also available in MATLAB file exchange website [157].

Chapter 8

An $\mathcal{O}(\sqrt{n}L)$ Infeasible Arc-Search Algorithm for LP

For interior-point methods using line search, we have discussed in previous chapters the dilemmas between theoretical analysis and computational experience. For example, higher-order algorithms that use second or higher-order derivatives have been proved to improve the computational efficiency [89, 82, 83] but higher-order algorithms have either poorer polynomial bound than the first-order algorithms [97] or do not have a polynomial bound at all [89]. In addition, some algorithm with the best polynomial bound performs poorly in a real computational test. For example, the short step interior-point algorithm, which searches for the optimizer in a narrow neighborhood, has the best polynomial bound [70] but performs very poorly in practical computation, while the long step interior-point algorithm [71], which searches for the optimizer in a larger neighborhood, performs better in numerical test but has inferior polynomial bound [144]. Even worse, Mehrotra's predictor-corrector (MPC) algorithm, which was widely regarded as the most efficient interior-point algorithm in computation before 2010 and is competitive to the simplex algorithm for large problems, has not been proved to be polynomial (it may not even be convergent [19]). It should be noted that lack of polynomiality was a serious concern for simplex method [65], therefore, motivated ellipsoid method for linear programming [59], and was one of the main arguments in the early development of interior-point algorithms [144, 57].

In Chapters 5, 6, and 7, we introduced arc-search method and show that the new strategy may solve the dilemmas. In Chapter 5, we proposed a feasible arc-search interior-point algorithm for linear programming which uses higher-order derivatives to construct an ellipse and approximate the central path. Intuitively, searching along this ellipse will generate a longer step size than searching along any straight line. Indeed, the arc-search (higher-order) algorithm in Chapter 5 achieved the best polynomial bound which has partially solved the dilemma. Some promising numerical test results were demonstrated.

In Chapter 6, we proposed an algorithm that uses infeasible starting point and showed the method has improved polynomial bound over algorithms that use line search. To demonstrate the superiority of the arc-search strategy proposed in Chapters 5 and 6 for practical problems, we devised an arc-search infeasible interior-point algorithm in Chapter 7, which allowed us to test a lot of more Netlib problems. The proposed algorithm is very similar to Mehrotra's algorithm but replaces search direction by an arc used in Chapters 5 and 6. The comprehensive numerical test for a larger pool of Netlib problems clearly shows the superiority of arc-search over traditional line-search for interior-point method.

Because the purpose of Chapter 7 is to demonstrate the computational merit of the arc-search method, the algorithm is a mimic of Mehrotra's algorithm and we have not shown its convergence. In fact, we noticed in Chapter 7 that Mehrotra's algorithm does not converge for several Netlib problems if the measure of handling degeneration is not applied. The purpose of this Chapter is to develop some infeasible interior-point algorithms which are both computationally competitive to simplex algorithms (i.e., at least as good as Mehrotra's algorithm) and theoretically polynomial. We first devise an algorithm slightly different from the one in Chapter 7 and we show that this algorithm is polynomial. We then propose a simplified version of the first algorithm. This simplified algorithm will search for optimizers in a neighborhood larger than those used in short step and long step path-following algorithms, thereby generating potentially a larger step size. Yet, we want to show that the modified algorithm has the best polynomial complexity bound, in particular, we want to show that the complexity bound is better than $\mathcal{O}(n^2L)$ in [171], which was established for an infeasible interior-point algorithm searching for optimizers in the long-step central-path neighborhood, and better than $\mathcal{O}(nL)$ in [92, 90, 161], which were obtained for infeasible interior-point algorithms using a narrow short-step central-path neighborhoods. As a matter of fact, the simplified algorithm achieves the complexity bound $\mathcal{O}(\sqrt{n}L)$, which is the same as the best polynomial bound for all interior-point algorithms.

To make these algorithms attractive in theory and efficient in numerical computation, we remove some unrealistic and unnecessary assumption made by infeasible interior-point algorithms using line search to prove some convergence results. First, we do not assume that the initial point has the form of $(\zeta\mathbf{e}, \mathbf{0}, \zeta\mathbf{e})$, where $\mathbf{e}$ is a vector of all ones, and ζ is a scalar which is an upper bound of the optimal solution as defined by $\|(\mathbf{x}^*, \mathbf{s}^*)\|_\infty \leq \zeta$, which is an unknown before an optimal solution is found. Computationally, some most efficient starting point selections in [89, 83] (see Chapters 5 and 6) do not meet this restriction.

Second, we remove a requirement[1] that $\|(\mathbf{r}_b^k, \mathbf{r}_c^k)\| \leq [\|(\mathbf{r}_b^0, \mathbf{r}_c^0)\|/\mu_0]\beta\mu_k$, which is required in the convergence analysis for infeasible interior-point algorithms in Chapter 4 (see also, for example, equation (6.2) of [144], equation (13) of [67], and Theorem 2.1 of [92]). Our extensive numerical experience shows that this unnecessary requirement is the major barrier in achieving a large step size for the infeasible interior-point methods developed before this author's work. This is the main reason that the infeasible interior-point algorithms using line search with proven convergence results do not perform well comparing to Mehrotra's algorithm which does not have any convergence result. We demonstrate the computational merits of the proposed algorithms by testing these algorithms along with Mehrotra's algorithm and the very efficient arc-search algorithm in Chapter 7 using Netlib problems and comparing the test results. To have a fair comparison, we use the same initial point, the same pre-process and post-process, and the same termination criteria for the four algorithms in all test problems. The result shows the superiority of these arc-search algorithms. These results solve all the dilemmas we discussed in Chapters 2, 3, and 4. The materials of this chapter is based on [159].

8.1 Preliminaries

Consider the linear programming in the standard form described in (1.4) and the dual programming that is also presented in the standard form in (1.5).

Interior-point algorithms require that all the iterates satisfy the conditions $\mathbf{x} > \mathbf{0}$ and $\mathbf{s} > \mathbf{0}$. Infeasible interior-point algorithms, however, allow the iterates to deviate from the equality constraints. Again, we denote the residuals of the equality constraints (the deviation from the feasibility) by

$$\mathbf{r}_b = \mathbf{A}\mathbf{x} - \mathbf{b}, \quad \mathbf{r}_c = \mathbf{A}^{\mathrm{T}}\boldsymbol{\lambda} + \mathbf{s} - \mathbf{c}. \tag{8.1}$$

the duality measure by

$$\mu = \frac{\mathbf{x}^{\mathrm{T}}\mathbf{s}}{n}. \tag{8.2}$$

The central-path $\mathcal{C}$ of the primal-dual linear programming problem is parametrized by a scalar $\tau > 0$ as follows. For each interior point $(\mathbf{x}, \boldsymbol{\lambda}, \mathbf{s}) \in \mathcal{C}$ on the central path, there is a $\tau > 0$ such that

$$\mathbf{A}\mathbf{x} = \mathbf{b} \tag{8.3a}$$

$$\mathbf{A}^{\mathrm{T}}\boldsymbol{\lambda} + \mathbf{s} = \mathbf{c} \tag{8.3b}$$

$$(\mathbf{x}, \mathbf{s}) > \mathbf{0} \tag{8.3c}$$

$$x_i s_i = \tau, \quad i = 1, \ldots, n. \tag{8.3d}$$

[1] In the requirement of $\|(\mathbf{r}_b^k, \mathbf{r}_c^k)\| \leq [\|(\mathbf{r}_b^0, \mathbf{r}_c^0)\|/\mu_0]\beta\mu_k$, $\beta \geq 1$ is a constant, k is the iteration count, $\mathbf{r}_b^k$ and $\mathbf{r}_c^k$, are residuals of primal and dual constraints, respectively, and μ_k is the duality measure. All of these notations will be given in the next section.

Therefore, the central path is an arc in $\mathbf{R}^{2n+m}$ parametrized as a function of τ and is denoted as

$$\mathcal{C} = \{(\mathbf{x}(\tau), \lambda(\tau), \mathbf{s}(\tau)) : \tau > 0\}. \tag{8.4}$$

As $\tau \to 0$, the central path $(\mathbf{x}(\tau), \lambda(\tau), \mathbf{s}(\tau))$ represented by (8.3) approaches to a solution of the linear programming problem represented by (1.4) because (8.3) reduces to the KKT conditions as $\tau \to 0$.

Because of the high cost of finding the feasible initial point and its corresponding central-path described in (8.3), we consider a modified problem which allows infeasible iterates on an arc $\mathcal{A}(\tau)$ which satisfies the following conditions.

$$\mathbf{A}\mathbf{x}(\tau) - \mathbf{b} = \tau \mathbf{r}_b^k := \mathbf{r}_b^k(\tau), \tag{8.5a}$$
$$\mathbf{A}^{\mathrm{T}}\lambda(\tau) + \mathbf{s}(\tau) - \mathbf{c} = \tau \mathbf{r}_c^k := \mathbf{r}_c^k(\tau), \tag{8.5b}$$
$$(\mathbf{x}(\tau), \mathbf{s}(\tau)) > \mathbf{0}, \tag{8.5c}$$
$$\mathbf{x}(\tau) \circ \mathbf{s}(\tau) = \tau \mathbf{x}^k \circ \mathbf{s}^k. \tag{8.5d}$$

where $\mathbf{x}(1) = \mathbf{x}^k$, $\mathbf{s}(1) = \mathbf{s}^k$, $\lambda(1) = \lambda^k$, $\mathbf{r}_b(1) = \mathbf{r}_b^k$, $\mathbf{r}_c(1) = \mathbf{r}_c^k$, $(\mathbf{r}_b(\tau), \mathbf{r}_c(\tau)) = \tau(\mathbf{r}_b^k, \mathbf{r}_c^k) \to \mathbf{0}$ as $\tau \to 0$. Clearly, as $\tau \to 0$, the arc defined as above will approach to an optimal solution of (1.4) because (8.5) reduces to KKT condition as $\tau \to 0$. We restrict the search for the optimizer in either the neighborhood $\mathcal{F}_1$ or the neighborhood $\mathcal{F}_2$ defined as follows:

$$\mathcal{F}_1 = \{(\mathbf{x}, \lambda, \mathbf{s}) : \quad (\mathbf{x}, \mathbf{s}) > \mathbf{0}, \quad x_i^k s_i^k > \theta \mu_k\}, \tag{8.6a}$$
$$\mathcal{F}_2 = \{(\mathbf{x}, \lambda, \mathbf{s}) : \quad (\mathbf{x}, \mathbf{s}) > \mathbf{0}\}, \tag{8.6b}$$

where $\theta \in (0,1)$ is a constant. The neighborhood (8.6b) is clearly the widest neighborhood used in all existing literatures. Throughout this chapter, we make the following assumption.

Assumption 1:

> **A** *is a full rank matrix.*

Assumption 1 is trivial and non-essential as **A** can always be reduced to meet this condition in polynomial operations. With this assumption, however, the mathematical treatment will be significantly simplified. In Section 8.4, we will describe a method based on [6] to check if a problem meets this assumption. If it is not, the method will reduce the problem to meet this assumption.

Assumption 2:

> *There is at least an optimal solution of (1.4), i.e., the KKT condition holds.*

This assumption is weaker than the assumption made for feasible interior-point method where an interior-point (or strictly feasible solution of (1.4)) is required. The assumption implies that there is at least one feasible solution of (1.4), which will be used in our convergence analysis.

Although the infeasible central path defined in (8.5) allows for infeasible initial point, the calculation of (8.5) is still not practical. We consider a simple approximation of (8.5) discussed in Chapters 5–7. Starting from any point $(\mathbf{x}^k, \lambda^k, \mathbf{s}^k)$ with $(\mathbf{x}^k, \mathbf{s}^k) > \mathbf{0}$, we consider a specific arc which is defined by the current iterate and $(\dot{\mathbf{x}}, \dot{\lambda}, \dot{\mathbf{s}})$ and $(\ddot{\mathbf{x}}, \ddot{\lambda}, \ddot{\mathbf{s}})$ as follows:

$$\begin{bmatrix} \mathbf{A} & \mathbf{0} & \mathbf{0} \\ \mathbf{0} & \mathbf{A}^{\mathrm{T}} & \mathbf{I} \\ \mathbf{S}^k & \mathbf{0} & \mathbf{X}^k \end{bmatrix} \begin{bmatrix} \dot{\mathbf{x}} \\ \dot{\lambda} \\ \dot{\mathbf{s}} \end{bmatrix} = \begin{bmatrix} \mathbf{r}_b^k \\ \mathbf{r}_c^k \\ \mathbf{x}^k \circ \mathbf{s}^k \end{bmatrix}, \tag{8.7}$$

$$\begin{bmatrix} \mathbf{A} & \mathbf{0} & \mathbf{0} \\ \mathbf{0} & \mathbf{A}^{\mathrm{T}} & \mathbf{I} \\ \mathbf{S}^k & \mathbf{0} & \mathbf{X}^k \end{bmatrix} \begin{bmatrix} \ddot{\mathbf{x}}(\sigma_k) \\ \ddot{\lambda}(\sigma_k) \\ \ddot{\mathbf{s}}(\sigma_k) \end{bmatrix} = \begin{bmatrix} \mathbf{0} \\ \mathbf{0} \\ -2\dot{\mathbf{x}} \circ \dot{\mathbf{s}} + \sigma_k \mu_k \mathbf{e} \end{bmatrix}, \tag{8.8}$$

where $\sigma_k \in [0,1]$ is the centering parameter discussed in [144, page 196], and the duality measure μ_k is evaluated at $(\mathbf{x}^k, \lambda^k, \mathbf{s}^k)$. We emphasize that the second-order derivatives are functions of σ_k which we will carefully select in the range of $0 < \sigma_{\min} \leq \sigma_k \leq 1$. A crucial but normally not emphasized fact is that, if $\mathbf{A}$ is full rank, and $\mathbf{X}^k$ and $\mathbf{S}^k$ are positive diagonal matrices, then the matrix

$$\begin{bmatrix} \mathbf{A} & \mathbf{0} & \mathbf{0} \\ \mathbf{0} & \mathbf{A}^{\mathrm{T}} & \mathbf{I} \\ \mathbf{S}^k & \mathbf{0} & \mathbf{X}^k \end{bmatrix}$$

is nonsingular. This guarantees that (8.7) and (8.8) have unique solutions, which is important not only in theory but also in computation, to all interior-point algorithms in linear programming. Therefore, we introduce the following assumption.

Assumption 3:

> $\mathbf{X}^k > \mathbf{0}$ *and* $\mathbf{S}^k > \mathbf{0}$ *are bounded below and away from zeros for all* k *iterations until the program is terminated.*

A simple trick of rescaling the step-length, introduced by Mehrotra [89] (see our implementation in Section 8.5.10), guarantees that this assumption holds.

Given the first and second derivatives defined by (8.7) and (8.8), an analytic expression of an ellipse, which is an approximation of the curve defined by (8.5), is derived in Chapter 5 (see also [151, 153]).

Theorem 8.1

Let $(\mathbf{x}(\alpha), \lambda(\alpha), \mathbf{s}(\alpha))$ *be an ellipse defined by* $(\mathbf{x}(\alpha), \lambda(\alpha), \mathbf{s}(\alpha))|_{\alpha=0} = (\mathbf{x}^k, \lambda^k, \mathbf{s}^k)$ *and its first and second derivatives* $(\dot{\mathbf{x}}, \dot{\lambda}, \dot{\mathbf{s}})$ *and* $(\ddot{\mathbf{x}}, \ddot{\lambda}, \ddot{\mathbf{s}})$ *which are defined by (8.7) and (8.8). Then, the ellipse is an approximation of* $\mathcal{A}(\tau)$ *and is given by*

$$\mathbf{x}(\alpha, \sigma) = \mathbf{x}^k - \dot{\mathbf{x}} \sin(\alpha) + \ddot{\mathbf{x}}(\sigma)(1 - \cos(\alpha)). \tag{8.9}$$

$$\lambda(\alpha,\sigma) = \lambda^k - \dot{\lambda}\sin(\alpha) + \ddot{\lambda}(\sigma)(1-\cos(\alpha)). \tag{8.10}$$

$$\mathbf{s}(\alpha,\sigma) = \mathbf{s}^k - \dot{\mathbf{s}}\sin(\alpha) + \ddot{\mathbf{s}}(\sigma)(1-\cos(\alpha)). \tag{8.11}$$

It is clear from the theorem that the search for the optimizer is carried out along an arc parametrized by α with an adjustable parameter σ. We will use several simple results that can easily be derived from (8.7) and (8.8). To simplify the notations, we will drop the superscript and subscript k unless a confusion might be introduced. We will also use the short notation $(\ddot{\mathbf{x}},\ddot{\lambda},\ddot{\mathbf{s}})$ instead of $(\ddot{\mathbf{x}}(\sigma),\ddot{\lambda}(\sigma),\ddot{\mathbf{s}}(\sigma))$. Using the relations

$$\dot{\mathbf{x}} = \mathbf{A}^{\mathrm{T}}(\mathbf{A}\mathbf{A}^{\mathrm{T}})^{-1}\mathbf{r}_b + \hat{\mathbf{A}}\mathbf{v}, \quad \mathbf{A}^{\mathrm{T}}\dot{\lambda} + \dot{\mathbf{s}} = \mathbf{r}_c, \quad \mathbf{X}^{-1}\dot{\mathbf{x}} + \mathbf{S}^{-1}\dot{\mathbf{s}} = \mathbf{e},$$

where $\hat{\mathbf{A}}$ is an orthonormal basis of the null space of $\mathbf{A}$ and using the similar derivation presented in Chapter 5 (see also [153, Lemma 3.5] or [154, Lemma 3.3]), we have

$$\begin{aligned}
& \mathbf{X}^{-1}\dot{\mathbf{x}} = \mathbf{X}^{-1}[\mathbf{A}^{\mathrm{T}}(\mathbf{A}\mathbf{A}^{\mathrm{T}})^{-1}\mathbf{r}_b + \hat{\mathbf{A}}\mathbf{v}], \\
& \mathbf{S}^{-1}(\mathbf{A}^{\mathrm{T}}\dot{\lambda} + \dot{\mathbf{s}}) = \mathbf{S}^{-1}\mathbf{r}_c, \quad \mathbf{X}^{-1}\dot{\mathbf{x}} + \mathbf{S}^{-1}\dot{\mathbf{s}} = \mathbf{e} \\
\Longleftrightarrow \quad & \begin{bmatrix} \mathbf{X}^{-1}\hat{\mathbf{A}}, & -\mathbf{S}^{-1}\mathbf{A}^{\mathrm{T}} \end{bmatrix} \begin{bmatrix} \mathbf{v} \\ \dot{\lambda} \end{bmatrix} = \mathbf{e} - \mathbf{X}^{-1}\mathbf{A}^{\mathrm{T}}(\mathbf{A}\mathbf{A}^{\mathrm{T}})^{-1}\mathbf{r}_b - \mathbf{S}^{-1}\mathbf{r}_c \\
\Longleftrightarrow \quad & \begin{bmatrix} \mathbf{v} \\ \dot{\lambda} \end{bmatrix} = \begin{bmatrix} (\hat{\mathbf{A}}^{\mathrm{T}}\mathbf{S}\mathbf{X}^{-1}\hat{\mathbf{A}})^{-1}\hat{\mathbf{A}}^{\mathrm{T}}\mathbf{S} \\ -(\mathbf{A}\mathbf{X}\mathbf{S}^{-1}\mathbf{A}^{\mathrm{T}})^{-1}\mathbf{A}\mathbf{X} \end{bmatrix} \left[\mathbf{e} - \mathbf{X}^{-1}\mathbf{A}^{\mathrm{T}}(\mathbf{A}\mathbf{A}^{\mathrm{T}})^{-1}\mathbf{r}_b - \mathbf{S}^{-1}\mathbf{r}_c\right].
\end{aligned} \tag{8.12}$$

This gives the following analytic solutions for (8.7).

$$\dot{\lambda} = -(\mathbf{A}\mathbf{X}\mathbf{S}^{-1}\mathbf{A}^{\mathrm{T}})^{-1}\mathbf{A}\mathbf{X}(\mathbf{e} - \mathbf{X}^{-1}\mathbf{A}^{\mathrm{T}}(\mathbf{A}\mathbf{A}^{\mathrm{T}})^{-1}\mathbf{r}_b - \mathbf{S}^{-1}\mathbf{r}_c), \tag{8.13a}$$

$$\dot{\mathbf{s}} = \mathbf{A}^{\mathrm{T}}(\mathbf{A}\mathbf{X}\mathbf{S}^{-1}\mathbf{A}^{\mathrm{T}})^{-1}\mathbf{A}\mathbf{X}(\mathbf{e} - \mathbf{X}^{-1}\mathbf{A}^{\mathrm{T}}(\mathbf{A}\mathbf{A}^{\mathrm{T}})^{-1}\mathbf{r}_b - \mathbf{S}^{-1}\mathbf{r}_c) + \mathbf{r}_c, \tag{8.13b}$$

$$\dot{\mathbf{x}} = \hat{\mathbf{A}}(\hat{\mathbf{A}}^{\mathrm{T}}\mathbf{S}\mathbf{X}^{-1}\hat{\mathbf{A}})^{-1}\hat{\mathbf{A}}^{\mathrm{T}}\mathbf{S}(\mathbf{e} - \mathbf{X}^{-1}\mathbf{A}^{\mathrm{T}}(\mathbf{A}\mathbf{A}^{\mathrm{T}})^{-1}\mathbf{r}_b - \mathbf{S}^{-1}\mathbf{r}_c) + \mathbf{A}^{\mathrm{T}}(\mathbf{A}\mathbf{A}^{\mathrm{T}})^{-1}\mathbf{r}_b. \tag{8.13c}$$

These relations can easily be reduced to some simpler formulas that will be used in Sections 8.5.6 and 8.5.8.

$$(\mathbf{A}\mathbf{X}\mathbf{S}^{-1}\mathbf{A}^{\mathrm{T}})\dot{\lambda} = \mathbf{A}\mathbf{X}\mathbf{S}^{-1}\mathbf{r}_c - \mathbf{b}, \tag{8.14a}$$

$$\dot{\mathbf{s}} = \mathbf{r}_c - \mathbf{A}^{\mathrm{T}}\dot{\lambda}, \tag{8.14b}$$

$$\dot{\mathbf{x}} = \mathbf{x} - \mathbf{X}\mathbf{S}^{-1}\dot{\mathbf{s}}. \tag{8.14c}$$

Several relations follow immediately from (8.7) and (8.8) (see also in [159]).

Lemma 8.1
Let $(\dot{\mathbf{x}},\dot{\lambda},\dot{\mathbf{s}})$ and $(\ddot{\mathbf{x}},\ddot{\lambda},\ddot{\mathbf{s}})$ be defined in (8.7) and (8.8). Then, the following relations hold.

$$\mathbf{s}\circ\dot{\mathbf{x}} + \mathbf{x}\circ\dot{\mathbf{s}} = \mathbf{x}\circ\mathbf{s}, \quad \mathbf{s}^{\mathrm{T}}\dot{\mathbf{x}} + \mathbf{x}^{\mathrm{T}}\dot{\mathbf{s}} = \mathbf{x}^{\mathrm{T}}\mathbf{s}, \quad \mathbf{s}^{\mathrm{T}}\ddot{\mathbf{x}} + \mathbf{x}^{\mathrm{T}}\ddot{\mathbf{s}} = -2\dot{\mathbf{x}}^{\mathrm{T}}\dot{\mathbf{s}} + \sigma\mu n, \quad \ddot{\mathbf{x}}^{\mathrm{T}}\ddot{\mathbf{s}} = 0. \tag{8.15}$$

Most popular interior-point algorithms of linear programming (e.g., Mehrotra's algorithm) use heuristics to select σ first and then select the step size. In Section 3.4 (see also [154]), it has been shown that a better strategy is to select both centering parameter σ and step size α at the same time. This requires $(\ddot{\mathbf{x}},\ddot{\lambda},\ddot{\mathbf{s}})$ to be represented explicitly in terms of σ_k. Applying the similar derivation in Section 3.4 to (8.8), we have the following explicit solution for $(\ddot{\mathbf{x}},\ddot{\lambda},\ddot{\mathbf{s}})$ in terms of σ.

$$\ddot{\mathbf{x}} = \hat{\mathbf{A}}(\hat{\mathbf{A}}^{\mathrm{T}}\mathbf{S}\mathbf{X}^{-1}\hat{\mathbf{A}})^{-1}\hat{\mathbf{A}}^{\mathrm{T}}\mathbf{X}^{-1}(-2\dot{\mathbf{x}}\circ\dot{\mathbf{s}}+\sigma\mu\mathbf{e}) := \mathbf{p}_x\sigma+\mathbf{q}_x, \tag{8.16a}$$

$$\ddot{\lambda} = -(\mathbf{A}\mathbf{X}\mathbf{S}^{-1}\mathbf{A}^{\mathrm{T}})^{-1}\mathbf{A}\mathbf{S}^{-1}(-2\dot{\mathbf{x}}\circ\dot{\mathbf{s}}+\sigma\mu\mathbf{e}) := \mathbf{p}_\lambda\sigma+\mathbf{q}_\lambda, \tag{8.16b}$$

$$\ddot{\mathbf{s}} = \mathbf{A}^{\mathrm{T}}(\mathbf{A}\mathbf{X}\mathbf{S}^{-1}\mathbf{A}^{\mathrm{T}})^{-1}\mathbf{A}\mathbf{S}^{-1}(-2\dot{\mathbf{x}}\circ\dot{\mathbf{s}}+\sigma\mu\mathbf{e}) := \mathbf{p}_s\sigma+\mathbf{q}_s. \tag{8.16c}$$

These relations can easily be reduced to some simpler formulas as follows:

$$(\mathbf{A}\mathbf{X}\mathbf{S}^{-1}\mathbf{A}^{\mathrm{T}})\ddot{\lambda} = \mathbf{A}\mathbf{S}^{-1}(2\dot{\mathbf{x}}\circ\dot{\mathbf{s}}-\sigma\mu\mathbf{e}), \tag{8.17a}$$

$$\ddot{\mathbf{s}} = -\mathbf{A}^{\mathrm{T}}\ddot{\lambda}, \tag{8.17b}$$

$$\ddot{\mathbf{x}} = \mathbf{S}^{-1}(\sigma\mu\mathbf{e}-\mathbf{X}\ddot{\mathbf{s}}-2\dot{\mathbf{x}}\circ\dot{\mathbf{s}}), \tag{8.17c}$$

or their equivalence to be used in Sections 8.5.6 and 8.5.8:

$$(\mathbf{A}\mathbf{X}\mathbf{S}^{-1}\mathbf{A}^{\mathrm{T}})\mathbf{p}_\lambda = -\mathbf{A}\mathbf{S}^{-1}\mu\mathbf{e}, \quad (\mathbf{A}\mathbf{X}\mathbf{S}^{-1}\mathbf{A}^{\mathrm{T}})\mathbf{q}_\lambda = 2\mathbf{A}\mathbf{S}^{-1}(\dot{\mathbf{x}}\circ\dot{\mathbf{s}}), \tag{8.18a}$$

$$\mathbf{p}_s = -\mathbf{A}^{\mathrm{T}}\mathbf{p}_\lambda, \quad \mathbf{q}_s = -\mathbf{A}^{\mathrm{T}}\mathbf{q}_\lambda, \tag{8.18b}$$

$$\mathbf{p}_x = \mathbf{S}^{-1}\mu\mathbf{e}-\mathbf{S}^{-1}\mathbf{X}\mathbf{p}_s, \quad \mathbf{q}_x = -\mathbf{S}^{-1}\mathbf{X}\mathbf{q}_s-2\mathbf{S}^{-1}(\dot{\mathbf{x}}\circ\dot{\mathbf{s}}). \tag{8.18c}$$

From (8.16) and (8.8), we also have

$$\begin{bmatrix} \mathbf{A} & \mathbf{0} & \mathbf{0} \\ \mathbf{0} & \mathbf{A}^{\mathrm{T}} & \mathbf{I} \\ \mathbf{S} & \mathbf{0} & \mathbf{X} \end{bmatrix} \begin{bmatrix} \mathbf{p}_x \\ \mathbf{p}_\lambda \\ \mathbf{p}_s \end{bmatrix} = \begin{bmatrix} \mathbf{0} \\ \mathbf{0} \\ \mu\mathbf{e} \end{bmatrix}, \tag{8.19}$$

$$\begin{bmatrix} \mathbf{A} & \mathbf{0} & \mathbf{0} \\ \mathbf{0} & \mathbf{A}^{\mathrm{T}} & \mathbf{I} \\ \mathbf{S} & \mathbf{0} & \mathbf{X} \end{bmatrix} \begin{bmatrix} \mathbf{q}_x \\ \mathbf{q}_\lambda \\ \mathbf{q}_s \end{bmatrix} = \begin{bmatrix} \mathbf{0} \\ \mathbf{0} \\ -2\dot{\mathbf{x}}\circ\dot{\mathbf{s}} \end{bmatrix}. \tag{8.20}$$

From these relations, it is straightforward to derive the following:

Lemma 8.2
Let $(\mathbf{p}_x,\mathbf{p}_\lambda,\mathbf{p}_s)$ and $(\mathbf{q}_x,\mathbf{q}_\lambda,\mathbf{q}_s)$ be defined in (8.19) and (8.20); $(\dot{\mathbf{x}},\dot{\lambda},\dot{\mathbf{s}})$ be defined in (8.7). Then, for every iteration k (k is omitted for the sake of notational simplicity), the following relations hold.

$$\mathbf{q}_x^{\mathrm{T}}\mathbf{p}_s = 0, \quad \mathbf{q}_s^{\mathrm{T}}\mathbf{p}_x = 0, \quad \mathbf{q}_x^{\mathrm{T}}\mathbf{q}_s = 0, \quad \mathbf{p}_x^{\mathrm{T}}\mathbf{p}_s = 0, \tag{8.21a}$$

$$\mathbf{s}^{\mathrm{T}}\mathbf{p}_x+\mathbf{x}^{\mathrm{T}}\mathbf{p}_s = n\mu, \quad \mathbf{s}^{\mathrm{T}}\mathbf{q}_x+\mathbf{x}^{\mathrm{T}}\mathbf{q}_s = -2\dot{\mathbf{x}}^{\mathrm{T}}\dot{\mathbf{s}}, \tag{8.21b}$$

$$\mathbf{s}\circ\mathbf{p}_x+\mathbf{x}\circ\mathbf{p}_s=\mu\mathbf{e},\quad \mathbf{s}\circ\mathbf{q}_x+\mathbf{x}\circ\mathbf{q}_s=-2\dot{\mathbf{x}}\circ\dot{\mathbf{s}}. \tag{8.21c}$$

Remark 8.1 Under Assumption 3, a simple but very important observation from (8.13) and (8.16) is that $\dot{\mathbf{x}}$, $\dot{\mathbf{s}}$, $\ddot{\mathbf{x}}$, and $\ddot{\mathbf{s}}$ are all bounded if $\mathbf{r}_b$ and $\mathbf{r}_c$ are bounded, which we will show later. ■

To prove the convergence of the first algorithm, the following condition is required in every iteration:

$$\mathbf{x}^k\circ\mathbf{s}^k\geq\theta\mu_k\mathbf{e} \tag{8.22}$$

where $\theta\in(0,1)$ is a constant.

Remark 8.2 Given $(\mathbf{x}^k,\lambda^k,\mathbf{s}^k,\dot{\mathbf{x}},\dot{\lambda},\dot{\mathbf{s}},\ddot{\mathbf{x}},\ddot{\lambda},\ddot{\mathbf{s}})$ with $(\mathbf{x}^k,\mathbf{s}^k)>(\mathbf{0},\mathbf{0})$, our strategy is to use the relations described in Lemmas 8.1 and 8.2 to find some appropriate $\alpha_k\in(0,\pi/2]$ and $\sigma_k\in[\sigma_{\min},1]$ such that

1. $\|(\mathbf{r}_b^{k+1},\mathbf{r}_c^{k+1})\|$ and μ_{k+1} decrease in every iteration and approach to zero as $k\to 0$.
2. $(\mathbf{x}^{k+1},\mathbf{s}^{k+1})>(\mathbf{0},\mathbf{0})$.
3. $\mathbf{x}^{k+1}\circ\mathbf{s}^{k+1}\geq\theta\mu_{k+1}\mathbf{e}$.

■

The next lemma to be used in the discussion is provided in Chapter 7.

Lemma 8.3
Let $\mathbf{r}_b^k=\mathbf{A}\mathbf{x}^k-\mathbf{b}$, $\mathbf{r}_c^k=\mathbf{A}^{\mathrm{T}}\lambda^k+\mathbf{s}^k-\mathbf{c}$, and $\nu_k=\prod_{j=0}^{k-1}(1-\sin(\alpha_j))$. Then, the following relations hold.

$$\mathbf{r}_b^k=\mathbf{r}_b^{k-1}(1-\sin(\alpha_{k-1}))=\cdots=\mathbf{r}_b^0\prod_{j=0}^{k-1}(1-\sin(\alpha_j))=\mathbf{r}_b^0\nu_k, \tag{8.23a}$$

$$\mathbf{r}_c^k=\mathbf{r}_c^{k-1}(1-\sin(\alpha_{k-1}))=\cdots=\mathbf{r}_c^0\prod_{j=0}^{k-1}(1-\sin(\alpha_j))=\mathbf{r}_c^0\nu_k. \tag{8.23b}$$

In a compact form, (8.23) can be rewritten as

$$(\mathbf{r}_b^k,\mathbf{r}_c^k)=\nu_k(\mathbf{r}_b^0,\mathbf{r}_c^0). \tag{8.24}$$

Lemma 8.3 indicates clearly that, in order to reduce $(\mathbf{r}_b^k,\mathbf{r}_c^k)$ quickly, we should take the largest possible step size $\alpha_k\to\pi/2$. If $\nu_k=0$, then $\mathbf{r}_b^k=\mathbf{0}=\mathbf{r}_c^k$. In this case, the problem has a feasible interior-point $(\mathbf{x}^k,\lambda^k,\mathbf{s}^k)$ and can be solved by using some efficient feasible interior-point algorithm. Therefore, in the remainder of this chapter, we use the following assumption:

Assumption 4:

$\nu_k > 0$ *for* $\forall k \geq 0$.

A rescale of α_k similar to the strategy discussed in [144] is suggested in Section 8.4.10, which guarantees that Assumptions 3 and 4 hold in every iteration. To examine the decreasing property of $\mu(\sigma_k, \alpha_k)$, we need the following result.

Lemma 8.4
Let α_k be the step length at kth iteration, and let $\mathbf{x}(\sigma_k, \alpha_k)$, $\mathbf{s}(\sigma_k, \alpha_k)$, and $\lambda(\sigma_k, \alpha_k)$ be defined in Theorem 8.1. Then, the updated duality measure after an iteration from k can be expressed as

$$\mu_{k+1} := \mu(\sigma_k, \alpha_k) = \frac{1}{n}\left[a_u(\alpha_k)\sigma_k + b_u(\alpha_k)\right], \tag{8.25}$$

where

$$a_u(\alpha_k) = n\mu_k(1-\cos(\alpha_k)) - (\dot{\mathbf{x}}^{\mathrm{T}}\mathbf{p}_s + \dot{\mathbf{s}}^{\mathrm{T}}\mathbf{p}_x)\sin(\alpha_k)(1-\cos(\alpha_k))$$

and

$$b_u(\alpha_k) = n\mu_k(1-\sin(\alpha_k)) - [\dot{\mathbf{x}}^{\mathrm{T}}\dot{\mathbf{s}}(1-\cos(\alpha_k))^2 + (\dot{\mathbf{s}}^{\mathrm{T}}\mathbf{q}_x + \dot{\mathbf{x}}^{\mathrm{T}}\mathbf{q}_s)\sin(\alpha_k)(1-\cos(\alpha_k))]$$

are coefficients which are functions of α_k.

Proof 8.1 Using (8.2), (8.16), and Lemmas 8.1 and 8.2, we have

$$\begin{aligned}
& n\mu(\sigma_k, \alpha_k) \\
= \; & \left(\mathbf{x}^k - \dot{\mathbf{x}}\sin(\alpha_k) + \ddot{\mathbf{x}}(1-\cos(\alpha_k))\right)^{\mathrm{T}} \left(\mathbf{s}^k - \dot{\mathbf{s}}\sin(\alpha_k) + \ddot{\mathbf{s}}(1-\cos(\alpha_k))\right) \\
= \; & \mathbf{x}^{k^{\mathrm{T}}}\mathbf{s}^k - \left(\mathbf{x}^{k^{\mathrm{T}}}\dot{\mathbf{s}} + \mathbf{s}^{k^{\mathrm{T}}}\dot{\mathbf{x}}\right)\sin(\alpha_k) + \left(\mathbf{x}^{k^{\mathrm{T}}}\ddot{\mathbf{s}} + \mathbf{s}^{k^{\mathrm{T}}}\ddot{\mathbf{x}}\right)(1-\cos(\alpha_k)) \\
& + \dot{\mathbf{x}}^{\mathrm{T}}\dot{\mathbf{s}}\sin^2(\alpha_k) - \left(\dot{\mathbf{x}}^{\mathrm{T}}\ddot{\mathbf{s}} + \dot{\mathbf{s}}^{\mathrm{T}}\ddot{\mathbf{x}}\right)\sin(\alpha_k)(1-\cos(\alpha_k)) \\
= \; & n\mu_k(1-\sin(\alpha_k)) + \left(\sigma_k\mu_k n - 2\dot{\mathbf{x}}^{\mathrm{T}}\dot{\mathbf{s}}\right)(1-\cos(\alpha_k)) + \dot{\mathbf{x}}^{\mathrm{T}}\dot{\mathbf{s}}\sin^2(\alpha_k) \\
& - \left(\dot{\mathbf{x}}^{\mathrm{T}}\ddot{\mathbf{s}} + \dot{\mathbf{s}}^{\mathrm{T}}\ddot{\mathbf{x}}\right)\sin(\alpha_k)(1-\cos(\alpha_k)) \\
= \; & n\mu_k(1-\sin(\alpha_k)) + n\sigma_k\mu_k(1-\cos(\alpha_k)) - \dot{\mathbf{x}}^{\mathrm{T}}\dot{\mathbf{s}}(1-\cos(\alpha_k))^2 \\
& - \left(\dot{\mathbf{x}}^{\mathrm{T}}\ddot{\mathbf{s}} + \dot{\mathbf{s}}^{\mathrm{T}}\ddot{\mathbf{x}}\right)\sin(\alpha_k)(1-\cos(\alpha_k)) \qquad (8.26) \\
= \; & \left[n\mu_k(1-\cos(\alpha_k)) - (\dot{\mathbf{x}}^{\mathrm{T}}\mathbf{p}_s + \dot{\mathbf{s}}^{\mathrm{T}}\mathbf{p}_x)\sin(\alpha_k)(1-\cos(\alpha_k))\right]\sigma_k \\
& + n\mu_k(1-\sin(\alpha_k)) \\
& - [\dot{\mathbf{x}}^{\mathrm{T}}\dot{\mathbf{s}}(1-\cos(\alpha_k))^2 + (\dot{\mathbf{s}}^{\mathrm{T}}\mathbf{q}_x + \dot{\mathbf{x}}^{\mathrm{T}}\mathbf{q}_s)\sin(\alpha_k)(1-\cos(\alpha_k))] \\
:= \; & a_u(\alpha_k)\sigma_k + b_u(\alpha_k).
\end{aligned}$$

This proves the lemma. ■

Since Assumption 3 implies that $\mu_k = \mathbf{x}^{k^{\mathrm{T}}}\mathbf{s}^k/n$ is bounded below and away from zero, in view of Lemma 5.4, it is easy to see from (8.26) the following proposition.

Proposition 8.1
For any fixed σ_k, if $\dot{\mathbf{x}}$, $\dot{\mathbf{s}}$, $\ddot{\mathbf{x}}$, and $\ddot{\mathbf{s}}$ are bounded, then there always exists $\alpha_k \in (0,1)$ bounded below and away from zero such that $\mu(\sigma_k,\alpha_k)$ decreases in every iteration. Moreover, $\mu_{k+1} := \mu(\sigma_k,\alpha_k) \to \mu_k(1-\sin(\alpha_k))$ as $\alpha_k \to 0$.

Now, we show that there exists α_k bounded below and away from zero such that the requirement 2 of Remark 8.2 holds. Let $\rho \in (0,1)$ be a constant, and

$$\underline{x}^k = \min_i x_i^k, \quad \underline{s}^k = \min_j s_j^k. \tag{8.27}$$

Denote ϕ_k and ψ_k such that

$$\phi_k = \min\{\rho\underline{x}^k, \nu_k\}, \quad \psi_k = \min\{\rho\underline{s}^k, \nu_k\}. \tag{8.28}$$

It is clear that

$$\mathbf{0} < \phi_k\mathbf{e} \le \rho\mathbf{x}^k, \quad \mathbf{0} < \phi_k\mathbf{e} < \nu_k\mathbf{e}, \tag{8.29a}$$

$$\mathbf{0} < \psi_k\mathbf{e} \le \rho\mathbf{s}^k, \quad \mathbf{0} < \psi_k\mathbf{e} \le \nu_k\mathbf{e}. \tag{8.29b}$$

Positivity of $\mathbf{x}(\sigma_k,\alpha_k)$ and $\mathbf{s}(\sigma_k,\alpha_k)$ is guaranteed if $(\mathbf{x}^0,\mathbf{s}^0) > \mathbf{0}$ and the following conditions hold.

$$\begin{aligned}
\mathbf{x}^{k+1} &= \mathbf{x}(\sigma_k,\alpha_k) = \mathbf{x}^k - \dot{\mathbf{x}}\sin(\alpha_k) + \ddot{\mathbf{x}}(1-\cos(\alpha_k)) \\
&= \mathbf{p}_x(1-\cos(\alpha_k))\sigma_k + [\mathbf{x}^k - \dot{\mathbf{x}}\sin(\alpha_k) + \mathbf{q}_x(1-\cos(\alpha_k))] \\
&:= a_x(\alpha_k)\sigma_k + b_x(\alpha_k) \ge \phi_k\mathbf{e}.
\end{aligned} \tag{8.30}$$

$$\begin{aligned}
\mathbf{s}^{k+1} &= \mathbf{s}(\sigma,\alpha_k) = \mathbf{s}^k - \dot{\mathbf{s}}\sin(\alpha_k) + \ddot{\mathbf{x}}(1-\cos(\alpha_k)) \\
&= \mathbf{p}_s(1-\cos(\alpha_k))\sigma_k + [\mathbf{s}^k - \dot{\mathbf{s}}\sin(\alpha_k) + \mathbf{q}_s(1-\cos(\alpha_k))] \\
&:= a_s(\alpha_k)\sigma_k + b_s(\alpha_k) \ge \psi_k\mathbf{e}.
\end{aligned} \tag{8.31}$$

If $\mathbf{x}^{k+1} = \mathbf{x}^k - \dot{\mathbf{x}}\sin(\alpha_k) + \ddot{\mathbf{x}}(1-\cos(\alpha_k)) \ge \rho\mathbf{x}^k$ holds, from (8.29a), we have $\mathbf{x}^{k+1} \ge \phi_k\mathbf{e}$. Therefore, inequality (8.30) will be satisfied if

$$(1-\rho)\mathbf{x}^k - \dot{\mathbf{x}}\sin(\alpha_k) + \ddot{\mathbf{x}}(1-\cos(\alpha_k)) \ge \mathbf{0}, \tag{8.32}$$

which holds for some $\alpha_k > 0$ bounded below and away from zero because $(1-\rho)\mathbf{x}^k > \mathbf{0}$ is bounded below and away from zero. Similarly, from (8.29b), inequality (8.31) will be satisfied if

$$(1-\rho)\mathbf{s}^k - \dot{\mathbf{s}}\sin(\alpha_k) + \ddot{\mathbf{s}}(1-\cos(\alpha_k)) \ge \mathbf{0}, \tag{8.33}$$

which holds for some $\alpha_k > 0$ bounded below and away from zero because $(1-\rho)\mathbf{s}^k > \mathbf{0}$ is bounded below and away from zero. We summarize the above discussion as the following proposition.

Proposition 8.2
There exists $\alpha_k > 0$ bounded below and away from zero such that $(\mathbf{x}^{k+1}, \mathbf{s}^{k+1}) > \mathbf{0}$ for all iteration k.

The next proposition addresses requirement 3 of Remark 8.2.

Proposition 8.3
There exist α_k bounded below and away from zero for all k such that (8.22) holds.

Proof 8.2 From (8.30) and (8.31), since $x_i^k s_i^k \geq \theta\mu_k$, we have

$$\begin{aligned}
& x_i^{k+1} s_i^{k+1} \\
= \; & [x_i^k - \dot{x}_i \sin(\alpha_k) + \ddot{x}_i(1-\cos(\alpha_k))][s_i^k - \dot{s}_i \sin(\alpha_k) + \ddot{s}_i(1-\cos(\alpha_k))] \\
= \; & x_i^k s_i^k - [\dot{x}_i s_i^k + x_i^k \dot{s}_i] \sin(\alpha_k) + [\ddot{x}_i s_i^k + x_i^k \ddot{s}_i](1-\cos(\alpha_k)) \\
& + \dot{x}_i \dot{s}_i \sin^2(\alpha_k) - [\ddot{x}_i \dot{s}_i + \dot{x}_i \ddot{s}_i] \sin(\alpha_k)(1-\cos(\alpha_k)) + \ddot{x}_i \ddot{s}_i (1-\cos(\alpha_k))^2 \\
= \; & x_i^k s_i^k (1-\sin(\alpha_k)) + \dot{x}_i \dot{s}_i [\sin^2(\alpha_k) - 2(1-\cos(\alpha_k))] + \sigma_k \mu_k (1-\cos(\alpha_k)) \\
& - [\ddot{x}_i \dot{s}_i + \dot{x}_i \ddot{s}_i] \sin(\alpha_k)(1-\cos(\alpha_k)) + \ddot{x}_i \ddot{s}_i (1-\cos(\alpha_k))^2 \\
= \; & x_i^k s_i^k (1-\sin(\alpha_k)) - \dot{x}_i \dot{s}_i (1-\cos(\alpha_k))^2 + \sigma_k \mu_k (1-\cos(\alpha_k)) \\
& - [\ddot{x}_i \dot{s}_i + \dot{x}_i \ddot{s}_i] \sin(\alpha_k)(1-\cos(\alpha_k)) + \ddot{x}_i \ddot{s}_i (1-\cos(\alpha_k))^2 \\
\geq \; & \theta\mu_k (1-\sin(\alpha_k)) + \sigma_k \mu_k (1-\cos(\alpha_k)) \\
& - [\ddot{x}_i \dot{s}_i + \dot{x}_i \ddot{s}_i] \sin(\alpha_k)(1-\cos(\alpha_k)) + (\ddot{x}_i \ddot{s}_i - \dot{x}_i \dot{s}_i)(1-\cos(\alpha_k))^2.
\end{aligned}$$

Therefore, using $\mu_{k+1} = \mu(\sigma_k, \alpha_k)$ and (8.26), we have

$$\begin{aligned}
& x_i^{k+1} s_i^{k+1} - \theta\mu_{k+1} \\
\geq \; & \theta\mu_k (1-\sin(\alpha_k)) + \theta\sigma_k \mu_k (1-\cos(\alpha_k)) + (1-\theta)\sigma_k \mu_k (1-\cos(\alpha_k)) \\
& - [\ddot{x}_i \dot{s}_i + \dot{x}_i \ddot{s}_i] \sin(\alpha_k)(1-\cos(\alpha_k)) + (\ddot{x}_i \ddot{s}_i - \dot{x}_i \dot{s}_i)(1-\cos(\alpha_k))^2. \\
& - \theta\mu_k (1-\sin(\alpha_k)) - \theta\sigma_k \mu_k (1-\cos(\alpha_k)) + \frac{\theta}{n} \dot{\mathbf{x}}^{\mathrm{T}} \dot{\mathbf{s}} (1-\cos(\alpha_k))^2 \\
& + \frac{\theta}{n} \left(\dot{\mathbf{x}}^{\mathrm{T}} \ddot{\mathbf{s}} + \dot{\mathbf{s}}^{\mathrm{T}} \ddot{\mathbf{x}}\right) \sin(\alpha_k)(1-\cos(\alpha_k)) \\
= \; & (1-\cos(\alpha_k)) \Bigg((1-\theta)\sigma_k \mu_k + \left(\ddot{x}_i \ddot{s}_i - \dot{x}_i \dot{s}_i + \frac{\theta}{n} \dot{\mathbf{x}}^{\mathrm{T}} \dot{\mathbf{s}}\right)(1-\cos(\alpha_k)) \\
& - \left(\ddot{x}_i \dot{s}_i + \dot{x}_i \ddot{s}_i - \frac{\theta}{n} \left(\dot{\mathbf{x}}^{\mathrm{T}} \ddot{\mathbf{s}} + \dot{\mathbf{s}}^{\mathrm{T}} \ddot{\mathbf{x}}\right)\right) \sin(\alpha_k) \Bigg) \\
:= \; & (1-\cos(\alpha_k)) p(\alpha).
\end{aligned} \tag{8.34}$$

Since Assumption 3 implies (a) μ_k is bounded below and away from zero, and (b) $\dot{\mathbf{x}}$, $\dot{\mathbf{s}}$, $\ddot{\mathbf{x}}$, and $\ddot{\mathbf{s}}$ are all bounded, $(1-\theta)\sigma_k\mu_k$ is bounded below and away from zero, there must be an α_k bounded below and away from zero such that $p(\alpha) \geq 0$. This proves the claim. ∎

Let $\sigma_{\min}$ and $\sigma_{\max}$ be constants, and $0 < \sigma_{\min} < \sigma_{\max} \leq 1$. From Propositions 8.1, and 8.2, and Lemma 8.3, we conclude:

Proposition 8.4
For any fixed σ_k such that $\sigma_{\min} \leq \sigma_k \leq \sigma_{\max}$, there is a constant $\delta > 0$ related to lower bound of α_k such that (a) $\mathbf{r}_b^k - \mathbf{r}_b^{k+1} \geq \delta\mathbf{e}$, (b) $\mathbf{r}_c^k - \mathbf{r}_c^{k+1} \geq \delta\mathbf{e}$, (c) $\mu_k - \mu_{k+1} \geq \delta$, and (d) $(\mathbf{x}^{k+1}, \mathbf{s}^{k+1}) > \mathbf{0}$.

8.2 A Basic Algorithm

This algorithm considers the search in the neighborhood (8.6a). Based on the discussion in the previous section, we will show in this section that the following arc-search infeasible interior-point algorithm is well-defined and converges in polynomial iterations.

Algorithm 8.1

Data: $\mathbf{A}$, $\mathbf{b}$, $\mathbf{c}$.
Parameter: $\varepsilon \in (0,1)$, $\sigma_{\min} \in (0,1)$, $\sigma_{\min} \leq \sigma_{max} \in (0,1)$, $\theta \in (0,1)$, *and* $\rho \in (0,1)$.
Initial point: $\lambda^0 = 0$ *and* $(\mathbf{x}^0, \mathbf{s}^0) > \mathbf{0}$.
for *iteration* $k = 0, 1, 2, \ldots$

Step 0: If $\|\mathbf{r}_b^0\| \leq \varepsilon$, $\|\mathbf{r}_c^0\| \leq \varepsilon$, *and* $\mu_k \leq \varepsilon$, *stop.*

Step 1: Calculate μ_k, $\mathbf{r}_b^k$, $\mathbf{r}_c^k$, $\dot{\lambda}$, $\dot{\mathbf{s}}$, $\dot{\mathbf{x}}$, $\mathbf{p}_x^k$, $\mathbf{p}_\lambda^k$, $\mathbf{p}_s^k$, $\mathbf{q}_x^k$, $\mathbf{q}_\lambda^k$, *and* $\mathbf{q}_s^k$.

Step 2: Find some appropriate $\alpha_k \in (0, \pi/2]$ *and* $\sigma_k \in [\sigma_{\min}, \sigma_{\max}]$ *to satisfy*

$$\mathbf{x}(\sigma_k,\alpha_k) \geq \phi_k\mathbf{e},\ \ \mathbf{s}(\sigma_k,\alpha_k) \geq \psi_k\mathbf{e},\ \ \mu_k > \mu(\sigma_k,\alpha_k),$$
$$\mathbf{x}(\sigma_k,\alpha_k)\mathbf{s}(\sigma_k,\alpha_k) \geq \theta\mu(\sigma_k,\alpha_k)\mathbf{e}.$$

Step 3: Set $(\mathbf{x}^{k+1}, \lambda^{k+1}, \mathbf{s}^{k+1}) = (\mathbf{x}(\sigma_k,\alpha_k), \lambda(\sigma_k,\alpha_k), \mathbf{s}(\sigma_k,\alpha_k))$ *and* $\mu_{k+1} = \mu(\sigma_k,\alpha_k)$.

Step 4: Set $k+1 \rightarrow k$. *Go back to Step 1.*

end (for)

The algorithm is well defined because of the three propositions in the previous section, i.e., there is a series of α_k bounded below and away from zero such that all conditions in Step 2 hold. Therefore, a constant $\rho \in (0,1)$ satisfying $\rho \geq (1-\sin(\alpha_k))$ does exist for all $k \geq 0$. Denote

$$\beta_k = \frac{\min\{\underline{x}^k, \underline{s}^k\}}{\nu_k} \geq 0, \tag{8.35}$$

and

$$\beta = \inf_k\{\beta_k\} \geq 0. \tag{8.36}$$

The next lemma shows that β is bounded below and away from zero.

Lemma 8.5
Assuming that $\rho \in (0,1)$ is a constant and for all $k \geq 0$, $\rho \geq (1-\sin(\alpha_k))$. Then, we have $\beta \geq \min\{\underline{x}^0, \underline{s}^0, 1\}$.

Proof 8.3 For $k=0$, $(\underline{x}^0, \underline{s}^0) > \mathbf{0}$, and $\nu_0 = 1$, therefore, $\beta_0 \geq \min\{\underline{x}^0, \underline{s}^0, 1\}$ holds. Assuming that $\beta_k \geq \min\{\underline{x}^0, \underline{s}^0, 1\}$ holds for $k > 0$, we would like to show that $\beta_{k+1} = \min\{\beta_k, 1\}$ holds for $k+1$. We divide our discussion into three cases.

Case 1: $\min\{\underline{x}^{k+1}, \underline{s}^{k+1}\} = \underline{x}^{k+1} \geq \rho \underline{x}^k \geq \underline{x}^k(1-\sin(\alpha_k))$. Then, we have

$$\beta_{k+1} = \frac{\min\{\underline{x}^{k+1}, \underline{s}^{k+1}\}}{\nu_{k+1}} \geq \frac{\underline{x}^k(1-\sin(\alpha_k))}{\nu_k(1-\sin(\alpha_k))} \geq \beta_k.$$

Case 2: $\min\{\underline{x}^{k+1}, \underline{s}^{k+1}\} = \underline{s}^{k+1} \geq \rho \underline{s}^k \geq \underline{s}^k(1-\sin(\alpha_k))$. Then, we have

$$\beta_{k+1} = \frac{\min\{\underline{x}^{k+1}, \underline{s}^{k+1}\}}{\nu_{k+1}} \geq \frac{\underline{s}^k(1-\sin(\alpha_k))}{\nu_k(1-\sin(\alpha_k))} \geq \beta_k.$$

Case 3: $\min\{\underline{x}^{k+1}, \underline{s}^{k+1}\} \geq \nu_k$. Then, we have

$$\beta_{k+1} = \frac{\min\{\underline{x}^{k+1}, \underline{s}^{k+1}\}}{\nu_{k+1}} \geq \frac{\nu_k}{\nu_k(1-\sin(\alpha_k))} \geq 1.$$

Adjoining these cases, we conclude $\beta \geq \min\{\beta_0, 1\}$. ■

The main purpose of the remaining section is to establish a polynomial bound for this algorithm. In view of (8.5) and (8.6a), to show the convergence of Algorithm 8.1, we need to show that there is a sequence of $\alpha_k \in (0, \pi/2]$ with $\sin(\alpha_k)$ being bounded by a polynomial of n, and $\sigma_k \in [\sigma_{\min}, \sigma_{max}]$, such that (a) $\mathbf{r}_b^k \to \mathbf{0}$, $\mathbf{r}_c^k \to \mathbf{0}$ (which can be shown by using Lemma 8.3), and $\mu_k \to 0$, (b) $(\mathbf{x}^k, \mathbf{s}^k) > \mathbf{0}$ for all $k \geq 0$, and (c) $\mathbf{x}(\sigma_k, \alpha_k) \circ \mathbf{s}(\sigma_k, \alpha_k) \geq \theta\mu(\sigma_k, \alpha_k)\mathbf{e}$ for all $k \geq 0$.

Although the strategy presented below is similar to the one used by Kojima [67], Kojima, Megiddo, and Mizuno [68], Wright [144], and Zhang [171], the convergence results in this chapter do not depend on some unrealistic and unnecessary restrictions assumed in those papers. We start with a simple but important observation. Let $\mathbf{D} = \mathbf{X}^{\frac{1}{2}}\mathbf{S}^{-\frac{1}{2}} = \mathrm{diag}(D_{ii})$.

Lemma 8.6
For Algorithm 8.1, there is a constant C_1 independent of n such that for $\forall i \in \{1,\ldots,n\}$

$$(D_{ii}^k)^{-1}\nu_k = \nu_k\sqrt{\frac{s_i^k}{x_i^k}} \le C_1\sqrt{n\mu_k}, \quad D_{ii}^k\nu_k = \nu_k\sqrt{\frac{x_i^k}{s_i^k}} \le C_1\sqrt{n\mu_k}. \tag{8.37}$$

Proof 8.4 We know that $\min\{\underline{x}^k, \underline{s}^k\} > 0$, $\nu_k > 0$, and $\beta > 0$ is a constant independent of n. By the definition of β_k, we have $x_i^k \ge \underline{x}^k \ge \beta_k\nu_k \ge \beta\nu_k$ and $s_j^k \ge \underline{s}^k \ge \beta_k\nu_k \ge \beta\nu_k$. This gives, for $\forall i \in \{1,\ldots,n\}$,

$$(D_{ii}^k)^{-1}\nu_k = \sqrt{\frac{s_i^k}{x_i^k}}\nu_k \le \frac{1}{\beta}\sqrt{s_i^k x_i^k} \le \frac{1}{\beta}\sqrt{n\mu_k} := C_1\sqrt{n\mu_k}.$$

Using a similar argument for $s_j^k \ge \beta\nu_k$, we can show that $D_{ii}\nu_k \le C_1\sqrt{n\mu_k}$. ■

The main idea in the proof is based on a crucial observation used in many literatures, for example, Mizuno [92] and Kojima [67]. Let $\bar{\mathcal{S}}$ be defined by (6.25). Let $(\bar{\mathbf{x}}, \bar{\lambda}, \bar{\mathbf{s}}) \in \bar{\mathcal{S}}$ be a feasible point satisfying $\mathbf{A}\bar{\mathbf{x}} = \mathbf{b}$ and $\mathbf{A}^{\mathrm{T}}\bar{\lambda} + \bar{\mathbf{s}} = \mathbf{c}$. The existence of $(\bar{\mathbf{x}}, \bar{\lambda}, \bar{\mathbf{s}})$ is guaranteed by Assumption 2. We will make an additional assumption in the remaining discussion.

Assumption 5:

> *There exists a big constant M which is independent to the problem size n and m such that* $\min_{(\bar{\mathbf{x}},\bar{\lambda},\bar{\mathbf{s}})\in\bar{\mathcal{S}}} \left\| \left(\mathbf{x}^0 - \bar{\mathbf{x}}, \mathbf{s}^0 - \bar{\mathbf{s}}\right) \right\| < M$.

For $(\bar{\mathbf{x}}, \bar{\lambda}, \bar{\mathbf{s}})$ meeting Assumption 5, since

$$\mathbf{A}\dot{\mathbf{x}} = \mathbf{r}_b^k = \nu_k\mathbf{r}_b^0 = \nu_k(\mathbf{A}\mathbf{x}^0 - \mathbf{b}) = \nu_k\mathbf{A}(\mathbf{x}^0 - \bar{\mathbf{x}}),$$

we have

$$\mathbf{A}(\dot{\mathbf{x}} - \nu_k(\mathbf{x}^0 - \bar{\mathbf{x}})) = \mathbf{0}.$$

Similarly, since

$$\mathbf{A}^{\mathrm{T}}\dot{\lambda} + \dot{\mathbf{s}} = \mathbf{r}_c^k = \nu_k\mathbf{r}_c^0 = \nu_k(\mathbf{A}^{\mathrm{T}}\lambda^0 + \mathbf{s}^0 - \mathbf{c}) = \nu_k(\mathbf{A}^{\mathrm{T}}(\lambda^0 - \bar{\lambda}) + (\mathbf{s}^0 - \bar{\mathbf{s}})),$$

we have

$$\mathbf{A}^{\mathrm{T}}(\dot{\lambda} - \nu_k(\lambda^0 - \bar{\lambda})) + (\dot{\mathbf{s}} - \nu_k(\mathbf{s}^0 - \bar{\mathbf{s}})) = \mathbf{0}.$$

Using Lemma 8.1, we have

$$\mathbf{s}\circ(\dot{\mathbf{x}}-\nu_k(\mathbf{x}^0-\bar{\mathbf{x}}))+\mathbf{x}\circ(\dot{\mathbf{s}}-\nu_k(\mathbf{s}^0-\bar{\mathbf{s}}))=\mathbf{x}\circ\mathbf{s}-\nu_k\mathbf{s}\circ(\mathbf{x}^0-\bar{\mathbf{x}})-\nu_k\mathbf{x}\circ(\mathbf{s}^0-\bar{\mathbf{s}}).$$

Thus, in matrix form, we have

$$\begin{bmatrix}\mathbf{A} & \mathbf{0} & \mathbf{0}\\ \mathbf{0} & \mathbf{A}^{\mathrm{T}} & \mathbf{I}\\ \mathbf{S} & \mathbf{0} & \mathbf{X}\end{bmatrix}\begin{bmatrix}\dot{\mathbf{x}}-\nu_k(\mathbf{x}^0-\bar{\mathbf{x}})\\ \dot{\lambda}-\nu_k(\lambda^0-\bar{\lambda})\\ \dot{\mathbf{s}}-\nu_k(\mathbf{s}^0-\bar{\mathbf{s}})\end{bmatrix}=\begin{bmatrix}\mathbf{0}\\ \mathbf{0}\\ \mathbf{x}\circ\mathbf{s}-\nu_k\mathbf{s}\circ(\mathbf{x}^0-\bar{\mathbf{x}})-\nu_k\mathbf{x}\circ(\mathbf{s}^0-\bar{\mathbf{s}})\end{bmatrix} \tag{8.38}$$

Denote $(\delta\mathbf{x},\delta\lambda,\delta\mathbf{s})=(\dot{\mathbf{x}}-\nu_k(\mathbf{x}^0-\bar{\mathbf{x}}),\dot{\lambda}-\nu_k(\lambda^0-\bar{\lambda}),\dot{\mathbf{s}}-\nu_k(\mathbf{s}^0-\bar{\mathbf{s}}))$ and $\mathbf{r}=\mathbf{r}^1+\mathbf{r}^2+\mathbf{r}^3$ with $(\mathbf{r}^1,\mathbf{r}^2,\mathbf{r}^3)=(\mathbf{x}\circ\mathbf{s},-\nu_k\mathbf{s}\circ(\mathbf{x}^0-\bar{\mathbf{x}}),-\nu_k\mathbf{x}\circ(\mathbf{s}^0-\bar{\mathbf{s}}))$. For $i=1,2,3$, let $(\delta\mathbf{x}^i,\delta\lambda^i,\delta\mathbf{s}^i)$ be the solution of

$$\begin{bmatrix}\mathbf{A} & \mathbf{0} & \mathbf{0}\\ \mathbf{0} & \mathbf{A}^{\mathrm{T}} & \mathbf{I}\\ \mathbf{S} & \mathbf{0} & \mathbf{X}\end{bmatrix}\begin{bmatrix}\delta\mathbf{x}^i\\ \delta\lambda^i\\ \delta\mathbf{s}^i\end{bmatrix}=\begin{bmatrix}\mathbf{0}\\ \mathbf{0}\\ \mathbf{r}^i\end{bmatrix} \tag{8.39}$$

Clearly, we have

$$\delta\mathbf{x}=\delta\mathbf{x}^1+\delta\mathbf{x}^2+\delta\mathbf{x}^3=\dot{\mathbf{x}}-\nu_k(\mathbf{x}^0-\bar{\mathbf{x}}), \tag{8.40a}$$
$$\delta\lambda=\delta\lambda^1+\delta\lambda^2+\delta\lambda^3=\dot{\lambda}-\nu_k(\lambda^0-\bar{\lambda}), \tag{8.40b}$$
$$\delta\mathbf{s}=\delta\mathbf{s}^1+\delta\mathbf{s}^2+\delta\mathbf{s}^3=\dot{\mathbf{s}}-\nu_k(\mathbf{s}^0-\bar{\mathbf{s}}). \tag{8.40c}$$

From the second row of (8.39), we have $(\mathbf{D}^{-1}\delta\mathbf{x}^i)^{\mathrm{T}}(\mathbf{D}\delta\mathbf{s}^i)=0$, for $i=1,2,3$, therefore,

$$\|\mathbf{D}^{-1}\delta\mathbf{x}^i\|^2,\|\mathbf{D}\delta\mathbf{s}^i\|^2\le\|\mathbf{D}^{-1}\delta\mathbf{x}^i\|^2+\|\mathbf{D}\delta\mathbf{s}^i\|^2=\|\mathbf{D}^{-1}\delta\mathbf{x}^i+\mathbf{D}\delta\mathbf{s}^i\|^2. \tag{8.41}$$

Applying $\mathbf{S}\delta\mathbf{x}^i+\mathbf{X}\delta\mathbf{s}^i=\mathbf{r}^i$ to (8.41) for $i=1,2,3$, respectively, we obtain the following relations

$$\|\mathbf{D}^{-1}\delta\mathbf{x}^1\|,\|\mathbf{D}\delta\mathbf{s}^1\|\le\|\mathbf{D}^{-1}\delta\mathbf{x}^1+\mathbf{D}\delta\mathbf{s}^1\|=\|(\mathbf{X}\mathbf{s})^{\frac{1}{2}}\|=\sqrt{\mathbf{x}^{\mathrm{T}}\mathbf{s}}=\sqrt{n\mu}, \tag{8.42a}$$
$$\|\mathbf{D}^{-1}\delta\mathbf{x}^2\|,\|\mathbf{D}\delta\mathbf{s}^2\|\le\|\mathbf{D}^{-1}\delta\mathbf{x}^2+\mathbf{D}\delta\mathbf{s}^2\|=\nu_k\|\mathbf{D}^{-1}(\mathbf{x}^0-\bar{\mathbf{x}})\|, \tag{8.42b}$$
$$\|\mathbf{D}^{-1}\delta\mathbf{x}^3\|,\|\mathbf{D}\delta\mathbf{s}^3\|\le\|\mathbf{D}^{-1}\delta\mathbf{x}^3+D\delta\mathbf{s}^3\|=\nu_k\|\mathbf{D}(\mathbf{s}^0-\bar{\mathbf{s}})\|. \tag{8.42c}$$

Considering (8.39) with $i=2$, we have

$$\mathbf{S}\delta\mathbf{x}^2+\mathbf{X}\delta\mathbf{s}^2=\mathbf{r}^2=-\nu_k\mathbf{S}(\mathbf{x}^0-\bar{\mathbf{x}}),$$

which is equivalent to

$$\delta\mathbf{x}^2=-\nu_k(\mathbf{x}^0-\bar{\mathbf{x}})-\mathbf{D}^2\delta\mathbf{s}^2. \tag{8.43}$$

Thus, from (8.40a), (8.43), and (8.42), we have

$$\|\mathbf{D}^{-1}\dot{\mathbf{x}}\| \quad=\quad \|\mathbf{D}^{-1}[\delta\mathbf{x}^1+\delta\mathbf{x}^2+\delta\mathbf{x}^3+\nu_k(\mathbf{x}^0-\bar{\mathbf{x}})]\|$$

$$
\begin{aligned}
&= \|\mathbf{D}^{-1}\delta\mathbf{x}^1 - \mathbf{D}\delta\mathbf{s}^2 + \mathbf{D}^{-1}\delta\mathbf{x}^3\| \\
&\le \|\mathbf{D}^{-1}\delta\mathbf{x}^1\| + \|\mathbf{D}\delta\mathbf{s}^2\| + \|\mathbf{D}^{-1}\delta\mathbf{x}^3\|.
\end{aligned} \tag{8.44}
$$

Considering (8.39) with $i = 3$, we have

$$
\mathbf{S}\delta\mathbf{x}^3 + \mathbf{X}\delta\mathbf{s}^3 = \mathbf{r}^3 = -\nu_k\mathbf{X}(\mathbf{s}^0 - \bar{\mathbf{s}}),
$$

which is equivalent to

$$
\delta\mathbf{s}^3 = -\nu_k(\mathbf{s}^0 - \bar{\mathbf{s}}) - \mathbf{D}^{-2}\delta\mathbf{x}^3. \tag{8.45}
$$

Thus, from (8.40c), (8.45), and (8.42), we have

$$
\begin{aligned}
\|\mathbf{D}\dot{\mathbf{s}}\| &= \|\mathbf{D}[\delta\mathbf{s}^1 + \delta\mathbf{s}^2 + \delta\mathbf{s}^3 + \nu_k(\mathbf{s}^0 - \bar{\mathbf{s}})]\| \\
&= \|\mathbf{D}\delta\mathbf{s}^1 + \mathbf{D}\delta\mathbf{s}^2 - \mathbf{D}^{-1}\delta\mathbf{x}^3\| \\
&\le \|\mathbf{D}\delta\mathbf{s}^1\| + \|\mathbf{D}\delta\mathbf{s}^2\| + \|\mathbf{D}^{-1}\delta\mathbf{x}^3\|.
\end{aligned} \tag{8.46}
$$

From (8.42a), we can summarize the above discussion as the following (see also [67]) lemma.

Lemma 8.7
Let $(\mathbf{x}^0, \lambda^0, \mathbf{s}^0)$ be the initial point of Algorithm 8.1, and $(\bar{\mathbf{x}}, \bar{\lambda}, \bar{\mathbf{s}}) \in \bar{\mathcal{S}}$ meet Assumption 5. Then,

$$
\|\mathbf{D}\dot{\mathbf{s}}\|, \|\mathbf{D}^{-1}\dot{\mathbf{x}}\| \le \sqrt{n\mu} + \|\mathbf{D}\delta\mathbf{s}^2\| + \|\mathbf{D}^{-1}\delta\mathbf{x}^3\|. \tag{8.47}
$$

Remark 8.3 If the initial point $(\mathbf{x}^0, \lambda^0, \mathbf{s}^0)$ is a feasible point satisfying $\mathbf{A}\mathbf{x}^0 = \mathbf{b}$ and $\mathbf{A}^{\mathrm{T}}\lambda^0 + \mathbf{s}^0 = \mathbf{c}$, then the problem is reduced to a feasible interior-point problem which has been discussed in Chapter 5. In this case, inequality (8.47) is reduced to $\|\mathbf{D}\dot{\mathbf{s}}\|, \|\mathbf{D}^{-1}\dot{\mathbf{x}}\| \le \sqrt{n\mu}$ because $\mathbf{x}^0 = \bar{\mathbf{x}}$, $\mathbf{s}^0 = \bar{\mathbf{s}}$, and $\|\mathbf{D}\delta\mathbf{s}^2\| = \|\mathbf{D}^{-1}\delta\mathbf{x}^3\| = 0$ from (8.42b) and (8.42c). Using $\|\mathbf{D}\dot{\mathbf{s}}\|, \|\mathbf{D}^{-1}\dot{\mathbf{x}}\| \le \sqrt{n\mu}$, we have proved in Chapter 5 that a feasible arc-search algorithm is polynomial with complexity bound $\mathcal{O}(\sqrt{n}\log(1/\varepsilon))$. In the remainder of the chapter, we will focus on the case that the initial point is infeasible. ▮

Lemma 8.8
Let $(\dot{\mathbf{x}}, \dot{\lambda}, \dot{\mathbf{s}})$ be defined in (8.7). Then, there is a constant C_2 independent of n such that in every iteration of Algorithm 8.1, the following inequality holds.

$$
\|\mathbf{D}\dot{\mathbf{s}}\|, \|\mathbf{D}^{-1}\dot{\mathbf{x}}\| \le C_2\sqrt{n\mu}. \tag{8.48}
$$

Proof 8.5 Since $\mathbf{D}$ and $\mathbf{D}^{-1}$ are diagonal matrices, in view of (8.39), we have

$$
(\mathbf{D}\delta\mathbf{s}^2)^{\mathrm{T}}(\mathbf{D}^{-1}\delta\mathbf{x}^3) = (\delta\mathbf{s}^2)^{\mathrm{T}}(\delta\mathbf{x}^3) = 0.
$$

Let $(\mathbf{x}^0, \lambda^0, \mathbf{s}^0)$ be the initial point of Algorithm 8.1, and $(\bar{\mathbf{x}}, \bar{\lambda}, \bar{\mathbf{s}}) \in \bar{\mathcal{S}}$ meet Assumption 5. Then, from (8.42b) and (8.42c), we have

$$\begin{aligned}
& \|\mathbf{D}\delta\mathbf{s}^2 - \mathbf{D}^{-1}\delta\mathbf{x}^3\|^2 \\
= \; & \|\mathbf{D}\delta\mathbf{s}^2\|^2 + \|\mathbf{D}^{-1}\delta\mathbf{x}^3\|^2 \\
= \; & \nu_k^2\|\mathbf{D}^{-1}(\mathbf{x}^0 - \bar{\mathbf{x}})\|^2 + \nu_k^2\|\mathbf{D}(\mathbf{s}^0 - \bar{\mathbf{s}})\|^2 \\
= \; & \nu_k^2 \left[(\mathbf{x}^0 - \bar{\mathbf{x}})^{\mathrm{T}} \operatorname{diag}\left(\frac{s_i^k}{x_i^k}\right) (\mathbf{x}^0 - \bar{\mathbf{x}}) + (\mathbf{s}^0 - \bar{\mathbf{s}})^{\mathrm{T}} \operatorname{diag}\left(\frac{x_i^k}{s_i^k}\right) (\mathbf{s}^0 - \bar{\mathbf{s}}) \right] \\
= \; & \nu_k^2 \left(\mathbf{x}^0 - \bar{\mathbf{x}}, \mathbf{s}^0 - \bar{\mathbf{s}}\right)^{\mathrm{T}} \begin{bmatrix} \operatorname{diag}\left(\frac{s_i^k}{x_i^k}\right) & \mathbf{0} \\ \mathbf{0} & \operatorname{diag}\left(\frac{x_i^k}{s_i^k}\right) \end{bmatrix} (\mathbf{x}^0 - \bar{\mathbf{x}}, \mathbf{s}^0 - \bar{\mathbf{s}}) \\
\leq \; & \nu_k^2 \max_i \left\{ \frac{s_i^k}{x_i^k}, \frac{x_i^k}{s_i^k} \right\} \| \left(\mathbf{x}^0 - \bar{\mathbf{x}}, \mathbf{s}^0 - \bar{\mathbf{s}}\right) \|^2 \\
= \; & \nu_k^2 \max_i \{D_{ii}^{-2}, D_{ii}^2\} \| \left(\mathbf{x}^0 - \bar{\mathbf{x}}, \mathbf{s}^0 - \bar{\mathbf{s}}\right) \|^2 \\
\leq \; & C_1^2 n\mu_k \| \left(\mathbf{x}^0 - \bar{\mathbf{x}}, \mathbf{s}^0 - \bar{\mathbf{s}}\right) \|^2, \qquad (8.49)
\end{aligned}$$

where the last inequality follows from Lemma 8.6. Adjoining this result with Lemma 8.7 and Assumption 5 gives

$$\|\mathbf{D}\dot{\mathbf{s}}\|, \|\mathbf{D}^{-1}\dot{\mathbf{x}}\| \leq \sqrt{n\mu_k} + \| \left(\mathbf{x}^0 - \bar{\mathbf{x}}, \mathbf{s}^0 - \bar{\mathbf{s}}\right) \| C_1\sqrt{n\mu_k} \leq C_2\sqrt{n\mu_k}.$$

This completes the proof. ■

From Lemma 8.8, we can obtain several inequalities that will be used in our convergence analysis. The first one is given as follows.

Lemma 8.9
Let $(\dot{\mathbf{x}}, \dot{\lambda}, \dot{\mathbf{s}})$ and $(\ddot{\mathbf{x}}, \ddot{\lambda}, \ddot{\mathbf{s}})$ be defined in (8.7) and (8.8). Then, there exists a constant $C_3 > 0$ independent of n such that the following relations hold.

$$\|\mathbf{D}^{-1}\ddot{\mathbf{x}}\|, \|\mathbf{D}\ddot{\mathbf{s}}\| \leq C_3 n\mu_k^{0.5}, \qquad (8.50a)$$

$$\|\mathbf{D}^{-1}\mathbf{p}_x\|, \|\mathbf{D}\mathbf{p}_s\| \leq \sqrt{\frac{n}{\theta}}\mu_k^{0.5}, \qquad (8.50b)$$

$$\|\mathbf{D}^{-1}\mathbf{q}_x\|, \|\mathbf{D}\mathbf{q}_s\| \leq \frac{2C_2^2}{\sqrt{\theta}} n\mu_k^{0.5}. \qquad (8.50c)$$

Proof 8.6 From the last row of (8.20), using the facts that $\mathbf{q}_x^{\mathrm{T}}\mathbf{q}_s = 0$, $x_i^k s_i^k > \theta\mu_k$, and Lemma 8.8, we have

$$\begin{aligned}
& \mathbf{S}\mathbf{q}_x + \mathbf{X}\mathbf{q}_s = -2\dot{\mathbf{x}} \circ \dot{\mathbf{s}} \\
\Longleftrightarrow \quad & \mathbf{D}^{-1}\mathbf{q}_x + \mathbf{D}\mathbf{q}_s = 2(\mathbf{X}\mathbf{S})^{-0.5}(-\dot{\mathbf{x}} \circ \dot{\mathbf{s}}) = 2(\mathbf{X}\mathbf{S})^{-0.5}(-\mathbf{D}^{-1}\dot{\mathbf{x}} \circ \mathbf{D}\dot{\mathbf{s}})
\end{aligned}$$

$$\begin{aligned}
\Longrightarrow \quad & ||\mathbf{D}^{-1}\mathbf{q}_x||^2, ||\mathbf{D}\mathbf{q}_s||^2 \le ||\mathbf{D}^{-1}\mathbf{q}_x||^2 + ||\mathbf{D}\mathbf{q}_s||^2 = ||\mathbf{D}^{-1}\mathbf{q}_x + \mathbf{D}\mathbf{q}_s||^2 \\
\le \quad & 4||(\mathbf{XS})^{-0.5}||^2 \left(||\mathbf{D}^{-1}\dot{\mathbf{x}}|| \cdot ||\mathbf{D}\dot{\mathbf{s}}||\right)^2 \\
\le \quad & \frac{4}{\theta\mu_k}\left(C_2^2 n\mu_k\right)^2 = \frac{(2C_2^2 n)^2}{\theta}\mu_k.
\end{aligned}$$

Taking the square root on both sides gives

$$||\mathbf{D}^{-1}\mathbf{q}_x||, ||\mathbf{D}\mathbf{q}_s|| \le \frac{2C_2^2}{\sqrt{\theta}} n\sqrt{\mu_k}. \tag{8.51}$$

From the last row of (8.19), using the facts that $\mathbf{p}_x^{\mathrm{T}}\mathbf{p}_s = 0$ and $x_i^k s_i^k \ge \theta\mu_k$, we have

$$\begin{aligned}
& \mathbf{S}\mathbf{p}_x + \mathbf{X}\mathbf{p}_s = \mu_k \mathbf{e} \\
\Longleftrightarrow \quad & \mathbf{D}^{-1}\mathbf{p}_x + \mathbf{D}\mathbf{p}_s = (\mathbf{XS})^{-0.5}\mu_k \mathbf{e} \\
\Longrightarrow \quad & ||\mathbf{D}^{-1}\mathbf{p}_x||^2, ||\mathbf{D}\mathbf{p}_s||^2 \le ||\mathbf{D}^{-1}\mathbf{p}_x||^2 + ||\mathbf{D}\mathbf{p}_s||^2 \\
& = ||\mathbf{D}^{-1}\mathbf{p}_x + \mathbf{D}\mathbf{p}_s||^2 \le ||(\mathbf{XS})^{-0.5}||^2 n(\mu_k)^2 \\
& \le n\mu_k/\theta.
\end{aligned}$$

Taking the square root on both sides gives

$$||\mathbf{D}^{-1}\mathbf{p}_x||, ||\mathbf{D}\mathbf{p}_s|| \le \sqrt{\frac{n}{\theta}}\sqrt{\mu_k}. \tag{8.52}$$

Combining (8.51) and (8.52) proves (8.50a). ■

The following inequalities are direct results of Lemmas 8.8 and 8.9.

Lemma 8.10
Let $(\dot{\mathbf{x}}, \dot{\lambda}, \dot{\mathbf{s}})$ and $(\ddot{\mathbf{x}}, \ddot{\lambda}, \ddot{\mathbf{s}})$ be defined in (8.7) and (8.8). Then, the following relations hold.

$$\frac{|\dot{\mathbf{x}}^{\mathrm{T}}\dot{\mathbf{s}}|}{n} \le C_2^2\mu_k, \quad \frac{|\ddot{\mathbf{x}}^{\mathrm{T}}\dot{\mathbf{s}}|}{n} \le C_2C_3\sqrt{n}\mu_k, \quad \frac{|\dot{\mathbf{x}}^{\mathrm{T}}\ddot{\mathbf{s}}|}{n} \le C_2C_3\sqrt{n}\mu_k. \tag{8.53}$$

Moreover,

$$|\dot{x}_i\dot{s}_i| \le C_2^2 n\mu_k, \quad |\ddot{x}_i\dot{s}_i| \le C_2C_3 n^{\frac{3}{2}}\mu_k, \quad |\dot{x}_i\ddot{s}_i| \le C_2C_3 n^{\frac{3}{2}}\mu_k, \quad |\ddot{x}_i\ddot{s}_i| \le C_3^2 n^2\mu_k. \tag{8.54}$$

Proof 8.7 The first relation of (8.53) is given as follows.

$$\frac{|\dot{\mathbf{x}}^{\mathrm{T}}\dot{\mathbf{s}}|}{n} = \frac{|(\mathbf{D}^{-1}\dot{\mathbf{x}})^{\mathrm{T}}(\mathbf{D}\dot{\mathbf{s}})|}{n} \le \frac{||\mathbf{D}^{-1}\dot{\mathbf{x}}|| \cdot ||\mathbf{D}\dot{\mathbf{s}}||}{n} \le C_2^2\mu_k. \tag{8.55}$$

Similarly, we have

$$\frac{|\ddot{\mathbf{x}}^{\mathrm{T}}\dot{\mathbf{s}}|}{n} = \frac{|(\mathbf{D}^{-1}\ddot{\mathbf{x}})^{\mathrm{T}}(\mathbf{D}\dot{\mathbf{s}})|}{n} \le \frac{||\mathbf{D}^{-1}\ddot{\mathbf{x}}|| \cdot ||\mathbf{D}\dot{\mathbf{s}}||}{n} \le C_2C_3\sqrt{n}\mu_k, \tag{8.56}$$

and

$$\frac{|\dot{\mathbf{x}}^{\mathrm{T}}\ddot{\mathbf{s}}|}{n} = \frac{|(\mathbf{D}^{-1}\dot{\mathbf{x}})^{\mathrm{T}}(\mathbf{D}\ddot{\mathbf{s}})|}{n} \leq \frac{\|\mathbf{D}^{-1}\dot{\mathbf{x}}\| \cdot \|\mathbf{D}\ddot{\mathbf{s}}\|}{n} \leq C_2C_3\sqrt{n}\mu_k. \tag{8.57}$$

The first relation of (8.54) is given as follows.

$$|\dot{x}_i\dot{s}_i| = |D_{ii}^{-1}\dot{x}_iD_{ii}\dot{s}_i| \leq |D_{ii}^{-1}\dot{x}_i| \cdot |D_{ii}\dot{s}_i| \leq \|\mathbf{D}^{-1}\dot{\mathbf{x}}\| \cdot \|\mathbf{D}\dot{\mathbf{s}}\| \leq C_2^2 n\mu_k. \tag{8.58}$$

Similar arguments can be used for the remaining inequalities of (8.54). ■

Now, we are ready to show that there exists a constant $\kappa_0 = \mathcal{O}(n^{\frac{3}{2}})$ such that for every iteration, for some $\sigma_k \in [\sigma_{\min}, \sigma_{\max}]$ and all $\sin(\alpha_k) \in (0, \frac{1}{\kappa_0}]$, all conditions in Step 2 of Algorithm 8.1 hold.

Lemma 8.11

There exists a positive constant C_4 independent of n and an $\bar{\alpha}$ defined by $\sin(\bar{\alpha}) \geq \frac{C_4}{\sqrt{n}}$ such that for $\forall k \geq 0$ and $\sin(\alpha_k) \in (0, \sin(\bar{\alpha})]$,

$$(x_i^{k+1}, s_i^{k+1}) := (x_i(\sigma_k, \alpha_k), s_i(\sigma_k, \alpha_k)) \geq (\phi_k, \psi_k) > \mathbf{0} \tag{8.59}$$

holds.

Proof 8.8 From (8.30) and (8.29a), a conservative estimation can be obtained by

$$x_i(\sigma_k, \alpha_k) = x_i^k - \dot{x}_i \sin(\alpha_k) + \ddot{x}_i(1 - \cos(\alpha_k)) \geq \rho x_i^k$$

which is equivalent to

$$x_i^k(1-\rho) - \dot{x}_i \sin(\alpha_k) + \ddot{x}_i(1 - \cos(\alpha_k)) \geq 0.$$

Multiplying D_{ii}^{-1} to this inequality and using Lemmas 8.8, 8.10, and 5.4, we have

$$\begin{aligned}
&(x_i^k s_i^k)^{0.5}(1-\rho) - D_{ii}^{-1}\dot{x}_i \sin(\alpha_k) + D_{ii}^{-1}\ddot{x}_i(1 - \cos(\alpha_k)) \\
&\geq \sqrt{\mu_k}\left(\sqrt{\theta}(1-\rho) - C_2\sqrt{n}\sin(\alpha_k) - C_3 n \sin^2(\alpha_k)\right).
\end{aligned}$$

Clearly, the last expression is greater than zero for all $\sin(\alpha_k) \leq \frac{C_4}{\sqrt{n}} \leq \sin(\bar{\alpha})$, where $C_4 = \frac{\sqrt{\theta}(1-\rho)}{2\max\{C_2, \sqrt{C_3}\}}$. This proves $x_i^{k+1} \geq \phi_k > 0$. Similarly, we have $s_i^{k+1} \geq \psi_k > 0$. This completes the proof. ■

Lemma 8.12

There exists a positive constant C_5 independent of n and an $\hat{\alpha}$ defined by $\sin(\hat{\alpha}) \geq \frac{C_5}{n^{\frac{1}{4}}}$ such that for $\forall k \geq 0$ and $\sin(\alpha) \in (0, \sin(\hat{\alpha})]$, the following relation

$$\mu_k(\sigma_k,\alpha_k) \le \mu_k\left(1-\frac{\sin(\alpha_k)}{4}\right) \le \mu_k\left(1-\frac{C_5}{4n^{\frac{1}{4}}}\right) \tag{8.60}$$

holds.

Proof 8.9 Using (8.26), Lemmas 5.4 and 8.10, we have

$$\begin{aligned}
\mu(\sigma_k,\alpha_k) &= \mu_k(1-\sin(\alpha_k))+\sigma_k\mu_k(1-\cos(\alpha_k))-\frac{\dot{\mathbf{x}}^{\mathrm{T}}\dot{\mathbf{s}}}{n}(1-\cos(\alpha_k))^2 \\
&\quad -\frac{\dot{\mathbf{x}}^{\mathrm{T}}\ddot{\mathbf{s}}+\dot{\mathbf{s}}^{\mathrm{T}}\ddot{\mathbf{x}}}{n}\sin(\alpha_k)(1-\cos(\alpha_k)) \\
&\le \mu_k\left[1-\sin(\alpha_k)+\sigma_k\sin^2(\alpha_k)\right] \\
&\quad +\left[\frac{|\dot{\mathbf{x}}^{\mathrm{T}}\dot{\mathbf{s}}|}{n}\sin^4(\alpha_k)+\left(\frac{|\dot{\mathbf{x}}^{\mathrm{T}}\ddot{\mathbf{s}}|}{n}+\frac{|\ddot{\mathbf{x}}^{\mathrm{T}}\dot{\mathbf{s}}|}{n}\right)\sin^3(\alpha_k)\right] \\
&\le \mu_k\left[1-\sin(\alpha_k)+\sigma_k\sin^2(\alpha_k)+C_2^2\sin^4(\alpha_k)+2C_2C_3\sqrt{n}\sin^3(\alpha_k)\right] \\
&= \mu_k\left[1-\sin(\alpha_k)\left(1-\sigma_k\sin(\alpha_k)-C_2^2\sin^3(\alpha_k)-2C_2C_3\sqrt{n}\sin^2(\alpha_k)\right)\right].
\end{aligned}$$

Let

$$C_5 = \frac{1}{4\max\{\sigma_k, C_2^{2/3}, \sqrt{2C_2C_3}\}}.$$

then, for all $\sin(\alpha_k)\in(0,\sin(\hat{\alpha})]$ and $\sigma_{\min} < \sigma_k \le \sigma_{\max}$, inequality (8.60) holds. This completes the proof. ■

Lemma 8.13

There exists a positive constant C_6 independent of n and an $\check{\alpha}$ defined by $\sin(\check{\alpha}) \ge \frac{C_6}{n^{\frac{3}{2}}}$ such that if $x_i^k s_i^k \ge \theta\mu_k$ holds, then for $\forall k\ge 0$, $\forall i\in\{1,\ldots,n\}$, and $\sin(\alpha)\in(0,\sin(\check{\alpha})]$, the following relation

$$x_i^{k+1}s_i^{k+1} \ge \theta\mu_{k+1} \tag{8.61}$$

holds.

Proof 8.10 Using (8.34) and (8.26), we have

$$\begin{aligned}
& x_i^{k+1}s_i^{k+1}-\theta\mu_{k+1} \\
\ge\ & \sigma_k\mu_k(1-\theta)(1-\cos(\alpha_k)) \\
& -\left[\ddot{x}_i\dot{s}_i+\dot{x}_i\ddot{s}_i-\frac{\theta(\dot{\mathbf{x}}^{\mathrm{T}}\ddot{\mathbf{s}}+\dot{\mathbf{s}}^{\mathrm{T}}\ddot{\mathbf{x}})}{n}\right]\sin(\alpha_k)(1-\cos(\alpha_k)) \\
& +\left[\ddot{x}_i\ddot{s}_i-\dot{x}_i\dot{s}_i+\frac{\theta(\dot{\mathbf{x}}^{\mathrm{T}}\dot{\mathbf{s}})}{n}\right](1-\cos(\alpha_k))^2
\end{aligned} \tag{8.62}$$

Therefore, if

$$\begin{aligned} \sigma_k\mu_k(1-\theta) &- \left[\ddot{x}_i\dot{s}_i+\dot{x}_i\ddot{s}_i-\frac{\theta(\dot{\mathbf{x}}^{\mathrm{T}}\ddot{\mathbf{s}}+\dot{\mathbf{s}}^{\mathrm{T}}\ddot{\mathbf{x}})}{n}\right]\sin(\alpha_k) \\ &+ \left[\ddot{x}_i\ddot{s}_i-\dot{x}_i\dot{s}_i+\frac{\theta(\dot{\mathbf{x}}^{\mathrm{T}}\dot{\mathbf{s}})}{n}\right](1-\cos(\alpha_k))\geq 0, \end{aligned} \tag{8.63}$$

then

$$x_i^{k+1}s_i^{k+1}\geq\theta\mu_{k+1}.$$

The inequality (8.63) holds if

$$\begin{aligned} \sigma_k\mu_k(1-\theta) &- \left(|\ddot{x}_i\dot{s}_i|+|\dot{x}_i\ddot{s}_i|+\left|\frac{\theta(\dot{\mathbf{x}}^{\mathrm{T}}\ddot{\mathbf{s}}+\dot{\mathbf{s}}^{\mathrm{T}}\ddot{\mathbf{x}})}{n}\right|\right)\sin(\alpha_k) \\ &- \left(|\ddot{x}_i\ddot{s}_i|+|\dot{x}_i\dot{s}_i|+\left|\frac{\theta(\dot{\mathbf{x}}^{\mathrm{T}}\dot{\mathbf{s}})}{n}\right|\right)\sin^2(\alpha_k)\geq 0. \end{aligned}$$

Using Lemma 8.10, we can easily find some $\breve{\alpha}$ defined by $\sin(\breve{\alpha})\geq\frac{C_6}{n^{\frac{3}{2}}}$ to meet the above inequality. ■

Now, the convergence result follows from the standard argument using Theorem 1.4, which is restated below for convenience.

Theorem 8.2

Let $\varepsilon\in(0,1)$ be given. Suppose that an algorithm generates a sequence of iterations $\{\chi_k\}$ that satisfies

$$\chi_{k+1}\leq\left(1-\frac{\delta}{n^{\omega}}\right)\chi_k,\quad k=0,1,2,\ldots, \tag{8.64}$$

for some positive constants δ and ω. Then, there exists an index K with

$$K=\mathcal{O}(n^{\omega}\log(\chi_0/\varepsilon))$$

such that

$$\chi_k\leq\varepsilon\quad\text{for }\ \forall k\geq K.$$

In view of Lemmas 8.3, 8.11, 8.12, 8.13, and Theorem 8.2, we can state our main result as the following:

Theorem 8.3

Algorithms 8.1 is a polynomial algorithm with polynomial complexity bound of $\mathcal{O}(n^{\frac{3}{2}}\max\{\log((\mathbf{x}^0)^{\mathrm{T}}\mathbf{s}^0/\varepsilon),\log(\|\mathbf{r}_b^0\|/\varepsilon),\log(\|\mathbf{r}_c^0\|/\varepsilon)\})$.

8.3 The $\mathcal{O}(\sqrt{n}L)$ Algorithm

Algorithm 8.2 is a simplified version of Algorithm 8.1. The only difference between the two algorithms is that this algorithm searches for optimizers in a larger neighborhood defined by (8.6b). From the discussion in Section 8.1, the following arc-search infeasible interior-point algorithm is well defined.

Algorithm 8.2

Data: $\mathbf{A}$, $\mathbf{b}$, $\mathbf{c}$.
Parameter: $\varepsilon \in (0,1)$, $\sigma_{\min} \in (0,1)$, $\sigma_{\max} \in (0,1)$, *and* $\rho \in (0,1)$.
Initial point: $\lambda^0 = 0$ *and* $(\mathbf{x}^0, \mathbf{s}^0) > 0$.
for *iteration* $k = 0, 1, 2, \ldots$

Step 0: If $\|\mathbf{r}_b^k\| \leq \varepsilon$, $\|\mathbf{r}_c^k\| \leq \varepsilon$, *and* $\mu_k \leq \varepsilon$, *stop.*

Step 1: Calculate μ_k, $\mathbf{r}_b^k$, $\mathbf{r}_c^k$, $\dot{\lambda}$, $\dot{\mathbf{s}}$, $\dot{\mathbf{x}}$, $\mathbf{p}_x^k$, $\mathbf{p}_\lambda^k$, $\mathbf{p}_s^k$, $\mathbf{q}_x^k$, $\mathbf{q}_\lambda^k$, *and* $\mathbf{q}_s^k$.

Step 2: Find some appropriate $\alpha_k \in (0, \pi/2]$ *and* $\sigma_k \in [\sigma_{\min}, \sigma_{\max}]$ *to satisfy*

$$\mathbf{x}(\sigma_k, \alpha_k) \geq \phi_k \mathbf{e}, \quad \mathbf{s}(\sigma_k, \alpha_k) \geq \psi_k \mathbf{e}, \quad \mu_k > \mu(\sigma_k, \alpha_k). \tag{8.65}$$

Step 3: Set $(\mathbf{x}^{k+1}, \lambda^{k+1}, \mathbf{s}^{k+1}) = (\mathbf{x}(\sigma_k, \alpha_k), \lambda(\sigma_k, \alpha_k), \mathbf{s}(\sigma_k, \alpha_k))$ *and* $\mu_{k+1} = \mu(\sigma_k, \alpha_k)$.

Step 4: Set $k+1 \to k$. *Go back to Step 1.*

end (for)

Remark 8.4 It is clear that the only difference between Algorithm 8.1 and Algorithm 8.2 is in Step 2, where Algorithm 8.2 does not require $x_i(\sigma_k, \alpha_k)s_i(\sigma_k, \alpha_k) \geq \theta\mu(\sigma_k, \alpha_k)$. We have seen from Lemma 8.13 that this requirement is the main barrier in achieving a better polynomial bound. ▮

Denote

$$\gamma = \min\{1, \rho\beta\}. \tag{8.66}$$

Lemma 8.14

Assume that Algorithms 8.2 terminates in finite iterations K and for some constant C, $\sin(\alpha_k) \leq \frac{C}{n^{1/4}}$ *holds. If* $\mu_0 > 1$, *then there is a positive constant* $\theta = \min_{k \in K}\{\frac{\gamma^2 \nu_k^2}{\mu_k}\}$ *such that* $\theta < 1$.

Proof 8.11 First , Assumption 3 implies that μ_k is bounded below and away from zero before the convergence. Assumption 4 indicates that $\nu_k > 0$. Since K is finite and $\gamma > 0$ is a constant, θ is an attainable positive constant. Setting $k=0$ in (8.23) yields $\nu_0 = 1$. This means that $\gamma^2\nu_0^2 \le 1$. Since $\mu_0 > 1$, we have $\theta \le \frac{\gamma^2\nu_0^2}{\mu_0} \le \frac{\nu_0^2}{\mu_0} < 1$. This verifies that the claim holds for $k=0$. Assume that $\frac{\nu_k^2}{\mu_k} < 1$ is true for $k \ge 0$, we show that the claim is true for $k+1$. In view of (8.26) and Lemmas 8.10 and 5.4, we have

$$
\begin{aligned}
\mu_{k+1} \ \ge\ & \mu_k(1-\sin(\alpha_k)) + \sigma_k\mu_k(1-\cos(\alpha_k)) - \left|\frac{\dot{\mathbf{x}}^{\mathrm{T}}\dot{\mathbf{s}}}{n}\right|(1-\cos(\alpha_k))^2 \\
& - \left|\frac{\dot{\mathbf{x}}^{\mathrm{T}}\ddot{\mathbf{s}} + \ddot{\mathbf{x}}^{\mathrm{T}}\dot{\mathbf{s}}}{n}\right| \sin(\alpha_k)(1-\cos(\alpha_k)) \\
\ge\ & \mu_k(1-\sin(\alpha_k)) + \frac{1}{2}\sigma_k\mu_k\sin^2(\alpha_k) - C_2^2\mu_k\sin^4(\alpha_k) \\
& -2C_2C_3\sqrt{n}\mu_k\sin^3(\alpha_k).
\end{aligned}
\tag{8.67}
$$

Therefore, using assumption of $\frac{\nu_k^2}{\mu_k} < 1$, we have

$$
\begin{aligned}
\frac{\nu_{k+1}^2}{\mu_{k+1}} &\le \frac{\nu_k^2(1-\sin(\alpha_k))^2}{\mu_k[1-\sin(\alpha_k)+\frac{1}{2}\sigma_k\sin^2(\alpha_k) - C_2^2\sin^4(\alpha_k) - 2C_2C_3\sqrt{n}\sin^3(\alpha_k)]} \\
&< \frac{1-2\sin(\alpha_k)+\sin^2(\alpha_k)}{1-\sin(\alpha_k)+\frac{1}{2}\sigma_k\sin^2(\alpha_k) - C_2^2\sin^4(\alpha_k) - 2C_2C_3\sqrt{n}\sin^3(\alpha_k)}.
\end{aligned}
\tag{8.68}
$$

To show that $\frac{\nu_{k+1}^2}{\mu_{k+1}} < 1$, we just need to show

$$
-\sin(\alpha_k)+\sin^2(\alpha_k) \le \frac{1}{2}\sigma_k\sin^2(\alpha_k) - C_2^2\sin^4(\alpha_k) - 2C_2C_3\sqrt{n}\sin^3(\alpha_k)
$$

or

$$
(1-\frac{1}{2}\sigma_k)\sin(\alpha_k) + 2C_2C_3\sqrt{n}\sin^2(\alpha_k) + C_2^2\sin^3(\alpha_k) \le 1.
$$

Clearly, there is a positive constant C and $\sin(\alpha_k) \le \frac{C}{n^{1/4}}$ such that the last inequality holds. This proves that $\frac{\nu_{k+1}^2}{\mu_{k+1}} < 1$, hence, $\theta = \min_{k\in K}\{\frac{\gamma^2\nu_k^2}{\mu_k}\} < 1$. ■

Lemma 8.15

If $\{\frac{\nu_k^2}{\mu_k}\} > 0$ is bounded below and away from zero before the convergence of Algorithm 8.2 and $\beta > 0$, then, for a nonnegative constant $\theta \ge 0$, the inequality $x_i^k s_i^k \ge \theta\mu_k$ holds for $\forall i \in \{1,2,\ldots,n\}$ and $\forall k \ge 0$.

Proof 8.12 First, Assumption 3 implies that μ_k is bounded below and away from zero before the convergence. Assumption 4 indicates that $\nu_k > 0$. By the definition of β, we have $\beta \le \frac{\underline{x}^k}{\nu_k} \le \frac{x_i^k}{\nu_k}$ and $\beta \le \frac{\underline{s}^k}{\nu_k} \le \frac{s_i^k}{\nu_k}$, which can be written as

$$
\underline{x}^k \ge \beta\nu_k > 0, \qquad \underline{s}^k \ge \beta\nu_k > 0.
$$

Since $\phi_k = \min\{\rho \underline{x}^k, \nu_k\}$, we have either $\phi_k = \rho \underline{x}^k \geq \rho\beta\nu_k$ or $\phi_k = \nu_k$, which means that

$$\phi_k \geq \min\{1, \rho\beta\}\nu_k = \gamma\,\nu_k. \tag{8.69}$$

Since $\psi_k = \min\{\rho \underline{s}^k, \nu_k\}$, with a similar argument, we can show

$$\psi_k \geq \min\{1, \rho\beta\}\nu_k = \gamma\,\nu_k. \tag{8.70}$$

Using (8.30) and (8.31), the definition of ϕ_k and ψ_k, and the above two formulas, we have

$$x_i^k s_i^k \geq \phi_{k-1}\psi_{k-1} \geq \gamma^2\nu_{k-1}^2 > \gamma^2\nu_{k-1}^2(1-\sin(\alpha_{k-1}))^2 = \gamma^2\nu_k^2 > 0.$$

Let $\theta = \inf_k\{\frac{\gamma^2\nu_k^2}{\mu_k}\}$, then, $\gamma^2\{\frac{\nu_k^2}{\mu_k}\} \geq \gamma^2 \inf_k\{\frac{\nu_k^2}{\mu_k}\} = \theta \geq 0$, and we have

$$x_i^k s_i^k \geq \frac{\gamma^2\nu_k^2}{\mu_k}\mu_k \geq \theta\mu_k.$$

This completes the proof. ■

Since $\nu_k > 0$ for all iteration k (see Assumption 4 and Section 8.4.10), we immediately have the following corollary.

Corollary 8.1

If Algorithm 8.2 terminates in finite iterations K, then we have $\theta = \min_{k \leq K}\{\frac{\gamma^2\nu_k^2}{\mu_k}\} > 0$, *where θ is a positive constant independent of n; in addition, $x_i^k s_i^k \geq \theta\mu_k$ holds for $\forall i \in \{1,2,\ldots,n\}$ and for $0 \leq k \leq K$.*

Proposition 8.5

Assume that $\|\mathbf{r}_b^0\|$, $\|\mathbf{r}_c^0\|$, and μ_0 are all finite. Then, Algorithm 8.2 terminates in finite iterations.

Proof 8.13 Since $\|\mathbf{r}_b^0\|$, $\|\mathbf{r}_c^0\|$, and μ_0 are finite, in view of Proposition 8.4, in every iteration, these variables decrease at least a constant. Therefore, the algorithm will terminate in finite steps. ■

Proposition 8.5 implies that $x_i^k s_i^k \geq \theta\mu_k$ holds for a positive constant $\theta, \forall i \in \{1,2,\ldots,n\}$ and for $0 \leq k \leq K$. Since $x_i^k s_i^k \geq \theta\mu_k$ is the most strict condition required in Lemma 8.13 ($\sin(\check{\alpha}) \geq \frac{C_6}{n^{\frac{3}{2}}}$ is smaller than both $\sin(\hat{\alpha}) \geq \frac{C_5}{n^{\frac{1}{4}}}$ and $\sin(\bar{\alpha}) \geq \frac{C_4}{\sqrt{n}}$) to show the polynormiality of Algorithm 8.1, and since Algorithm 8.2 does not check the condition $x_i^k s_i^k \geq \theta\mu_k$, and it checks only (8.65), from Lemmas 8.11, 8.12, and Theorem 8.2, we conclude

Theorem 8.4
If Algorithm 8.2 terminates in finite iterations, it converges in a number of iterations bounded by a polynomial of the order

$$\mathcal{O}(n^{\frac{1}{2}} \max\{\log((\mathbf{x}^0)^{\mathrm{T}}\mathbf{s}^0/\varepsilon), \log(\|\mathbf{r}_b^0\|/\varepsilon), \log(\|\mathbf{r}_c^0\|/\varepsilon)\}).$$

8.4 Implementation Details

The two proposed algorithms are very similar to the one in Chapter 7, in that they all solve standard form linear programming using arc-search techniques. Many implementation details for the four algorithms (two proposed in this chapter, one proposed in Chapter 7, and Mehrotra's algorithm described in Chapter 4) are in common. However, some algorithm-specific parameters (Section 8.5.1), selection of α_k (Section 8.5.8 for arc-search), section of σ_k (Section 8.5.9), and rescale α_k (Section 8.5.10) for the two algorithms proposed in this chapter are different from the ones discussed in Chapter 7. Most implementation details have been thoroughly discussed in Chapter 7. Since all these details affect the numerical efficiency, we summarize all details implemented for the algorithms discussed in this chapter and explain the reasons why some implementations in Chapter 7 are adopted and some are not.

8.4.1 Default parameters

Several parameters are used in Algorithms 8.1 and 8.2. In our implementation, the following defaults are used without a serious effort to optimize the results for all test problems: $\theta = \min\{10^{-6}, 0.1 * \min\{\mathbf{x}^0 \circ \mathbf{s}^0\}/\mu_0\}$, $\sigma_{\min} = 10^{-6}$, $\sigma_{\max} = 0.4$ for Algorithm 8.1, $\sigma_{\max} = 0.3$ for Algorithm 8.2, $\rho = 0.01$, and $\varepsilon = 10^{-8}$. Note that θ is used only in Algorithm 8.1 and this θ selection guarantees $x_i^0 s_i^0 \geq \theta\mu_0$.

8.4.2 Initial point selection

Initial point selection has been known to be an important factor in the computational efficiency for most infeasible interior-point algorithms [23, 172]. We use the methods proposed in [89, 83] to generate candidate initial points. We then compare

$$\max\{\|\mathbf{A}\mathbf{x}^0 - \mathbf{b}\|, \|\mathbf{A}^{\mathrm{T}}\lambda^0 + \mathbf{s}^0 - \mathbf{c}\|, \mu_0\} \tag{8.71}$$

obtained by these two methods and select the initial point with smaller value of (8.71) as we guess this selection may reduce the number of iterations (see Chapter 7 for detail).

8.4.3 Pre-process and post-process

Pre-solver strategies for the standard linear programming problems represented in the form of (1.4) and solved in normal equations were fully investigated in Chapter 7. Five of them were determined to be effective and efficient in application. The same set of the pre-solver is used for algorithms described in this chapter. The post-process is also the same as in Chapter 7.

8.4.4 Matrix scaling

Based on the test and analysis of Chapter 7, it is determined that matrix scaling does not improve efficiency in general. Therefore, we will not use scaling in the implementation. However, the ratio

$$\frac{\max |A_{i,j}|}{\min \{|A_{k,l}| \ : \ A_{k,l} \neq 0\}} \tag{8.72}$$

is used to determine if pre-process rule 9 of Chapter 7 is used.

8.4.5 Removing row dependency from A

Removing row dependency from **A** is studied in [6], Andersen reported an efficient method that removes row dependency of **A**. Based on the study in Chapter 7, we choose to not use this function unless we feel it is necessary when it is used as part of handling degenerate solutions, discussed below. To have a fair comparison of all tested algorithms, we will clarify in Section 8.5 which algorithms and/or problems use this function and which algorithms and/or problems do not use this function.

8.4.6 Linear algebra for sparse Cholesky matrix

Similar to Mehrotra's algorithm, the majority of the computational cost of the proposed algorithms is related to solving sparse Cholesky systems (8.14) and (8.18), which can be expressed as an abstract problem as follows.

$$\mathbf{AD}^2\mathbf{A}^{\mathrm{T}}\mathbf{u} = \mathbf{L}\Lambda\mathbf{L}^{\mathrm{T}}\mathbf{u} = \mathbf{v}, \tag{8.73}$$

where $\mathbf{D} = \mathbf{X}^{\frac{1}{2}}\mathbf{S}^{-\frac{1}{2}}$ is identical in (8.14) and (8.18), $\mathbf{L}$ is a lower triangle matrix, Λ is a diagonal matrix; but $\mathbf{u}$ and $\mathbf{v}$ are different vectors. Many popular LP solvers [23, 172] call a software package [104] which uses some linear algebra specifically developed for the ill-conditioned sparse Cholesky decomposition [77]. However, MATLAB® has not yet implemented the function with the features for ill-conditioned matrices. We implement the same method as in Chapter 7.

8.4.7 Handling degenerate solutions

Difficulty caused by degenerate solutions in interior-point methods for linear programming has been an issue for a long time [48]. Similar observation was also reported in [44]. In our implementation, we have an option for handling degenerate solutions, as described in Chapter 7.

8.4.8 Analytic solution of α_k

Given σ_k, we know that α_k can be calculated in analytic form [151, 157]. Since Algorithms 8.1 and 8.2 are slightly different from the ones in Chapter 7, the formulas to calculate α_k are slightly different too. We provide these formulas without giving proofs (the proof is very similar to the ones in Chapter 7). For each $i \in \{1,\dots,n\}$, we can select the largest α_{x_i} such that for any $\alpha \in [0, \alpha_{x_i}]$, the ith inequality of (8.30) holds, and the largest α_{s_i} such that for any $\alpha \in [0, \alpha_{s_i}]$ the ith inequality of (8.31) holds. We then define

$$\alpha^x = \min_{i\in\{1,\dots,n\}} \{\alpha_{x_i}\}, \tag{8.74}$$

$$\alpha^s = \min_{i\in\{1,\dots,n\}} \{\alpha_{s_i}\}, \tag{8.75}$$

$$\alpha_k = \min\{\alpha^x, \alpha^s\}, \tag{8.76}$$

where α_{x_i} and α_{s_i} can be obtained (using a similar argument as in Chapter 7) in analytical forms represented by ϕ_k, $\dot{\mathbf{x}}_i$, $\ddot{\mathbf{x}}_i$ $(= \mathbf{p}_{x_i}\sigma + \mathbf{q}_{x_i})$, ψ_k, $\dot{\mathbf{s}}_i$, and $\ddot{\mathbf{s}}_i$ $(= \mathbf{p}_{s_i}\sigma + \mathbf{q}_{s_i})$. In every iteration, several σ may be tried to find the best σ_k and α_k while ϕ_k, ψ_k, $\dot{\mathbf{x}}$, $\dot{\mathbf{s}}$, $\mathbf{p}_x$, $\mathbf{p}_s$, $\mathbf{q}_x$ and $\mathbf{q}_s$ are fixed (see details in the next section).

Case 1 ($\dot{x}_i = 0$ and $\ddot{x}_i \neq 0$):

$$\alpha_{x_i} = \begin{cases} \frac{\pi}{2} & \text{if } x_i - \phi_k + \ddot{x}_i \geq 0 \\ \cos^{-1}\left(\frac{x_i-\phi_k+\ddot{x}_i}{\ddot{x}_i}\right) & \text{if } x_i - \phi_k + \ddot{x}_i \leq 0. \end{cases} \tag{8.77}$$

Case 2 ($\ddot{x}_i = 0$ and $\dot{x}_i \neq 0$):

$$\alpha_{x_i} = \begin{cases} \frac{\pi}{2} & \text{if } \dot{x}_i \leq x_i - \phi_k \\ \sin^{-1}\left(\frac{x_i-\phi_k}{\dot{x}_i}\right) & \text{if } \dot{x}_i \geq x_i - \phi_k \end{cases} \tag{8.78}$$

Case 3 ($\dot{x}_i > 0$ and $\ddot{x}_i > 0$):

Let

$$\beta = \sin^{-1}\left(\frac{\ddot{x}_i}{\sqrt{\dot{x}_i^2 + \ddot{x}_i^2}}\right). \tag{8.79}$$

$$\alpha_{x_i} = \begin{cases} \frac{\pi}{2} & \text{if } x_i - \phi_k + \ddot{x}_i \geq \sqrt{\dot{x}_i^2 + \ddot{x}_i^2} \\ \sin^{-1}\left(\frac{x_i-\phi_k+\ddot{x}_i}{\sqrt{\dot{x}_i^2+\ddot{x}_i^2}}\right) - \sin^{-1}\left(\frac{\ddot{x}_i}{\sqrt{\dot{x}_i^2+\ddot{x}_i^2}}\right) & \text{if } x_i - \phi_k + \ddot{x}_i \leq \sqrt{\dot{x}_i^2 + \ddot{x}_i^2} \end{cases} \tag{8.80}$$

Case 4 ($\dot{x}_i > 0$ and $\ddot{x}_i < 0$):
Let

$$\beta = \sin^{-1}\left(\frac{-\ddot{x}_i}{\sqrt{\dot{x}_i^2 + \ddot{x}_i^2}}\right). \tag{8.81}$$

$$\alpha_{x_i} = \begin{cases} \frac{\pi}{2} & \text{if } x_i - \phi_k + \ddot{x}_i \geq \sqrt{\dot{x}_i^2 + \ddot{x}_i^2} \\ \sin^{-1}\left(\frac{x_i - \phi_k + \ddot{x}_i}{\sqrt{\dot{x}_i^2 + \ddot{x}_i^2}}\right) + \sin^{-1}\left(\frac{-\ddot{x}_i}{\sqrt{\dot{x}_i^2 + \ddot{x}_i^2}}\right) & \text{if } x_i - \phi_k + \ddot{x}_i \leq \sqrt{\dot{x}_i^2 + \ddot{x}_i^2} \end{cases} \tag{8.82}$$

Case 5 ($\dot{x}_i < 0$ and $\ddot{x}_i < 0$):
Let

$$\beta = \sin^{-1}\left(\frac{-\ddot{x}_i}{\sqrt{\dot{x}_i^2 + \ddot{x}_i^2}}\right). \tag{8.83}$$

$$\alpha_{x_i} = \begin{cases} \frac{\pi}{2} & \text{if } x_i - \phi_k + \ddot{x}_i \geq 0 \\ \pi - \sin^{-1}\left(\frac{-(x_i - \phi_k + \ddot{x}_i)}{\sqrt{\dot{x}_i^2 + \ddot{x}_i^2}}\right) - \sin^{-1}\left(\frac{-\ddot{x}_i}{\sqrt{\dot{x}_i^2 + \ddot{x}_i^2}}\right) & \text{if } x_i - \phi_k + \ddot{x}_i \leq 0 \end{cases} \tag{8.84}$$

Case 6 ($\dot{x}_i < 0$ and $\ddot{x}_i > 0$):

$$\alpha_{x_i} = \frac{\pi}{2}. \tag{8.85}$$

Case 7 ($\dot{x}_i = 0$ and $\ddot{x}_i = 0$):

$$\alpha_{x_i} = \frac{\pi}{2}. \tag{8.86}$$

Case 1a ($\dot{s}_i = 0$, $\ddot{s}_i \neq 0$):

$$\alpha_{s_i} = \begin{cases} \frac{\pi}{2} & \text{if } s_i - \psi_k + \ddot{s}_i \geq 0 \\ \cos^{-1}\left(\frac{s_i - \psi_k + \ddot{s}_i}{\ddot{s}_i}\right) & \text{if } s_i - \psi_k + \ddot{s}_i \leq 0. \end{cases} \tag{8.87}$$

Case 2a ($\ddot{s}_i = 0$ and $\dot{s}_i \neq 0$):

$$\alpha_{s_i} = \begin{cases} \frac{\pi}{2} & \text{if } \dot{s}_i \leq s_i - \psi_k \\ \sin^{-1}\left(\frac{s_i - \psi_k}{\dot{s}_i}\right) & \text{if } \dot{s}_i \geq s_i - \psi_k \end{cases} \tag{8.88}$$

Case 3a ($\dot{s}_i > 0$ and $\ddot{s}_i > 0$):

$$\alpha_{s_i} = \begin{cases} \frac{\pi}{2} & \text{if } s_i - \psi_k + \ddot{s}_i \geq \sqrt{\dot{s}_i^2 + \ddot{s}_i^2} \\ \sin^{-1}\left(\frac{s_i - \psi_k + \ddot{s}_i}{\sqrt{\dot{s}_i^2 + \ddot{s}_i^2}}\right) - \sin^{-1}\left(\frac{\ddot{s}_i}{\sqrt{\dot{s}_i^2 + \ddot{s}_i^2}}\right) & \text{if } s_i - \psi_k + \ddot{s}_i < \sqrt{\dot{s}_i^2 + \ddot{s}_i^2} \end{cases} \tag{8.89}$$

Case 4a ($\dot{s}_i > 0$ and $\ddot{s}_i < 0$):

$$\alpha_{s_i} = \begin{cases} \frac{\pi}{2} & \text{if } s_i - \psi_k + \ddot{s}_i \geq \sqrt{\dot{s}_i^2 + \ddot{s}_i^2} \\ \sin^{-1}\left(\frac{s_i - \psi_k + \ddot{s}_i}{\sqrt{\dot{s}_i^2 + \ddot{s}_i^2}}\right) + \sin^{-1}\left(\frac{-\ddot{s}_i}{\sqrt{\dot{s}_i^2 + \ddot{s}_i^2}}\right) & \text{if } s_i - \psi_k + \ddot{s}_i \leq \sqrt{\dot{s}_i^2 + \ddot{s}_i^2} \end{cases} \tag{8.90}$$

Case 5a ($\dot{s}_i < 0$ and $\ddot{s}_i < 0$):

$$\alpha_{s_i} = \begin{cases} \frac{\pi}{2} & \text{if } s_i - \psi_k + \ddot{s}_i \geq 0 \\ \pi - \sin^{-1}\left(\frac{-(s_i - \psi_k + \ddot{s}_i)}{\sqrt{\dot{s}_i^2 + \ddot{s}_i^2}}\right) - \sin^{-1}\left(\frac{-\ddot{s}_i}{\sqrt{\dot{s}_i^2 + \ddot{s}_i^2}}\right) & \text{if } s_i - \psi_k + \ddot{s}_i \leq 0 \end{cases} \tag{8.91}$$

Case 6a ($\dot{s}_i < 0$ and $\ddot{s}_i > 0$):

$$\alpha_{s_i} = \frac{\pi}{2}. \tag{8.92}$$

Case 7a ($\dot{s}_i = 0$ and $\ddot{s}_i = 0$):

$$\alpha_{s_i} = \frac{\pi}{2}. \tag{8.93}$$

8.4.9 Selection of centering parameter σ_k

From the convergence analysis, it is clear that the best strategy to achieve a large step size is to select both α_k and σ_k at the same time, similar to the idea of Section 3.4. Therefore, we will find a σ_k which maximizes the step size α_k. The problem can be expressed as

$$\max_{\sigma \in [\sigma_{\min}, \sigma_{\max}]} \min_{i \in \{1,\ldots,n\}} \{\alpha_{x_i}(\sigma), \alpha_{s_i}(\sigma)\}, \tag{8.94}$$

where $0 < \sigma_{\min} < \sigma_{\max} < 1$, $\alpha_{x_i}(\sigma)$ and $\alpha_{s_i}(\sigma)$ are calculated using (8.77)-(8.93) for a fixed $\sigma \in [\sigma_{\min}, \sigma_{\max}]$. Problem (8.94) is a minimax problem without regularity conditions involving derivatives. Golden section search [34] seems to be a reasonable method for solving this problem. However, given the fact from (8.30) that $\alpha_{x_i}(\sigma)$ is a monotonic increasing function of σ if $p_{x_i} > 0$ and $\alpha_{x_i}(\sigma)$ is a monotonic decreasing function of σ if $p_{x_i} < 0$ (and similar properties hold for $\alpha_{s_i}(\sigma)$), we can use the condition

$$\min\{\min_{i \in p_{x_i} < 0} \alpha_{x_i}(\sigma), \min_{i \in p_{s_i} < 0} \alpha_{s_i}(\sigma)\} > \min\{\min_{i \in p_{x_i} > 0} \alpha_{x_i}(\sigma), \min_{i \in p_{s_i} > 0} \alpha_{s_i}(\sigma)\}, \tag{8.95}$$

to devise an efficient bisection search to solve (8.94).

Algorithm 8.3

Data: $(\dot{\mathbf{x}},\dot{\mathbf{s}})$, $(\mathbf{p}_x,\mathbf{p}_s)$, $(\mathbf{q}_x,\mathbf{q}_s)$, $(\mathbf{x}^k,\mathbf{s}^k)$, ϕ_k, *and* ψ_k.
Parameter: $\varepsilon\in(0,1)$, $\sigma_{lb}=\sigma_{\min}$, $\sigma_{ub}=\sigma_{\max}\le 1$.
for *iteration* $k=0,1,2,\ldots$

Step 0: If $\sigma_{ub}-\sigma_{lb}\le\varepsilon$, *set* $\alpha=\min\limits_{i\in\{1,\ldots,n\}}\{\alpha_{x_i}(\sigma),\alpha_{s_i}(\sigma)\}$, *stop.*

Step 1: Set $\sigma=\sigma_{lb}+0.5(\sigma_{ub}-\sigma_{lb})$.

Step 2: Calculate $\alpha_{x_i}(\sigma)$ *and* $\alpha_{s_i}(\sigma)$ *using (8.77)-(8.93).*

Step 3: If (8.95) holds, set $\sigma_{lb}=\sigma$, *else set* $\sigma_{ub}=\sigma$.

Step 4: Set $k+1\to k$. *Go back to Step 1.*

end (for)

It is known that Golden section search yields a new interval whose length is 0.618 of the previous interval in all iterations [80], while the proposed algorithm yields a new interval whose length is 0.5 of the previous interval and, therefore, is more efficient than Golden section.

For Algorithm 8.1, after executing Algorithm 8.3, we may still need to further reduce α_k using Golden section or bisection to satisfy

$$\mu(\sigma_k,\alpha_k)<\mu_k, \tag{8.96a}$$
$$\mathbf{x}(\sigma_k,\alpha_k)\mathbf{s}(\sigma_k,\alpha_k)\ge\theta\mu(\sigma_k,\alpha_k)\mathbf{e}. \tag{8.96b}$$

For Algorithm 8.2, after executing Algorithm 8.3, we need to check only (8.96a) and decide if further reduction of α_k is needed.

Remark 8.5 Comparing Lemmas 8.11 and 8.12, we guess that the restriction of satisfying the condition of $\mu(\sigma_k,\alpha_k)<\mu_k$ is weaker than the restriction of satisfying the conditions of $\mathbf{x}(\sigma_k,\alpha_k)\ge\phi_k$ and $\mathbf{s}(\sigma_k,\alpha_k)\ge\psi_k$, which are solved in (8.94). Indeed, we observed that σ_k and α_k obtained by solving (8.94) always satisfy the weaker restriction. Nevertheless, we keep this check for the safety concern. ■

8.4.10 Rescaling α_k

To maintain $\nu_k>0$, in each iteration, after an α_k is found as in the above process, we rescale $\alpha_k=\min\{0.9999\alpha_k, 0.99\pi/2\}<0.99\pi/2$ so that $\nu_k=\nu_{k-1}(1-\sin(\alpha_k))>0$ holds in every iteration. Therefore, Assumption 4 is always satisfied. We notice that the rescaling also prevents $\mathbf{x}^k$ and $\mathbf{s}^k$ from getting too close to zero in early iterations, which may cause problems while solving (8.73).

8.4.11 Terminate criteria

The main stopping criterion used in the implementations is slightly deviated from the one described in previous sections but follows the convention used by most infeasible interior-point software implementations, such as LIPSOL [172]

$$\frac{\|\mathbf{r}_b^k\|}{\max\{1,\|\mathbf{b}\|\}} + \frac{\|\mathbf{r}_c^k\|}{\max\{1,\|\mathbf{c}\|\}} + \frac{\mu_k}{\max\{1,\|\mathbf{c}^{\mathrm{T}}\mathbf{x}^k\|,\|\mathbf{b}^{\mathrm{T}}\lambda^k\|\}} < 10^{-8}.$$

In case the algorithms fail to find a good search direction, the program also stops if step sizes $\alpha_k^x < 10^{-8}$ and $\alpha_k^s < 10^{-8}$.

Finally, if (a) due to the numerical problem, $\|\mathbf{r}_b^k\|$ or $\|\mathbf{r}_c^k\|$ does not decrease but $10\|\mathbf{r}_b^{k-1}\| < \|\mathbf{r}_b^k\|$ or $10\|\mathbf{r}_c^{k-1}\| < \|\mathbf{r}_c^k\|$, or (b) if $\mu < 10^{-8}$, the program stops.

8.5 Numerical Tests

The two algorithms proposed in this chapter are implemented in MATLAB functions and named as `arclp1.m` and `arclp2.m`. These two algorithms are compared with two efficient algorithms, the arc-search algorithm proposed in Chapter 7 (implemented as `curvelp.m`) and the well-known Mehrotra's algorithm discussed in Chapters 4 and 7 (implemented as `mehrotra.m`).

The main cost of the four algorithms in each iteration is the same, involving the linear algebra for sparse Cholesky decomposition, which is the first equation of (8.73). The cost of Cholesky decomposition is $\mathcal{O}(n^3)$, which is much higher than the next expensive task $\mathcal{O}(n^2)$, the cost of solving the second equation of (8.73). Therefore, we conclude that the iteration count is a good measure to compare the performance for these four algorithms. All MATLAB codes of the above four algorithms are tested against to each other using the benchmark Netlib problems. The four MATLAB codes use exactly the same initial point, the same stopping criteria, the same pre-process and the same post-process, so that the comparison of the performance of the four algorithms is reasonable. Numerical tests for all algorithms have been performed for all Netlib linear programming problems that are presented in standard form, except `Osa_60` ($m = 10281$ and $n = 232966$) because the PC computer used for the testing does not have enough memory to handle this problem. The iteration numbers used to solve these problems are listed in Table 8.1.

We noted, in Chapter 7, that `curvelp.m` and `mehrotra.m` have difficulty with some problems because of the degenerate solutions, but the proposed Algorithms 8.1 and 8.2 implemented as `arclp1.m` and `arclp2.m` have no difficulty with all test problems. Although we have the option of handling degenerate solutions implemented in `arclp1.m` and `arclp2.m`, this option is not used for all the test problems. However, `curvelp.m` and `mehrotra.m` have to use this option because these two codes reached some degenerate solutions for several problems which make them difficult to solve or need significantly more iterations. For problems marked with '+', this option is called only for Mehrotra's method. For problems marked with '*', both `curvelp.m` and `mehrotra.m` need to call this option for better results. For problems with '**', this option is called for both `curvelp.m` and `mehrotra.m` but `curvelp.m` does not need to call this feature, however, calling this feature reduces iteration count. We need to keep in mind that, although using the option described in Section 8.4.7 reduces the iteration count significantly, these iterations are significantly more expensive [161]. Therefore, simply comparing iteration counts for problems marked with '+', '*', and '**' will lead to a conclusion in favor of `curvelp.m` and `mehrotra.m` (which is what we will do in the following discussions).

Table 8.1: Comparison of arclp1.m, arclp2.m, curvelp.m, and mehrotra.m for problems in Netlib.

Problem	algorithm	iter	obj	infeasibility
Adlittle	curvelp.m	15	2.2549e+05	1.0e-07
	mehrotra.m	15	2.2549e+05	3.4e-08
	arclp1.m	16	2.2549e+05	3.0e-11
	arclp2.m	17	2.2549e+05	8.0e-11
Afiro	curvelp.m	9	-464.7531	1.0e-11
	mehrotra.m	9	-464.7531	8.0e-12
	arclp1.m	9	-464.7531	6.2e-13
	arclp2.m	9	-464.7531	1.0e-12
Agg	curvelp.m	18	-3.5992e+07	5.0e-06
	mehrotra.m	22	-3.5992e+07	5.2e-05
	arclp1.m	20	-3.5992e+07	3.7e-06
	arclp2.m	20	-3.5992e+07	7.0e-06
Agg2	curvelp.m	18	-2.0239e+07	4.6e-07
	mehrotra.m	20	-2.0239e+07	5.2e-07
	arclp1.m	21	-2.0239e+07	3.1e-08
	arclp2.m	21	-2.0239e+07	2.6e-08
Agg3	curvelp.m	17	1.0312e+07	3.1e-08
	mehrotra.m	18	1.0312e+07	8.8e-09
	arclp1.m	20	1.0312e+07	1.5e-08
	arclp2.m	20	1.0312e+07	1.8e-08
Bandm	curvelp.m	19	-158.6280	3.2e-11
	mehrotra.m	22	-158.6280	8.3e-10
	arclp1.m	20	-158.6280	3.6e-11
	arclp2.m	20	-158.6280	3.4e-11
Beaconfd	curvelp.m	10	3.3592e+04	1.4e-12
	mehrotra.m	11	3.3592e+04	1.4e-10
	arclp1.m	11	3.3592e+04	1.8e-12
	arclp2.m	11	3.3592e+04	6.0e-12
Blend	curvelp.m	12	-30.8121	1.0e-09
	mehrotra.m	14	-30.8122	4.9e-11
	arclp1.m	14	-30.8122	1.6e-12
	arclp2.m	14	-30.8121	2.5e-12
Bnl1	curvelp.m	32	1.9776e+03	2.7e-09
	mehrotra.m	35	1.9776e+03	3.4e-09

	arclp1.m	34	1.9776e+03	2.9e-09
	arclp2.m	34	1.9776e+03	7.8e-10
Bnl2+	curvelp.m	31	1.8112e+03	5.4e-10
	mehrotra.m	38	1.8112e+03	9.3e-07
	arclp1.m	35	1.8112e+03	3.5e-06
	arclp2.m	35	1.8112e+03	1.9e-07
Brandy	curvelp.m	21	1.5185e+03	3.0e-06
	mehrotra.m	19	1.5185e+03	6.2e-08
	arclp1.m	24	1.5185e+03	2.4e-06
	arclp2.m	23	1.5185e+03	1.8e-07
Degen2*	curvelp.m	16	-1.4352e+03	1.9e-08
	mehrotra.m	17	-1.4352e+03	2.0e-10
	arclp1.m	19	-1.4352e+03	5.9e-10
	arclp2.m	19	-1.4352e+03	1.5e-08
Degen3*	curvelp.m	22	-9.8729e+02	7.0e-05
	mehrotra.m	22	-9.8729e+02	1.2e-09
	arclp1.m	35	-9.8729e+02	8.6e-08
	arclp2.m	26	-9.8729e+02	1.2e-08
fffff800	curve	26	5.5568e+005	4.3e-05
	mehrotra.m	31	5.5568e+05	7.7e-04
	arclp1.m	28	5.5568e+05	3.7e-09
	arclp2.m	28	5.5568e+05	4.9e-09
Israel	curvelp.m	23	-8.9664e+05	7.4e-08
	mehrotra.m	29	-8.9665e+05	1.8e-08
	arclp1.m	27	-8.9664e+05	3.4e-08
	arclp2.m	25	-8.9664e+05	6.3e-08
Lotfi	curvelp.m	14	-25.2647	3.5e-10
	mehrotra.m	18	-25.2647	2.7e-07
	arclp1.m	16	-25.2646	7.8e-09
	arclp2.m	16	-25.2647	6.5e-09
Maros_r7	curvelp.m	18	1.4972e+06	1.6e-08
	mehrotra.m	21	1.4972e+06	6.4e-09
	arclp1.m	20	1.4972e+06	1.7e-09
	arclp2.m	20	1.4972e+06	1.8e-09
Osa_07*	curvelp.m	37	5.3574e+05	4.2e-07
	mehrotra.m	35	5.3578e+05	1.5e-07
	arclp1.m	32	5.3578e+05	8.4e-10
	arclp2.m	30	5.3578e+05	5.7e-5
Osa_14	curvelp.m	35	1.1065e+06	2.0e-09
	mehrotra.m	37	1.1065e+06	3.0e-08
	arclp1.m	42	1.1065e+06	5.2e-09
	arclp2.m	42	1.1065e+06	8.9e-09
Osa_30	curvelp.m	32	2.1421e+06	1.0e-08
	mehrotra.m	36	2.1421e+06	1.3e-08
	arclp1.m	42	2.1421e+06	1.3e-08
	arclp2.m	39	2.1421e+06	1.6e-08
Qap12	curvelp.m	22	5.2289e+02	1.9e-08
	mehrotra.m	24	5.2289e+02	6.2e-09
	arclp1.m	23	5.2289e+02	2.9e-10
	arclp2.m	22	5.2289e+02	2.5e-09
Qap15*	curvelp.m	27	1.0411e+03	3.9e-07
	mehrotra.m	44	1.0410e+03	1.5e-05
	arclp1.m	28	1.0410e+03	8.4e-08

	arclp2.m	27	1.0410e+03	1.4e-08
Qap8*	curvelp.m	12	2.0350e+02	1.2e-12
	mehrotra.m	13	2.0350e+02	7.1e-09
	arclp1.m	12	2.0350e+02	6.2e-11
	arclp2.m	12	2.0350e+02	1.1e-10
Sc105	curvelp.m	10	-52.2021	3.8e-12
	mehrotra.m	11	-52.2021	9.8e-11
	arclp1.m	11	-52.2021	2.2e-12
	arclp2.m	11	-52.2021	5.6e-12
Sc205	curvelp.m	13	-52.2021	3.7e-10
	mehrotra.m	12	-52.2021	8.8e-11
	arclp1.m	12	-52.2021	4.4e-11
	arclp2.m	12	-52.2021	4.5e-11
Sc50a	curvelp.m	10	-64.5751	3.4e-12
	mehrotra.m	9	-64.5751	8.3e-08
	arclp1.m	10	-64.5751	8.5e-13
	arclp2.m	10	-64.5751	5.9e-13
Sc50b	curvelp.m	8	-70.0000	1.0e-10
	mehrotra.m	8	-70.0000	9.1e-07
	arclp1.m	10	-70.0000	3.6e-12
	arclp2.m	10	-70.0000	1.8373e-12
Scagr25	curvelp.m	19	-1.4753e+07	5.0e-07
	mehrotra.m	18	-1.4753e+07	4.6e-09
	arclp1.m	19	-1.4753e+07	1.7e-08
	arclp2.m	19	-1.4753e+07	2.1e-08
Scagr7	curvelp.m	15	-2.3314e+06	2.7e-09
	mehrotra.m	17	-2.3314e+06	1.1e-07
	arclp1.m	17	-2.3314e+06	7.0e-10
	arclp2.m	17	-2.3314e+06	9.2e-10
Scfxm1+	curvelp.m	20	1.8417e+04	3.1e-07
	mehrotra.m	22	1.8417e+04	1.6e-08
	arclp1.m	21	1.8417e+04	3.3e-05
	arclp2.m	21	1.8417e+04	6.8e-06
Scfxm2	curvelp.m	23	3.6660e+04	2.3e-06
	mehrotra.m	26	3.6660e+04	2.6e-08
	arclp1.m	24	3.6660e+04	4.8e-05
	arclp2.m	24	3.6661e+04	1.1e-05
Scfxm3+	curvelp.m	24	5.4901e+04	1.9e-06
	mehrotra.m	23	5.4901e+04	9.8e-08
	arclp1.m	23	5.4901e+04	1.2e-04
	arclp2.m	25	5.4901e+04	4.0988e-04
Scrs8	curvelp.m	23	9.0430e+02	1.2e-11
	mehrotra.m	30	9.0430e+02	1.8e-10
	arclp1.m	28	9.0430e+02	1.0e-10
	arclp2.m	27	9.0429e+2	1.2e-08
Scsd1	curvelp.m	12	8.6666	1.0e-10
	mehrotra.m	13	8.6666	8.7e-14
	arclp1.m	11	8.6666	3.3e-15
	arclp2.m	11	8.6667	5.3e-15
Scsd6	curvelp.m	14	50.5000	1.5e-13
	mehrotra.m	16	50.5000	8.6e-15
	arclp1.m	16	50.5000	2.6e-13
	arclp2.m	16	50.5000	4.8e-13

Scsd8	curvelp.m	13	9.0500e+02	6.7e-10
	mehrotra.m	14	9.0500e+02	1.3e-10
	arclp1.m	15	9.0500e+02	2.6e-13
	arclp2.m	15	9.0500e+02	3.7e-13
Sctap1	curvelp.m	20	1.4122e+03	2.6e-10
	mehrotra.m	27	1.4123e+03	0.0031
	arclp1.m	20	1.4123e+03	1.4e-11
	arclp2.m	20	1.4123e+03	1.8e-11
Sctap2	curvelp.m	21	1.7248e+03	2.1e-10
	mehrotra.m	21	1.7248e+03	4.4e-07
	arclp1.m	22	1.7248e+03	1.4e-12
	arclp2.m	20	1.7248e+03	1.1e-12
Sctap3	curvelp.m	20	1.4240e+03	5.7e-08
	mehrotra.m	22	1.4240e+03	5.9e-07
	arclp1.m	21	1.4240e+03	1.9e-12
	arclp2.m	21	1.4240e+03	2.5e-12
Share1b	curvelp.m	22	-7.6589e+04	6.5e-08
	mehrotra.m	25	-7.6589e+04	1.5e-06
	arclp1.m	26	-7.6589e+04	1.9e-07
	arclp2.m	26	-7.6589e+04	2.3e-07
Share2b	curvelp.m	13	-4.1573e+02	4.9e-11
	mehrotra.m	15	-4.1573e+02	7.9e-10
	arclp1.m	15	-4.1573e+02	1.4e-10
	arclp2.m	15	-4.1573e+02	8.9e-11
Ship04l	curvelp.m	17	1.7933e+06	5.2e-11
	mehrotra.m	18	1.7933e+06	2.9e-11
	arclp1.m	19	1.7933e+06	1.3e-10
	arclp2.m	18	1.7933e+06	5.9e-11
Ship04s	curvelp.m	17	1.7987e+06	2.2e-11
	mehrotra.m	20	1.7987e+06	4.5e-09
	arclp1.m	19	1.7987e+06	3.1e-10
	arclp2.m	19	1.7987e+06	3.1e-09
Ship08l	curvelp.m	19	1.9090e+06	1.6e-07
	mehrotra.m	22	1.9091e+06	1.0e-10
	arclp1.m	20	1.9090e+06	1.8e-11
	arclp2.m	20	1.9091e+06	1.2e-11
Ship08s	curvelp.m	17	1.9201e+06	3.7e-08
	mehrotra.m	20	1.9201e+06	4.5e-12
	arclp1.m	19	1.9201e+06	1.7e-09
	arclp2.m	19	1.9201e+06	3.2e-11
Ship12l	curvelp.m	21	1.4702e+06	4.7e-13
	mehrotra.m	21	1.4702e+06	1.0e-08
	arclp1.m	21	1.4702e+06	3.0e-10
	arclp2.m	21	1.4702e+06	6.5e-11
Ship12s	curvelp.m	17	1.4892e+06	1.0e-10
	mehrotra.m	19	1.4892e+06	2.1e-13
	arclp1.m	21	1.4892e+06	5.0e-11
	arclp2.m	20	1.4892e+06	1.4e-10
Stocfor1**	curvelp.m	20/14	-4.1132e+04	2.8e-10
	mehrotra.m	14	-4.1132e+04	1.1e-10
	arclp1.m	13	-4.1132e+04	8.6890e-11
	arclp2.m	14	-4.1132e+04	1.1e-10

Stocfor2	curvelp.m	22	-3.9024e+04	2.1e-09
	mehrotra.m	22	-3.9024e+04	1.6e-09
	arclp1.m	22	-3.9024e+04	4.3e-09
	arclp2.m	22	-3.9024e+04	4.3e-09
Stocfor3	curvelp.m	34	-3.9976e+04	4.7e-08
	mehrotra.m	38	-3.9976e+04	6.4e-08
	arclp1.m	37	-3.9977e+04	7.7e-08
	arclp2.m	37	-3.9976e+04	6.8e-08
Truss	curvelp.m	25	4.5882e+05	1.7e-07
	mehrotra.m	26	4.5882e+05	9.5e-06
	arclp1.m	24	4.5882e+05	5.2e-07
	arclp2.m	24	4.5882e+05	1.7e-09

Performance profile[2] is used to compare the efficiency of the four algorithms. Figure 8.1 is the performance profile of iteration numbers of the four algorithms. It is clear that `curvelp.m` is the most efficient algorithm of the four algorithms; `arclp2.m` is slightly better than `arclp1.m` and `mehrotra.m`; and the efficiencies of `arclp1.m` and `mehrotra.m` are roughly the same. Overall, the efficiency difference of the four algorithms is not very significant. Given the fact that `arclp1.m` and `arclp2.m` are convergent in theory and more stable in numerical test, we believe that these two algorithms are better choices in practical applications.

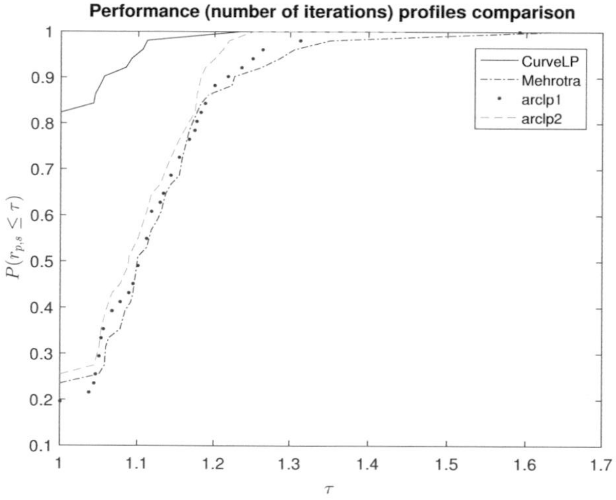

Figure 8.1: Performance profile comparison of the four algorithms.

[2]To our best knowledge, performance profile was first used in [132] to compare the performance of different algorithms. The method has been becoming very popular after its merit was carefully analyzed in [29].

8.6 Concluding Remarks

In this chapter, we propose two computationally efficient polynomial interior-point algorithms. These algorithms search the optimizer along ellipses that approximate the central path. The first algorithm is proved to be polynomial and its simplified version, the second algorithm, has a better complexity bound than all existing infeasible interior-point algorithms and achieves the best complexity bound for all existing, feasible or infeasible, interior-point algorithms. Numerical test results for all the Netlib standard linear programming problems show that these two algorithms are competitive to the state-of-the-art Mehrotra's Algorithm which has no convergence result. The results obtained in this chapter solve all the dilemmas discussed in Section 4.4.

III

ARC-SEARCH INTERIOR-POINT METHODS: EXTENSIONS

Chapter 9

An Arc-Search Algorithm for Convex Quadratic Programming

Interior-point algorithms for convex quadratic programming have been proposed for decades by many researchers. One of the earliest and well-known algorithms was proposed by Dikin [27]. After Karmarkar's pioneer work on interior-point polynomial algorithm for linear programming [57], interior-point polynomial algorithms for convex quadratic programming have been investigated by many researchers. For example, Ye and Tse [166] extended Karmarkar's algorithm and proved that their algorithm has polynomial complexity bound $O(n\log(1/\varepsilon))$. Monteiro and Adler proposed a different algorithm [96] and improved the complexity bound to $O(\sqrt{n}\log(1/\varepsilon))$. However, these algorithms do not use higher-order information, which is believed intuitively (and demonstrated by numerical test) to be useful in practical implementation [8, 89]. Gondzio [41] also considered multiple centrality corrections for linear programming and provided numerical results to demonstrate the potential benefit of higher-order method, but [89, 41] did not show the polynomiality of their methods.

The first polynomial higher-order algorithm can probably be attributed to Monteiro, Adler, and Resende [97], who derived the complexity bound $O(n^{\frac{1}{2}(1+\frac{1}{r})}\log(1/\varepsilon))$ for their algorithm, where $r \geq 1$ is the order of derivatives. Clearly, this bound is not as good as the one for the first-order algorithm [96].

This chapter extends the arc-search technique discussed in Chapter 5 to convex quadratic problem. A polynomial algorithm with complexity bound $O(\sqrt{n}\log(1/\varepsilon))$ is proposed. A MATLAB® code is implemented for the algorithm. A simple example is used to demonstrate how this algorithm works. The proposed algorithm is also tested on all convex quadratic programming problems listed in [53]. The result is compared with the one obtained by LOQO [47]. This preliminary result shows that the proposed algorithm is promising because it uses fewer iterations in all these tested problems than the iterations LOQO uses.

9.1 Problem Descriptions

This chapter considers the convex quadratic programming (QP) in the standard form:

$$\begin{aligned} (QP) \qquad & \min && \frac{1}{2}\mathbf{x}^{\mathrm{T}}\mathbf{H}\mathbf{x}+\mathbf{c}^{\mathrm{T}}\mathbf{x}, \\ & \text{subject to} && \mathbf{A}\mathbf{x}=\mathbf{b}, \quad \mathbf{x}\geq \mathbf{0}, \end{aligned} \tag{9.1}$$

where $\mathbf{0} \preceq \mathbf{H} \in \mathbb{R}^{n\times n}$ is a positive semidefinite matrix, $\mathbf{A} \in \mathbb{R}^{m\times n}$, $\mathbf{b} \in \mathbb{R}^m$, $\mathbf{c} \in \mathbb{R}^n$ are given, and $\mathbf{x} \in \mathbb{R}^n$ is the vector to be optimized. Associated with the quadratic programming is the dual programming (DQP) that is also presented in the standard form:

$$\begin{aligned} (DQP) \qquad & \max && -\frac{1}{2}\mathbf{x}^{\mathrm{T}}\mathbf{H}\mathbf{x}+\mathbf{b}^{\mathrm{T}}\boldsymbol{\lambda}, \\ & \text{subject to} && -\mathbf{H}\mathbf{x}+\mathbf{A}^{\mathrm{T}}\boldsymbol{\lambda}+\mathbf{s}=\mathbf{c}, \quad \mathbf{s}\geq \mathbf{0}, \quad \mathbf{x}\geq \mathbf{0}, \end{aligned} \tag{9.2}$$

where $\boldsymbol{\lambda} \in \mathbb{R}^m$ is the dual variable vector, and $\mathbf{s} \in \mathbb{R}^n$ is the dual slack vector.

Denote the feasible set $\mathcal{F}$ as a collection of all points that meet the constraints of (QP) and (DQP).

$$\mathcal{F} = \{(\mathbf{x},\boldsymbol{\lambda},\mathbf{s}) : \mathbf{A}\mathbf{x}=\mathbf{b}, \mathbf{A}^{\mathrm{T}}\boldsymbol{\lambda}+\mathbf{s}-\mathbf{H}\mathbf{x}=\mathbf{c}, (\mathbf{x},\mathbf{s})\geq \mathbf{0}\}, \tag{9.3}$$

and the strictly feasible set $\mathcal{F}^o$ as a collection of all points that meet the constraints of (QP) and (DQP) and are strictly positive

$$\mathcal{F}^o = \{(\mathbf{x},\boldsymbol{\lambda},\mathbf{s}) : \mathbf{A}\mathbf{x}=\mathbf{b}, \mathbf{A}^{\mathrm{T}}\boldsymbol{\lambda}+\mathbf{s}-\mathbf{H}\mathbf{x}=\mathbf{c}, (\mathbf{x},\mathbf{s})> \mathbf{0}\}. \tag{9.4}$$

Throughout this chapter, we make the following assumptions.

Assumptions:

1. $\mathbf{A}$ is a full rank matrix.
2. $\mathcal{F}^o$ is not empty.

Assumption 2 implies the existence of a central path. Since $1 \leq m$ and $m < n$, assumption 2 implies $2 \leq n$. It is well known that $\mathbf{x} \in \mathbb{R}^n$ is an optimal solution of (9.1) if and only if $\mathbf{x}$, $\boldsymbol{\lambda}$, and $\mathbf{s}$ meet the following KKT conditions

$$\mathbf{A}\mathbf{x}=\mathbf{b} \tag{9.5a}$$

$$\mathbf{A}^{\mathrm{T}}\boldsymbol{\lambda}+\mathbf{s}-\mathbf{H}\mathbf{x}=\mathbf{c} \tag{9.5b}$$

$$x_i s_i = 0, \quad i=1,\ldots,n \tag{9.5c}$$

$$(\mathbf{x},\mathbf{s}) \geq \mathbf{0}. \tag{9.5d}$$

For convex (QP) problem, KKT condition is also sufficient for $\mathbf{x}$ to be a global optimal solution. Similar to the linear programming, the central path following algorithm proposed here tries to search the optimizers along an arc that is an approximation of the central path $\mathcal{C} \in \mathcal{F}^o \subset \mathcal{F}$, where the central path $\mathcal{C}$ is parametrized by a scalar $\tau > 0$ as follows. For each interior point $(\mathbf{x},\boldsymbol{\lambda},\mathbf{s})$ on the central path, there is a $\tau > 0$ such that

$$\mathbf{A}\mathbf{x}=\mathbf{b} \tag{9.6a}$$

$$\mathbf{A}^{\mathrm{T}}\boldsymbol{\lambda}+\mathbf{s}-\mathbf{H}\mathbf{x}=\mathbf{c} \tag{9.6b}$$

$$x_i s_i = \tau, \quad i=1,\ldots,n \tag{9.6c}$$

$$(\mathbf{x},\mathbf{s}) > \mathbf{0}. \tag{9.6d}$$

Therefore, the central path is an arc in $\mathbb{R}^{2n+m}$ parametrized as a function of τ and is denoted as

$$\mathcal{C} = \{(\mathbf{x}(\tau),\boldsymbol{\lambda}(\tau),\mathbf{s}(\tau)) : \tau > 0\}. \tag{9.7}$$

As $\tau \to 0$, the moving point $(\mathbf{x}(\tau),\boldsymbol{\lambda}(\tau),\mathbf{s}(\tau))$ on the central path represented by (9.6) approaches to the solution of (QP) represented by (9.1).

9.2 An Arc-search Algorithm for Convex Quadratic Programming

Let $1 > \theta > 0$, and the duality measure be given by (5.17) which is repeated below.

$$\mu = \frac{\mathbf{x}^{\mathrm{T}}\mathbf{s}}{n}.$$

We denote

$$\mathcal{N}_2(\theta) = \{(\mathbf{x},\boldsymbol{\lambda},\mathbf{s}) \mid \mathbf{A}\mathbf{x}=\mathbf{b},\, -\mathbf{H}\mathbf{x}+\mathbf{A}^{\mathrm{T}}\boldsymbol{\lambda}+\mathbf{s}=\mathbf{c},\, (\mathbf{x},\mathbf{s}) > \mathbf{0},\, \|\mathbf{x}\circ\mathbf{s}-\mu\mathbf{e}\| \leq \theta\mu\}. \tag{9.8}$$

It is worthwhile to note $\mathcal{N}_2(\theta) \subset \mathcal{F}^o$. For $(\mathbf{x},\boldsymbol{\lambda},\mathbf{s}) \in \mathcal{N}_2(\theta)$, since $(1-\theta)\mu \leq x_i s_i \leq (1+\theta)\mu$, we have

$$\frac{x_i s_i}{1+\theta} \leq \frac{\max_i(x_i s_i)}{1+\theta} \leq \mu \leq \frac{\min_i(x_i s_i)}{1-\theta} \leq \frac{x_i s_i}{1-\theta}. \tag{9.9}$$

The idea of arc-search proposed in this chapter is very simple. The algorithm starts from a feasible point in $\mathcal{N}_2(\theta)$ close to the central path, constructs an arc that passes through the point and approximates the central path, searches along the arc to a new point in a larger area $\mathcal{N}_2(2\theta)$ that reduces the duality gap $\mathbf{x}^{\mathrm{T}}\mathbf{s}$ and meets (9.6a), (9.6b), and (9.6d). The process is repeated by finding a better point close to the central path or on the central path in $\mathcal{N}_2(\theta)$ that simultaneously meets (9.6a), (9.6b), and (9.6d). As the duality measure or duality gap approaches zero, condition (9.6c) is met and the optimal solution is then found.

9.2.1 Predictor step

We will use an ellipse $\mathcal{E}$ in $2n+m$ dimensional space to approximate the central path C described by (9.6), where

$$\mathcal{E} = \{(\mathbf{x}(\alpha),\lambda(\alpha),\mathbf{s}(\alpha)) : (\mathbf{x}(\alpha),\lambda(\alpha),\mathbf{s}(\alpha)) = \vec{\mathbf{a}}\cos(\alpha)+\vec{\mathbf{b}}\sin(\alpha)+\vec{\mathbf{c}}\}, \tag{9.10}$$

$\vec{\mathbf{a}} \in \mathbb{R}^{2n+m}$ and $\vec{\mathbf{b}} \in \mathbb{R}^{2n+m}$ are the axes of the ellipse, and $\vec{\mathbf{c}} \in \mathbb{R}^{2n+m}$ is the center of the ellipse. Given a point $\mathbf{y} = (\mathbf{x},\lambda,\mathbf{s}) = (\mathbf{x}(\alpha_0),\lambda(\alpha_0),\mathbf{s}(\alpha_0)) \in \mathcal{E}$, which is close to or on the central path, $\vec{\mathbf{a}}$, $\vec{\mathbf{b}}$, $\vec{\mathbf{c}}$ are functions of α, $\mathbf{y}$, $\dot{\mathbf{y}}$, and $\ddot{\mathbf{y}}$, where $\dot{\mathbf{y}}$ and $\ddot{\mathbf{y}}$ are defined as

$$\begin{bmatrix} \mathbf{A} & \mathbf{0} & \mathbf{0} \\ -\mathbf{H} & \mathbf{A}^{\mathrm{T}} & \mathbf{I} \\ \mathbf{S} & \mathbf{0} & \mathbf{X} \end{bmatrix} \begin{bmatrix} \dot{\mathbf{x}} \\ \dot{\lambda} \\ \dot{\mathbf{s}} \end{bmatrix} = \begin{bmatrix} \mathbf{0} \\ \mathbf{0} \\ \mathbf{x}\circ\mathbf{s} \end{bmatrix}, \tag{9.11}$$

$$\begin{bmatrix} \mathbf{A} & \mathbf{0} & \mathbf{0} \\ -\mathbf{H} & \mathbf{A}^{\mathrm{T}} & \mathbf{I} \\ \mathbf{S} & \mathbf{0} & \mathbf{X} \end{bmatrix} \begin{bmatrix} \ddot{\mathbf{x}} \\ \ddot{\lambda} \\ \ddot{\mathbf{s}} \end{bmatrix} = \begin{bmatrix} \mathbf{0} \\ \mathbf{0} \\ -2\dot{\mathbf{x}}\circ\dot{\mathbf{s}} \end{bmatrix}. \tag{9.12}$$

It has been shown in Chapter 5 that the calculation of $\vec{\mathbf{a}}$, $\vec{\mathbf{b}}$, and $\vec{\mathbf{c}}$ in the expression of the ellipse can be avoided. The following formulas are used instead.

Theorem 9.1
Let $(\mathbf{x}(\alpha),\lambda(\alpha),\mathbf{s}(\alpha))$ be an arc defined by (9.10) passing through a point $(\mathbf{x},\lambda,\mathbf{s}) \in \mathcal{E}$, and its first and second derivatives at $(\mathbf{x},\lambda,\mathbf{s})$ be $(\dot{\mathbf{x}},\dot{\lambda},\dot{\mathbf{s}})$ and $(\ddot{\mathbf{x}},\ddot{\lambda},\ddot{\mathbf{s}})$ which are defined by (9.11) and (9.12). Then, an ellipse approximation of the central path is given by

$$\mathbf{x}(\alpha) = \mathbf{x} - \dot{\mathbf{x}}\sin(\alpha) + \ddot{\mathbf{x}}(1-\cos(\alpha)). \tag{9.13}$$

$$\lambda(\alpha) = \lambda - \dot{\lambda}\sin(\alpha) + \ddot{\lambda}(1-\cos(\alpha)). \tag{9.14}$$

$$\mathbf{s}(\alpha) = \mathbf{s} - \dot{\mathbf{s}}\sin(\alpha) + \ddot{\mathbf{s}}(1-\cos(\alpha)). \tag{9.15}$$

Assuming $(\mathbf{x},\mathbf{s}) > \mathbf{0}$, one can easily see that, if $\frac{\dot{\mathbf{x}}}{\mathbf{x}}$, $\frac{\ddot{\mathbf{x}}}{\mathbf{x}}$, $\frac{\dot{\mathbf{s}}}{\mathbf{s}}$, and $\frac{\ddot{\mathbf{s}}}{\mathbf{s}}$ are bounded (we will show that this claim is true), and if α is small enough, then $\mathbf{x}(\alpha) > \mathbf{0}$ and $\mathbf{s}(\alpha) > \mathbf{0}$. We will also show that searching along this arc will reduce the duality measure, i.e., $\mu(\alpha) = \frac{\mathbf{x}(\alpha)^{\mathrm{T}}\mathbf{s}(\alpha)}{n} < \mu$.

Lemma 9.1
Let $(\mathbf{x},\lambda,\mathbf{s})$ be a strictly feasible point of (QP) and (DQP), $(\dot{\mathbf{x}},\dot{\lambda},\dot{\mathbf{s}})$ and $(\ddot{\mathbf{x}},\ddot{\lambda},\ddot{\mathbf{s}})$ meet (9.11) and (9.12), $(\mathbf{x}(\alpha),\lambda(\alpha),\mathbf{s}(\alpha))$ be calculated using (9.13), (9.14), and (9.15), then the following conditions hold.

$$\mathbf{A}\mathbf{x}(\alpha)=\mathbf{b},\quad \mathbf{A}^{\mathrm{T}}\lambda(\alpha)+\mathbf{s}(\alpha)-\mathbf{H}\mathbf{x}(\alpha)=\mathbf{c}.$$

Proof 9.1 Since $(\mathbf{x},\lambda,\mathbf{s})$ is a strict feasible point, the result follows from direct calculation by using (9.4), (9.11), (9.12), and Theorem 9.1. ■

Lemma 9.2
Let $\dot{\mathbf{x}}$, $\dot{\mathbf{s}}$, $\ddot{\mathbf{x}}$, and $\ddot{\mathbf{s}}$ be defined in (9.11) and (9.12), and $\mathbf{H}$ be positive semidefinite matrix. Then, the following relations hold.

$$\dot{\mathbf{x}}^{\mathrm{T}}\dot{\mathbf{s}}=\dot{\mathbf{x}}^{\mathrm{T}}\mathbf{H}\dot{\mathbf{x}}\geq\mathbf{0}, \tag{9.16}$$

$$\ddot{\mathbf{x}}^{\mathrm{T}}\ddot{\mathbf{s}}=\ddot{\mathbf{x}}^{\mathrm{T}}\mathbf{H}\ddot{\mathbf{x}}\geq\mathbf{0}. \tag{9.17}$$

$$\ddot{\mathbf{x}}^{\mathrm{T}}\dot{\mathbf{s}}=\dot{\mathbf{x}}^{\mathrm{T}}\ddot{\mathbf{s}}=\dot{\mathbf{x}}^{\mathrm{T}}\mathbf{H}\ddot{\mathbf{x}}. \tag{9.18}$$

$$\begin{aligned}&-(\dot{\mathbf{x}}^{\mathrm{T}}\dot{\mathbf{s}})(1-\cos(\alpha))^2-(\ddot{\mathbf{x}}^{\mathrm{T}}\ddot{\mathbf{s}})\sin^2(\alpha)\\ \leq\;&(\ddot{\mathbf{x}}^{\mathrm{T}}\dot{\mathbf{s}}+\dot{\mathbf{x}}^{\mathrm{T}}\ddot{\mathbf{s}})\sin(\alpha)(1-\cos(\alpha))\\ \leq\;&(\dot{\mathbf{x}}^{\mathrm{T}}\dot{\mathbf{s}})(1-\cos(\alpha))^2+(\ddot{\mathbf{x}}^{\mathrm{T}}\ddot{\mathbf{s}})\sin^2(\alpha).\end{aligned} \tag{9.19}$$

$$\begin{aligned}&-(\dot{\mathbf{x}}^{\mathrm{T}}\dot{\mathbf{s}})\sin^2(\alpha)-(\ddot{\mathbf{x}}^{\mathrm{T}}\ddot{\mathbf{s}})(1-\cos(\alpha))^2\\ \leq\;&(\ddot{\mathbf{x}}^{\mathrm{T}}\dot{\mathbf{s}}+\dot{\mathbf{x}}^{\mathrm{T}}\ddot{\mathbf{s}})\sin(\alpha)(1-\cos(\alpha))\\ \leq\;&(\dot{\mathbf{x}}^{\mathrm{T}}\dot{\mathbf{s}})\sin^2(\alpha)+(\ddot{\mathbf{x}}^{\mathrm{T}}\ddot{\mathbf{s}})(1-\cos(\alpha))^2.\end{aligned} \tag{9.20}$$

For $\alpha=\frac{\pi}{2}$, (9.19) and (9.20) reduce to

$$-\left(\dot{\mathbf{x}}^{\mathrm{T}}\dot{\mathbf{s}}+\ddot{\mathbf{x}}^{\mathrm{T}}\ddot{\mathbf{s}}\right)\leq(\ddot{\mathbf{x}}^{\mathrm{T}}\dot{\mathbf{s}}+\dot{\mathbf{x}}^{\mathrm{T}}\ddot{\mathbf{s}})\leq\dot{\mathbf{x}}^{\mathrm{T}}\dot{\mathbf{s}}+\ddot{\mathbf{x}}^{\mathrm{T}}\ddot{\mathbf{s}}. \tag{9.21}$$

Proof 9.2 Pre-multiplying $\dot{\mathbf{x}}^{\mathrm{T}}$ to the second rows of (9.11) and pre-multiplying $\ddot{\mathbf{x}}^{\mathrm{T}}$ to the second rows of (9.12), then using the first rows of (9.11) and (9.12), gives $\dot{\mathbf{x}}^{\mathrm{T}}\dot{\mathbf{s}}=\dot{\mathbf{x}}^{\mathrm{T}}\mathbf{H}\dot{\mathbf{x}}$ and $\ddot{\mathbf{x}}^{\mathrm{T}}\ddot{\mathbf{s}}=\ddot{\mathbf{x}}^{\mathrm{T}}\mathbf{H}\ddot{\mathbf{x}}$. (9.16) and (9.17) follow from the fact that $\mathbf{H}$ is positive semidefinite. Pre-multiply $\ddot{\mathbf{x}}^{\mathrm{T}}$ in the second row of (9.11) and, using the first row of (9.12), we have

$$\ddot{\mathbf{x}}^{\mathrm{T}}\mathbf{H}\dot{\mathbf{x}}=\ddot{\mathbf{x}}^{\mathrm{T}}\dot{\mathbf{s}}.$$

Pre-multiply $\dot{\mathbf{x}}^{\mathrm{T}}$ in the second row of (9.12) and, using the first row of (9.11), we have

$$\dot{\mathbf{x}}^{\mathrm{T}}\ddot{\mathbf{s}}=\dot{\mathbf{x}}^{\mathrm{T}}\mathbf{H}\ddot{\mathbf{x}}=\ddot{\mathbf{x}}^{\mathrm{T}}\dot{\mathbf{s}}.$$

This gives,

$$(\dot{\mathbf{x}}(1-\cos(\alpha))+\ddot{\mathbf{x}}\sin(\alpha))^{\mathrm{T}}\mathbf{H}(\dot{\mathbf{x}}(1-\cos(\alpha))+\ddot{\mathbf{x}}\sin(\alpha))$$

$$\begin{aligned}
&= (\dot{\mathbf{x}}^{\mathrm{T}}\mathbf{H}\dot{\mathbf{x}})(1-\cos(\alpha))^{\mathbf{2}} + \mathbf{2}(\dot{\mathbf{x}}^{\mathrm{T}}\mathbf{H}\ddot{\mathbf{x}})\sin(\alpha)(1-\cos(\alpha)) + (\ddot{\mathbf{x}}^{\mathrm{T}}\mathbf{H}\ddot{\mathbf{x}})\sin^{\mathbf{2}}(\alpha) \\
&= (\dot{\mathbf{x}}^{\mathrm{T}}\dot{\mathbf{s}})(1-\cos(\alpha))^2 + (\ddot{\mathbf{x}}^{\mathrm{T}}\ddot{\mathbf{s}})\sin^2(\alpha) + (\ddot{\mathbf{x}}^{\mathrm{T}}\dot{\mathbf{s}}+\dot{\mathbf{x}}^{\mathrm{T}}\ddot{\mathbf{s}})\sin(\alpha)(1-\cos(\alpha)) \geq 0,
\end{aligned}$$

which is the first inequality of (9.19), and

$$\begin{aligned}
& (\dot{\mathbf{x}}(1-\cos(\alpha)) - \ddot{\mathbf{x}}\sin(\alpha))^{\mathrm{T}}\mathbf{H}(\dot{\mathbf{x}}(1-\cos(\alpha)) - \ddot{\mathbf{x}}\sin(\alpha)) \\
&= (\dot{\mathbf{x}}^{\mathrm{T}}\mathbf{H}\dot{\mathbf{x}})(1-\cos(\alpha))^{\mathbf{2}} - \mathbf{2}(\dot{\mathbf{x}}^{\mathrm{T}}\mathbf{H}\ddot{\mathbf{x}})\sin(\alpha)(1-\cos(\alpha)) + (\ddot{\mathbf{x}}^{\mathrm{T}}\mathbf{H}\ddot{\mathbf{x}})\sin^{\mathbf{2}}(\alpha) \\
&= (\dot{\mathbf{x}}^{\mathrm{T}}\dot{\mathbf{s}})(1-\cos(\alpha))^2 + (\ddot{\mathbf{x}}^{\mathrm{T}}\ddot{\mathbf{s}})\sin^2(\alpha) - (\ddot{\mathbf{x}}^{\mathrm{T}}\dot{\mathbf{s}}+\dot{\mathbf{x}}^{\mathrm{T}}\ddot{\mathbf{s}})\sin(\alpha)(1-\cos(\alpha)) \geq 0,
\end{aligned}$$

which is the second inequality of (9.19). Replacing $\dot{\mathbf{x}}(1-\cos(\alpha))$ and $\ddot{\mathbf{x}}\sin(\alpha)$ by $\dot{\mathbf{x}}\sin(\alpha)$ and $\ddot{\mathbf{x}}(1-\cos(\alpha))$, and following the same derivation, we can obtain (9.20). This finishes the proof. ■

A simple result, which was provided in (6.54) and restated here (see also [95]), will be used.

Lemma 9.3

Let $\mathbf{u}$, $\mathbf{v}$, *and* $\mathbf{w}$ *be real vectors of same size satisfying* $\mathbf{u}+\mathbf{v}=\mathbf{w}$ *and* $\mathbf{u}^{\mathrm{T}}\mathbf{v}\geq 0$. *Then,*

$$2\|\mathbf{u}\|\cdot\|\mathbf{v}\| \leq \|\mathbf{u}\|^2+\|\mathbf{v}\|^2 \leq \|\mathbf{u}\|^2+\|\mathbf{v}\|^2+2\mathbf{u}^{\mathrm{T}}\mathbf{v} = \|\mathbf{u}+\mathbf{v}\|^2 = \|\mathbf{w}\|^2. \tag{9.22}$$

Using Lemmas 9.2, 5.3, and 9.3, we can show that $\frac{\dot{\mathbf{x}}}{\mathbf{x}}$, $\frac{\ddot{\mathbf{x}}}{\mathbf{x}}$, $\frac{\dot{\mathbf{s}}}{\mathbf{s}}$, and $\frac{\ddot{\mathbf{s}}}{\mathbf{s}}$ are bounded, as claimed in the following two Lemmas.

Lemma 9.4

Let $(\mathbf{x},\lambda,\mathbf{s})\in\mathcal{N}_2(\theta)$ *and* $(\dot{\mathbf{x}},\dot{\lambda},\dot{\mathbf{s}})$ *meet (9.11). Then,*

$$\left\|\frac{\dot{\mathbf{x}}}{\mathbf{x}}\right\|^2+\left\|\frac{\dot{\mathbf{s}}}{\mathbf{s}}\right\|^2 \leq \frac{n}{1-\theta}, \tag{9.23}$$

$$\left\|\frac{\dot{\mathbf{x}}}{\mathbf{x}}\right\|^2\left\|\frac{\dot{\mathbf{s}}}{\mathbf{s}}\right\|^2 \leq \left(\frac{n}{2(1-\theta)}\right)^2, \tag{9.24}$$

$$0 \leq \frac{\dot{\mathbf{x}}^{\mathrm{T}}\dot{\mathbf{s}}}{\mu} \leq \frac{n(1+\theta)}{2(1-\theta)} := \delta_1 n. \tag{9.25}$$

Proof 9.3 From the last row of (9.11), we have

$$\begin{aligned}
& \mathbf{S}\dot{\mathbf{x}}+\mathbf{X}\dot{\mathbf{s}} = \mathbf{X}\mathbf{S}\mathbf{e} \\
\Longleftrightarrow\quad & \mathbf{X}^{-\frac{1}{2}}\mathbf{S}^{\frac{1}{2}}\dot{\mathbf{x}}+\mathbf{X}^{\frac{1}{2}}\mathbf{S}^{-\frac{1}{2}}\dot{\mathbf{s}} = \mathbf{X}^{\frac{1}{2}}\mathbf{S}^{\frac{1}{2}}\mathbf{e}.
\end{aligned}$$

Let $\mathbf{u}=\mathbf{X}^{-\frac{1}{2}}\mathbf{S}^{\frac{1}{2}}\dot{\mathbf{x}}$, $\mathbf{v}=\mathbf{X}^{\frac{1}{2}}\mathbf{S}^{-\frac{1}{2}}\dot{\mathbf{s}}$, and $\mathbf{w}=\mathbf{X}^{\frac{1}{2}}\mathbf{S}^{\frac{1}{2}}\mathbf{e}$, from Lemma 9.2, $\mathbf{u}^{\mathrm{T}}\mathbf{v}\geq 0$. Using Lemma 9.3, we have

$$\|\mathbf{u}\|^2+\|\mathbf{v}\|^2 = \sum_{i=1}^{n}\frac{\dot{x}_i^2 s_i}{x_i} + \sum_{i=1}^{n}\frac{\dot{s}_i^2 x_i}{s_i} \leq \|\mathbf{X}^{\frac{1}{2}}\mathbf{S}^{\frac{1}{2}}\mathbf{e}\|^2 = n\mu.$$

Dividing both sides of the inequality by $\min_j s_j x_j$ and using (9.9) gives

$$\sum_{i=1}^{n} \frac{\dot{x}_i^2}{x_i^2} + \sum_{i=1}^{n} \frac{\dot{s}_i^2}{s_i^2} \le \frac{n\mu}{\min_j(s_j x_j)} \le \frac{n}{1-\theta},$$

or equivalently

$$\left\|\frac{\dot{\mathbf{x}}}{\mathbf{x}}\right\|^2 + \left\|\frac{\dot{\mathbf{s}}}{\mathbf{s}}\right\|^2 \le \frac{n}{1-\theta}.$$

This proves (9.23). Combining (9.23) and Lemma 5.3 yields

$$\left\|\frac{\dot{\mathbf{x}}}{\mathbf{x}}\right\|^2 \left\|\frac{\dot{\mathbf{s}}}{\mathbf{s}}\right\|^2 \le \left(\frac{n}{2(1-\theta)}\right)^2.$$

This leads to

$$\left\|\frac{\dot{\mathbf{x}}}{\mathbf{x}}\right\| \left\|\frac{\dot{\mathbf{s}}}{\mathbf{s}}\right\| \le \frac{n}{2(1-\theta)}. \tag{9.26}$$

Therefore, using (9.9) and Cauchy-Schwarz inequality yields

$$\begin{aligned} \frac{\dot{\mathbf{x}}^{\mathrm{T}}\dot{\mathbf{s}}}{\mu} &\le \frac{|\dot{\mathbf{x}}|^{\mathrm{T}}|\dot{\mathbf{s}}|}{\mu} \le (1+\theta)\frac{|\dot{\mathbf{x}}|^{\mathrm{T}}|\dot{\mathbf{s}}|}{\max_i(x_i s_i)} \le (1+\theta)\left(\frac{|\dot{\mathbf{x}}|}{\mathbf{x}}\right)^{\mathrm{T}}\left(\frac{|\dot{\mathbf{s}}|}{\mathbf{s}}\right) \\ &\le (1+\theta)\left\|\frac{\dot{\mathbf{x}}}{\mathbf{x}}\right\| \left\|\frac{\dot{\mathbf{s}}}{\mathbf{s}}\right\| \le \frac{n(1+\theta)}{2(1-\theta)}, \end{aligned} \tag{9.27}$$

which is the second inequality of (9.25). From Lemma 9.2, $\dot{\mathbf{x}}^{\mathrm{T}}\dot{\mathbf{s}} = \dot{\mathbf{x}}^{\mathrm{T}}\mathbf{H}\dot{\mathbf{x}} \ge 0$, we have the first inequality of (9.25). ∎

Lemma 9.5
Let $(\mathbf{x},\lambda,\mathbf{s}) \in \mathcal{N}_2(\theta)$, $(\dot{\mathbf{x}},\dot{\lambda},\dot{\mathbf{s}})$ *and* $(\ddot{\mathbf{x}},\ddot{\lambda},\ddot{\mathbf{s}})$ *meet (9.11) and (9.12). Then,*

$$\left\|\frac{\ddot{\mathbf{x}}}{\mathbf{x}}\right\|^2 + \left\|\frac{\ddot{\mathbf{s}}}{\mathbf{s}}\right\|^2 \le \frac{(1+\theta)n^2}{(1-\theta)^3}. \tag{9.28}$$

$$0 \le \frac{\ddot{\mathbf{x}}^{\mathrm{T}}\ddot{\mathbf{s}}}{\mu} \le \frac{n^2(1+\theta)^2}{2(1-\theta)^3} := \delta_2 n^2. \tag{9.29}$$

$$\left|\frac{\dot{\mathbf{x}}^{\mathrm{T}}\ddot{\mathbf{s}}}{\mu}\right| \le \frac{n^{\frac{3}{2}}(1+\theta)^{\frac{3}{2}}}{(1-\theta)^2} := \delta_3 n^{\frac{3}{2}}, \quad \left|\frac{\ddot{\mathbf{x}}^{\mathrm{T}}\dot{\mathbf{s}}}{\mu}\right| \le \frac{n^{\frac{3}{2}}(1+\theta)^{\frac{3}{2}}}{(1-\theta)^2} := \delta_3 n^{\frac{3}{2}}. \tag{9.30}$$

Proof 9.4 Similar to the proof of Lemma 9.4, from the last row of (9.12), we have

$$\begin{aligned} & \mathbf{S}\ddot{\mathbf{x}} + \mathbf{X}\ddot{\mathbf{s}} = -2(\dot{\mathbf{x}} \circ \dot{\mathbf{s}}) \\ \iff\ & \mathbf{X}^{-\frac{1}{2}}\mathbf{S}^{\frac{1}{2}}\ddot{\mathbf{x}} + \mathbf{X}^{\frac{1}{2}}\mathbf{S}^{-\frac{1}{2}}\ddot{\mathbf{s}} = -2\mathbf{X}^{-\frac{1}{2}}\mathbf{S}^{-\frac{1}{2}}(\dot{\mathbf{x}} \circ \dot{\mathbf{s}}), \end{aligned}$$

Let $\mathbf{u} = \mathbf{X}^{-\frac{1}{2}}\mathbf{S}^{\frac{1}{2}}\ddot{\mathbf{x}}$, $\mathbf{v} = \mathbf{X}^{\frac{1}{2}}\mathbf{S}^{-\frac{1}{2}}\ddot{\mathbf{s}}$, and $\mathbf{w} = -2\mathbf{X}^{-\frac{1}{2}}\mathbf{S}^{-\frac{1}{2}}(\dot{\mathbf{x}} \circ \dot{\mathbf{s}})$, from Lemma 9.2, $\mathbf{u}^{\mathrm{T}}\mathbf{v} \geq 0$. Using Lemma 9.3, we have

$$\|\mathbf{u}\|^2 + \|\mathbf{v}\|^2 = \sum_{i=1}^{n} \frac{\ddot{x}_i^2 s_i}{x_i} + \sum_{i=1}^{n} \frac{\ddot{s}_i^2 x_i}{s_i} \leq \left\| -2\mathbf{X}^{-\frac{1}{2}}\mathbf{S}^{-\frac{1}{2}}(\dot{\mathbf{x}} \circ \dot{\mathbf{s}}) \right\|^2 = 4\sum_{i=1}^{n} \left(\frac{\dot{x}_i^2}{x_i} \frac{\dot{s}_i^2}{s_i} \right)$$

Dividing both sides of the inequality by μ and using (9.9) gives

$$(1-\theta)\left(\sum_{i=1}^{n} \frac{\ddot{x}_i^2}{x_i^2} + \sum_{i=1}^{n} \frac{\ddot{s}_i^2}{s_i^2} \right) \leq 4(1+\theta)\left(\sum_{i=1}^{n} \left(\frac{\dot{x}_i^2}{x_i^2} \frac{\dot{s}_i^2}{s_i^2} \right) \right),$$

or equivalently

$$\left\| \frac{\ddot{\mathbf{x}}}{\mathbf{x}} \right\|^2 + \left\| \frac{\ddot{\mathbf{s}}}{\mathbf{s}} \right\|^2 \leq 4\frac{1+\theta}{1-\theta} \left\| \frac{\dot{\mathbf{x}}}{\mathbf{x}} \circ \frac{\dot{\mathbf{s}}}{\mathbf{s}} \right\|^2 \leq \frac{(1+\theta)n^2}{(1-\theta)^3}.$$

Therefore, using Lemma 5.3, we have

$$\left\| \frac{\ddot{\mathbf{x}}}{\mathbf{x}} \right\|^2 \left\| \frac{\ddot{\mathbf{s}}}{\mathbf{s}} \right\|^2 \leq \frac{1}{4} \left(\frac{(1+\theta)n^2}{(1-\theta)^3} \right)^2,$$

and

$$\begin{aligned} \frac{\ddot{\mathbf{x}}^{\mathrm{T}}\ddot{\mathbf{s}}}{\mu} &\leq \frac{|\ddot{\mathbf{x}}|^{\mathrm{T}}|\ddot{\mathbf{s}}|}{\mu} \leq (1+\theta)\frac{|\ddot{\mathbf{x}}|^{\mathrm{T}}|\ddot{\mathbf{s}}|}{\max_i(x_i s_i)} \leq (1+\theta)\left(\frac{|\ddot{\mathbf{x}}|}{\mathbf{x}} \right)^{\mathrm{T}} \left(\frac{|\ddot{\mathbf{s}}|}{\mathbf{s}} \right) \\ &\leq (1+\theta)\left\| \frac{\ddot{\mathbf{x}}}{\mathbf{x}} \right\| \left\| \frac{\ddot{\mathbf{s}}}{\mathbf{s}} \right\| \leq \frac{n^2(1+\theta)^2}{2(1-\theta)^3}, \end{aligned}$$

which is the second inequality of (9.29). From Lemma 9.2, $\ddot{\mathbf{x}}^{\mathrm{T}}\ddot{\mathbf{s}} = \ddot{\mathbf{x}}^{\mathrm{T}}\mathbf{H}\ddot{\mathbf{x}} \geq 0$, we have the first inequality of (9.29). Similarly, it is easy to show (9.30). ∎

Using the bounds established in Lemmas 9.2, 9.4, 9.5, and 5.4, we can obtain the lower bound and upper bound for $\mu(\alpha)$.

Lemma 9.6
Let $(\mathbf{x}, \lambda, \mathbf{s}) \in \mathcal{N}_2(\theta)$, $(\dot{\mathbf{x}}, \dot{\lambda}, \dot{\mathbf{s}})$ and $(\ddot{\mathbf{x}}, \ddot{\lambda}, \ddot{\mathbf{s}})$ meet (9.11) and (9.12). Let $\mathbf{x}(\alpha)$ and $\mathbf{s}(\alpha)$ be defined by (9.13) and (9.15). Then,

$$\begin{aligned} & \mu(1-\sin(\alpha)) - \frac{1}{n}\left((\ddot{\mathbf{x}}^{\mathrm{T}}\dot{\mathbf{s}})\sin^4(\alpha) + (\dot{\mathbf{x}}^{\mathrm{T}}\dot{\mathbf{s}})\sin^2(\alpha) \right) \\ \leq\ & \mu(\alpha) = \mu(1-\sin(\alpha)) \\ & + \frac{1}{n}\left(\left(\ddot{\mathbf{x}}^{\mathrm{T}}\ddot{\mathbf{s}} - \dot{\mathbf{x}}^{\mathrm{T}}\dot{\mathbf{s}} \right)(1-\cos(\alpha))^2 - \left(\dot{\mathbf{x}}^{\mathrm{T}}\ddot{\mathbf{s}} + \dot{\mathbf{s}}^{\mathrm{T}}\ddot{\mathbf{x}} \right)\sin(\alpha)(1-\cos(\alpha)) \right) \\ \leq\ & \mu(1-\sin(\alpha)) + \frac{1}{n}\left((\ddot{\mathbf{x}}^{\mathrm{T}}\ddot{\mathbf{s}})\sin^4(\alpha) + (\ddot{\mathbf{x}}^{\mathrm{T}}\dot{\mathbf{s}})\sin^2(\alpha) \right). \end{aligned} \tag{9.31}$$

Proof 9.5 Using (9.13) and (9.15), we have

$$
\begin{aligned}
n\mu(\alpha) = \mathbf{x}(\alpha)^{\mathrm{T}}\mathbf{s}(\alpha) \quad &= \left(\mathbf{x}^{\mathrm{T}} - \dot{\mathbf{x}}^{\mathrm{T}}\sin(\alpha) + \ddot{\mathbf{x}}^{\mathrm{T}}(1-\cos(\alpha))\right) \\
&\quad \left(\mathbf{s} - \dot{\mathbf{s}}\sin(\alpha) + \ddot{\mathbf{s}}(1-\cos(\alpha))\right) \\
&= \mathbf{x}^{\mathrm{T}}\mathbf{s} - \mathbf{x}^{\mathrm{T}}\dot{\mathbf{s}}\sin(\alpha) + \mathbf{x}^{\mathrm{T}}\ddot{\mathbf{s}}(1-\cos(\alpha)) \\
&\quad -\dot{\mathbf{x}}^{\mathrm{T}}\mathbf{s}\sin(\alpha) + \dot{\mathbf{x}}^{\mathrm{T}}\dot{\mathbf{s}}\sin^2(\alpha) - \dot{\mathbf{x}}^{\mathrm{T}}\ddot{\mathbf{s}}\sin(\alpha)(1-\cos(\alpha)) \\
&\quad +\ddot{\mathbf{x}}^{\mathrm{T}}\mathbf{s}(1-\cos(\alpha)) - \ddot{\mathbf{x}}^{\mathrm{T}}\dot{\mathbf{s}}\sin(\alpha)(1-\cos(\alpha)) \\
&\quad +\ddot{\mathbf{x}}^{\mathrm{T}}\ddot{\mathbf{s}}(1-\cos(\alpha))^2 \\
&= \mathbf{x}^{\mathrm{T}}\mathbf{s} - (\mathbf{x}^{\mathrm{T}}\dot{\mathbf{s}} + \mathbf{s}^{\mathrm{T}}\dot{\mathbf{x}})\sin(\alpha) + (\mathbf{x}^{\mathrm{T}}\ddot{\mathbf{s}} + \mathbf{s}^{\mathrm{T}}\ddot{\mathbf{x}})(1-\cos(\alpha)) \\
&\quad -(\dot{\mathbf{x}}^{\mathrm{T}}\ddot{\mathbf{s}} + \dot{\mathbf{s}}^{\mathrm{T}}\ddot{\mathbf{x}})\sin(\alpha)(1-\cos(\alpha)) + \dot{\mathbf{x}}^{\mathrm{T}}\dot{\mathbf{s}}\sin^2(\alpha) \\
&\quad +\ddot{\mathbf{x}}^{\mathrm{T}}\ddot{\mathbf{s}}(1-\cos(\alpha))^2 \\
\text{use last rows of} \quad &= n\mu(1-\sin(\alpha)) - 2\dot{\mathbf{x}}^{\mathrm{T}}\dot{\mathbf{s}}(1-\cos(\alpha)) \\
\text{(9.11) and (9.12)} \quad &\quad -(\dot{\mathbf{x}}^{\mathrm{T}}\ddot{\mathbf{s}} + \dot{\mathbf{s}}^{\mathrm{T}}\ddot{\mathbf{x}})\sin(\alpha)(1-\cos(\alpha)) \\
&\quad +\dot{\mathbf{x}}^{\mathrm{T}}\dot{\mathbf{s}}(1-\cos^2(\alpha)) + \ddot{\mathbf{x}}^{\mathrm{T}}\ddot{\mathbf{s}}(1-\cos(\alpha))^2 \\
&= n\mu(1-\sin(\alpha)) + (\ddot{\mathbf{x}}^{\mathrm{T}}\ddot{\mathbf{s}} - \dot{\mathbf{x}}^{\mathrm{T}}\dot{\mathbf{s}})(1-\cos(\alpha))^2 \\
&\quad -(\dot{\mathbf{x}}^{\mathrm{T}}\ddot{\mathbf{s}} + \dot{\mathbf{s}}^{\mathrm{T}}\ddot{\mathbf{x}})\sin(\alpha)(1-\cos(\alpha)) \\
&\leq n\mu(1-\sin(\alpha)) + (\ddot{\mathbf{x}}^{\mathrm{T}}\ddot{\mathbf{s}} - \dot{\mathbf{x}}^{\mathrm{T}}\dot{\mathbf{s}})(1-\cos(\alpha))^2 \\
\text{use (9.19)} \quad &\quad +(\dot{\mathbf{x}}^{\mathrm{T}}\dot{\mathbf{s}})(1-\cos(\alpha))^2 + (\ddot{\mathbf{x}}^{\mathrm{T}}\ddot{\mathbf{s}})\sin^2(\alpha) \\
\text{use Lemma 5.4} \quad &\leq n\mu\left(1-\sin(\alpha)\right) + (\ddot{\mathbf{x}}^{\mathrm{T}}\ddot{\mathbf{s}})\sin^4(\alpha) + (\ddot{\mathbf{x}}^{\mathrm{T}}\ddot{\mathbf{s}})\sin^2(\alpha) \\
\text{use Lemma 9.5} \quad &\leq n\mu\left(1-\sin(\alpha) + \delta_2 n\sin^2(\alpha) + \delta_2 n\sin^4(\alpha)\right).
\end{aligned}
$$

This proves the second inequality of the lemma. Combining the last equality of the above formulas and (9.20) proves the first inequality of the lemma. ■

To keep all iterates of the algorithm inside the feasible set, we need $(\mathbf{x}(\alpha), \mathbf{s}(\alpha)) > \mathbf{0}$ in all the iterations. We will prove that this is guaranteed if $\mu(\alpha) > 0$ holds. The following corollary states the condition for $\mu(\alpha) > 0$ to hold.

Corollary 9.1
If $\mu > 0$, *then for any fixed* $\theta \in (0,1)$, *there is an* $\bar{\alpha}$ *depending on* θ, *such that for any* $\sin(\alpha) \leq \sin(\bar{\alpha})$, $\mu(\alpha) > 0$. *In particular, if* $\theta = 0.148$, $\sin(\bar{\alpha}) \geq 0.6286$.

Proof 9.6 Since

$$
\begin{aligned}
\mu(\alpha) &\geq \mu\left((1-\sin(\alpha) - \frac{1}{n\mu}\left((\dot{\mathbf{x}}^{\mathrm{T}}\dot{\mathbf{s}})\sin^4(\alpha) + (\dot{\mathbf{x}}^{\mathrm{T}}\dot{\mathbf{s}})\sin^2(\alpha)\right)\right) \\
&\geq \mu\left((1-\sin(\alpha) - \frac{(1+\theta)}{2(1-\theta)}\left(\sin^4(\alpha) + \sin^2(\alpha)\right)\right) := \mu r(\alpha),
\end{aligned}
$$

$\mu > 0$, and $r(\alpha)$ is a monotonic decreasing function in $[0, \frac{\pi}{2}]$ with $r(0) > 0$, $r(\frac{\pi}{2}) < 0$, there is a unique real solution $\sin(\bar{\alpha}) \in (0,1)$ of $r(\alpha) = 0$ such that for all $\sin(\alpha) <$

$\sin(\bar{\alpha})$, $r(\alpha) > 0$, or $\mu(\alpha) > 0$. It is easy to verify that if $\theta = 0.148$, $\sin(\bar{\alpha}) = 0.6286$ is the solution of $r(\alpha) = 0$. ∎

Remark 9.1 Intuitively, to search in a wider region will generate a longer step. Therefore, the larger the θ is, the better. However, in order to derive the convergence result, $\theta \leq 0.148$ is imposed in Lemma 9.12. ∎

To reduce the duality measure in an iteration, we need to have $\mu(\alpha) \leq \mu$. For linear programming, it has been shown in Chapter 5 that $\mu(\alpha) \leq \mu$ for $\alpha \in [0, \hat{\alpha}]$ with $\hat{\alpha} = \frac{\pi}{2}$, and the larger the α in the interval is, the smaller the $\mu(\alpha)$ will be. This claim may not be true for the convex quadratic programming and it needs to be modified. We first introduce the famous Cardano's formula, which can be found in [113].

Lemma 9.7
Let p and q be the real numbers that are related to the following cubic algebra equation

$$x^3 + px + q = 0.$$

If

$$\Delta = \left(\frac{q}{2}\right)^2 + \left(\frac{p}{3}\right)^3 > 0,$$

then the cubic equation has one real root that is given by

$$x = \sqrt[3]{-\frac{q}{2} + \sqrt{\left(\frac{q}{2}\right)^2 + \left(\frac{p}{3}\right)^3}} + \sqrt[3]{-\frac{q}{2} - \sqrt{\left(\frac{q}{2}\right)^2 + \left(\frac{p}{3}\right)^3}}.$$

Lemma 9.8
Let $(\mathbf{x}, \lambda, \mathbf{s}) \in \mathcal{N}_2(\theta)$, $(\dot{\mathbf{x}}, \dot{\lambda}, \dot{\mathbf{s}})$ and $(\ddot{\mathbf{x}}, \ddot{\lambda}, \ddot{\mathbf{s}})$ meet (9.11) and (9.12). Let $\mathbf{x}(\alpha)$ and $\mathbf{s}(\alpha)$ be defined by (9.13) and (9.15). Then, there exists

$$\hat{\alpha} = \begin{cases} \frac{\pi}{2}, & \text{if } \frac{\ddot{\mathbf{x}}^{\mathrm{T}}\ddot{\mathbf{s}}}{n\mu} \leq \frac{1}{2} \\ \sin^{-1}(g), & \text{if } \frac{\ddot{\mathbf{x}}^{\mathrm{T}}\ddot{\mathbf{s}}}{n\mu} > \frac{1}{2} \end{cases} \tag{9.32}$$

where

$$g = \sqrt[3]{\frac{n\mu}{2\ddot{\mathbf{x}}^{\mathrm{T}}\ddot{\mathbf{s}}} + \sqrt{\left(\frac{n\mu}{2\ddot{\mathbf{x}}^{\mathrm{T}}\ddot{\mathbf{s}}}\right)^2 + \left(\frac{1}{3}\right)^3}} + \sqrt[3]{\frac{n\mu}{2\ddot{\mathbf{x}}^{\mathrm{T}}\ddot{\mathbf{s}}} - \sqrt{\left(\frac{n\mu}{2\ddot{\mathbf{x}}^{\mathrm{T}}\ddot{\mathbf{s}}}\right)^2 + \left(\frac{1}{3}\right)^3}},$$

such that for every $\alpha \in [0, \hat{\alpha}]$, $\mu(\alpha) \leq \mu$.

Proof 9.7 From the second inequality of (9.31), we have

$$\mu(\alpha)-\mu \le \mu \sin(\alpha)\left(-1+\frac{\ddot{\mathbf{x}}^{\mathrm{T}}\ddot{\mathbf{s}}}{n\mu}\sin(\alpha)+\frac{\ddot{\mathbf{x}}^{\mathrm{T}}\ddot{\mathbf{s}}}{n\mu}\sin^3(\alpha)\right).$$

Clearly, if $\frac{\ddot{\mathbf{x}}^{\mathrm{T}}\ddot{\mathbf{s}}}{n\mu} \le \frac{1}{2}$, for any $\alpha \in [0, \frac{\pi}{2}]$, the function

$$f(\alpha) := \left(-1+\frac{\ddot{\mathbf{x}}^{\mathrm{T}}\ddot{\mathbf{s}}}{n\mu}\sin(\alpha)+\frac{\ddot{\mathbf{x}}^{\mathrm{T}}\ddot{\mathbf{s}}}{n\mu}\sin^3(\alpha)\right) \le 0,$$

and $\mu(\alpha) \le \mu$. If $\frac{\ddot{\mathbf{x}}^{\mathrm{T}}\ddot{\mathbf{s}}}{n\mu} > \frac{1}{2}$, using Cardano's formula, we conclude that the function f has one real solution $\sin(\alpha) \in (0,1)$. The solution is given as

$$\sin(\hat{\alpha}) = \sqrt[3]{\frac{n\mu}{2\ddot{\mathbf{x}}^{\mathrm{T}}\ddot{\mathbf{s}}}+\sqrt{\left(\frac{n\mu}{2\ddot{\mathbf{x}}^{\mathrm{T}}\ddot{\mathbf{s}}}\right)^2+\left(\frac{1}{3}\right)^3}}+\sqrt[3]{\frac{n\mu}{2\ddot{\mathbf{x}}^{\mathrm{T}}\ddot{\mathbf{s}}}-\sqrt{\left(\frac{n\mu}{2\ddot{\mathbf{x}}^{\mathrm{T}}\ddot{\mathbf{s}}}\right)^2+\left(\frac{1}{3}\right)^3}}.$$

This proves the Lemma. ■

According to Theorem 9.1, Lemmas 9.1, 9.4, 9.5, 9.6, and 9.8, if α is small enough, then $(\mathbf{x}(\alpha),\mathbf{s}(\alpha)) > \mathbf{0}$, and $\mu(\alpha) < \mu$, i.e., the search along the arc defined by Theorem 9.1, will generate a strict feasible point with smaller duality measure. Since $(\mathbf{x},\mathbf{s}) > \mathbf{0}$ holds in all iterations, reducing the duality measure to zero means approaching to the solution of the convex quadratic programming. We will apply a similar idea used in Chapters 3 and 5, i.e., starting with an iterate in $\mathcal{N}_2(\theta)$, searching along the approximated central path in order to reduce the duality measure and to keep the iterate in $\mathcal{N}_2(2\theta)$, and then making a correction to move the iterate back to $\mathcal{N}_2(\theta)$. Let

$$a_0 = -\theta\mu < 0,$$

$$a_1 = \theta\mu > 0,$$

$$a_2 = 2\theta\frac{\dot{\mathbf{x}}^{\mathrm{T}}\dot{\mathbf{s}}}{n} = 2\theta\frac{\dot{\mathbf{x}}^{\mathrm{T}}\mathbf{H}\dot{\mathbf{x}}}{n} \ge 0,$$

$$a_3 = \left\|\dot{\mathbf{x}}\circ\ddot{\mathbf{s}}+\dot{\mathbf{s}}\circ\ddot{\mathbf{x}}-\frac{1}{n}(\dot{\mathbf{x}}^{\mathrm{T}}\ddot{\mathbf{s}}+\dot{\mathbf{s}}^{\mathrm{T}}\ddot{\mathbf{x}})\mathbf{e}\right\| \ge 0,$$

$$a_4 = \left\|\ddot{\mathbf{x}}\circ\ddot{\mathbf{s}}-\dot{\mathbf{s}}\circ\dot{\mathbf{x}}-\frac{1}{n}(\ddot{\mathbf{x}}^{\mathrm{T}}\ddot{\mathbf{s}}-\dot{\mathbf{s}}^{\mathrm{T}}\dot{\mathbf{x}})\mathbf{e}\right\| + 2\theta\frac{\dot{\mathbf{x}}^{\mathrm{T}}\dot{\mathbf{s}}}{n} \ge 0.$$

We define a quartic polynomial in terms of $\sin(\alpha)$ as follows:

$$q(\alpha) = a_4\sin^4(\alpha)+a_3\sin^3(\alpha)+a_2\sin^2(\alpha)+a_1\sin(\alpha)+a_0 = 0. \quad (9.33)$$

Since $q(\alpha)$ is a monotonic increasing function of $\alpha \in [0, \frac{\pi}{2}]$, $q(0) < 0$ and $q(\frac{\pi}{2}) > 0$, the polynomial has exactly one positive root in $[0, \frac{\pi}{2}]$. Moreover, since (9.33) is a quartic equation, all the solutions are analytical and the computational cost is independent of the size of $\mathbf{A}$ (n and m) and is negligible [35].

Lemma 9.9
Let $\theta \le 0.148$ *and* $(\mathbf{x}^k, \lambda^k, \mathbf{s}^k) \in \mathcal{N}_2(\theta)$, $(\dot{\mathbf{x}}, \dot{\lambda}, \dot{\mathbf{s}})$ *and* $(\ddot{\mathbf{x}}, \ddot{\lambda}, \ddot{\mathbf{s}})$ *be calculated from (9.11) and (9.12). Let* $\sin(\tilde{\alpha})$ *be the only positive real solution of (9.33) in* $[0,1]$. *Assume* $\sin(\alpha) \le \min\{\sin(\tilde{\alpha}), \sin(\bar{\alpha})\}$, *let* $(\mathbf{x}(\alpha), \lambda(\alpha), \mathbf{s}(\alpha))$ *and* $\mu(\alpha)$ *be updated as follows*

$$(\mathbf{x}(\alpha), \lambda(\alpha), \mathbf{s}(\alpha)) = (\mathbf{x}^k, \lambda^k, \mathbf{s}^k) - (\dot{\mathbf{x}}, \dot{\lambda}, \dot{\mathbf{s}}) \sin(\alpha) + (\ddot{\mathbf{x}}, \ddot{\lambda}, \ddot{\mathbf{s}})(1 - \cos(\alpha)), \quad (9.34)$$

$$\begin{aligned} \mu(\alpha) &= \mu_k(1 - \sin(\alpha)) \\ &+ \frac{1}{n}\left[(\ddot{\mathbf{x}}^{\mathrm{T}}\ddot{\mathbf{s}} - \dot{\mathbf{x}}^{\mathrm{T}}\dot{\mathbf{s}})(1 - \cos(\alpha))^2 - (\dot{\mathbf{x}}^{\mathrm{T}}\ddot{\mathbf{s}} + \dot{\mathbf{s}}^{\mathrm{T}}\ddot{\mathbf{x}}) \sin(\alpha)(1 - \cos(\alpha))\right]. \end{aligned} \quad (9.35)$$

Then, $(\mathbf{x}(\alpha), \lambda(\alpha), \mathbf{s}(\alpha)) \in \mathcal{N}_2(2\theta)$.

Proof 9.8 Since $\sin(\tilde{\alpha})$ is the only positive real solution of (9.33) in $[0,1]$ and $q(0) < 0$, substituting a_0, a_1, a_2, a_3 and a_4 into (9.33), we have, for all $\sin(\alpha) \le \sin(\tilde{\alpha})$,

$$\begin{aligned} & \left(\left\|\ddot{\mathbf{x}} \circ \ddot{\mathbf{s}} - \dot{\mathbf{s}} \circ \dot{\mathbf{x}} - \frac{1}{n}(\ddot{\mathbf{x}}^{\mathrm{T}}\ddot{\mathbf{s}} - \dot{\mathbf{s}}^{\mathrm{T}}\dot{\mathbf{x}})\mathbf{e}\right\|\right) \sin^4(\alpha) \\ & + \left(\left\|\dot{\mathbf{x}} \circ \ddot{\mathbf{s}} + \dot{\mathbf{s}} \circ \ddot{\mathbf{x}} - \frac{1}{n}(\dot{\mathbf{x}}^{\mathrm{T}}\ddot{\mathbf{s}} + \dot{\mathbf{s}}^{\mathrm{T}}\ddot{\mathbf{x}})\mathbf{e}\right\|\right) \sin^3(\alpha) \\ \le & -\left(2\theta \frac{\dot{\mathbf{x}}^{\mathrm{T}}\dot{\mathbf{s}}}{n}\right) \sin^4(\alpha) - \left(2\theta \frac{\dot{\mathbf{x}}^{\mathrm{T}}\dot{\mathbf{s}}}{n}\right) \sin^2(\alpha) + \theta\mu_k(1 - \sin(\alpha)). \end{aligned} \quad (9.36)$$

From (9.11), (9.12), (9.34) and (9.35), using Lemmas 5.4, 9.6 and (9.36), we have

$$\begin{aligned} & \left\|\mathbf{x}(\alpha) \circ \mathbf{s}(\alpha) - \mu(\alpha)\mathbf{e}\right\| \\ = & \left\|(\mathbf{x}^k \circ \mathbf{s}^k - \mu_k\mathbf{e})(1 - \sin(\alpha))\right. \\ & + \left(\ddot{\mathbf{x}} \circ \ddot{\mathbf{s}} - \dot{\mathbf{x}} \circ \dot{\mathbf{s}} - \frac{1}{n}(\ddot{\mathbf{x}}^{\mathrm{T}}\ddot{\mathbf{s}} - \dot{\mathbf{x}}^{\mathrm{T}}\dot{\mathbf{s}})\mathbf{e}\right)(1 - \cos(\alpha))^2 \\ & \left. - \left(\dot{\mathbf{x}} \circ \ddot{\mathbf{s}} + \dot{\mathbf{s}} \circ \ddot{\mathbf{x}} - \frac{1}{n}(\dot{\mathbf{x}}^{\mathrm{T}}\ddot{\mathbf{s}} + \ddot{\mathbf{x}}^{\mathrm{T}}\dot{\mathbf{s}})\mathbf{e}\right) \sin(\alpha)(1 - \cos(\alpha))\right\| \\ \le & (1 - \sin(\alpha))\left\|\mathbf{x}^k \circ \mathbf{s}^k - \mu_k\mathbf{e}\right\| \\ & + \left\|(\ddot{\mathbf{x}} \circ \ddot{\mathbf{s}} - \dot{\mathbf{x}} \circ \dot{\mathbf{s}} - \frac{1}{n}(\ddot{\mathbf{x}}^{\mathrm{T}}\ddot{\mathbf{s}} - \dot{\mathbf{x}}^{\mathrm{T}}\dot{\mathbf{s}}))\mathbf{e}\right\|(1 - \cos(\alpha))^2 \end{aligned}$$

$$
\begin{aligned}
& +\left\|(\dot{\mathbf{x}} \circ \ddot{\mathbf{s}}+\dot{\mathbf{s}} \circ \ddot{\mathbf{x}}-\frac{1}{n}(\dot{\mathbf{x}}^{\mathrm{T}} \ddot{\mathbf{s}}+\ddot{\mathbf{x}}^{\mathrm{T}} \dot{\mathbf{s}}) \mathbf{e}\right\| \sin(\alpha)(1-\cos(\alpha)) \\
\leq\ & \theta \mu_k(1-\sin(\alpha)) \\
& +\left\|(\ddot{\mathbf{x}} \circ \ddot{\mathbf{s}}-\dot{\mathbf{x}} \circ \dot{\mathbf{s}}-\frac{1}{n}(\ddot{\mathbf{x}}^{\mathrm{T}} \ddot{\mathbf{s}}-\dot{\mathbf{x}}^{\mathrm{T}} \dot{\mathbf{s}})) \mathbf{e}\right\| \sin^4(\alpha)+a_3 \sin^3(\alpha) \\
\leq\ & 2 \theta \mu_k(1-\sin(\alpha))-\left(2 \theta \frac{\dot{\mathbf{x}}^{\mathrm{T}} \dot{\mathbf{s}}}{n}\right)(\sin^4(\alpha)+\sin^2(\alpha)) \\
\leq\ & 2 \theta \mu(\alpha). \qquad \text{[use Lemma 9.6]}
\end{aligned} \tag{9.37}
$$

Hence, the point $(\mathbf{x}(\alpha), \lambda(\alpha), \mathbf{s}(\alpha))$ satisfies the proximity condition for $\mathcal{N}_2(2\theta)$. To check the positivity condition $(\mathbf{x}(\alpha), \mathbf{s}(\alpha)) > \mathbf{0}$, note that the initial condition $(\mathbf{x}, \mathbf{s}) > 0$. It follows from (9.37) and Corollary 9.1 that, for $\theta \leq 0.148$ and $\sin(\alpha) \leq \sin(\bar{\alpha})$,

$$
x_i(\alpha) s_i(\alpha) \geq (1-2\theta)\mu(\alpha) > 0. \tag{9.38}
$$

Therefore, we cannot have $x_i(\alpha) = 0$ or $s_i(\alpha) = 0$ for any index i when $\alpha \in [0, \sin^{-1}(\bar{\alpha})]$. This proves $(\mathbf{x}(\alpha), \mathbf{s}(\alpha)) > \mathbf{0}$. ∎

Remark 9.2 It is worthwhile to note, by examining the proof of Lemma 9.9, that $\sin(\tilde{\alpha})$ is selected for the proximity condition (9.37) to hold, and $\sin(\bar{\alpha})$ is selected for $\mu(\alpha) > 0$, thereby assuring the positivity condition (9.38) to hold. ∎

The lower bound of $\sin(\bar{\alpha})$ is estimated in Corollary 9.1. To estimate the lower bound of $\sin(\tilde{\alpha})$, we need the following lemma.

Lemma 9.10
Let $(\mathbf{x}, \lambda, \mathbf{s}) \in \mathcal{N}_2(\theta)$, $(\dot{\mathbf{x}}, \dot{\lambda}, \dot{\mathbf{s}})$ *and* $(\ddot{\mathbf{x}}, \ddot{\lambda}, \ddot{\mathbf{s}})$ *meet (9.11) and (9.12). Then,*

$$
\left\|\dot{\mathbf{x}} \circ \dot{\mathbf{s}}\right\| \leq \frac{\mu(1+\theta)}{2(1-\theta)} n, \tag{9.39}
$$

$$
\left\|\ddot{\mathbf{x}} \circ \ddot{\mathbf{s}}\right\| \leq \frac{(1+\theta)^2 \mu}{2(1-\theta)^3} n^2, \tag{9.40}
$$

$$
\left\|\ddot{\mathbf{x}} \circ \dot{\mathbf{s}}\right\| \leq \frac{(1+\theta)^{\frac{3}{2}} \mu}{(1-\theta)^2} n^{\frac{3}{2}}, \tag{9.41}
$$

$$
\left\|\dot{\mathbf{x}} \circ \ddot{\mathbf{s}}\right\| \leq \frac{(1+\theta)^{\frac{3}{2}} \mu}{(1-\theta)^2} n^{\frac{3}{2}}. \tag{9.42}
$$

Proof 9.9 Since

$$
\left\|\frac{\dot{\mathbf{x}}}{\mathbf{x}}\right\|^2 = \sum_{i=1}^{n} \left(\frac{\dot{x}_i}{x_i}\right)^2, \quad \left\|\frac{\dot{\mathbf{s}}}{\mathbf{s}}\right\|^2 = \sum_{i=1}^{n} \left(\frac{\dot{s}_i}{s_i}\right)^2,
$$

From Lemma 9.4, we have

$$
\begin{aligned}
& \left(\frac{n}{2(1-\theta)}\right)^2 \\
\geq \; & \left\|\frac{\dot{\mathbf{x}}}{\mathbf{x}}\right\|^2\left\|\frac{\dot{\mathbf{s}}}{\mathbf{s}}\right\|^2=\left(\sum_{i=1}^n\left(\frac{\dot{x}_i}{x_i}\right)^2\right)\left(\sum_{i=1}^n\left(\frac{\dot{s}_i}{s_i}\right)^2\right) \\
\geq \; & \sum_{i=1}^n\left(\frac{\dot{x}_i}{x_i}\frac{\dot{s}_i}{s_i}\right)^2=\left\|\frac{\dot{\mathbf{x}}}{\mathbf{x}}\circ\frac{\dot{\mathbf{s}}}{\mathbf{s}}\right\|^2 \\
\geq \; & \sum_{i=1}^n\left(\frac{\dot{x}_i\dot{s}_i}{(1+\theta)\mu}\right)^2=\frac{1}{(1+\theta)^2\mu^2}\left\|\dot{\mathbf{x}}\circ\dot{\mathbf{s}}\right\|^2.
\end{aligned}
$$

This proves (9.39). Using

$$
\left\|\frac{\ddot{\mathbf{x}}}{\mathbf{x}}\right\|^2=\sum_{i=1}^n\left(\frac{\ddot{x}_i}{x_i}\right)^2, \quad \left\|\frac{\ddot{\mathbf{s}}}{\mathbf{s}}\right\|^2=\sum_{i=1}^n\left(\frac{\ddot{s}_i}{s_i}\right)^2,
$$

and Lemmas 9.5 and 5.3, then, following the same procedure, it is easy to verify (9.40). From (9.23) and (9.28), we have

$$
\begin{aligned}
& \left(\frac{n}{(1-\theta)}\right)\left(\frac{(1+\theta)n^2}{(1-\theta)^3}\right) \\
\geq \; & \left\|\frac{\ddot{\mathbf{x}}}{\mathbf{x}}\right\|^2\left\|\frac{\dot{\mathbf{s}}}{\mathbf{s}}\right\|^2+\left\|\frac{\dot{\mathbf{x}}}{\mathbf{x}}\right\|^2\left\|\frac{\ddot{\mathbf{s}}}{\mathbf{s}}\right\|^2 \\
= \; & \left(\sum_{i=1}^n\left(\frac{\ddot{x}_i}{x_i}\right)^2\right)\left(\sum_{i=1}^n\left(\frac{\dot{s}_i}{s_i}\right)^2\right)+\left(\sum_{i=1}^n\left(\frac{\dot{x}_i}{x_i}\right)^2\right)\left(\sum_{i=1}^n\left(\frac{\ddot{s}_i}{s_i}\right)^2\right) \\
\geq \; & \sum_{i=1}^n\left(\frac{\ddot{x}_i\dot{s}_i}{x_is_i}\right)^2+\sum_{i=1}^n\left(\frac{\dot{x}_i\ddot{s}_i}{x_is_i}\right)^2 \\
\geq \; & \sum_{i=1}^n\left(\frac{\ddot{x}_i\dot{s}_i}{(1+\theta)\mu}\right)^2+\sum_{i=1}^n\left(\frac{\dot{x}_i\ddot{s}_i}{(1+\theta)\mu}\right)^2 \\
= \; & \frac{1}{(1+\theta)^2\mu^2}\left(\left\|\ddot{\mathbf{x}}\circ\dot{\mathbf{s}}\right\|^2+\left\|\dot{\mathbf{x}}\circ\ddot{\mathbf{s}}\right\|^2\right).
\end{aligned}
\tag{9.43}
$$

This proves the lemma. ■

The next technical lemma will be used a few times in the rest of the book.

Lemma 9.11
Let $\mathbf{u}$ *and* $\mathbf{v}$ *be the n-dimensional vectors. Then,*

$$
\left\|\mathbf{u}\circ\mathbf{v}-\frac{1}{n}\left(\mathbf{u}^{\mathrm{T}}\mathbf{v}\right)\mathbf{e}\right\|\leq\left\|\mathbf{u}\circ\mathbf{v}\right\|.
$$

Proof 9.10 Simple calculation gives

$$\begin{aligned}
&\left\|\mathbf{u}\circ\mathbf{v}-\frac{1}{n}\left(\mathbf{u}^{\mathrm{T}}\mathbf{v}\right)\mathbf{e}\right\|^2 \\
&= \sum_{i=1}^{n}\left(u_iv_i-\frac{1}{n}\sum_{i=1}^{n}u_iv_i\right)^2 \\
&= \sum_{i=1}^{n}\left(u_i^2v_i^2-\frac{2u_iv_i}{n}\sum_{i=1}^{n}u_iv_i+\frac{1}{n^2}\left(\sum_{i=1}^{n}u_iv_i\right)^2\right) \\
&= \sum_{i=1}^{n}\left(u_i^2v_i^2\right)-\frac{2}{n}\left(\sum_{i=1}^{n}u_iv_i\right)^2+\frac{1}{n}\left(\sum_{i=1}^{n}u_iv_i\right)^2 \\
&= \sum_{i=1}^{n}\left(u_i^2v_i^2\right)-\frac{1}{n}\left(\sum_{i=1}^{n}u_iv_i\right)^2 \le \|\mathbf{u}\circ\mathbf{v}\|^2.
\end{aligned}$$

This completes the proof. ■

This result will be used in the proof of the following lemma.

Lemma 9.12
Let $\theta \le 0.148$. Then, $\sin(\tilde{\alpha}) \ge \frac{2\theta}{\sqrt{n}}$ for $n \ge 2$.

Proof 9.11 First, notice that $q(\sin(\alpha))$ is a monotonic increasing function of $\sin(\alpha)$ for $\alpha \in [0, \frac{\pi}{2}]$ and $q(\sin(0)) < 0$, therefore, we need only to show that $q(\frac{2\theta}{\sqrt{n}}) < 0$ for $\theta \le 0.148$ and $n \ge 2$. From Lemma 9.11, we have

$$\left\|\dot{\mathbf{x}}\circ\ddot{\mathbf{s}}+\dot{\mathbf{s}}\circ\ddot{\mathbf{x}}-\frac{1}{n}(\dot{\mathbf{x}}^{\mathrm{T}}\ddot{\mathbf{s}}+\dot{\mathbf{s}}^{\mathrm{T}}\ddot{\mathbf{x}})\right\| \le \left\|\dot{\mathbf{x}}\circ\ddot{\mathbf{s}}\right\|+\left\|\dot{\mathbf{s}}\circ\ddot{\mathbf{x}}\right\|,$$

and

$$\left\|\ddot{\mathbf{x}}\circ\dot{\mathbf{s}}-\dot{\mathbf{s}}\circ\ddot{\mathbf{x}}-\frac{1}{n}(\ddot{\mathbf{x}}^{\mathrm{T}}\dot{\mathbf{s}}-\dot{\mathbf{s}}^{\mathrm{T}}\ddot{\mathbf{x}})\right\| \le \left\|\ddot{\mathbf{x}}\circ\dot{\mathbf{s}}\right\|+\left\|\dot{\mathbf{s}}\circ\ddot{\mathbf{x}}\right\|.$$

Using these results and Lemmas 9.10, 9.4, and 9.5 for the quartic polynomial (9.33), we have, for $\alpha \in [0, \frac{\pi}{2}]$,

$$\begin{aligned}
q(\sin(\alpha)) \le &\left(\left\|\ddot{\mathbf{x}}\circ\ddot{\mathbf{s}}\right\|+\left\|\dot{\mathbf{s}}\circ\dot{\mathbf{x}}\right\|+2\theta\frac{\dot{\mathbf{x}}^{\mathrm{T}}\dot{\mathbf{s}}}{n}\right)\sin^4(\alpha)+\left(\left\|\dot{\mathbf{x}}\circ\ddot{\mathbf{s}}\right\|+\left\|\dot{\mathbf{s}}\circ\ddot{\mathbf{x}}\right\|\right)\sin^3(\alpha) \\
&+2\theta\frac{\dot{\mathbf{x}}^{\mathrm{T}}\dot{\mathbf{s}}}{n}\sin^2(\alpha)+\theta\mu\sin(\alpha)-\theta\mu \\
\le &\mu\left(\left(\frac{(1+\theta)^2}{2(1-\theta)^3}n^2+\frac{n(1+\theta)}{2(1-\theta)}+\frac{\theta(1+\theta)}{(1-\theta)}\right)\sin^4(\alpha)\right.
\end{aligned}$$

$$+2\frac{(1+\theta)^{\frac{3}{2}}}{(1-\theta)^2}n^{\frac{3}{2}}\sin^3(\alpha)+\frac{\theta(1+\theta)}{(1-\theta)}\sin^2(\alpha)+\theta\sin(\alpha)-\theta\Big).$$

Substituting $\sin(\alpha)=\frac{2\theta}{\sqrt{n}}$ gives

$$\begin{aligned}q\Big(\frac{2\theta}{\sqrt{n}}\Big)\le&\mu\Big(\Big(\frac{(1+\theta)^2}{2(1-\theta)^3}n^2+\frac{n(1+\theta)}{2(1-\theta)}+\frac{\theta(1+\theta)}{(1-\theta)}\Big)\frac{16\theta^4}{n^2}+2\frac{(1+\theta)^{\frac{3}{2}}n^{\frac{3}{2}}}{(1-\theta)^2}\frac{8\theta^3}{n^{\frac{3}{2}}}\\&+\frac{\theta(1+\theta)}{(1-\theta)}\frac{4\theta^2}{n}+\theta\frac{2\theta}{\sqrt{n}}-\theta\Big)\\=&\theta\mu\Big(\frac{8\theta^3(1+\theta)^2}{(1-\theta)^3}+\frac{8\theta^3(1+\theta)}{n(1-\theta)}+\frac{16\theta^4(1+\theta)}{(1-\theta)n^2}\\&+\frac{16\theta^2(1+\theta)^{\frac{3}{2}}}{(1-\theta)^2}+\frac{4\theta^2(1+\theta)}{n(1-\theta)}+\frac{2\theta}{\sqrt{n}}-1\Big):=\theta\mu p(\theta).\end{aligned}\tag{9.44}$$

Since $p(\theta)$ is monotonic increasing function of θ, $p(0)<0$, $n\ge 2$, and it is easy to verify that $p(0.148)<0$ for $n=2$, this proves the lemma. ∎

9.2.2 Corrector step

Corollary 9.1, Lemmas 9.1, 9.9, and 9.12 prove the feasibility of searching for the optimizer along the ellipse. To move the iterate back to $\mathcal{N}_2(\theta)$, we use the direction defined by

$$\begin{bmatrix}\mathbf{A} & \mathbf{0} & \mathbf{0}\\ -\mathbf{H} & \mathbf{A}^{\mathrm{T}} & \mathbf{I}\\ \mathbf{S}(\alpha) & \mathbf{0} & \mathbf{X}(\alpha)\end{bmatrix}\begin{bmatrix}\Delta\mathbf{x}\\ \Delta\lambda\\ \Delta\mathbf{s}\end{bmatrix}=\begin{bmatrix}\mathbf{0}\\ \mathbf{0}\\ \mu(\alpha)\mathbf{e}-\mathbf{x}(\alpha)\circ\mathbf{s}(\alpha)\end{bmatrix},\tag{9.45}$$

and we update $(\mathbf{x}^{k+1},\lambda^{k+1},\mathbf{s}^{k+1})$ and μ_{k+1} by

$$(\mathbf{x}^{k+1},\lambda^{k+1},\mathbf{s}^{k+1})=(\mathbf{x}(\alpha),\lambda(\alpha),\mathbf{s}(\alpha))+(\Delta\mathbf{x},\Delta\lambda,\Delta\mathbf{s}),\tag{9.46}$$

$$\mu_{k+1}=\frac{\mathbf{x}^{k+1^{\mathrm{T}}}\mathbf{s}^{k+1}}{n}.\tag{9.47}$$

Next, we show that the combined step (searching along the arc in $\mathcal{N}_2(2\theta)$ and moving back to $\mathcal{N}_2(\theta)$) will reduce the duality gap of the iterate, i.e., $\mu_{k+1}<\mu_k$, if we select some appropriate θ and α. We introduce the following Lemma before we prove this result.

Lemma 9.13
Let $(\mathbf{x}(\alpha),\lambda(\alpha),\mathbf{s}(\alpha))\in\mathcal{N}_2(2\theta)$ and $(\Delta\mathbf{x},\Delta\lambda,\Delta\mathbf{s})$ be defined as in (9.45). Then,

$$0\le\frac{\Delta\mathbf{x}^{\mathrm{T}}\Delta\mathbf{s}}{n}\le\frac{2\theta^2(1+2\theta)}{n(1-2\theta)^2}\mu(\alpha):=\frac{\delta_0}{n}\mu(\alpha).\tag{9.48}$$

Proof 9.12 First, pre-multiply $\Delta\mathbf{x}^{\mathrm{T}}$ in the second row of (9.45), then, applying the first row of (9.45) gives $0 \le \Delta\mathbf{x}^{\mathrm{T}}\mathbf{H}\Delta\mathbf{x} = \Delta\mathbf{x}^{\mathrm{T}}\Delta\mathbf{s}$, which is the first inequality of (9.48). From the third row of (9.45), we have

$$\mathbf{S}(\alpha)\Delta\mathbf{x} + \mathbf{X}(\alpha)\Delta\mathbf{s} = \mu(\alpha)\mathbf{e} - \mathbf{X}(\alpha)\mathbf{S}(\alpha)\mathbf{e}.$$

Multiplying both sides by $\mathbf{X}^{-\frac{1}{2}}(\alpha)\mathbf{S}^{-\frac{1}{2}}(\alpha)$ gives

$$\mathbf{X}^{-\frac{1}{2}}(\alpha)\mathbf{S}^{\frac{1}{2}}(\alpha)\Delta\mathbf{x} + \mathbf{X}^{\frac{1}{2}}(\alpha)\mathbf{S}^{-\frac{1}{2}}(\alpha)\Delta\mathbf{s} = \mathbf{X}^{-\frac{1}{2}}(\alpha)\mathbf{S}^{-\frac{1}{2}}(\alpha)\Big(\mu(\alpha)\mathbf{e} - \mathbf{X}(\alpha)\mathbf{S}(\alpha)\mathbf{e}\Big).$$

Let $\mathbf{D} = \mathbf{X}^{\frac{1}{2}}(\alpha)\mathbf{S}^{-\frac{1}{2}}(\alpha)$, $\mathbf{u} = \mathbf{D}^{-1}\Delta\mathbf{x}$, and $\mathbf{v} = \mathbf{D}\Delta\mathbf{s}$. It is easy to see that $\mathbf{u}^{\mathrm{T}}\mathbf{v} = \Delta\mathbf{x}^{\mathrm{T}}\Delta\mathbf{s} = \Delta\mathbf{x}^{\mathrm{T}}\mathbf{H}\Delta\mathbf{x} \ge 0$. Using Lemma 9.3 and the assumption of $(\mathbf{x}(\alpha), \lambda(\alpha), \mathbf{s}(\alpha)) \in \mathcal{N}_2(2\theta)$, we have

$$\begin{aligned}
&\sum_{i=1}^{n}\left(\frac{(\Delta x_i)^2 s_i(\alpha)}{x_i(\alpha)} + \frac{(\Delta s_i)^2 x_i(\alpha)}{s_i(\alpha)}\right) \\
=&\|\mathbf{D}^{-1}\Delta\mathbf{x}\|^2 + \|\mathbf{D}\Delta\mathbf{s}\|^2 \\
\le&\|\mathbf{X}^{-\frac{1}{2}}(\alpha)\mathbf{S}^{-\frac{1}{2}}(\alpha)[\mu(\alpha)\mathbf{e} - \mathbf{X}(\alpha)\mathbf{S}(\alpha)\mathbf{e}]\|^2 \\
\le&\sum_{i=1}^{n}\frac{(\mu(\alpha) - x_i(\alpha)s_i(\alpha))^2}{x_i(\alpha)s_i(\alpha)} \\
\le&\frac{\sum_{i=1}^{n}(\mu(\alpha) - x_i(\alpha)s_i(\alpha))^2}{\min_i(x_i(\alpha)s_i(\alpha))} \\
\le&\frac{(2\theta)^2\mu^2(\alpha)}{(1-2\theta)\mu(\alpha)} = \frac{(2\theta)^2\mu(\alpha)}{(1-2\theta)}.
\end{aligned} \tag{9.49}$$

Dividing both sides by $\mu(\alpha)$ and using $x_i(\alpha)s_i(\alpha) \ge \mu(\alpha)(1-2\theta)$ yields

$$\begin{aligned}
&\sum_{i=1}^{n}(1-2\theta)\left(\frac{(\Delta x_i)^2}{x_i^2(\alpha)} + \frac{(\Delta s_i)^2}{s_i^2(\alpha)}\right) \\
=&(1-2\theta)\left(\left\|\frac{\Delta\mathbf{x}}{\mathbf{x}(\alpha)}\right\|^2 + \left\|\frac{\Delta\mathbf{s}}{\mathbf{s}(\alpha)}\right\|^2\right) \\
\le&\frac{(2\theta)^2}{(1-2\theta)},
\end{aligned} \tag{9.50}$$

i.e.,

$$\left\|\frac{\Delta\mathbf{x}}{\mathbf{x}(\alpha)}\right\|^2 + \left\|\frac{\Delta\mathbf{s}}{\mathbf{s}(\alpha)}\right\|^2 \le \left(\frac{2\theta}{1-2\theta}\right)^2. \tag{9.51}$$

Invoking Lemma 5.3, we have

$$\left\|\frac{\Delta \mathbf{x}}{\mathbf{x}(\alpha)}\right\|^2 \cdot \left\|\frac{\Delta \mathbf{s}}{\mathbf{s}(\alpha)}\right\|^2 \leq \frac{1}{4}\left(\frac{2\theta}{1-2\theta}\right)^4. \tag{9.52}$$

This gives

$$\left\|\frac{\Delta \mathbf{x}}{\mathbf{x}(\alpha)}\right\| \cdot \left\|\frac{\Delta \mathbf{s}}{\mathbf{s}(\alpha)}\right\| \leq \frac{2\theta^2}{(1-2\theta)^2}. \tag{9.53}$$

Using Cauchy-Schwarz inequality, we have

$$\begin{aligned}
&\frac{(\Delta \mathbf{x})^{\mathrm{T}}(\Delta \mathbf{s})}{\mu(\alpha)} \\
\leq &\sum_{i=1}^{n} \frac{|\Delta x_i||\Delta s_i|}{\mu(\alpha)} \\
\leq &(1+2\theta)\sum_{i=1}^{n} \frac{|\Delta x_i|}{x_i(\alpha)} \frac{|\Delta s_i|}{s_i(\alpha)} \\
= &(1+2\theta)\left|\frac{\Delta \mathbf{x}}{\mathbf{x}(\alpha)}\right|^{\mathrm{T}} \left|\frac{\Delta \mathbf{s}}{\mathbf{s}(\alpha)}\right| \\
\leq &(1+2\theta)\left\|\frac{\Delta \mathbf{x}}{\mathbf{x}(\alpha)}\right\| \cdot \left\|\frac{\Delta \mathbf{s}}{\mathbf{s}(\alpha)}\right\| \\
\leq &\frac{2\theta^2(1+2\theta)}{(1-2\theta)^2}.
\end{aligned} \tag{9.54}$$

Therefore,

$$\frac{(\Delta \mathbf{x})^{\mathrm{T}}(\Delta \mathbf{s})}{n} \leq \frac{2\theta^2(1+2\theta)}{n(1-2\theta)^2}\mu(\alpha). \tag{9.55}$$

This proves the lemma. ■

For linear programming, it is known (see Chapter 3) that $\mu_{k+1} = \mu(\alpha)$. This claim is not always true for the convex quadratic programming, as is pointed out in Lemma 9.14. Therefore, some extra work is needed in order to make sure that the duality measure will be reduced in every iteration.

Lemma 9.14

Let $(\mathbf{x}(\alpha), \lambda(\alpha), \mathbf{s}(\alpha)) \in \mathcal{N}_2(2\theta)$ and $(\Delta \mathbf{x}, \Delta \lambda, \Delta \mathbf{s})$ be defined as in (9.45). Let $(\mathbf{x}^{k+1}, \lambda^{k+1}, \mathbf{s}^{k+1})$ be defined as in (9.46). Then,

$$\mu(\alpha) \leq \mu_{k+1} := \frac{\mathbf{x}^{k+1^{\mathrm{T}}}\mathbf{s}^{k+1}}{n} \leq \mu(\alpha)\left(1 + \frac{2\theta^2(1+2\theta)}{n(1-2\theta)^2}\right) = \mu(\alpha)\left(1 + \frac{\delta_0}{n}\right)$$

Proof 9.13 Using the third row of (9.45), we have $\frac{\mathbf{x}(\alpha)^{\mathrm{T}}\Delta\mathbf{s}+\mathbf{s}(\alpha)^{\mathrm{T}}\Delta\mathbf{x}}{n}=0$. From Lemma 9.13, it is, therefore, straightforward to obtain

$$\begin{aligned}\mu(\alpha) &\leq \frac{\mathbf{x}(\alpha)^{\mathrm{T}}\mathbf{s}(\alpha)}{n}+\frac{1}{n}\Delta\mathbf{x}^{\mathrm{T}}\Delta\mathbf{s}=\frac{(\mathbf{x}(\alpha)+\Delta\mathbf{x})^{\mathrm{T}}(\mathbf{s}(\alpha)+\Delta\mathbf{s})}{n}\\ &= \mu_{k+1}\leq\mu(\alpha)+\frac{2\theta^2(1+2\theta)}{n(1-2\theta)^2}\mu(\alpha).\end{aligned}$$

This proves the lemma. ■

Now, we show that the correction step brings the iterate from $\mathcal{N}_2(2\theta)$ back to $\mathcal{N}_2(\theta)$.

Lemma 9.15

Let $(\mathbf{x}(\alpha),\lambda(\alpha),\mathbf{s}(\alpha))\in\mathcal{N}_2(2\theta)$ and $(\Delta\mathbf{x},\Delta\lambda,\Delta\mathbf{s})$ be defined as in (9.45). Let $(\mathbf{x}^{k+1},\lambda^{k+1},\mathbf{s}^{k+1})$ be updated by using (9.46). Then, for $\theta\leq 0.148$ and $\sin(\alpha)\leq\sin(\bar{\alpha})$, $(\mathbf{x}^{k+1},\lambda^{k+1},\mathbf{s}^{k+1})\in\mathcal{N}_2(\theta)$.

Proof 9.14 From Lemma 9.11, we have

$$0\leq\left\|\Delta\mathbf{x}\circ\Delta\mathbf{s}-\frac{1}{n}(\Delta\mathbf{x}^{\mathrm{T}}\Delta\mathbf{s})\mathbf{e}\right\|^2=\|\Delta\mathbf{x}\circ\Delta\mathbf{s}\|^2-n\left(\frac{1}{n}\Delta\mathbf{x}^{\mathrm{T}}\Delta\mathbf{s}\right)^2\leq\|\Delta\mathbf{x}\circ\Delta\mathbf{s}\|^2. \tag{9.56}$$

Let $\mathbf{D}=\mathbf{X}^{\frac{1}{2}}(\alpha)\mathbf{S}^{-\frac{1}{2}}(\alpha)$. Pre-multiplying $\left(\mathbf{X}(\alpha)\mathbf{S}(\alpha)\right)^{\frac{1}{2}}$ in the last row of (9.45) yields

$$\mathbf{D}\Delta\mathbf{s}+\mathbf{D}^{-1}\Delta\mathbf{x}=\left(\mathbf{X}(\alpha)\mathbf{S}(\alpha)\right)^{-\frac{1}{2}}\left(\mu(\alpha)\mathbf{e}-\mathbf{X}(\alpha)\mathbf{S}(\alpha)\mathbf{e}\right).$$

Let $\mathbf{u}=\mathbf{D}\Delta\mathbf{s}$, $\mathbf{v}=\mathbf{D}^{-1}\Delta\mathbf{x}$, use the technical Lemma 3.2 and the assumption of $(\mathbf{x}(\alpha),\lambda(\alpha),\mathbf{s}(\alpha))\in\mathcal{N}_2(2\theta)$, we have

$$\begin{aligned}\left\|\Delta\mathbf{x}\circ\Delta\mathbf{s}\right\| &= \left\|\mathbf{u}\circ\mathbf{v}\right\|\leq 2^{-\frac{3}{2}}\left\|\left(\mathbf{X}(\alpha)\mathbf{S}(\alpha)\right)^{-\frac{1}{2}}\left(\mu(\alpha)\mathbf{e}-\mathbf{X}(\alpha)\mathbf{S}(\alpha)\mathbf{e}\right)\right\|^2\\ &= 2^{-\frac{3}{2}}\sum_{i=1}^{n}\frac{(\mu(\alpha)-x_i(\alpha)s_i(\alpha))^2}{x_i(\alpha)s_i(\alpha)}\\ &\leq 2^{-\frac{3}{2}}\frac{\|\mu(\alpha)\mathbf{e}-\mathbf{x}(\alpha)\circ\mathbf{s}(\alpha)\|^2}{\min_i(x_i(\alpha)s_i(\alpha))}\\ &\leq 2^{-\frac{3}{2}}\frac{(2\theta)^2\mu(\alpha)^2}{(1-2\theta)\mu(\alpha)}=2^{\frac{1}{2}}\frac{\theta^2\mu(\alpha)}{(1-2\theta)}.\end{aligned} \tag{9.57}$$

Define $(\mathbf{x}^{k+1}(t),\mathbf{s}^{k+1}(t))=(\mathbf{x}(\alpha),\mathbf{s}(\alpha))+t(\Delta\mathbf{x},\Delta\mathbf{s})$. Using the last row of (9.45) and $n\mu(\alpha)=\mathbf{x}(\alpha)^{\mathrm{T}}\mathbf{s}(\alpha)$, we have

$$\mu_{k+1}(t) = \frac{\left(\mathbf{x}(\alpha)+t\Delta\mathbf{x}\right)^{\mathrm{T}}\left(\mathbf{s}(\alpha)+t\Delta\mathbf{s}\right)}{n}$$

$$= \frac{\mathbf{x}(\alpha)^{\mathrm{T}}\mathbf{s}(\alpha)+t^2\Delta\mathbf{x}^{\mathrm{T}}\Delta\mathbf{s}}{n} = \mu(\alpha)+t^2\frac{\Delta\mathbf{x}^{\mathrm{T}}\Delta\mathbf{s}}{n}. \tag{9.58}$$

Using this relation and (9.56), (9.57), and Lemmas 9.9 and 9.14, we have

$$\begin{aligned}
& \left\|\mathbf{x}^{k+1}(t)\circ\mathbf{s}^{k+1}(t)-\mu_{k+1}(t)\mathbf{e}\right\| \\
= & \left\|[\mathbf{x}(\alpha)+t\Delta\mathbf{x}]\circ[\mathbf{s}(\alpha)+t\Delta\mathbf{s}]-\mu(\alpha)\mathbf{e}-\frac{t^2}{n}\Delta\mathbf{x}^{\mathrm{T}}\Delta\mathbf{s}\mathbf{e}\right\| \\
= & \left\|\mathbf{x}(\alpha)\circ\mathbf{s}(\alpha)+t[\mu(\alpha)\mathbf{e}-\mathbf{x}(\alpha)\circ\mathbf{s}(\alpha)]+t^2\Delta\mathbf{x}\circ\Delta\mathbf{s}-\mu(\alpha)\mathbf{e}-\frac{t^2}{n}\Delta\mathbf{x}^{\mathrm{T}}\Delta\mathbf{s}\mathbf{e}\right\| \\
= & \left\|(1-t)\left[\mathbf{x}(\alpha)\circ\mathbf{s}(\alpha)-\mu(\alpha)\mathbf{e}\right]+t^2\left(\Delta\mathbf{x}\circ\Delta\mathbf{s}-\frac{1}{n}\Delta\mathbf{x}^{\mathrm{T}}\Delta\mathbf{s}\mathbf{e}\right)\right\| \\
\le & (1-t)(2\theta)\mu(\alpha)+t^2\frac{2^{\frac{1}{2}}\theta^2}{(1-2\theta)}\mu(\alpha) \qquad \text{[use Lemma 9.9, (9.56), and (9.57)]} \\
\le & \left((1-t)(2\theta)+t^2\frac{2^{\frac{1}{2}}\theta^2}{(1-2\theta)}\right)\mu_{k+1} \qquad \text{[use Lemma 9.14]} \\
:= & f(t,\theta)\mu_{k+1}.
\end{aligned} \tag{9.59}$$

Therefore, taking $t=1$ gives $\left\|\mathbf{x}^{k+1}\circ\mathbf{s}^{k+1}-\mu_{k+1}\mathbf{e}\right\| \le \frac{2^{\frac{1}{2}}\theta^2}{(1-2\theta)}\mu_{k+1}$. It is easy to see that, for $\theta \le 0.29$,

$$\frac{2^{\frac{1}{2}}\theta^2}{(1-2\theta)} \le \theta.$$

For $\theta \le 0.148$ and $t\in[0,1]$, noticing that $0\le f(t,\theta)\le f(t,0.148)\le 0.296(1-t)+0.044t^2<1$, and using Corollary 9.1, we have, for an additional condition $\sin(\alpha)\le\sin(\bar{\alpha})$,

$$\begin{aligned}
x_i^{k+1}(t)s_i^{k+1}(t) &\ge (1-f(t,\theta))\,\mu_{k+1}(t) \\
&= (1-f(t,\theta))\left(\mu(\alpha)+\frac{t^2}{n}\Delta\mathbf{x}^{\mathrm{T}}\Delta\mathbf{s}\right) \\
&\ge (1-f(t,\theta))\,\mu(\alpha) \\
&> 0,
\end{aligned} \tag{9.60}$$

Therefore, $(\mathbf{x}^{k+1}(t),\mathbf{s}^{k+1}(t))>\mathbf{0}$ for $t\in[0,1]$, i.e., $(\mathbf{x}^{k+1},\mathbf{s}^{k+1})>\mathbf{0}$. This finishes the proof. ■

Lemma 9.16
For $\theta \le 0.148$, if

$$\sin(\alpha)=\frac{\theta}{\sqrt{n}}, \tag{9.61}$$

then $\mu_{k+1} < \mu_k$. *Moreover, for* $\sin(\alpha) = \frac{\theta}{\sqrt{n}}$,

$$\mu_{k+1} \le \mu_k \left(1 - \frac{0.148\theta}{\sqrt{n}}\right). \tag{9.62}$$

Proof 9.15 From Lemmas 9.14, 9.6, 5.4, 9.2, 9.4, and 9.5, we have

$$\mu_{k+1} \le \mu(\alpha)\left(1 + \frac{2\theta^2(1+2\theta)}{n(1-2\theta)^2}\right) \tag{9.63a}$$

$$\le \mu_k \left(1 - \sin(\alpha) + \left(\frac{\ddot{\mathbf{x}}^{\mathrm{T}}\ddot{\mathbf{s}}}{n\mu} - \frac{\dot{\mathbf{x}}^{\mathrm{T}}\dot{\mathbf{s}}}{n\mu}\right)\sin^4(\alpha) - \left(\frac{\dot{\mathbf{x}}^{\mathrm{T}}\ddot{\mathbf{s}}}{n\mu} + \frac{\dot{\mathbf{s}}^{\mathrm{T}}\ddot{\mathbf{x}}}{n\mu}\right)\sin^3(\alpha)\right)\left(1 + \frac{\delta_0}{n}\right)$$

$$\le \mu_k \left(1 - \sin(\alpha) + \frac{n(1+\theta)^2}{2(1-\theta)^3}\sin^4(\alpha) + \frac{2n^{\frac{1}{2}}(1+\theta)^{\frac{3}{2}}}{(1-\theta)^2}\sin^3(\alpha)\right)\left(1 + \frac{\delta_0}{n}\right) \tag{9.63b}$$

$$= \mu_k \left[1 + \frac{\delta_0}{n} - \left(1 + \frac{\delta_0}{n}\right)\sin(\alpha) + \delta_2 n\left(1 + \frac{\delta_0}{n}\right)\sin^4(\alpha) + 2\delta_3 n^{\frac{1}{2}}\left(1 + \frac{\delta_0}{n}\right)\sin^3(\alpha)\right]. \tag{9.63c}$$

Substituting $\sin(\alpha) = \frac{\theta}{\sqrt{n}}$ into (9.63c) gives

$$\begin{aligned}
&\mu_{k+1} \\
\le\ & \mu_k \Big\{1 + \frac{2\theta^2(1+2\theta)}{n(1-2\theta)^2} - \frac{\theta}{\sqrt{n}} - \frac{2\theta^3(1+2\theta)}{n^{\frac{3}{2}}(1-2\theta)^2} + \frac{\theta^4(1+\theta)^2}{2n(1-\theta)^3} \\
& + \frac{\theta^6(1+\theta)^2(1+2\theta)}{n^2(1-\theta)^3(1-2\theta)^2} + \frac{2\theta^3(1+\theta)^{\frac{3}{2}}}{n(1-\theta)^2} + \frac{4\theta^5(1+\theta)^{\frac{3}{2}}(1+2\theta)}{n^2(1-2\theta)^2(1-\theta)^2}\Big\} \\
=\ & \mu_k \Big\{1 - \theta\Big[\frac{1}{\sqrt{n}} - \frac{1}{n}\left(\frac{2\theta(1+2\theta)}{(1-2\theta)^2} + \frac{\theta^3(1+\theta)^2}{2(1-\theta)^3} + \frac{2\theta^2(1+\theta)^{\frac{3}{2}}}{(1-\theta)^2}\right)\Big] \\
& - \theta\Big[\frac{2\theta^2(1+2\theta)}{n^{\frac{3}{2}}(1-2\theta)^2} - \frac{1}{n^2}\left(\frac{\theta^5(1+\theta)^2(1+2\theta)}{(1-\theta)^3(1-2\theta)^2} + \frac{4\theta^4(1+\theta)^{\frac{3}{2}}(1+2\theta)}{(1-2\theta)^2(1-\theta)^2}\right)\Big]\Big\}
\end{aligned} \tag{9.64}$$

For $\theta \le 0.148$, we have

$$\frac{2\theta^2(1+2\theta)}{(1-2\theta)^2} > \frac{\theta^5(1+\theta)^2(1+2\theta)}{(1-\theta)^3(1-2\theta)^2} + \frac{4\theta^4(1+\theta)^{\frac{3}{2}}(1+2\theta)}{(1-2\theta)^2(1-\theta)^2},$$

and

$$\frac{2\theta(1+2\theta)}{(1-2\theta)^2} + \frac{\theta^3(1+\theta)^2}{2(1-\theta)^3} + \frac{2\theta^2(1+\theta)^{\frac{3}{2}}}{(1-\theta)^2} < 0.852,$$

therefore,

$$\mu_{k+1} < \mu_k \left\{ 1 - \theta \left(\frac{1}{\sqrt{n}} - \frac{0.852}{\sqrt{n}} \right) \right\} = \mu_k \left(1 - \frac{0.148\theta}{\sqrt{n}} \right).$$

This proves (9.62). ■

We summarize all the results in this section as the following theorem.

Theorem 9.2
Let $\theta = 0.148$, $n \geq 2$, and $(\mathbf{x}^k, \lambda^k, \mathbf{s}^k) \in \mathcal{N}_2(\theta)$. Then, for $\sin(\alpha) = \frac{\theta}{\sqrt{n}}$, it holds that $(\mathbf{x}(\alpha), \lambda(\alpha), \mathbf{s}(\alpha)) \in \mathcal{N}_2(2\theta)$; $(\mathbf{x}^{k+1}, \lambda^{k+1}, \mathbf{s}^{k+1}) \in \mathcal{N}_2(\theta)$; and $\mu_{k+1} \leq \mu_k \left(1 - \frac{0.148\theta}{\sqrt{n}}\right)$.

Proof 9.16 From Corollary 9.1 and Lemma 9.12, we have $\sin(\alpha) \leq \min\{\sin(\tilde{\alpha}), \sin(\bar{\alpha})\}$. Therefore, Lemma 9.9 holds, i.e., $(\mathbf{x}(\alpha), \lambda(\alpha), \mathbf{s}(\alpha)) \in \mathcal{N}_2(2\theta)$. Since $\sin(\alpha) \leq \sin(\bar{\alpha})$ and $(\mathbf{x}(\alpha), \lambda(\alpha), \mathbf{s}(\alpha)) \in \mathcal{N}_2(2\theta)$, Lemma 9.15 states $(\mathbf{x}^{k+1}, \lambda^{k+1}, \mathbf{s}^{k+1}) \in \mathcal{N}_2(\theta)$. Since $\sin(\alpha) = \frac{\theta}{\sqrt{n}}$, Lemma 9.16 states $\mu_{k+1} \leq \mu_k \left(1 - \frac{0.148\theta}{\sqrt{n}}\right)$. This finishes the proof. ■

We present the proposed algorithm as follows:

Algorithm 9.1
(Arc-search path-following)
Data: $\mathbf{A}$, $\mathbf{H} \succeq \mathbf{0}$, $\mathbf{b}$, $\mathbf{c}$, $\theta = 0.148$, $\varepsilon > 0$, *initial point* $(\mathbf{x}^0, \lambda^0, \mathbf{s}^0) \in \mathcal{N}_2(\theta)$, *and* $\mu_0 = \frac{\mathbf{x}^{0^{\mathrm{T}}}\mathbf{s}^0}{n}$.
for *iteration* $k = 0, 1, 2, \ldots$

Step 1: Solve the linear systems of equations (9.11) and (9.12) to get $(\dot{\mathbf{x}}, \dot{\lambda}, \dot{\mathbf{s}})$ *and* $(\ddot{\mathbf{x}}, \ddot{\lambda}, \ddot{\mathbf{s}})$.

Step 2: Let $\sin(\alpha) = \frac{\theta}{\sqrt{n}}$. *Update* $(\mathbf{x}(\alpha), \lambda(\alpha), \mathbf{s}(\alpha))$ *and* $\mu(\alpha)$ *by (9.34) and (9.35).*

Step 3: Calculate $(\Delta\mathbf{x}, \Delta\lambda, \Delta\mathbf{s})$ *by solving (9.45), update* $(\mathbf{x}^{k+1}, \lambda^{k+1}, \mathbf{s}^{k+1})$ *and* μ_{k+1} *by using (9.46) and (9.47). Set* $k+1 \to k$. *Go back to Step 1.*

end (for)

9.3 Convergence Analysis

The first result in this section extends a result of linear programming (Theorem 2.1) to convex quadratic programming.

Lemma 9.17
Suppose Assumption 2 holds, i.e., $\mathcal{F}^o \neq \emptyset$. Then, for each $K \geq 0$, the set

$$\{(\mathbf{x},\mathbf{s}) \mid (\mathbf{x},\lambda,\mathbf{s}) \in \mathcal{F}, \quad \mathbf{x}^{\mathrm{T}}\mathbf{s} \leq K\}$$

is bounded.

Proof 9.17 The proof is almost identical to the proof in Theorem 2.1. It is given here for completeness. Let $(\bar{\mathbf{x}},\bar{\lambda},\bar{\mathbf{s}})$ be any fixed vector in $\mathcal{F}^o$, and $(\mathbf{x},\lambda,\mathbf{s})$ be any vector in $\mathcal{F}$ with $\mathbf{x}^{\mathrm{T}}\mathbf{s} \leq K$. Then,

$$\mathbf{A}^{\mathrm{T}}(\bar{\lambda}-\lambda)+(\bar{\mathbf{s}}-\mathbf{s})-\mathbf{H}(\bar{\mathbf{x}}-\mathbf{x})=\mathbf{0},$$

therefore,

$$(\bar{\mathbf{x}}-\mathbf{x})^{\mathrm{T}}\left(\mathbf{A}^{\mathrm{T}}(\bar{\lambda}-\lambda)+(\bar{\mathbf{s}}-\mathbf{s})-\mathbf{H}(\bar{\mathbf{x}}-\mathbf{x})\right)=0.$$

Since

$$\mathbf{A}(\bar{\mathbf{x}}-\mathbf{x})=\mathbf{0},$$

this means

$$(\bar{\mathbf{x}}-\mathbf{x})^{\mathrm{T}}(\bar{\mathbf{s}}-\mathbf{s})=(\bar{\mathbf{x}}-\mathbf{x})^{\mathrm{T}}\mathbf{H}(\bar{\mathbf{x}}-\mathbf{x})\geq 0.$$

This leads to

$$\bar{\mathbf{x}}^{\mathrm{T}}\bar{\mathbf{s}}+K \geq \bar{\mathbf{x}}^{\mathrm{T}}\bar{\mathbf{s}}+\mathbf{x}^{\mathrm{T}}\mathbf{s} \geq \bar{\mathbf{x}}^{\mathrm{T}}\mathbf{s}+\mathbf{x}^{\mathrm{T}}\bar{\mathbf{s}}.$$

Since $(\bar{\mathbf{x}},\bar{\mathbf{s}})>\mathbf{0}$ is fixed, let

$$\xi = \min_{i=1,\cdots,n} \min\{\bar{x}_i,\bar{s}_i\}.$$

Then,

$$\bar{\mathbf{x}}^{\mathrm{T}}\bar{\mathbf{s}}+K \geq \xi\mathbf{e}^{\mathrm{T}}(\mathbf{x}+\mathbf{s}) \geq \max_{i=1,\cdots,n} \max\{\xi x_i,\xi s_i\},$$

i.e., for $i \in \{1,\cdots,n\}$,

$$0 \leq x_i \leq \frac{1}{\xi}(K+\bar{\mathbf{x}}^{\mathrm{T}}\bar{\mathbf{s}}), \qquad 0 \leq s_i \leq \frac{1}{\xi}(K+\bar{\mathbf{x}}^{\mathrm{T}}\bar{\mathbf{s}}).$$

This proves the lemma. ∎

The following theorem is a direct result of Lemmas 9.17, 9.1, Theorem 9.2, KKT conditions, Theorem 1.5.

Theorem 9.3
Suppose Assumptions 1 and 2 hold, then the sequence generated by Algorithm 9.1 converges to a set of accumulation points, and all these accumulation points are global optimal solutions of the convex quadratic programming.

Let $(\mathbf{x}^*, \lambda^*, \mathbf{s}^*)$ be any solution of (9.5), where $\mathbf{x}^*$ is a solution of the primary quadratic programming and $(\lambda^*, \mathbf{s}^*)$ is a solution of the dual quadratic programming, following the notation of [10], we denote index sets $\mathcal{B}$, $\mathcal{S}$, and $\mathcal{T}$ as

$$\mathcal{B} = \{j \in \{1, \dots, n\} \mid x_j^* \neq 0\}. \tag{9.65}$$

$$\mathcal{S} = \{j \in \{1, \dots, n\} \mid s_j^* \neq 0\}. \tag{9.66}$$

$$\mathcal{T} = \{j \in \{1, \dots, n\} \mid s_j^* = x_j^* = 0\}. \tag{9.67}$$

Goldman-Tucker demonstrated [40] (see Theorem 1.3) that if $\mathbf{H} = \mathbf{0}$, then there is at least one primal-dual solution $(\mathbf{x}^*, \lambda^*, \mathbf{s}^*)$ such that the corresponding $\mathcal{B} \cap \mathcal{S} = \emptyset = \mathcal{T}$ and $\mathcal{B} \cup \mathcal{S} = \{1, \dots, n\}$. A solution with this property is called strictly complementary. This property has been used in many papers to prove the locally super-linear convergence of interior-point algorithms in linear programming. However, it is pointed out in [49] that this partition does not hold for general quadratic programming problems.

A simple example of convex quadratic program is given by

$$\min \frac{1}{2} x^2, \quad x \geq 0.$$

Its dual program is

$$\min \frac{1}{2} s^2, \quad s \geq 0.$$

Clearly, the optimal solution $x^* = 0$ and $s^* = 0$ is not a strictly complementary solution.

We will show that as long as a convex quadratic programming has strictly complementary solution(s), an interior-point algorithm will generate a sequence to approach strict complementary solution(s). As a matter of fact, from Lemma 9.17, we can extend the result of [144, Lemma 5.13] to the case of convex quadratic programming, and obtain the following lemma which is independent of any algorithm.

Lemma 9.18

Let $\mu_0 > 0$, and $\gamma \in (0,1)$. Assume that the convex QP has strictly complementary solution(s). Then, for all points $(\mathbf{x}, \lambda, \mathbf{s}) \in \mathcal{F}^o$, $x_i s_i > \gamma \mu$, and $\mu < \mu_0$, there are constants M, C_1, and C_2 such that

$$\|(\mathbf{x}, \mathbf{s})\| \leq M, \tag{9.68}$$

$$0 < x_i \leq \mu / C_1 \quad (i \in \mathcal{S}), \qquad 0 < s_i \leq \mu / C_1 \quad (i \in \mathcal{B}). \tag{9.69}$$

$$s_i \geq C_2 \gamma \quad (i \in \mathcal{S}), \qquad x_i \geq C_2 \gamma \quad (i \in \mathcal{B}). \tag{9.70}$$

Proof 9.18 The first result follows immediately from Lemma 9.17 by setting $K = n\mu_0$. Let $(\mathbf{x}^*, \lambda^*, \mathbf{s}^*)$ be any primal-dual strictly complementary solution. Since $(\mathbf{x}^*, \lambda^*, \mathbf{s}^*)$ and $(\mathbf{x}, \lambda, \mathbf{s})$ are both feasible,

$$\mathbf{A}(\mathbf{x}-\mathbf{x}^*) = \mathbf{0}, \qquad \mathbf{A}^{\mathrm{T}}(\lambda - \lambda^*) + (\mathbf{s}-\mathbf{s}^*) - \mathbf{H}(\mathbf{x}-\mathbf{x}^*) = \mathbf{0}.$$

$$(\mathbf{x}-\mathbf{x}^*)^{\mathrm{T}}(\mathbf{s}-\mathbf{s}^*) = (\mathbf{x}-\mathbf{x}^*)^{\mathrm{T}}\mathbf{H}(\mathbf{x}-\mathbf{x}^*) \geq 0. \tag{9.71}$$

Since $(\mathbf{x}^*, \lambda^*, \mathbf{s}^*)$ is strictly complementary solution, $\mathcal{T} = \emptyset$, $x_i^* = 0$ for $i \in \mathcal{S}$, and $s_i^* = 0$ for $i \in \mathcal{B}$. Since $\mathbf{x}^{\mathrm{T}}\mathbf{s} = n\mu$, $(\mathbf{x}^*)^{\mathrm{T}}\mathbf{s}^* = 0$, from (9.71), we have

$$n\mu \geq \mathbf{x}^{\mathrm{T}}\mathbf{s}^* + \mathbf{s}^{\mathrm{T}}\mathbf{x}^* = \sum_{i\in\mathcal{S}} x_i s_i^* + \sum_{i\in\mathcal{B}} x_i^* s_i.$$

Since each term in the summations is positive and bounded above by $n\mu$, we have for any $i \in \mathcal{S}$, $s_i^* > 0$, therefore,

$$0 < x_i \leq \frac{n\mu}{s_i^*}.$$

Denote $D^* = \{(\lambda^*, \mathbf{s}^*) \mid s_i^* > 0\}$ and $P^* = \{(\mathbf{x}^*) \mid x_i^* > 0\}$, we have

$$0 < x_i \leq \frac{n\mu}{\sup_{(\lambda^*,\mathbf{s}^*)\in D^*} s_i^*}.$$

This leads to

$$\max_{i\in\mathcal{S}} x_i \leq \frac{n\mu}{\min_{i\in\mathcal{S}} \sup_{(\lambda^*,\mathbf{s}^*)\in D^*} s_i^*}.$$

Similarly,

$$\max_{i\in\mathcal{B}} s_i \leq \frac{n\mu}{\min_{i\in\mathcal{B}} \sup_{\mathbf{x}^*\in P^*} x_i^*}.$$

Combining these two inequalities gives

$$\begin{aligned}\max\{\max_{i\in\mathcal{S}} x_i, \max_{i\in\mathcal{B}} s_i\} &\leq \frac{n\mu}{\min\{\min_{i\in\mathcal{S}} \sup_{(\lambda^*,\mathbf{s}^*)\in D^*} s_i^*, \min_{i\in\mathcal{B}} \sup_{\mathbf{x}^*\in P^*} x_i^*\}} \\ &= \frac{\mu}{C_1}.\end{aligned}$$

This proves (9.69). Finally, $x_i s_i \geq \gamma\mu$, hence, for any $i \in \mathcal{S}$,

$$s_i \geq \frac{\gamma\mu}{x_i} \geq \frac{\gamma\mu}{\mu/C_1} = C_2\gamma.$$

Similarly, for any $i \in \mathcal{B}$,

$$x_i \geq \frac{\gamma\mu}{s_i} \geq \frac{\gamma\mu}{\mu/C_1} = C_2\gamma.$$

This completes the proof. ■

Lemma 9.18 leads to the following

Theorem 9.4
Let $(\mathbf{x}^k, \lambda^k, \mathbf{s}^k) \in \mathcal{N}_2(\theta)$ *be generated by Algorithm 9.1. Assume that the convex QP has strictly complementary solution(s). Then, every limit point of the sequence is a strictly complementary primary-dual solution of the convex quadratic programming, i.e.,*

$$s_i^* \geq C_2\gamma \quad (i \in \mathcal{S}), \qquad x_i^* \geq C_2\gamma \quad (i \in \mathcal{B}). \tag{9.72}$$

Proof 9.19 From Lemma 9.18, $(\mathbf{x}^k, \mathbf{s}^k)$ is bounded, therefore, there is at least one limit point $(\mathbf{x}^*, \mathbf{s}^*)$. Since (x_i^k, s_i^k) is in the neighborhood of the central path, i.e., $x_i^k s_i^k > \gamma\mu_k = (1-\theta)\mu_k$,

$$s_i^k \geq C_2\gamma \quad (i \in \mathcal{S}), \qquad x_i^k \geq C_2\gamma \quad (i \in \mathcal{B}),$$

every limit point will meet (9.72) due to the fact that $C_2\gamma$ is a constant. ■

We now show that the complexity bound of Algorithm 9.1 is $O(\sqrt{n}\log(1/\varepsilon))$. For this purpose, we need Theorem 1.4, which is restated below for convenience.

Theorem 9.5
Let $\varepsilon \in (0,1)$ *be given. Suppose that an algorithm for solving (9.5) generates a sequence of iterations that satisfies*

$$\mu_{k+1} \leq \left(1 - \frac{\delta}{n^{\omega}}\right)\mu_k, \quad k = 0,1,2,\ldots, \tag{9.73}$$

for some positive constants δ *and* ω. *Suppose that the starting point* $(\mathbf{x}^0, \lambda^0, \mathbf{s}^0)$ *satisfies* $\mu_0 \leq 1/\varepsilon$. *Then, there exists an index* K *with*

$$K = O(n^{\omega}\log(1/\varepsilon))$$

such that

$$\mu_k \leq \varepsilon \quad \text{for} \quad \forall k \geq K.$$

Combining Theorems 9.2 and 9.5 gives:

Theorem 9.6
The complexity of Algorithm 9.1 is bounded by $O(\sqrt{n}\log(1/\varepsilon))$.

9.4 Implementation Details

Algorithm 9.1 is presented in a form that is convenient for convergence analysis. Some implementation details that make the algorithm effective and efficient are discussed in this section.

9.4.1 Termination criterion

Algorithm 9.1 needs a termination criterion in real implementation. One can use

$$\mu_k \leq \varepsilon, \tag{9.74a}$$

$$\|\mathbf{r}_B\| = \|\mathbf{A}\mathbf{x}^k - \mathbf{b}\| \leq \varepsilon, \tag{9.74b}$$

$$\|\mathbf{r}_C\| = \|\mathbf{A}^{\mathrm{T}}\lambda^k + \mathbf{s}^k - \mathbf{H}\mathbf{x}^{\mathbf{k}} - \mathbf{c}\| \leq \varepsilon, \tag{9.74c}$$

$$(x^k, s^k) > 0. \tag{9.74d}$$

An alternative criterion is given in *linprog* [172]

$$\frac{\|\mathbf{r}_B\|}{\max\{1, \|\mathbf{b}\|\}} + \frac{\|\mathbf{r}_C\|}{\max\{1, \|\mathbf{c}\|\}} + \frac{\mu}{\max\{1, \|\mathbf{c}^{\mathrm{T}}\mathbf{x}\|, \|\mathbf{b}^{\mathrm{T}}\lambda\|\}} \leq \varepsilon. \tag{9.75}$$

9.4.2 Finding initial $(\mathbf{x}^0, \lambda^0, \mathbf{s}^0) \in \mathcal{N}_2(\theta)$

Algorithm 9.1 requires an initial point $(\mathbf{x}^0, \lambda^0, \mathbf{s}^0) \in \mathcal{N}_2(\theta)$. We use a modified algorithm of [20] to provide such an initial point in our implementation. Denote

$$\mathbf{X} = \mathrm{diag}(x_1, \ldots, x_n), \quad \mathbf{S} = \mathrm{diag}(s_1, \ldots, s_n).$$

Starting from any point $(\mathbf{x}, \lambda, \mathbf{s})$ with $(\mathbf{x}, \mathbf{s}) > \mathbf{0}$ that may or may not be in $\mathcal{F}^o$, moving the point to a point close to or on the central path amounts to approximately solving

$$F(\mathbf{x}(t), \lambda(t), \mathbf{s}(t)) = \begin{pmatrix} \mathbf{A}\mathbf{x} - \mathbf{b} \\ \mathbf{A}^{\mathrm{T}}\lambda + \mathbf{s} - \mathbf{H}\mathbf{x} - \mathbf{c} \\ \mathbf{X}\mathbf{S}\mathbf{e} - t\mu\mathbf{e} \end{pmatrix} = \mathbf{0}, \qquad (\mathbf{x}, \mathbf{s}) > \mathbf{0}. \tag{9.76}$$

(9.76) can be solved by repeatedly searching along Newton directions while keeping $(\mathbf{x}, \mathbf{s}) > \mathbf{0}$. In each step, the Newton direction $(d\mathbf{x}, d\lambda, d\mathbf{s})$ can be calculated by

$$\begin{bmatrix} \mathbf{A} & \mathbf{0} & \mathbf{0} \\ -\mathbf{H} & \mathbf{A}^{\mathrm{T}} & \mathbf{I} \\ \mathbf{S} & \mathbf{0} & \mathbf{X} \end{bmatrix} \begin{bmatrix} d\mathbf{x} \\ d\lambda \\ d\mathbf{s} \end{bmatrix} = \begin{bmatrix} \mathbf{0} \\ \mathbf{0} \\ \mu\mathbf{e} - \mathbf{x} \circ \mathbf{s} \end{bmatrix}. \tag{9.77}$$

This process is described in the following

Algorithm 9.2
(Find Initial $(\mathbf{x}^0, \lambda^0, \mathbf{s}^0) \in \mathcal{N}_2(\theta)$**)**
Data: $\mathbf{A}$, $\mathbf{H} \succeq \mathbf{0}$, $\mathbf{b}$, $\mathbf{c}$, $\varepsilon > 0$, *and initial point* $(\mathbf{x}^0, \lambda^0, \mathbf{s}^0)$ *with* $(\mathbf{x}^0, \mathbf{s}^0) > \mathbf{0}$.
for *iteration* $k = 0, 1, 2, \ldots$

Check conditions

$$\|\mathbf{r}_B\| = \|\mathbf{A}\mathbf{x}^k - \mathbf{b}\| \leq \varepsilon, \tag{9.78a}$$

$$\|\mathbf{r}_C\| = \|\mathbf{A}^{\mathrm{T}}\lambda^k + \mathbf{s}^k - \mathbf{H}\mathbf{x}^{\mathbf{k}} - \mathbf{c}\| \le \varepsilon, \tag{9.78b}$$

$$\|\mathbf{r}_t\| = \|\mathbf{X}^k\mathbf{S}^k\mathbf{e} - \mu\mathbf{e}\| \le \theta\mu, \tag{9.78c}$$

$$(\mathbf{x}^k, \mathbf{s}^k) > \mathbf{0}. \tag{9.78d}$$

If (9.78) holds, $(\mathbf{x}^k, \lambda^k, \mathbf{s}^k)$ *is a point in* $\mathcal{N}_2(\theta)$*. Set the solution* $(\mathbf{x}^0, \lambda^0, \mathbf{s}^0) = (\mathbf{x}^k, \lambda^k, \mathbf{s}^k)$ *and stop.*

If (9.78) does not hold, calculate the Newton direction $(d\mathbf{x}^k, d\lambda^k, d\mathbf{s}^k)$ *from (9.77). Carry out line search along the Newton direction*

$$(\mathbf{x}^{k+1}, \lambda^{k+1}, \mathbf{s}^{k+1}) = (\mathbf{x}^k + \alpha d\mathbf{x}^k, \lambda^k + \alpha d\lambda^k, \mathbf{s}^k + \alpha d\mathbf{s}^k) \tag{9.79}$$

such that the α *satisfies* $(\mathbf{x}^k + \alpha d\mathbf{x}^k, \mathbf{s}^k + \alpha d\mathbf{s}^k) > \mathbf{0}$ *and*

$$\begin{aligned} & \left(\|\mathbf{A}\mathbf{x}^{k+1} - \mathbf{b}\|^2 + \|\mathbf{A}^{\mathrm{T}}\lambda^{k+1} + \mathbf{s}^{k+1} - \mathbf{H}\mathbf{x}^{\mathbf{k+1}} - \mathbf{c}\|^2 + \|\mathbf{X}^{\mathbf{k+1}}\mathbf{S}^{\mathbf{k+1}}\mathbf{e} - \mu\mathbf{e}\|^2\right) \\ < \ & \left(\|\mathbf{A}\mathbf{x}^k - \mathbf{b}\|^2 + \|\mathbf{A}^{\mathrm{T}}\lambda^k + \mathbf{s}^k - \mathbf{H}\mathbf{x}^{\mathbf{k}} - \mathbf{c}\|^2 + \|\mathbf{X}^{\mathbf{k}}\mathbf{S}^{\mathbf{k}}\mathbf{e} - \mu\mathbf{e}\|^2\right) \end{aligned} \tag{9.80}$$

end (for)

Remark 9.3 Since $(\mathbf{x}, \mathbf{s}) > \mathbf{0}$, and the search direction is toward the central path, it is not surprising that we observe in all of our test examples that the condition $(\mathbf{x}^k + \alpha d\mathbf{x}^k, \mathbf{s}^k + \alpha d\mathbf{s}^k) > \mathbf{0}$ always holds. However, a rigorous analysis is needed or an alternative algorithm with rigorous analysis should be used if one wants to find initial points for general convex quadratic programming problems. ∎

9.4.3 Solving linear systems of equations

In Algorithm 9.1, the majority computational operation in each iteration is to solve linear systems of equations (9.11), (9.12), and (9.45). Directly solving each of these linear systems of equations requires $O(2n+m)^3$ operation count. The following Theorem and its corollaries provide a more efficient way to solve these linear systems of equations.

Theorem 9.7
Let $\hat{\mathbf{A}} \in \mathbb{R}^{n\times(n-m)}$ *be a base of the null space of* $\mathbf{A}$*. Let* $(\mathbf{x}, \lambda, \mathbf{s}) \in \mathcal{F}^o$ *and* $(\dot{\mathbf{x}}, \dot{\lambda}, \dot{\mathbf{s}})$ *meet (9.11). Then,*

$$\frac{\dot{\mathbf{x}}}{\mathbf{x}} = \mathbf{X}^{-1}\hat{\mathbf{A}}(\hat{\mathbf{A}}^{\mathrm{T}}(\mathbf{S}\mathbf{X}^{-1} + \mathbf{H})\hat{\mathbf{A}})^{-1}\hat{\mathbf{A}}^{\mathrm{T}}\mathbf{S}\mathbf{e},$$

$$\frac{\dot{\mathbf{s}}}{\mathbf{s}} = \mathbf{e} - \frac{\dot{\mathbf{x}}}{\mathbf{x}},$$

and

$$\dot{\lambda} = \left(\mathbf{A}\mathbf{A}^{\mathrm{T}}\right)^{-1}\mathbf{A}\left(\mathbf{H}\dot{\mathbf{x}} - \dot{\mathbf{s}}\right).$$

Proof 9.20 Pre-multiplying $\left(\mathbf{A}\mathbf{A}^{\mathrm{T}}\right)^{-1}\mathbf{A}$ in $\mathbf{A}^{\mathrm{T}}\dot{\lambda}+\dot{\mathbf{s}}-\mathbf{H}\dot{\mathbf{x}}=\mathbf{0}$ gives the last equation. Since $\mathbf{A}\dot{\mathbf{x}}=\mathbf{0}$, we have $\mathbf{A}\mathbf{X}\frac{\dot{\mathbf{x}}}{\mathbf{x}}=\mathbf{0}$, this means that there exists a vector $\mathbf{v}$ such that $\mathbf{X}\frac{\dot{\mathbf{x}}}{\mathbf{x}}=\hat{\mathbf{A}}\mathbf{v}$, i.e.,

$$\frac{\dot{\mathbf{x}}}{\mathbf{x}}=\mathbf{X}^{-1}\hat{\mathbf{A}}\mathbf{v}. \tag{9.81}$$

From the last row of (9.11)

$$\frac{\dot{\mathbf{s}}}{\mathbf{s}}=\mathbf{e}-\frac{\dot{\mathbf{x}}}{\mathbf{x}}=\mathbf{e}-\mathbf{X}^{-1}\hat{\mathbf{A}}\mathbf{v}. \tag{9.82}$$

Similarly, $\mathbf{A}^{\mathrm{T}}\dot{\lambda}+\dot{\mathbf{s}}-\mathbf{H}\dot{\mathbf{x}}=\mathbf{0}$ is equivalent to

$$\mathbf{S}^{-1}\mathbf{A}^{\mathrm{T}}\dot{\lambda}+\frac{\dot{\mathbf{s}}}{\mathbf{s}}-\mathbf{S}^{-1}\mathbf{H}\hat{\mathbf{A}}\mathbf{v}=\mathbf{0}. \tag{9.83}$$

Substituting (9.82) into (9.83) gives $\mathbf{S}^{-1}\mathbf{A}^{\mathrm{T}}\dot{\lambda}+\mathbf{e}-\mathbf{X}^{-1}\hat{\mathbf{A}}\mathbf{v}-\mathbf{S}^{-1}\mathbf{H}\hat{\mathbf{A}}\mathbf{v}=\mathbf{0}$, or in matrix form

$$\left[(\mathbf{X}^{-1}+\mathbf{S}^{-1}\mathbf{H})\hat{\mathbf{A}},-\mathbf{S}^{-1}\mathbf{A}^{\mathrm{T}}\right]\begin{bmatrix}\mathbf{v}\\ \dot{\lambda}\end{bmatrix}=\mathbf{e}. \tag{9.84}$$

Since $\mathbf{H}$ is positive semidefinite, $(\mathbf{S}\mathbf{X}^{-1}+\mathbf{H})$ is positive definite and invertible, hence, $(\mathbf{X}^{-1}+\mathbf{S}^{-1}\mathbf{H})=\mathbf{S}^{-1}(\mathbf{S}\mathbf{X}^{-1}+\mathbf{H})$ is invertible. Since $(\mathbf{X}^{-1}+\mathbf{S}^{-1}\mathbf{H})^{-1}\mathbf{S}^{-1}=(\mathbf{S}\mathbf{X}^{-1}+\mathbf{H})^{-1}$ is positive definite, $\mathbf{A}$ and $\hat{\mathbf{A}}$ are full rank matrices, we have that $\left(\mathbf{A}(\mathbf{X}^{-1}+\mathbf{S}^{-1}\mathbf{H})^{-1}\mathbf{S}^{-1}\mathbf{A}^{\mathrm{T}}\right)$ and $\hat{\mathbf{A}}^{\mathrm{T}}\mathbf{S}(\mathbf{X}^{-1}+\mathbf{S}^{-1}\mathbf{H})\hat{\mathbf{A}}$ are positive definite and invertible. It is easy to verify that

$$\begin{bmatrix}\left(\hat{\mathbf{A}}^{\mathrm{T}}\mathbf{S}(\mathbf{X}^{-1}+\mathbf{S}^{-1}\mathbf{H})\hat{\mathbf{A}}\right)^{-1}\hat{\mathbf{A}}^{\mathrm{T}}\mathbf{S}\\ -\left(\mathbf{A}(\mathbf{X}^{-1}+\mathbf{S}^{-1}\mathbf{H})^{-1}\mathbf{S}^{-1}\mathbf{A}^{\mathrm{T}}\right)^{-1}\mathbf{A}(\mathbf{X}^{-1}+\mathbf{S}^{-1}\mathbf{H})^{-1}\end{bmatrix}\left[(\mathbf{X}^{-1}+\mathbf{S}^{-1}\mathbf{H})\hat{\mathbf{A}},-\mathbf{S}^{-1}\mathbf{A}^{\mathrm{T}}\right]=\mathbf{I}.$$

Taking the inverse in (9.84) gives

$$\begin{bmatrix}\mathbf{v}\\ \dot{\lambda}\end{bmatrix}=\begin{bmatrix}\left(\hat{\mathbf{A}}^{\mathrm{T}}\mathbf{S}(\mathbf{X}^{-1}+\mathbf{S}^{-1}\mathbf{H})\hat{\mathbf{A}}\right)^{-1}\hat{\mathbf{A}}^{\mathrm{T}}\mathbf{S}\\ -\left(\mathbf{A}(\mathbf{X}^{-1}+\mathbf{S}^{-1}\mathbf{H})^{-1}\mathbf{S}^{-1}\mathbf{A}^{\mathrm{T}}\right)^{-1}\mathbf{A}(\mathbf{X}^{-1}+\mathbf{S}^{-1}\mathbf{H})^{-1}\end{bmatrix}\mathbf{e}. \tag{9.85}$$

Substituting (9.85) into (9.81) proves the result. ■

Remark 9.4 $\hat{\mathbf{A}}$ will be a sparse matrix if $\mathbf{A}$ is a sparse matrix and if sparse QR decomposition [26] is used. This feature is important for large size problems. $\left(\mathbf{A}\mathbf{A}^{\mathrm{T}}\right)^{-1}\mathbf{A}$ and $\mathbf{H}$ are constants and independent on iterations, therefore, $\left(\mathbf{A}\mathbf{A}^{\mathrm{T}}\right)^{-1}\mathbf{A}$ can be stored, and the computation of $\dot{\lambda}$ can be very efficient. The exactly same idea can be extended to solve (9.12) and (9.45). ■

Corollary 9.2
Let $\hat{\mathbf{A}} \in \mathbb{R}^{n\times(n-m)}$ *be a base of the null space of* $\mathbf{A}$. *Let* $(\mathbf{x},\lambda,\mathbf{s}) \in \mathcal{F}^o$, $(\dot{\mathbf{x}},\dot{\lambda},\dot{\mathbf{s}})$ *and* $(\ddot{\mathbf{x}},\ddot{\lambda},\ddot{\mathbf{s}})$ *meet (9.11) and (9.12). Define* $\mathbf{f} = -2\frac{\dot{\mathbf{x}}}{\mathbf{x}} \circ \frac{\dot{\mathbf{s}}}{\mathbf{s}}$. *Then,*

$$\frac{\ddot{\mathbf{x}}}{\mathbf{x}} = \mathbf{X}^{-1}\hat{\mathbf{A}}\left(\hat{\mathbf{A}}^{\mathrm{T}}\left(\mathbf{S}\mathbf{X}^{-1}+\mathbf{H}\right)\hat{\mathbf{A}}\right)^{-1}\hat{\mathbf{A}}^{\mathrm{T}}\mathbf{S}\mathbf{f},$$

$$\frac{\ddot{\mathbf{s}}}{\mathbf{s}} = \mathbf{f} - \frac{\ddot{\mathbf{x}}}{\mathbf{x}},$$

and

$$\ddot{\lambda} = \left(\mathbf{A}\mathbf{A}^{\mathrm{T}}\right)^{-1}\mathbf{A}\left(\mathbf{H}\ddot{\mathbf{x}} - \ddot{\mathbf{s}}\right).$$

Corollary 9.3
Let $\hat{\mathbf{A}} \in \mathbb{R}^{n\times(n-m)}$ *be a base of the null space of* $\mathbf{A}$. *Let* $(\mathbf{x},\lambda,\mathbf{s}) \in \mathcal{F}^o$, $(\dot{\mathbf{x}},\dot{\lambda},\dot{\mathbf{s}})$ *and* $(\ddot{\mathbf{x}},\ddot{\lambda},\ddot{\mathbf{s}})$ *meet (9.11) and (9.12). Define* $\mathbf{f}_1 = \frac{\mu\mathbf{e}}{\mathbf{x}(\alpha)\circ\mathbf{s}(\alpha)} - \mathbf{e}$. *Then,*

$$\frac{\Delta\mathbf{x}}{\mathbf{x}(\alpha)} = \mathbf{X}^{-1}(\alpha)\hat{\mathbf{A}}\left(\hat{\mathbf{A}}^{\mathrm{T}}\left(\mathbf{S}(\alpha)\mathbf{X}^{-1}(\alpha)+\mathbf{H}\right)\hat{\mathbf{A}}\right)^{-1}\hat{\mathbf{A}}^{\mathrm{T}}\mathbf{S}(\alpha)\mathbf{f}_1,$$

$$\frac{\Delta\mathbf{s}}{\mathbf{s}(\alpha)} = \mathbf{f}_1 - \frac{\Delta\mathbf{x}}{\mathbf{x}(\alpha)},$$

and,

$$\Delta\lambda = \left(\mathbf{A}\mathbf{A}^{\mathrm{T}}\right)^{-1}\mathbf{A}\left(\mathbf{H}\Delta\mathbf{x} - \Delta\mathbf{s}\right).$$

Theorem 9.7, Corollaries 9.2 and 9.3 provide efficient formulas to calculate $(\dot{\mathbf{x}},\dot{\lambda},\dot{\mathbf{s}})$, $(\ddot{\mathbf{x}},\ddot{\lambda},\ddot{\mathbf{s}})$, and $(\Delta\mathbf{x},\Delta\lambda,\Delta\mathbf{s})$. Let $\hat{\mathbf{A}}$ be obtained by *QR* decomposition $\mathbf{A}^{\mathrm{T}} = [\mathbf{Q}_1,\hat{\mathbf{A}}]\begin{bmatrix}\mathbf{R}\\ \mathbf{0}\end{bmatrix}$. The computational procedure for $\dot{\mathbf{x}}$, $\ddot{\mathbf{x}}$, $\dot{\mathbf{s}}$, $\ddot{\mathbf{s}}$, $\dot{\lambda}$, and $\ddot{\lambda}$ is summarized as follows.

Algorithm 9.3
(compute $\dot{\mathbf{x}}$, $\ddot{\mathbf{x}}$, $\dot{\mathbf{s}}$, $\ddot{\mathbf{s}}$, $\dot{\lambda}$, and $\ddot{\lambda}$)

Data: Matrices $\mathbf{A}$, $\hat{\mathbf{A}}$, $\mathbf{H} \succeq \mathbf{0}$, $\mathbf{X}$, $\mathbf{S}$, *and* $\mathbf{e}$.

Compute $\hat{\mathbf{A}}^{\mathrm{T}}\mathbf{S}\mathbf{e}$.
Compute $\mathbf{P} = \hat{\mathbf{A}}^{\mathrm{T}}\left(\mathbf{S}\mathbf{X}^{-1}+\mathbf{H}\right)\hat{\mathbf{A}}$.
Compute $\mathbf{R} = \mathbf{P}^{-1}$.
Compute $\dot{\mathbf{x}} = \hat{\mathbf{A}}\mathbf{R}\hat{\mathbf{A}}^{\mathrm{T}}\mathbf{S}\mathbf{e}$, $\frac{\dot{\mathbf{x}}}{\mathbf{x}} = \mathbf{X}^{-1}\dot{\mathbf{x}}$, $\frac{\dot{\mathbf{s}}}{\mathbf{s}} = \left(\mathbf{e} - \frac{\dot{\mathbf{x}}}{\mathbf{x}}\right)$, *and* $\dot{\mathbf{s}} = \mathbf{S}\frac{\dot{\mathbf{s}}}{\mathbf{s}}$.
Compute $\mathbf{f} = -2\frac{\dot{\mathbf{x}}}{\mathbf{x}} \circ \frac{\dot{\mathbf{s}}}{\mathbf{s}}$.
Compute $\ddot{\mathbf{x}} = \hat{\mathbf{A}}\mathbf{R}\hat{\mathbf{A}}^{\mathrm{T}}\mathbf{S}\mathbf{f}$ *and* $\ddot{\mathbf{s}} = \mathbf{S}\mathbf{f} - \mathbf{S}\mathbf{X}^{-1}\ddot{\mathbf{x}}$.
Compute $\dot{\lambda} = \left(\mathbf{A}\mathbf{A}^{\mathrm{T}}\right)^{-1}\mathbf{A}\left(\mathbf{H}\dot{\mathbf{x}} - \dot{\mathbf{s}}\right)$ *and* $\ddot{\lambda} = \left(\mathbf{A}\mathbf{A}^{\mathrm{T}}\right)^{-1}\mathbf{A}\left(\mathbf{H}\ddot{\mathbf{x}} - \ddot{\mathbf{s}}\right)$.

The computational counts for $(\dot{\mathbf{x}},\dot{\lambda},\dot{\mathbf{s}})$ and $(\ddot{\mathbf{x}},\ddot{\lambda},\ddot{\mathbf{s}})$ are estimated as follows.

- Computing $\hat{\mathbf{A}}^{\mathrm{T}}\mathbf{S}\mathbf{e}$ requires $O((n-m)n)$;

- Computing $\mathbf{R} = \left(\hat{\mathbf{A}}^{\mathrm{T}}\left(\mathbf{S}\mathbf{X}^{-1}+\mathbf{H}\right)\hat{\mathbf{A}}\right)^{-1}$ requires $O((n-m)n^2+(n-m)^2n+(n-m)^3)$;
- Computing $\dot{\mathbf{x}} = \hat{\mathbf{A}}\mathbf{R}\hat{\mathbf{A}}^{\mathrm{T}}\mathbf{S}\mathbf{e}$, $\frac{\dot{\mathbf{x}}}{\mathbf{x}} = \mathbf{X}^{-1}\dot{\mathbf{x}}$, $\frac{\dot{\mathbf{s}}}{\mathbf{s}} = \left(\mathbf{e}-\frac{\dot{\mathbf{x}}}{\mathbf{x}}\right)$, and $\dot{\mathbf{s}} = \mathbf{S}\frac{\dot{\mathbf{s}}}{\mathbf{s}}$ requires $O(n(n-m)+n)$;
- Computing $\mathbf{f} = -2\frac{\dot{\mathbf{x}}}{\mathbf{x}} \circ \frac{\dot{\mathbf{s}}}{\mathbf{s}}$ requires $O(n)$;
- Computing $\ddot{\mathbf{x}} = \hat{\mathbf{A}}\mathbf{R}\hat{\mathbf{A}}^{\mathrm{T}}\mathbf{S}\mathbf{f}$ and $\ddot{\mathbf{s}} = \mathbf{S}\mathbf{f}-\mathbf{S}\mathbf{X}^{-1}\ddot{\mathbf{x}}$ requires $O(n(n-m)+n)$.
- Computing $\dot{\lambda} = \left(\mathbf{A}\mathbf{A}^{\mathrm{T}}\right)^{-1}\mathbf{A}\left(\mathbf{H}\dot{\mathbf{x}}-\dot{\mathbf{s}}\right)$ and $\ddot{\lambda} = \left(\mathbf{A}\mathbf{A}^{\mathrm{T}}\right)^{-1}\mathbf{A}\left(\mathbf{H}\ddot{\mathbf{x}}-\ddot{\mathbf{s}}\right)$ requires $O(n(n+m)+n)$.

Clearly, this implementation is much cheaper than directly solving the systems of equations. A similar algorithm is used to calculate $(\Delta\mathbf{x}, \Delta\lambda, \Delta\mathbf{s})$.

Algorithm 9.4
(compute $\Delta\mathbf{x}, \Delta\lambda, \Delta\mathbf{s}$)

Data: Matrices $\mathbf{A}$, $\hat{\mathbf{A}}$, $\mathbf{H} \succeq \mathbf{0}$, $\mathbf{X}(\alpha)$, $\mathbf{S}(\alpha)$, *and* $\mathbf{e}$.
Compute $\mathbf{f}_1 = \mathbf{X}^{-1}(\alpha)\mathbf{S}^{-1}(\alpha)\mu\mathbf{e}-\mathbf{e}$.
Compute $\hat{\mathbf{A}}^{\mathrm{T}}\mathbf{S}(\alpha)\mathbf{f}_1$.
Compute $\mathbf{P} = \hat{\mathbf{A}}^{\mathrm{T}}\left(\mathbf{S}(\alpha)\mathbf{X}^{-1}(\alpha)+\mathbf{H}\right)\hat{\mathbf{A}}$.
Compute $\mathbf{R} = \mathbf{P}^{-1}$.
Compute $\Delta\mathbf{x} = \hat{\mathbf{A}}\mathbf{R}\hat{\mathbf{A}}^{\mathrm{T}}\mathbf{S}(\alpha)\mathbf{f}_1$ *and* $\Delta\mathbf{s} = \mathbf{S}(\alpha)\mathbf{f}_1 - \mathbf{S}(\alpha)\mathbf{X}^{-1}(\alpha)\Delta\mathbf{x}$.
Compute $\Delta\lambda = \left(\mathbf{A}\mathbf{A}^{\mathrm{T}}\right)^{-1}\mathbf{A}\left(\mathbf{H}\Delta\mathbf{x}-\Delta\mathbf{s}\right)$.

9.4.4 Duality measure reduction

Directly using $\sin(\alpha) = \frac{\theta}{\sqrt{n}}$ in Algorithm 9.1 provides a convenient formula to prove the polynomiality. However, this choice of $\sin(\alpha)$ is too conservative in practice, because the step size in $\mathcal{N}_2(2\theta)$ may be small and the duality measure may not be reduced fast enough. A better choice of $\sin(\alpha)$ should have a larger step in every iteration so that the polynomiality is reserved and fast convergence is achieved.

From analysis in Section 9.2, conditions that restrict step size are proximity conditions, positivity conditions, and duality reduction condition. Assuming $\theta \le 0.148$, the proximity condition (9.59) holds for $(\mathbf{x}^{k+1}, \mathbf{s}^{k+1})$ without other restriction; however, three more factors restrict the search step length. First, proximity condition (9.37) is met for $\sin(\alpha) \in [0, \sin(\tilde{\alpha})]$, where $\sin(\tilde{\alpha})$ is the smallest positive solution of (9.33) and it is estimated very conservatively in Lemma 9.12. However, *an efficient implementation should use* $\sin(\tilde{\alpha})$, *the smallest positive solution of (9.33)*. Since (9.33) is a quartic function of $\sin(\alpha)$, the cost of finding the smallest positive solution is negligible [35]. Second, from (9.38) and (9.60), $\mu(\alpha) > 0$ is required for positivity conditions $(\mathbf{x}(\alpha), \mathbf{s}(\alpha)) > \mathbf{0}$ and

$(\mathbf{x}^{k+1},\mathbf{s}^{k+1}) > \mathbf{0}$ to hold. Since $\sin(\bar{\alpha})$ estimated in Corollary 9.1 may be a little conservative, *we directly calculate* $\sin(\bar{\alpha})$, *which is the smallest positive solution of*

$$\mu(\alpha) \geq \mu(1-\sin(\alpha)) - \frac{1}{n}\Big((\dot{\mathbf{x}}^{\mathrm{T}}\dot{\mathbf{s}})\sin^4(\alpha) + (\dot{\mathbf{x}}^{\mathrm{T}}\dot{\mathbf{s}})\sin^2(\alpha)\Big) = \sigma, \tag{9.86}$$

where $\sigma > 0$ is a small number. The positivity conditions are guaranteed for all $\sin(\alpha) \in [0, \sin(\bar{\alpha})]$. Since (9.86) is a quartic function of $\sin(\alpha)$, the cost of finding the smallest positive solution is negligible. Third, from (9.63a) and Lemma 9.6, for $\mu_{k+1} \leq \mu_k$ to hold, we need

$$\begin{aligned} & \frac{2\theta^2(1+2\theta)}{n(1-2\theta)^2} - \left(1 + \frac{2\theta^2(1+2\theta)}{n(1-2\theta)^2}\right)\sin(\alpha) \\ & + \left(1 + \frac{2\theta^2(1+2\theta)}{n(1-2\theta)^2}\right)\frac{\ddot{\mathbf{x}}^{\mathrm{T}}\ddot{\mathbf{s}}}{n\mu}\left(\sin^2(\alpha)+\sin^4(\alpha)\right) \\ \leq\ & 0. \end{aligned}$$

For the sake of convergence analysis, Lemma 9.16 is used. For efficient implementation, the following solution should be adopted. Denote $p = \frac{2\theta^2(1+2\theta)}{n(1-2\theta)^2} > 0$, $q = \frac{\ddot{\mathbf{x}}^{\mathrm{T}}\ddot{\mathbf{s}}}{n\mu} > 0$, $z = \sin(\alpha) \in [0,1]$, and

$$F(z) = (1+p)qz^4 + (1+p)qz^2 - (1+p)z + p.$$

For $z \in [0,1]$ and $q \leq \frac{1}{6}$, $F'(z) = (1+p)(4qz^3 + 2qz - 1) \leq 0$, therefore, the upper bound of the duality measure is a monotonic decreasing function of $\sin(\alpha)$ for $\alpha \in [0, \frac{\pi}{2}]$. The larger the α is, the smaller the upper bound of the duality gap will be. For $q > \frac{1}{6}$, to minimize the upper bound of the duality gap, we can find the solution of $F'(z) = 0$. It is easy to check from discriminator of Lemma 9.7 (see also [113]) that the cubic polynomial $F'(z)$ has only one real solution, which is given by

$$\begin{aligned} \sin(\breve{\alpha}) &= \sqrt[3]{\frac{n\mu}{8\ddot{\mathbf{x}}^{\mathrm{T}}\ddot{\mathbf{s}}} + \sqrt{\left(\frac{n\mu}{8\ddot{\mathbf{x}}^{\mathrm{T}}\ddot{\mathbf{s}}}\right)^2 + \left(\frac{1}{6}\right)^3}} + \sqrt[3]{\frac{n\mu}{8\ddot{\mathbf{x}}^{\mathrm{T}}\ddot{\mathbf{s}}} - \sqrt{\left(\frac{n\mu}{8\ddot{\mathbf{x}}^{\mathrm{T}}\ddot{\mathbf{s}}}\right)^2 + \left(\frac{1}{6}\right)^3}} \\ &:= \chi. \end{aligned} \tag{9.87}$$

Since $F''(\sin(\breve{\alpha}) = (1+p)(12q\sin^2(\breve{\alpha}) + 2q) > 0$, at $\sin(\breve{\alpha}) \in [0,1)$, the upper bound of the duality gap is minimized. Therefore, we can define

$$\breve{\alpha} = \begin{cases} \frac{\pi}{2}, & \text{if } \frac{\ddot{\mathbf{x}}^{\mathrm{T}}\ddot{\mathbf{s}}}{n\mu} \leq \frac{1}{6} \\ \sin^{-1}(\chi), & \text{if } \frac{\ddot{\mathbf{x}}^{\mathrm{T}}\ddot{\mathbf{s}}}{n\mu} > \frac{1}{6}. \end{cases} \tag{9.88}$$

It is worthwhile to note that for $\alpha < \breve{\alpha}$, $F'(\sin(\alpha)) < 0$, i.e., $F(\sin(\alpha))$ is a monotonic decreasing function of $\alpha \in [0, \breve{\alpha}]$. As we can see from the above discussion, *$\bar{\alpha}$ and $\breve{\alpha}$ are used for satisfying positivity conditions and minimizing the upper bound of the duality gap*, and there is little room to improve these values. To further minimize the duality gap in each iteration, we may select the final step size $\sin(\alpha)$ as follows.

Algorithm 9.5
(Select step size)
Data: Fixed iteration number ℓ, $\sin(\breve{\alpha})$, $\sin(\bar{\alpha})$, and $\sin(\tilde{\alpha})$.

Step 1: If $\sin(\tilde{\alpha}) = \min\{\sin(\breve{\alpha}), \sin(\bar{\alpha}), \sin(\tilde{\alpha})\} < \min\{\sin(\breve{\alpha}), \sin(\bar{\alpha})\} = \sin(\breve{\alpha})$, using golden section search (ℓ iterations) to get an α in the interval $[\sin(\tilde{\alpha}), \sin(\breve{\alpha})]$ such that

$$\left\| \mathbf{x}(\alpha) \circ \mathbf{s}(\alpha) - \mu(\alpha)\mathbf{e} \right\| \leq 2\theta\mu(\alpha).$$

Step 2: Otherwise, select $\alpha = \breve{\alpha}$.

Therefore, Algorithm 9.1 can be implemented as follows.

Algorithm 9.6
(Arc-search path-following)
Data: $\mathbf{A}$, $\mathbf{H} \succeq \mathbf{0}$, $\mathbf{b}$, $\mathbf{c}$, $\theta = 0.148$, $\varepsilon > \sigma > 0$.
Step 0: Find initial point $(\mathbf{x}^0, \lambda^0, \mathbf{s}^0) \in \mathcal{N}_2(\theta)$ using Algorithm 9.2, and $\mu_0 = \frac{\mathbf{x}^{0^{\mathrm{T}}}\mathbf{s}^0}{n}$.
for *iteration $k = 0, 1, 2, \ldots$*

Step 1: If (9.75) holds or $\mu \leq \sigma$, stop. Otherwise continue.

Step 2: Compute $(\dot{\mathbf{x}}, \dot{\lambda}, \dot{\mathbf{s}})$ and $(\ddot{\mathbf{x}}, \ddot{\lambda}, \ddot{\mathbf{s}})$ using Algorithm 9.3.

Step 3: Find $\sin(\tilde{\alpha})$, the smallest positive solution of the quartic polynomial of (9.33), $\sin(\bar{\alpha})$, the smallest positive solution of the quartic polynomial of (9.86), and $\sin(\breve{\alpha})$ from (9.88). Select α by Algorithm 9.5. Update $(\mathbf{x}(\alpha), \lambda(\alpha), \mathbf{s}(\alpha))$ and $\mu(\alpha)$ by (9.34) and (9.35).

Step 4: Compute $(\Delta\mathbf{x}, \Delta\lambda, \Delta\mathbf{s})$ using Algorithm 9.4, update $(\mathbf{x}^{k+1}, \lambda^{k+1}, \mathbf{s}^{k+1})$ and μ_{k+1} by using (9.46) and (9.47). Set $k+1 \rightarrow k$. Go back to Step 1.

end (for)

Remark 9.5 The condition $\mu > \sigma$ guarantees that the equation (9.86) has a positive solution before terminate criterion is met. ∎

9.5 Numerical Examples

In this section, we will first use a simple example to demonstrate how the algorithm works. Then, we will test QP examples originating from [53] and compare the result with the one reported in [47].

9.5.1 A simple example

In this subsection, we use a problem in [106, page 464] to illustrate the difference between the active set method and the arc-search interior-point method developed in this chapter.

The problem is given as follows:

$$\min_{x} \quad f(x) = (x_1-1)^2+(x_2-2.5)^2 \tag{9.89a}$$
$$\text{subject to} \quad x_1-2x_2+2 \geq 0, \tag{9.89b}$$
$$-x_1-2x_2+6 \geq 0, \tag{9.89c}$$
$$-x_1+2x_2+2 \geq 0, \tag{9.89d}$$
$$x_1 \geq 0, \quad x_2 \geq 0. \tag{9.89e}$$

The optimal solution is $\mathbf{x}^* = (1.4, 1.7)$. Having assumed that every search provides an accurate solution and the initial point is $\mathbf{x}^0 = (2,0)$, the active set algorithm will find the initial active set, constraints 3 and 5, then searches to the point $\mathbf{x}^1 = (1,0)$, then searches to the point $\mathbf{x}^2 = (1,1.5)$, then finds the optimal solution. The detail of the search procedure is provided in [106, pages 464-465]. The search path is depicted in the red line in Figure 9.1.

The problem is converted to the standard form suitable for interior point algorithms as follows.

$$\min_{x} \quad f(x) = \frac{1}{2}\mathbf{x}^{\mathrm{T}}\mathbf{H}\mathbf{x}+\mathbf{c}^{\mathrm{T}}\mathbf{x}+7.25 \tag{9.90a}$$
$$\text{subject to} \quad x_1-2x_2-x_3+2=0, \tag{9.90b}$$
$$-x_1-2x_2-x_4+6=0, \tag{9.90c}$$
$$-x_1+2x_2-x_5+2=0, \tag{9.90d}$$
$$x_1 \geq 0, \quad x_2 \geq 0, \quad x_3 \geq 0, \quad x_4 \geq 0, \quad x_5 \geq 0, \tag{9.90e}$$

where $\mathbf{H} = diag(2,2,0,0,0)$, $\mathbf{c}^{\mathrm{T}} = (-2,-5,0,0,0)$, and $\mathbf{b}^{\mathrm{T}} = (-2,-6,-2)$. In standard form, the optimal solution is $\mathbf{x}^* = (1.4,\ 1.7,\ 0.0,\ 1.2,\ 4.0)$. The initial point $\mathbf{x}^0 = (2,\ 0.01,\ 0.01,\ 0.01,\ 0.01)$ and $\mathbf{s}^0 = (0.5,\ 100,\ 100,\ 100,\ 100)$ is used so that the initial $(\mathbf{x},\mathbf{s})$ is an interior point and close to the initial point of the active set method. This initial point satisfies $\mathbf{x}^0 \circ \mathbf{s}^0 = \mathbf{e}$. $\varepsilon = 0.000001$ is used in the termination criterion. $\sigma = 0.000001\varepsilon^3$ is used in (9.86). The central path projected in the (x_1,x_2) plane in Figures 9.1, 9.2, and 9.3 is a dot line in black. The

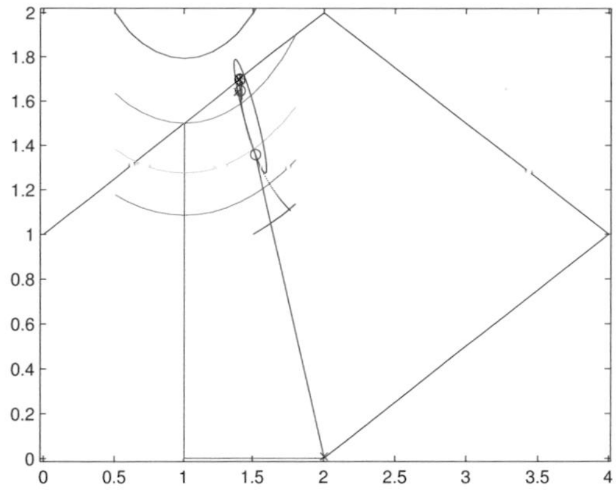

Figure 9.1: Arc-search for the example in [106, page 464].

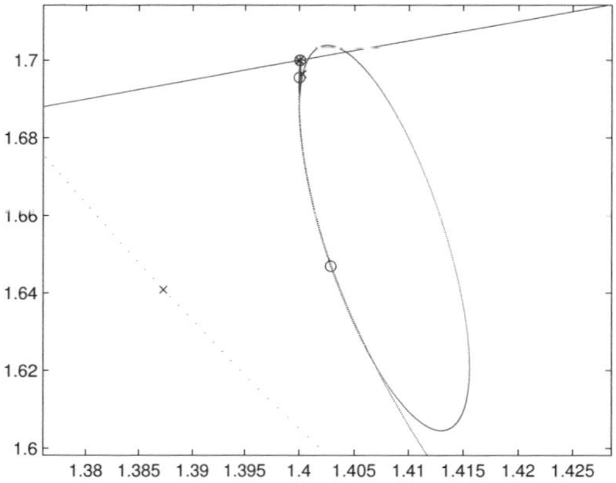

Figure 9.2: Arc-search approaches optimal solution for the example.

approximations of the central path are ellipses that are projected to the (x_1, x_2) plane in Figures 9.1, 9.2, and 9.3. They are dot lines in blue. Unlike its counterpart in linear programming, the central path is not close to a straight line, but the ellipse approximation of the central path in quadratic programming is still very good. The corrector step is also very efficient in bringing the iterate back to the central path. Figures 9.2 and 9.3 are magnified parts of Figure 9.1. They provide more detailed information when the arc-search approaches the optimal solution. In the plots, the red 'x' is the initial point of the test problem; the red circle 'o' are the points obtained by either Algorithm 9.2 (move to the central path initially) or the correct steps (Step 4); the black 'x' are the points obtained by searching along the ellipses (Step 3); the red '*' is the optimal solution of the problem. After five

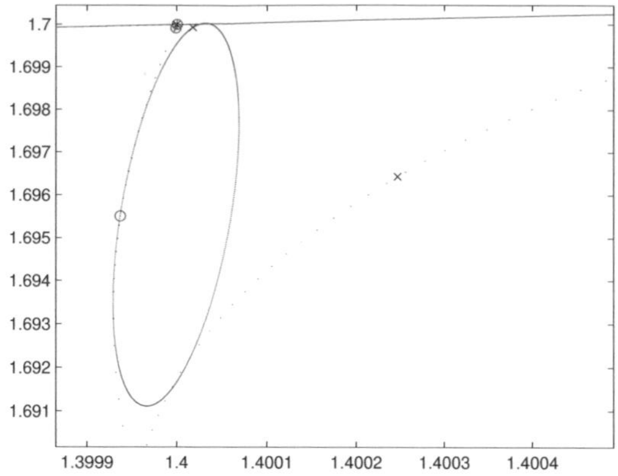

Figure 9.3: Arc-search approaches optimal solution for the example.

(5) iterations, the algorithm finds the optimal solution of this problem. At the convergence point, the slack variable is $\mathbf{s}^* = (0,\ 0,\ 0.8,\ 0,\ 0)$, therefore, the algorithm converges to a strict complementary solution. The duality gap values in the iterations are $(\mu_0,\ \mu_1,\ \mu_2,\ \mu_3,\ \mu_4,\ \mu_5) = (1,\ 0.45454,\ 0.03578,\ 6.38692e-004,\ 3.08053e-007,\ 7.63864e-014)$. From these figures, we can see intuitively that searching along an ellipse that approximates the central path is attractive. The convergence rate appears to be super-linear.

9.5.2 Test on problems in [53]

In [47], a quadratic programming software LOQO is presented and numerical test were conducted for many quadratic programming problems. Among these problems, some are from a set of nonlinear programming test problems published in a book [53] which is open to access, easy to convert to MATLAB format, and widely used by researchers to test nonlinear programming algorithms. Algorithm 9.1, with some improvements described in Section 9.4, is implemented in MATLAB and tested for all 7 convex quadratic problems collected in [53]. The result is compared to the one obtained by LOQO [47]. The polynomial algorithm proposed in this chapter uses the same number iterations as or fewer iterations than LOQO in every problem, and converges to some equally good or better points in all problems. The comparison is listed in Table 9.1. For these 7 problems, LOQO uses 67 iterations, while the proposed algorithm uses 49 iterations, 27% fewer in total iterations than LOQO; and the latter converges to points which have equal or higher accuracy in objective functions.

Table 9.1: Iteration counts for test problems in [53].

Problem	iteration numbers using arc-search	objective value obtained by arc-search	iteration numbers using LOQO	objective value obtained by LOGO
hs021	12	4.0000001e-2	13	4.0000001e-2
hs035	6	-8.8888889e+0	8	-8.8888889e+0
hs035mod	5	-8.7500000e+0	5	-8.7499999e+0
hs051	4	-6.0000000e+0	8	-6.0000000e+0
hs052	8	-6.7335244e-1	9	-6.7335244e-1
hs051	8	-1.9069767e+0	11	-1.9069767e+0
hs076	6	-4.6818182e+0	13	-4.6818182e+0

9.6 Concluding Remarks

This chapter proposed an arc-search interior-point path-following algorithm for convex quadratic programming that searches for the optimizers along ellipses that approximate the central path. The algorithm is proved to be polynomial with the complexity bound $O(\sqrt{n}\log(1/\varepsilon))$, which is the best bound for linear programming. A simple example is provided in order to demonstrate how the algorithm works. Preliminary test on quadratic programming problems originating from [53] shows that the proposed algorithm is promising.

As we already know, infeasible starting point and wide neighborhood are two other strategies that can also improve the computational efficiency. However, these strategies have not yet been investigated.

Chapter 10

An Arc-Search Algorithm for QP with Box Constraints

To avoid the cost of finding a feasible initial point for general inequality constrained optimization problems, infeasible interior-point method becomes very popular. However, if a feasible initial point is available, intuitively, a feasible interior-point method is more attractive than the infeasible interior-point method, because the effort to bring the iterates gradually from infeasible region to feasible region is not needed anymore. Another advantage of the feasible interior-point method is that all iterates are feasible, which is desired by many engineering problems where real-time optimization code may have to be terminated before an optimizer is found. In this case, the latest available iterate is still feasible and can be applied as a nearly optimal solution. In this chapter, we show that the convex quadratic programming problem subject only to bound (box) constraints does have an explicit feasible initial point. This problem has been studied in [51], where a first-order interior-point algorithm is considered. In this chapter, we will consider an arc-search higher-order interior-point algorithm, which starts at a feasible initial point. We believe that these two features will improve the computational efficiency. We propose an algorithm that is specially designed for this problem and show that the algorithm is also more efficient than the algorithm in the previous chapter because of two improvements due to the special structure of the constraints: (1) the enlarged searching neighborhood, and (2) an explicit feasible initial interior point. Moreover, we show that this algorithm has a very desirable theoretical property, the best polynomial complexity. Although the idea of the proof of polynomiality is somewhat similar to that used in Chapter 9, we provided a complete proof for several reasons: (i) the proof shows that the searching neighborhood is larger than the general convex quadratic programming with linear constraints, which supports the claim of the efficiency of the newly proposed algorithm, (ii) the proof is fairly different from the one in Chapter 9 because of some special structure of the box constraints and the proof carefully takes care of these differences, and (iii) the proof provides complete results. The algorithm is implemented in MATLAB®. Some numerical test is presented in order to demonstrate the effectiveness and efficiency of the proposed algorithm. The materials of this chapter are based on [155].

10.1 Problem Descriptions

We will consider a convex quadratic problem with box constraints in a standard form:

$$(BQP) \quad \min \ \tfrac{1}{2}\mathbf{x}^{\mathrm{T}}\mathbf{H}\mathbf{x}+\mathbf{c}^{\mathrm{T}}\mathbf{x}, \quad \text{subject to} \quad -\mathbf{e}\le\mathbf{x}\le\mathbf{e}, \tag{10.1}$$

where $\mathbf{0} \preceq \mathbf{H} \in \mathbf{R}^{n\times n}$ is a positive definite matrix, $\mathbf{c} \in \mathbf{R}^n$ is given, and $\mathbf{x} \in \mathbf{R}^n$ is the variable vector to be optimized.

In view of the KKT conditions (1.49), since $\mathbf{H}$ is positive definite matrix, $\mathbf{x}$ is an optimal solution of (9.1) if and only if $\mathbf{x}$, λ, and γ satisfy

$$-\lambda+\gamma-\mathbf{H}\mathbf{x}=\mathbf{c}, \tag{10.2a}$$

$$-\mathbf{e}\le\mathbf{x}\le\mathbf{e}, \tag{10.2b}$$

$$(\lambda,\gamma)\ge\mathbf{0}, \tag{10.2c}$$

$$\lambda_i(e_i-x_i)=0, \quad \gamma_i(e_i+x_i)=0, \quad i=1,\ldots,n. \tag{10.2d}$$

Denote $\mathbf{y}=\mathbf{e}-\mathbf{x}\ge\mathbf{0}$, $\mathbf{z}=\mathbf{e}+\mathbf{x}\ge\mathbf{0}$. The KKT conditions can be rewritten as

$$\mathbf{H}\mathbf{x}+\mathbf{c}+\lambda-\gamma=\mathbf{0}, \tag{10.3a}$$

$$\mathbf{x}+\mathbf{y}=\mathbf{e}, \quad \mathbf{x}-\mathbf{z}=-\mathbf{e}, \tag{10.3b}$$

$$(\mathbf{y},\mathbf{z},\lambda,\gamma)\ge\mathbf{0}, \tag{10.3c}$$

$$\lambda_i y_i=0, \quad \gamma_i z_i=0, \quad i=1,\ldots,n. \tag{10.3d}$$

For the convex (QP) problem, the KKT conditions are also sufficient for $\mathbf{x}$ to be a global optimal solution (see Chapter 1). Denote the *feasible set* $\mathcal{F}$ as a collection of all points that meet the constraints (10.3a), (10.3b), (10.3c)

$$\mathcal{F}=\{(\mathbf{x},\mathbf{y},\mathbf{z},\lambda,\gamma): \mathbf{H}\mathbf{x}+\mathbf{c}+\lambda-\gamma=\mathbf{0}, (\mathbf{y},\mathbf{z},\lambda,\gamma)\ge\mathbf{0}, \mathbf{x}+\mathbf{y}=\mathbf{e}, \mathbf{x}-\mathbf{z}=-\mathbf{e}\}, \tag{10.4}$$

and the *strictly feasible* set $\mathcal{F}^o$ as a collection of all points that meet the constraints (10.3a), (10.3b), and are strictly positive in (10.3c)

$$\mathcal{F}^o = \{(\mathbf{x},\mathbf{y},\mathbf{z},\boldsymbol{\lambda},\boldsymbol{\gamma}) : \mathbf{H}\mathbf{x}+\mathbf{c}+\boldsymbol{\lambda}-\boldsymbol{\gamma}=\mathbf{0}, (\mathbf{y},\mathbf{z},\boldsymbol{\lambda},\boldsymbol{\gamma}) > \mathbf{0}, \mathbf{x}+\mathbf{y}=\mathbf{e}, \mathbf{x}-\mathbf{z}=-\mathbf{e}\}. \tag{10.5}$$

Similar to the linear programming, the *central path* $\mathcal{C} \in \mathcal{F}^o \subset \mathcal{F}$ is defined as a curve in finite dimensional space parametrized by a scalar $\tau > 0$ as follows. For each interior point $(\mathbf{x},\mathbf{y},\mathbf{z},\boldsymbol{\lambda},\boldsymbol{\gamma}) \in \mathcal{F}^o$ on the central path, there is a $\tau > 0$ such that

$$\mathbf{H}\mathbf{x}+\mathbf{c}+\boldsymbol{\lambda}-\boldsymbol{\gamma}=\mathbf{0}, \tag{10.6a}$$

$$\mathbf{x}+\mathbf{y}=\mathbf{e}, \quad \mathbf{x}-\mathbf{z}=-\mathbf{e}, \tag{10.6b}$$

$$(\mathbf{y},\mathbf{z},\boldsymbol{\lambda},\boldsymbol{\gamma}) > \mathbf{0}, \tag{10.6c}$$

$$\lambda_i y_i = \tau, \quad \gamma_i z_i = \tau, \quad i=1,\ldots,n. \tag{10.6d}$$

Therefore, the central path is an arc that is parametrized as a function of τ and is denoted as

$$\mathcal{C} = \{(\mathbf{x}(\tau),\mathbf{y}(\tau),\mathbf{z}(\tau),\boldsymbol{\lambda}(\tau),\boldsymbol{\gamma}(\tau)) : \tau > 0\}. \tag{10.7}$$

As $\tau \to 0$, the moving point $(\mathbf{x}(\tau),\mathbf{y}(\tau),\mathbf{z}(\tau),\boldsymbol{\lambda}(\tau),\boldsymbol{\gamma}(\tau))$ on the central path represented by (10.6) approaches the solution of (QP) represented by (9.1). Throughout the rest of this chapter, the following assumption is made.

Assumption:

1. $\mathcal{F}^o$ is not empty.

Assumption 1 implies the existence of a central path. This assumption is always true for the convex quadratic programming problem with box constraints. An explicit initial interior point will be provided later in this chapter.

Let $1 > \theta > 0$, denote $\mathbf{p} = (\mathbf{y},\mathbf{z})$, $\boldsymbol{\omega} = (\boldsymbol{\lambda},\boldsymbol{\gamma})$, and the *duality measure*

$$\mu = \frac{\boldsymbol{\lambda}^{\mathrm{T}}\mathbf{y}+\boldsymbol{\gamma}^{\mathrm{T}}\mathbf{z}}{2n} = \frac{\mathbf{p}^{\mathrm{T}}\boldsymbol{\omega}}{2n}. \tag{10.8}$$

A set of neighborhood of the central path is defined as

$$\mathcal{N}_2(\theta) = \{(\mathbf{x},\mathbf{y},\mathbf{z},\boldsymbol{\lambda},\boldsymbol{\gamma}) \in \mathcal{F}^o : \|\mathbf{p}\circ\boldsymbol{\omega}-\mu\mathbf{e}\| \le \theta\mu\} \subset \mathcal{F}^o. \tag{10.9}$$

As the duality measure is reduced to zero, the neighborhood of $\mathcal{N}_2(\theta)$ will be a neighborhood of the central path that approaches the optimizer(s) of the BQP problem, therefore, all points inside $\mathcal{N}_2(\theta)$ will approach the optimizer(s) of the BQP problem. For $(\mathbf{x},\mathbf{y},\mathbf{z},\boldsymbol{\lambda},\boldsymbol{\gamma}) \in \mathcal{N}_2(\theta)$, since $(1-\theta)\mu \le \omega_i p_i \le (1+\theta)\mu$, where ω_i are either λ_i or γ_i, and p_i are either y_i or z_i, it must have

$$\frac{\omega_i p_i}{1+\theta} \le \frac{\max_i \omega_i p_i}{1+\theta} \le \mu \le \frac{\min_i \omega_i p_i}{1-\theta} \le \frac{\omega_i p_i}{1-\theta}. \tag{10.10}$$

10.2 An Interior-Point Algorithm for Convex QP with Box Constraints

The idea of the *arc-search* algorithm proposed in this section is very simple. The algorithm starts from a feasible point in $\mathcal{N}_2(\theta)$ close to the central path, constructs an arc that passes through the point and approximates the central path, searches along the arc to a new point in a larger area $\mathcal{N}_2(2\theta)$ that reduces the duality measure $\mathbf{p}^{\mathrm{T}}\omega$ and meets (10.6a), (10.6b), and (10.6c). The process is repeated by finding a better point close to the central path or on the central path in $\mathcal{N}_2(\theta)$ that simultaneously meets (10.6a), (10.6b), and (10.6c) and reduces the duality measure. When the duality measure is reduced to zero, the optimal solution is found.

Following the idea used in Chapter 9, an *ellipse* $\mathcal{E}$ in an appropriate dimensional space introduced in Chapter 5 will be used to approximate the central path $\mathcal{C}$ described by (10.6), where

$$\begin{aligned}\mathcal{E} \;=\; & \{(\mathbf{x}(\alpha),\mathbf{y}(\alpha),\mathbf{z}(\alpha),\lambda(\alpha),\gamma(\alpha)) : \\ & (\mathbf{x}(\alpha),\mathbf{y}(\alpha),\mathbf{z}(\alpha),\lambda(\alpha),\gamma(\alpha)) = \vec{\mathbf{a}}\cos(\alpha)+\vec{\mathbf{b}}\sin(\alpha)+\vec{\mathbf{c}}\},\end{aligned} \tag{10.11}$$

$\vec{\mathbf{a}} \in \mathbb{R}^{5n}$ and $\vec{\mathbf{b}} \in \mathbb{R}^{5n}$ are the axes of the ellipse, $\vec{\mathbf{c}} \in \mathbb{R}^{5n}$ is the center of the ellipse. Given a point $(\mathbf{x},\mathbf{y},\mathbf{z},\lambda,\gamma) = (\mathbf{x}(\alpha_0),\mathbf{y}(\alpha_0),\mathbf{z}(\alpha_0),\lambda(\alpha_0),\gamma(\alpha_0)) \in \mathcal{E}$ which is close to or on the central path, $\vec{\mathbf{a}}$, $\vec{\mathbf{b}}$, $\vec{\mathbf{c}}$ are functions of α, $(\mathbf{x},\lambda,\gamma,\mathbf{y},\mathbf{z})$, $(\dot{\mathbf{x}},\dot{\mathbf{y}},\dot{\mathbf{z}},\dot{\lambda},\dot{\gamma})$, and $(\ddot{\mathbf{x}},\ddot{\mathbf{y}},\ddot{\mathbf{z}},\ddot{\lambda},\ddot{\gamma})$, where $(\dot{\mathbf{x}},\dot{\mathbf{y}},\dot{\mathbf{z}},\dot{\lambda},\dot{\gamma})$ and $(\ddot{\mathbf{x}},\ddot{\mathbf{y}},\ddot{\mathbf{z}},\ddot{\lambda},\ddot{\gamma})$ are defined by

$$\begin{bmatrix} \mathbf{H} & \mathbf{0} & \mathbf{0} & \mathbf{I} & -\mathbf{I} \\ \mathbf{I} & \mathbf{I} & \mathbf{0} & \mathbf{0} & \mathbf{0} \\ \mathbf{I} & \mathbf{0} & -\mathbf{I} & \mathbf{0} & \mathbf{0} \\ \mathbf{0} & \Lambda & \mathbf{0} & \mathbf{Y} & \mathbf{0} \\ \mathbf{0} & \mathbf{0} & \Gamma & \mathbf{0} & \mathbf{Z} \end{bmatrix} \begin{bmatrix} \dot{\mathbf{x}} \\ \dot{\mathbf{y}} \\ \dot{\mathbf{z}} \\ \dot{\lambda} \\ \dot{\gamma} \end{bmatrix} = \begin{bmatrix} \mathbf{0} \\ \mathbf{0} \\ \mathbf{0} \\ \lambda \circ \mathbf{y} \\ \gamma \circ \mathbf{z} \end{bmatrix}, \tag{10.12}$$

and

$$\begin{bmatrix} \mathbf{H} & \mathbf{0} & \mathbf{0} & \mathbf{I} & -\mathbf{I} \\ \mathbf{I} & \mathbf{I} & \mathbf{0} & \mathbf{0} & \mathbf{0} \\ \mathbf{I} & \mathbf{0} & -\mathbf{I} & \mathbf{0} & \mathbf{0} \\ \mathbf{0} & \Lambda & \mathbf{0} & \mathbf{Y} & \mathbf{0} \\ \mathbf{0} & \mathbf{0} & \Gamma & \mathbf{0} & \mathbf{Z} \end{bmatrix} \begin{bmatrix} \ddot{\mathbf{x}} \\ \ddot{\mathbf{y}} \\ \ddot{\mathbf{z}} \\ \ddot{\lambda} \\ \ddot{\gamma} \end{bmatrix} = \begin{bmatrix} \mathbf{0} \\ \mathbf{0} \\ \mathbf{0} \\ -2\dot{\lambda} \circ \dot{\mathbf{y}} \\ -2\dot{\gamma} \circ \dot{\mathbf{z}} \end{bmatrix}, \tag{10.13}$$

where $\Lambda = \mathrm{diag}(\lambda)$, $\Gamma = \mathrm{diag}(\gamma)$, $\mathbf{Y} = \mathrm{diag}(\mathbf{y})$, and $\mathbf{Z} = \mathrm{diag}(\mathbf{z})$. The first rows of (10.12) and (10.13) are equivalent to

$$\mathbf{H}\dot{\mathbf{x}} = \dot{\gamma} - \dot{\lambda}, \qquad \mathbf{H}\ddot{\mathbf{x}} = \ddot{\gamma} - \ddot{\lambda}. \tag{10.14}$$

The next 2 rows of (10.12) and (10.13) are equivalent to

$$\dot{\mathbf{x}} = -\dot{\mathbf{y}}, \qquad \dot{\mathbf{x}} = \dot{\mathbf{z}}, \qquad \ddot{\mathbf{x}} = -\ddot{\mathbf{y}}, \qquad \ddot{\mathbf{x}} = \ddot{\mathbf{z}}. \tag{10.15}$$

The last 2 rows of (10.12) and (10.13) are equivalent to

$$\mathbf{p} \circ \dot{\omega} + \dot{\mathbf{p}} \circ \omega = \mathbf{p} \circ \omega, \tag{10.16}$$

$$\mathbf{p} \circ \ddot{\omega} + \ddot{\mathbf{p}} \circ \omega = -2\dot{\mathbf{p}} \circ \dot{\omega}, \tag{10.17}$$

where $\circ$ denotes the Hadamard product.

It has been shown in Chapter 5 that one can avoid the calculation of $\vec{\mathbf{a}}$, $\vec{\mathbf{b}}$, and $\vec{\mathbf{c}}$ in the expression of the ellipse. The following formulas are used instead.

Theorem 10.1

Let the 5-tuple $(\mathbf{x}(\alpha), \mathbf{y}(\alpha), \mathbf{z}(\alpha), \lambda(\alpha), \gamma(\alpha))$ be an arc defined by (10.11) passing through a point $(\mathbf{x}, \mathbf{y}, \mathbf{z}, \lambda, \gamma) \in \mathcal{E}$, and its first and second derivatives at $(\mathbf{x}, \mathbf{y}, \mathbf{z}, \lambda, \gamma)$ be $(\dot{\mathbf{x}}, \dot{\mathbf{y}}, \dot{\mathbf{z}}, \dot{\lambda}, \dot{\gamma})$ and $(\ddot{\mathbf{x}}, \ddot{\mathbf{y}}, \ddot{\mathbf{z}}, \ddot{\lambda}, \ddot{\gamma})$ which are defined by (10.12) and (10.13). Then, an ellipse approximation of the central path is given by

$$\mathbf{x}(\alpha) = \mathbf{x} - \dot{\mathbf{x}} \sin(\alpha) + \ddot{\mathbf{x}}(1 - \cos(\alpha)), \tag{10.18}$$

$$\mathbf{y}(\alpha) = \mathbf{y} - \dot{\mathbf{y}} \sin(\alpha) + \ddot{\mathbf{y}}(1 - \cos(\alpha)), \tag{10.19}$$

$$\mathbf{z}(\alpha) = \mathbf{z} - \dot{\mathbf{z}} \sin(\alpha) + \ddot{\mathbf{z}}(1 - \cos(\alpha)), \tag{10.20}$$

$$\lambda(\alpha) = \lambda - \dot{\lambda} \sin(\alpha) + \ddot{\lambda}(1 - \cos(\alpha)), \tag{10.21}$$

$$\gamma(\alpha) = \gamma - \dot{\gamma} \sin(\alpha) + \ddot{\gamma}(1 - \cos(\alpha)). \tag{10.22}$$

Two compact representations for $\mathbf{p}(\alpha) = (\mathbf{y}(\alpha), \mathbf{z}(\alpha))$ and $\omega(\alpha) = (\lambda(\alpha), \gamma(\alpha))$ are given below:

$$\mathbf{p}(\alpha) = \mathbf{p} - \dot{\mathbf{p}} \sin(\alpha) + \ddot{\mathbf{p}}(1 - \cos(\alpha)), \tag{10.23}$$

$$\omega(\alpha) = \omega - \dot{\omega} \sin(\alpha) + \ddot{\omega}(1 - \cos(\alpha)). \tag{10.24}$$

The duality measure at point $(\mathbf{x}(\alpha), \mathbf{p}(\alpha), \omega(\alpha))$ is defined as:

$$\mu(\alpha) = \frac{\lambda(\alpha)^{\mathrm{T}}\mathbf{y}(\alpha) + \gamma(\alpha)^{\mathrm{T}}\mathbf{z}(\alpha)}{2n} = \frac{\mathbf{p}(\alpha)^{\mathrm{T}}\omega(\alpha)}{2n}. \tag{10.25}$$

Assuming $(\mathbf{y}, \mathbf{z}, \lambda, \gamma) > \mathbf{0}$, one can easily see that if $\frac{\dot{\mathbf{y}}}{\mathbf{y}}, \frac{\dot{\mathbf{z}}}{\mathbf{z}}, \frac{\dot{\lambda}}{\lambda}, \frac{\dot{\gamma}}{\gamma}, \frac{\ddot{\mathbf{y}}}{\mathbf{y}}, \frac{\ddot{\mathbf{z}}}{\mathbf{z}}, \frac{\ddot{\lambda}}{\lambda}, \frac{\ddot{\gamma}}{\gamma}$ are bounded (this will be shown to be true), and if α is small enough, then, $\mathbf{y}(\alpha) > \mathbf{0}$, $\mathbf{z}(\alpha) > \mathbf{0}$, $\lambda(\alpha) > \mathbf{0}$, and $\gamma(\alpha) > \mathbf{0}$. It will also be shown that searching along this ellipse will reduce the duality measure, i.e., $\mu(\alpha) < \mu$.

Lemma 10.1

Let $(\mathbf{x}, \mathbf{y}, \mathbf{z}, \lambda, \gamma)$ be a strictly feasible point of (BQP), $(\dot{\mathbf{x}}, \dot{\mathbf{y}}, \dot{\mathbf{z}}, \dot{\lambda}, \dot{\gamma})$ and $(\ddot{\mathbf{x}}, \ddot{\mathbf{y}}, \ddot{\mathbf{z}}, \ddot{\lambda}, \ddot{\gamma})$ meet (10.12) and (10.13), $(\mathbf{x}(\alpha), \mathbf{y}(\alpha), \mathbf{z}(\alpha), \lambda(\alpha), \gamma(\alpha))$ be calculated using (10.18), (10.19), (10.20), (10.21), and (10.22), then the following conditions hold.

$$\mathbf{x}(\alpha) + \mathbf{y}(\alpha) = \mathbf{e}, \quad \mathbf{x}(\alpha) - \mathbf{z}(\alpha) = -\mathbf{e}, \quad \mathbf{H}\mathbf{x}(\alpha) + \mathbf{c} + \lambda(\alpha) - \gamma(\alpha) = \mathbf{0}.$$

Proof 10.1 Since $(\mathbf{x},\mathbf{y},\mathbf{z},\lambda,\gamma)$ is a strictly feasible point, the result follows from direct calculation by using (10.5), (10.12), (10.13), and Theorem 10.1. ∎

Lemma 10.2
Let $(\dot{\mathbf{x}},\dot{\mathbf{p}},\dot{\omega})$ be defined by (10.12), $(\ddot{\mathbf{x}},\ddot{\mathbf{p}},\ddot{\omega})$ be defined by (10.13), and $\mathbf{H}$ be positive semidefinite matrix. Then, the following relations hold:

$$\dot{\mathbf{p}}^{\mathrm{T}}\dot{\omega} = \dot{\mathbf{x}}^{\mathrm{T}}(\dot{\gamma}-\dot{\lambda}) = \dot{\mathbf{x}}^{\mathrm{T}}\mathbf{H}\dot{\mathbf{x}} \geq 0, \tag{10.26}$$

$$\ddot{\mathbf{p}}^{\mathrm{T}}\ddot{\omega} = \ddot{\mathbf{x}}^{\mathrm{T}}(\ddot{\gamma}-\ddot{\lambda}) = \ddot{\mathbf{x}}^{\mathrm{T}}\mathbf{H}\ddot{\mathbf{x}} \geq 0, \tag{10.27}$$

$$\ddot{\mathbf{p}}^{\mathrm{T}}\dot{\omega} = \ddot{\mathbf{x}}^{\mathrm{T}}(\dot{\gamma}-\dot{\lambda}) = \dot{\mathbf{x}}^{\mathrm{T}}(\ddot{\gamma}-\ddot{\lambda}) = \dot{\mathbf{p}}^{\mathrm{T}}\ddot{\omega} = \dot{\mathbf{x}}^{\mathrm{T}}\mathbf{H}\ddot{\mathbf{x}}; \tag{10.28}$$

$$\begin{aligned} & -(\dot{\mathbf{x}}^{\mathrm{T}}\mathbf{H}\dot{\mathbf{x}})(1-\cos(\alpha))^2 - (\ddot{\mathbf{x}}^{\mathrm{T}}\mathbf{H}\ddot{\mathbf{x}})\sin^2(\alpha) \\ \leq\ & (\ddot{\mathbf{x}}^{\mathrm{T}}(\dot{\gamma}-\dot{\lambda}) + \dot{\mathbf{x}}^{\mathrm{T}}(\ddot{\gamma}-\ddot{\lambda}))\sin(\alpha)(1-\cos(\alpha)) \\ \leq\ & (\dot{\mathbf{x}}^{\mathrm{T}}\mathbf{H}\dot{\mathbf{x}})(1-\cos(\alpha))^2 + (\ddot{\mathbf{x}}^{\mathrm{T}}\mathbf{H}\ddot{\mathbf{x}})\sin^2(\alpha); \end{aligned} \tag{10.29}$$

and

$$\begin{aligned} & -(\dot{\mathbf{x}}^{\mathrm{T}}\mathbf{H}\dot{\mathbf{x}})\sin^2(\alpha) - (\ddot{\mathbf{x}}^{\mathrm{T}}\mathbf{H}\ddot{\mathbf{x}})(1-\cos(\alpha))^2 \\ \leq\ & (\ddot{\mathbf{x}}^{\mathrm{T}}(\dot{\gamma}-\dot{\lambda}) + \dot{\mathbf{x}}^{\mathrm{T}}(\ddot{\gamma}-\ddot{\lambda}))\sin(\alpha)(1-\cos(\alpha)) \\ \leq\ & (\dot{\mathbf{x}}^{\mathrm{T}}\mathbf{H}\dot{\mathbf{x}})\sin^2(\alpha) + (\ddot{\mathbf{x}}^{\mathrm{T}}\mathbf{H}\ddot{\mathbf{x}})(1-\cos(\alpha))^2. \end{aligned} \tag{10.30}$$

For $\alpha = \frac{\pi}{2}$, (10.29) and (10.30) reduce to

$$-\left(\dot{\mathbf{x}}^{\mathrm{T}}\mathbf{H}\dot{\mathbf{x}} + \ddot{\mathbf{x}}^{\mathrm{T}}\mathbf{H}\ddot{\mathbf{x}}\right) \leq (\ddot{\mathbf{x}}^{\mathrm{T}}\mathbf{H}\dot{\mathbf{x}} + \dot{\mathbf{x}}^{\mathrm{T}}\mathbf{H}\ddot{\mathbf{x}}) \leq \dot{\mathbf{x}}^{\mathrm{T}}\mathbf{H}\dot{\mathbf{x}} + \ddot{\mathbf{x}}^{\mathrm{T}}\mathbf{H}\ddot{\mathbf{x}}. \tag{10.31}$$

Proof 10.2 From (10.15), we have

$$\dot{\mathbf{x}}^{\mathrm{T}}(\dot{\gamma}-\dot{\lambda}) = \dot{\mathbf{z}}^{\mathrm{T}}\dot{\gamma} + \dot{\mathbf{y}}^{\mathrm{T}}\dot{\lambda} = \dot{\mathbf{p}}^{\mathrm{T}}\dot{\omega},$$

$$\ddot{\mathbf{x}}^{\mathrm{T}}(\ddot{\gamma}-\ddot{\lambda}) = \ddot{\mathbf{z}}^{\mathrm{T}}\ddot{\gamma} + \ddot{\mathbf{y}}^{\mathrm{T}}\ddot{\lambda} = \ddot{\mathbf{p}}^{\mathrm{T}}\ddot{\omega},$$

$$\ddot{\mathbf{x}}^{\mathrm{T}}(\dot{\gamma}-\dot{\lambda}) = \ddot{\mathbf{p}}^{\mathrm{T}}\dot{\omega},$$

and

$$\dot{\mathbf{x}}^{\mathrm{T}}(\ddot{\gamma}-\ddot{\lambda}) = \dot{\mathbf{p}}^{\mathrm{T}}\ddot{\omega}.$$

Pre-multiplying $\dot{\mathbf{x}}^{\mathrm{T}}$ and $\ddot{\mathbf{x}}^{\mathrm{T}}$ to (10.14) gives

$$\dot{\mathbf{x}}^{\mathrm{T}}(\dot{\gamma}-\dot{\lambda}) = \dot{\mathbf{x}}^{\mathrm{T}}\mathbf{H}\dot{\mathbf{x}},$$

$$\ddot{\mathbf{x}}^{\mathrm{T}}(\ddot{\gamma}-\ddot{\lambda}) = \ddot{\mathbf{x}}^{\mathrm{T}}\mathbf{H}\ddot{\mathbf{x}},$$

$$\ddot{\mathbf{x}}^{\mathrm{T}}(\dot{\gamma}-\dot{\lambda}) = \ddot{\mathbf{x}}^{\mathrm{T}}\mathbf{H}\dot{\mathbf{x}} = \dot{\mathbf{x}}^{\mathrm{T}}\mathbf{H}\ddot{\mathbf{x}} = \dot{\mathbf{x}}^{\mathrm{T}}(\ddot{\gamma}-\ddot{\lambda}).$$

Equations (10.26) and (10.27) follow from the first two equations and the fact that $\mathbf{H}$ is positive semidefinite. The last equation is equivalent to (10.28). Using (10.26), (10.27), and (10.28) gives

$$\begin{aligned}
& (\dot{\mathbf{x}}(1-\cos(\alpha))+\ddot{\mathbf{x}}\sin(\alpha))^{\mathrm{T}}\mathbf{H}(\dot{\mathbf{x}}(1-\cos(\alpha))+\ddot{\mathbf{x}}\sin(\alpha)) \\
= & (\dot{\mathbf{x}}^{\mathrm{T}}\mathbf{H}\dot{\mathbf{x}})(1-\cos(\alpha))^2+2(\dot{\mathbf{x}}^{\mathrm{T}}\mathbf{H}\ddot{\mathbf{x}})\sin(\alpha)(1-\cos(\alpha))+(\ddot{\mathbf{x}}^{\mathrm{T}}\mathbf{H}\ddot{\mathbf{x}})\sin^2(\alpha) \\
= & (\dot{\mathbf{x}}^{\mathrm{T}}\mathbf{H}\dot{\mathbf{x}})(1-\cos(\alpha))^2+(\ddot{\mathbf{x}}^{\mathrm{T}}\mathbf{H}\ddot{\mathbf{x}})\sin^2(\alpha) \\
+ & (\ddot{\mathbf{x}}^{\mathrm{T}}(\dot{\gamma}-\dot{\lambda})+\dot{\mathbf{x}}^{\mathrm{T}}(\ddot{\gamma}-\ddot{\lambda}))\sin(\alpha)(1-\cos(\alpha)) \geq 0,
\end{aligned}$$

which is the first inequality of (10.29). Using (10.26), (10.27), and (10.28) also gives

$$\begin{aligned}
& (\dot{\mathbf{x}}(1-\cos(\alpha))-\ddot{\mathbf{x}}\sin(\alpha))^{\mathrm{T}}\mathbf{H}(\dot{\mathbf{x}}(1-\cos(\alpha))-\ddot{\mathbf{x}}\sin(\alpha)) \\
= & (\dot{\mathbf{x}}^{\mathrm{T}}\mathbf{H}\dot{\mathbf{x}})(1-\cos(\alpha))^2-2(\dot{\mathbf{x}}^{\mathrm{T}}\mathbf{H}\ddot{\mathbf{x}})\sin(\alpha)(1-\cos(\alpha))+(\ddot{\mathbf{x}}^{\mathrm{T}}\mathbf{H}\ddot{\mathbf{x}})\sin^2(\alpha) \\
= & (\dot{\mathbf{x}}^{\mathrm{T}}\mathbf{H}\dot{\mathbf{x}})(1-\cos(\alpha))^2+(\ddot{\mathbf{x}}^{\mathrm{T}}\mathbf{H}\ddot{\mathbf{x}})\sin^2(\alpha) \\
& -(\ddot{\mathbf{x}}^{\mathrm{T}}(\dot{\gamma}-\dot{\lambda})+\dot{\mathbf{x}}^{\mathrm{T}}(\ddot{\gamma}-\ddot{\lambda}))\sin(\alpha)(1-\cos(\alpha)) \geq 0,
\end{aligned}$$

which is the second inequality of (10.29). Replacing $\dot{\mathbf{x}}(1-\cos(\alpha))$ and $\ddot{\mathbf{x}}\sin(\alpha)$ by $\dot{\mathbf{x}}\sin(\alpha)$ and $\ddot{\mathbf{x}}(1-\cos(\alpha))$, and using the same method, one can obtain equation (10.30). ■

From Lemmas 10.2, 5.3, and 9.3, it can be shown that $\frac{\dot{\mathbf{p}}}{\mathbf{p}} := \left(\frac{\dot{\mathbf{y}}}{\mathbf{y}}, \frac{\dot{\mathbf{z}}}{\mathbf{z}}\right)$, $\frac{\dot{\omega}}{\omega} := \left(\frac{\dot{\lambda}}{\lambda}, \frac{\dot{\gamma}}{\gamma}\right)$, $\frac{\ddot{\mathbf{p}}}{\mathbf{p}} := \left(\frac{\ddot{\mathbf{y}}}{\mathbf{y}}, \frac{\ddot{\mathbf{z}}}{\mathbf{z}}\right)$, and $\frac{\ddot{\omega}}{\omega} := \left(\frac{\ddot{\lambda}}{\lambda}, \frac{\ddot{\gamma}}{\gamma}\right)$ are all bounded as claimed in the following two Lemmas.

Lemma 10.3
Let $(\mathbf{x},\mathbf{p},\omega) = (\mathbf{x},\mathbf{y},\mathbf{z},\lambda,\gamma) \in \mathcal{N}_2(\theta)$ *and* $(\dot{\mathbf{x}},\dot{\mathbf{p}},\dot{\omega}) = (\dot{\mathbf{x}},\dot{\mathbf{y}},\dot{\mathbf{z}},\dot{\lambda},\dot{\gamma})$ *meet equation (10.12). Then,*

$$\left\|\frac{\dot{\mathbf{p}}}{\mathbf{p}}\right\|^2+\left\|\frac{\dot{\omega}}{\omega}\right\|^2 \leq \frac{2n}{1-\theta}, \tag{10.32}$$

$$\left\|\frac{\dot{\mathbf{p}}}{\mathbf{p}}\right\|^2\left\|\frac{\dot{\omega}}{\omega}\right\|^2 \leq \left(\frac{n}{1-\theta}\right)^2, \tag{10.33}$$

$$0 \leq \frac{\dot{\mathbf{p}}^{\mathrm{T}}\dot{\omega}}{\mu} \leq \frac{1+\theta}{1-\theta}n := \delta_1 n. \tag{10.34}$$

Proof 10.3 From the last two rows of (10.12), or equivalently (10.16), it must have

$$\begin{aligned}
\Lambda\dot{\mathbf{y}}+\mathbf{Y}\dot{\lambda} &= \Lambda\mathbf{Y}\mathbf{e}, \\
\Gamma\dot{\mathbf{z}}+\mathbf{Z}\dot{\gamma} &= \Gamma\mathbf{Z}\mathbf{e}.
\end{aligned}$$

Pre-multiplying $\mathbf{Y}^{-\frac{1}{2}}\Lambda^{-\frac{1}{2}}$ on both sides of the first equality gives

$$\mathbf{Y}^{-\frac{1}{2}}\Lambda^{\frac{1}{2}}\dot{\mathbf{y}}+\mathbf{Y}^{\frac{1}{2}}\Lambda^{-\frac{1}{2}}\dot{\lambda} = \mathbf{Y}^{\frac{1}{2}}\Lambda^{\frac{1}{2}}\mathbf{e}.$$

Pre-multiplying $\mathbf{Z}^{-\frac{1}{2}}\Gamma^{-\frac{1}{2}}$ on both sides of the second equality gives

$$\mathbf{Z}^{-\frac{1}{2}}\Gamma^{\frac{1}{2}}\dot{\mathbf{z}}+\mathbf{Z}^{\frac{1}{2}}\Gamma^{-\frac{1}{2}}\dot{\gamma}=\mathbf{Z}^{\frac{1}{2}}\Gamma^{\frac{1}{2}}\mathbf{e}. \tag{10.35}$$

Let $\mathbf{u}=\begin{bmatrix}\mathbf{Y}^{-\frac{1}{2}}\Lambda^{\frac{1}{2}}\dot{\mathbf{y}}\\ \mathbf{Z}^{-\frac{1}{2}}\Gamma^{\frac{1}{2}}\dot{\mathbf{z}}\end{bmatrix}$, $\mathbf{v}=\begin{bmatrix}\mathbf{Y}^{\frac{1}{2}}\Lambda^{-\frac{1}{2}}\dot{\lambda}\\ \mathbf{Z}^{\frac{1}{2}}\Gamma^{-\frac{1}{2}}\dot{\gamma}\end{bmatrix}$, and $\mathbf{w}=\begin{bmatrix}\mathbf{Y}^{\frac{1}{2}}\Lambda^{\frac{1}{2}}\mathbf{e}\\ \mathbf{Z}^{\frac{1}{2}}\Gamma^{\frac{1}{2}}\mathbf{e}\end{bmatrix}$, using (10.15) and Lemma 10.2 yields $\mathbf{u}^{\mathrm{T}}\mathbf{v}=\dot{\mathbf{y}}^{\mathrm{T}}\dot{\lambda}+\dot{\mathbf{z}}^{\mathrm{T}}\dot{\gamma}=\dot{\mathbf{x}}^{\mathrm{T}}(\dot{\gamma}-\dot{\lambda})\geq 0$. Using Lemma 9.3 and (10.8) yields

$$\begin{aligned}\|\mathbf{u}\|^2+\|\mathbf{v}\|^2 &= \sum_{i=1}^{n}\left(\frac{\dot{y}_i^2\lambda_i}{y_i}+\frac{\dot{z}_i^2\gamma_i}{z_i}\right)+\sum_{i=1}^{n}\left(\frac{\dot{\lambda}_i^2 y_i}{\lambda_i}+\frac{\dot{\gamma}_i^2 z_i}{\gamma_i}\right)\\ &\leq \sum_{i=1}^{n}(y_i\lambda_i+z_i\gamma_i)=\sum_{i=1}^{2n}p_i\omega_i=2n\mu.\end{aligned} \tag{10.36}$$

Since $p_i>0$ and $\omega_i>0$, dividing both sides of the inequality by $\min_i(p_i\omega_i)$ and using (10.10) gives

$$\sum_{i=1}^{n}\left(\frac{\dot{y}_i^2}{y_i^2}+\frac{\dot{z}_i^2}{z_i^2}\right)+\sum_{i=1}^{n}\left(\frac{\dot{\gamma}_i^2}{\gamma_i^2}+\frac{\dot{\lambda}_i^2}{\lambda_i^2}\right)=\left\|\frac{\dot{\mathbf{p}}}{\mathbf{p}}\right\|^2+\left\|\frac{\dot{\omega}}{\omega}\right\|^2\leq\frac{2n\mu}{\min_i(p_i\omega_i)}\leq\frac{2n}{1-\theta}. \tag{10.37}$$

This proves (10.32). Combining (10.32) and Lemma 5.3 yields

$$\left\|\frac{\dot{\mathbf{p}}}{\mathbf{p}}\right\|^2\left\|\frac{\dot{\omega}}{\omega}\right\|^2\leq\left(\frac{n}{(1-\theta)}\right)^2.$$

This leads to

$$\left\|\frac{\dot{\mathbf{p}}}{\mathbf{p}}\right\|\left\|\frac{\dot{\omega}}{\omega}\right\|\leq\frac{n}{(1-\theta)}. \tag{10.38}$$

Therefore, using (10.10) and Cauchy-Schwarz inequality yields

$$\begin{aligned}\frac{\dot{\mathbf{p}}^{\mathrm{T}}\dot{\omega}}{\mu} &\leq \frac{|\dot{\mathbf{p}}|^{\mathrm{T}}|\dot{\omega}|}{\mu}\leq(1+\theta)\frac{|\dot{\mathbf{p}}|^{\mathrm{T}}|\dot{\omega}|}{\max_i(p_i\omega_i)}\leq(1+\theta)\left(\frac{|\dot{\mathbf{p}}|}{\mathbf{p}}\right)^{\mathrm{T}}\left(\frac{|\dot{\omega}|}{\omega}\right)\\ &\leq (1+\theta)\left\|\frac{\dot{\mathbf{p}}}{\mathbf{p}}\right\|\left\|\frac{\dot{\omega}}{\omega}\right\|\leq\frac{1+\theta}{1-\theta}n,\end{aligned} \tag{10.39}$$

which is the second inequality of (10.34). From Lemma 10.2, $\dot{\mathbf{p}}^{\mathrm{T}}\dot{\omega}=\dot{\mathbf{x}}^{\mathrm{T}}(\dot{\gamma}-\dot{\lambda})=\dot{\mathbf{x}}^{\mathrm{T}}\mathbf{H}\dot{\mathbf{x}}\geq 0$, the first inequality of (10.34) follows. ■

Lemma 10.4
Let $(\mathbf{x},\mathbf{p},\omega)=(\mathbf{x},\mathbf{y},\mathbf{z},\lambda,\gamma)\in\mathcal{N}_2(\theta)$, $(\dot{\mathbf{x}},\dot{\mathbf{y}},\dot{\mathbf{z}},\dot{\lambda},\dot{\gamma})$ *and* $(\ddot{\mathbf{x}},\ddot{\mathbf{y}},\ddot{\mathbf{z}},\ddot{\lambda},\ddot{\gamma})$ *meet equations (10.12) and (10.13). Then*

$$\left\|\frac{\ddot{\mathbf{p}}}{\mathbf{p}}\right\|^2+\left\|\frac{\ddot{\omega}}{\omega}\right\|^2\leq\frac{4(1+\theta)n^2}{(1-\theta)^3}, \tag{10.40}$$

$$\left\|\frac{\ddot{\mathbf{p}}}{\mathbf{p}}\right\|^2\left\|\frac{\ddot{\omega}}{\omega}\right\|^2 \le \left(\frac{2(1+\theta)n^2}{(1-\theta)^3}\right)^2, \tag{10.41}$$

$$0 \le \frac{\ddot{\mathbf{p}}^{\mathrm{T}}\ddot{\omega}}{\mu} \le \frac{2(1+\theta)^2}{(1-\theta)^3}n^2 := \delta_2 n^2, \tag{10.42}$$

$$\left|\frac{\dot{\mathbf{p}}^{\mathrm{T}}\ddot{\omega}}{\mu}\right| \le \frac{(2n(1+\theta))^{\frac{3}{2}}}{(1-\theta)^2} := \delta_3 n^{\frac{3}{2}}, \quad \left|\frac{\ddot{\mathbf{p}}^{\mathrm{T}}\dot{\omega}}{\mu}\right| \le \frac{(2n(1+\theta))^{\frac{3}{2}}}{(1-\theta)^2} := \delta_3 n^{\frac{3}{2}}. \tag{10.43}$$

Proof 10.4 Similar to the proof of Lemma 10.3, from (10.13) and (10.17), it must have

$$\begin{aligned} & \Lambda\ddot{\mathbf{y}} + \mathbf{Y}\ddot{\lambda} = -2\left(\dot{\mathbf{y}}\circ\dot{\lambda}\right) \\ \Longleftrightarrow\quad & \mathbf{Y}^{-\frac{1}{2}}\Lambda^{\frac{1}{2}}\ddot{\mathbf{y}} + \mathbf{Y}^{\frac{1}{2}}\Lambda^{-\frac{1}{2}}\ddot{\lambda} = -2\mathbf{Y}^{-\frac{1}{2}}\Lambda^{-\frac{1}{2}}\left(\dot{\mathbf{y}}\circ\dot{\lambda}\right), \end{aligned}$$

and

$$\begin{aligned} & \Gamma\ddot{\mathbf{z}} + \mathbf{Z}\ddot{\gamma} = -2\left(\dot{\mathbf{z}}\circ\dot{\gamma}\right) \\ \Longleftrightarrow\quad & \mathbf{Z}^{-\frac{1}{2}}\Gamma^{\frac{1}{2}}\ddot{\mathbf{z}} + \mathbf{Z}^{\frac{1}{2}}\Gamma^{-\frac{1}{2}}\ddot{\gamma} = -2\mathbf{Z}^{-\frac{1}{2}}\Gamma^{-\frac{1}{2}}\left(\dot{\mathbf{z}}\circ\dot{\gamma}\right). \end{aligned}$$

Let $\mathbf{u} = \begin{bmatrix} \mathbf{Y}^{-\frac{1}{2}}\Lambda^{\frac{1}{2}}\ddot{\mathbf{y}} \\ \mathbf{Z}^{-\frac{1}{2}}\Gamma^{\frac{1}{2}}\ddot{\mathbf{z}} \end{bmatrix}$, $\mathbf{v} = \begin{bmatrix} \mathbf{Y}^{\frac{1}{2}}\Lambda^{-\frac{1}{2}}\ddot{\lambda} \\ \mathbf{Z}^{\frac{1}{2}}\Gamma^{-\frac{1}{2}}\ddot{\gamma} \end{bmatrix}$, and $\mathbf{w} = \begin{bmatrix} -2\mathbf{Y}^{-\frac{1}{2}}\Lambda^{-\frac{1}{2}}\left(\dot{\mathbf{y}}\circ\dot{\lambda}\right) \\ -2\mathbf{Z}^{-\frac{1}{2}}\Gamma^{-\frac{1}{2}}\left(\dot{\mathbf{z}}\circ\dot{\gamma}\right) \end{bmatrix}$, using (10.15) and Lemma 10.2 yields $\mathbf{u}^{\mathrm{T}}\mathbf{v} = \ddot{\mathbf{y}}^{\mathrm{T}}\ddot{\lambda} + \ddot{\mathbf{z}}^{\mathrm{T}}\ddot{\gamma} = \ddot{\mathbf{x}}^{\mathrm{T}}(\ddot{\gamma} - \ddot{\lambda}) \ge 0$. Using Lemma 9.3 yields

$$\begin{aligned} \|\mathbf{u}\|^2 + \|\mathbf{v}\|^2 &= \sum_{i=1}^{n}\left(\frac{\ddot{y}_i^2\lambda_i}{y_i} + \frac{\ddot{z}_i^2\gamma_i}{z_i}\right) + \sum_{i=1}^{n}\left(\frac{\ddot{\lambda}_i^2 y_i}{\lambda_i} + \frac{\ddot{\gamma}_i^2 z_i}{\gamma_i}\right) \\ &\le \left\|-2\mathbf{Y}^{-\frac{1}{2}}\Lambda^{-\frac{1}{2}}\left(\dot{\mathbf{y}}\circ\dot{\lambda}\right)\right\|^2 + \left\|-2\mathbf{Z}^{-\frac{1}{2}}\Gamma^{-\frac{1}{2}}\left(\dot{\mathbf{z}}\circ\dot{\gamma}\right)\right\|^2 \\ &= 4\sum_{i=1}^{n}\left(\frac{\dot{y}_i^2}{y_i}\frac{\dot{\lambda}_i^2}{\lambda_i} + \frac{\dot{z}_i^2}{z_i}\frac{\dot{\gamma}_i^2}{\gamma_i}\right). \end{aligned}$$

Dividing both sides of the inequality by μ and using (10.10), we have

$$\begin{aligned} & (1-\theta)\left(\sum_{i=1}^{n}\left(\frac{\ddot{y}_i^2}{y_i^2} + \frac{\ddot{z}_i^2}{z_i^2}\right) + \sum_{i=1}^{n}\left(\frac{\ddot{\lambda}_i^2}{\lambda_i^2} + \frac{\ddot{\gamma}_i^2}{\gamma_i^2}\right)\right) \\ = \; & (1-\theta)\left(\left\|\frac{\ddot{\mathbf{p}}}{\mathbf{p}}\right\|^2 + \left\|\frac{\ddot{\omega}}{\omega}\right\|^2\right) \\ \le \; & 4(1+\theta)\left(\sum_{i=1}^{n}\left(\frac{\dot{y}_i^2}{y_i^2}\frac{\dot{\lambda}_i^2}{\lambda_i^2} + \frac{\dot{z}_i^2}{z_i^2}\frac{\dot{\gamma}_i^2}{\gamma_i^2}\right)\right), \end{aligned}$$

in view of Lemma 10.3, this leads to

$$\left\|\frac{\ddot{\mathbf{p}}}{\mathbf{p}}\right\|^2 + \left\|\frac{\ddot{\omega}}{\omega}\right\|^2 \le 4\frac{1+\theta}{1-\theta}\left\|\frac{\dot{\mathbf{p}}}{\mathbf{p}}\circ\frac{\dot{\omega}}{\omega}\right\|^2 \le 4\frac{1+\theta}{1-\theta}\left\|\frac{\dot{\mathbf{p}}}{\mathbf{p}}\right\|^2\left\|\frac{\dot{\omega}}{\omega}\right\|^2 \le \frac{4(1+\theta)n^2}{(1-\theta)^3}. \tag{10.44}$$

This proves (10.40). Combining (10.40) and Lemma 5.3 yields

$$\left\|\frac{\ddot{\mathbf{p}}}{\mathbf{p}}\right\|^2\left\|\frac{\ddot{\omega}}{\omega}\right\|^2 \le \left(\frac{2(1+\theta)n^2}{(1-\theta)^3}\right)^2.$$

Using (10.10) and Cauchy-Schwarz inequality yields

$$\begin{aligned}\frac{\ddot{\mathbf{p}}^{\mathrm{T}}\ddot{\omega}}{\mu} &\le \frac{|\ddot{\mathbf{p}}|^{\mathrm{T}}|\ddot{\omega}|}{\mu} \le (1+\theta)\frac{|\ddot{\mathbf{p}}|^{\mathrm{T}}|\ddot{\omega}|}{\max_i(p_i\omega_i)} \le (1+\theta)\left(\frac{|\ddot{\mathbf{p}}|}{\mathbf{p}}\right)^{\mathrm{T}}\left(\frac{|\ddot{\omega}|}{\omega}\right)\\ &\le (1+\theta)\left\|\frac{\ddot{\mathbf{p}}}{\mathbf{p}}\right\|\left\|\frac{\ddot{\omega}}{\omega}\right\| \le \frac{2n^2(1+\theta)^2}{(1-\theta)^3},\end{aligned}$$

which is the second inequality of (10.42). Using (10.15) and Lemma 10.2, one must have $\ddot{\mathbf{p}}^{\mathrm{T}}\ddot{\omega} = \ddot{\mathbf{y}}^{\mathrm{T}}\ddot{\lambda} + \ddot{\mathbf{z}}^{\mathrm{T}}\ddot{\gamma} = \ddot{\mathbf{x}}^{\mathrm{T}}(\ddot{\gamma}-\ddot{\lambda}) = \ddot{\mathbf{x}}^{\mathrm{T}}\mathbf{H}\ddot{\mathbf{x}} \ge 0$. This proves the first inequality of (10.42). Finally, using (10.10), Cauchy-Schwarz inequality, (10.32), and (10.40) yields

$$\begin{aligned}&\frac{|\dot{\mathbf{p}}^{\mathrm{T}}\ddot{\omega}|}{\mu} \le \frac{|\dot{\mathbf{p}}|^{\mathrm{T}}|\ddot{\omega}|}{\mu} \le (1+\theta)\frac{|\dot{\mathbf{p}}|^{\mathrm{T}}|\ddot{\omega}|}{\max_i(p_i\omega_i)} \le (1+\theta)\left(\frac{|\dot{\mathbf{p}}|}{\mathbf{p}}\right)^{\mathrm{T}}\left(\frac{|\ddot{\omega}|}{\omega}\right)\\ &\le (1+\theta)\left\|\frac{\dot{\mathbf{p}}}{\mathbf{p}}\right\|\left\|\frac{\ddot{\omega}}{\omega}\right\| \le (1+\theta)\left(\frac{2n}{1-\theta}\right)^{\frac{1}{2}}\left(\frac{4(1+\theta)n^2}{(1-\theta)^3}\right)^{\frac{1}{2}} \le \frac{(2n(1+\theta))^{\frac{3}{2}}}{(1-\theta)^2}.\end{aligned}$$

This proves the first inequality of (10.43). Replacing $\dot{\mathbf{p}}$ by $\ddot{\mathbf{p}}$ and $\ddot{\omega}$ by $\dot{\omega}$, then using the same reasoning, one can prove the second inequality of (10.43). ∎

From the bounds established in Lemmas 10.2, 10.3, 10.4, and 5.4, the lower bound and upper bound for $\mu(\alpha)$ can be obtained.

Lemma 10.5

Let $(\mathbf{x},\mathbf{p},\omega) = (\mathbf{x},\mathbf{y},\mathbf{z},\lambda,\gamma) \in \mathcal{N}_2(\theta)$, $(\dot{\mathbf{x}},\dot{\mathbf{y}},\dot{\mathbf{z}},\dot{\lambda},\dot{\gamma})$ *and* $(\ddot{\mathbf{x}},\ddot{\mathbf{y}},\ddot{\mathbf{z}},\ddot{\lambda},\ddot{\gamma})$ *meet equations (10.12) and (10.13). Let* $\mathbf{x}(\alpha)$, $\mathbf{y}(\alpha)$, $\mathbf{z}(\alpha)$, $\lambda(\alpha)$, *and* $\gamma(\alpha)$ *be defined by (10.18), (10.19), (10.20), (10.21), and (10.22). Then,*

$$\begin{aligned}&\mu(1-\sin(\alpha)) - \frac{1}{2n}\ddot{\mathbf{x}}^{\mathrm{T}}\mathbf{H}\ddot{\mathbf{x}}\left((1-\cos(\alpha))^2+\sin^2(\alpha)\right)\\ \le&\mu(\alpha) = \mu(1-\sin(\alpha)) + \frac{1}{2n}\left(\ddot{\mathbf{x}}^{\mathrm{T}}(\ddot{\gamma}-\ddot{\lambda}) - \dot{\mathbf{x}}^{\mathrm{T}}(\dot{\gamma}-\dot{\lambda})\right)(1-\cos(\alpha))^2\\ &-\frac{1}{2n}\left(\dot{\mathbf{x}}^{\mathrm{T}}(\ddot{\gamma}-\ddot{\lambda}) + \ddot{\mathbf{x}}^{\mathrm{T}}(\dot{\gamma}-\dot{\lambda})\right)\sin(\alpha)(1-\cos(\alpha))\\ \le&\mu(1-\sin(\alpha)) + \frac{1}{2n}\ddot{\mathbf{x}}^{\mathrm{T}}\mathbf{H}\ddot{\mathbf{x}}\left((1-\cos(\alpha))^2+\sin^2(\alpha)\right). \end{aligned} \tag{10.45}$$

Proof 10.5 Using (10.19), (10.21), (10.16), and (10.17), one must have

$$\mathbf{y}^{\mathrm{T}}(\alpha)\lambda(\alpha)$$

$$
\begin{aligned}
&=\left(\mathbf{y}^{\mathrm{T}}-\dot{\mathbf{y}}^{\mathrm{T}}\sin(\alpha)+\ddot{\mathbf{y}}^{\mathrm{T}}(1-\cos(\alpha))\right)\left(\lambda-\dot{\lambda}\sin(\alpha)+\ddot{\lambda}(1-\cos(\alpha))\right)\\
&=\mathbf{y}^{\mathrm{T}}\lambda-\mathbf{y}^{\mathrm{T}}\dot{\lambda}\sin(\alpha)+\mathbf{y}^{\mathrm{T}}\ddot{\lambda}(1-\cos(\alpha))\\
&\quad-\dot{\mathbf{y}}^{\mathrm{T}}\lambda\sin(\alpha)+\dot{\mathbf{y}}^{\mathrm{T}}\dot{\lambda}\sin^2(\alpha)-\dot{\mathbf{y}}^{\mathrm{T}}\ddot{\lambda}\sin(\alpha)(1-\cos(\alpha))\\
&\quad+\ddot{\mathbf{y}}^{\mathrm{T}}\lambda(1-\cos(\alpha))-\ddot{\mathbf{y}}^{\mathrm{T}}\dot{\lambda}\sin(\alpha)(1-\cos(\alpha))+\ddot{\mathbf{y}}^{\mathrm{T}}\ddot{\lambda}(1-\cos(\alpha))^2\\
&=\mathbf{y}^{\mathrm{T}}\lambda-(\mathbf{y}^{\mathrm{T}}\dot{\lambda}+\lambda^{\mathrm{T}}\dot{\mathbf{y}})\sin(\alpha)+(\mathbf{y}^{\mathrm{T}}\ddot{\lambda}+\lambda^{\mathrm{T}}\ddot{\mathbf{y}})(1-\cos(\alpha))\\
&\quad-(\dot{\mathbf{y}}^{\mathrm{T}}\ddot{\lambda}+\dot{\lambda}^{\mathrm{T}}\ddot{\mathbf{y}})\sin(\alpha)(1-\cos(\alpha))+\dot{\mathbf{y}}^{\mathrm{T}}\dot{\lambda}\sin^2(\alpha)+\ddot{\mathbf{y}}^{\mathrm{T}}\ddot{\lambda}(1-\cos(\alpha))^2\\
&=\mathbf{y}^{\mathrm{T}}\lambda(1-\sin(\alpha))-2\dot{\mathbf{y}}^{\mathrm{T}}\dot{\lambda}(1-\cos(\alpha))\\
&\quad-(\dot{\mathbf{y}}^{\mathrm{T}}\ddot{\lambda}+\dot{\lambda}^{\mathrm{T}}\ddot{\mathbf{y}})\sin(\alpha)(1-\cos(\alpha))\\
&\quad+\dot{\mathbf{y}}^{\mathrm{T}}\dot{\lambda}(1-\cos^2(\alpha))+\ddot{\mathbf{y}}^{\mathrm{T}}\ddot{\lambda}(1-\cos(\alpha))^2\\
&=\mathbf{y}^{\mathrm{T}}\lambda(1-\sin(\alpha))+(\ddot{\mathbf{y}}^{\mathrm{T}}\ddot{\lambda}-\dot{\mathbf{y}}^{\mathrm{T}}\dot{\lambda})(1-\cos(\alpha))^2\\
&\quad-(\dot{\mathbf{y}}^{\mathrm{T}}\ddot{\lambda}+\dot{\lambda}^{\mathrm{T}}\ddot{\mathbf{y}})\sin(\alpha)(1-\cos(\alpha)). \qquad (10.46)
\end{aligned}
$$

Using (10.20), (10.22), (10.16), (10.17), and a similar derivation of (10.46), one gets

$$
\begin{aligned}
\mathbf{z}^{\mathrm{T}}(\alpha)\gamma(\alpha) &= \mathbf{z}^{\mathrm{T}}\gamma(1-\sin(\alpha))+(\ddot{\mathbf{z}}^{\mathrm{T}}\ddot{\gamma}-\dot{\mathbf{z}}^{\mathrm{T}}\dot{\gamma})(1-\cos(\alpha))^2\\
&\quad-(\dot{\mathbf{z}}^{\mathrm{T}}\ddot{\gamma}+\dot{\gamma}^{\mathrm{T}}\ddot{\mathbf{z}})\sin(\alpha)(1-\cos(\alpha)). \qquad (10.47)
\end{aligned}
$$

Combining (10.46) and (10.47), then using (10.15) and (10.29) yields

$$
\begin{aligned}
&2n\mu(\alpha)=\mathbf{p}^{\mathrm{T}}(\alpha)\omega(\alpha)\\
&=\mathbf{y}^{\mathrm{T}}(\alpha)\lambda(\alpha)+\mathbf{z}^{\mathrm{T}}(\alpha)\gamma(\alpha)\\
&=(\mathbf{y}^{\mathrm{T}}\lambda+\mathbf{z}^{\mathrm{T}}\gamma)(1-\sin(\alpha))+(\ddot{\mathbf{y}}^{\mathrm{T}}\ddot{\lambda}+\ddot{\mathbf{z}}^{\mathrm{T}}\ddot{\gamma}-\dot{\mathbf{y}}^{\mathrm{T}}\dot{\lambda}-\dot{\mathbf{z}}^{\mathrm{T}}\dot{\gamma})(1-\cos(\alpha))^2\\
&\quad-(\dot{\mathbf{y}}^{\mathrm{T}}\ddot{\lambda}+\dot{\mathbf{z}}^{\mathrm{T}}\ddot{\gamma}+\ddot{\mathbf{y}}^{\mathrm{T}}\dot{\lambda}+\ddot{\mathbf{z}}^{\mathrm{T}}\dot{\gamma})\sin(\alpha)(1-\cos(\alpha))\\
&=(\mathbf{y}^{\mathrm{T}}\lambda+\mathbf{z}^{\mathrm{T}}\gamma)(1-\sin(\alpha))+(\ddot{\mathbf{x}}^{\mathrm{T}}(\ddot{\gamma}-\ddot{\lambda})-\dot{\mathbf{x}}^{\mathrm{T}}(\dot{\gamma}-\dot{\lambda}))(1-\cos(\alpha))^2\\
&\quad-(\dot{\mathbf{x}}^{\mathrm{T}}(\ddot{\gamma}-\ddot{\lambda})+\ddot{\mathbf{x}}^{\mathrm{T}}(\dot{\gamma}-\dot{\lambda}))\sin(\alpha)(1-\cos(\alpha)) \qquad (10.48)\\
&\le(\mathbf{y}^{\mathrm{T}}\lambda+\mathbf{z}^{\mathrm{T}}\gamma)(1-\sin(\alpha))+(\ddot{\mathbf{x}}^{\mathrm{T}}\mathbf{H}\ddot{\mathbf{x}}-\dot{\mathbf{x}}^{\mathrm{T}}\mathbf{H}\dot{\mathbf{x}})(1-\cos(\alpha))^2\\
&\quad+\dot{\mathbf{x}}^{\mathrm{T}}\mathbf{H}\dot{\mathbf{x}}(1-\cos(\alpha))^2+\ddot{\mathbf{x}}^{\mathrm{T}}\mathbf{H}\ddot{\mathbf{x}}\sin^2(\alpha)\\
&=(\mathbf{y}^{\mathrm{T}}\lambda+\mathbf{z}^{\mathrm{T}}\gamma)(1-\sin(\alpha))+\ddot{\mathbf{x}}^{\mathrm{T}}\mathbf{H}\ddot{\mathbf{x}}(1-\cos(\alpha))^2+\ddot{\mathbf{x}}^{\mathrm{T}}\mathbf{H}\ddot{\mathbf{x}}\sin^2(\alpha).
\end{aligned}
$$

Dividing both sides by $2n$ proves the second inequality of the lemma. Combining (10.48) and (10.30) proves the first inequality of the lemma. ∎

To keep all the iterates of the algorithm inside the strictly feasible set, $(\mathbf{p}(\alpha),\omega(\alpha))>\mathbf{0}$ for all iterations is required. This is guaranteed when $\mu(\alpha)>0$ holds. The following corollary states the condition for $\mu(\alpha)>0$ to hold.

Corollary 10.1

If $\mu > 0$, then, for any fixed $\theta \in (0,1)$, there is an $\bar{\alpha} > 0$ depending on θ, such that for any $\sin(\alpha) \leq \sin(\bar{\alpha})$, $\mu(\alpha) > 0$. In particular, if $\theta = 0.19$, $\sin(\bar{\alpha}) \geq 0.6158$.

Proof 10.6 Using the formulas $\dot{\mathbf{x}}^{\mathrm{T}}\mathbf{H}\dot{\mathbf{x}}^{\mathrm{T}} = \dot{\mathbf{x}}^{\mathrm{T}}(\dot{\gamma} - \dot{\lambda}) = \dot{\mathbf{p}}^{\mathrm{T}}\dot{\omega}$ and $((1-\cos(\alpha))^2 \leq \sin^4(\alpha)$ presented in Lemmas 10.2 and 5.4, we can rewrite the first inequality of (10.45) as

$$\begin{aligned}
\mu(\alpha) &\geq \mu\left(1 - \sin(\alpha) - \frac{1}{2n\mu}\dot{\mathbf{p}}^{\mathrm{T}}\dot{\omega}\left(\sin^4(\alpha) + \sin^2(\alpha)\right)\right) \\
&\geq \mu\left(1 - \sin(\alpha) - \frac{(1+\theta)}{2(1-\theta)}\left(\sin^4(\alpha) + \sin^2(\alpha)\right)\right) := \mu r(\alpha)
\end{aligned}$$

where (10.34) is used for the second inequality. Since $\mu > 0$, and $r(\alpha)$ is a monotonic decreasing function in $[0, \frac{\pi}{2}]$ with $r(0) > 0$ and $r(\frac{\pi}{2}) < 0$, there is a unique real solution $\sin(\bar{\alpha}) \in (0,1)$ of $r(\alpha) = 0$ such that for all $\sin(\alpha) < \sin(\bar{\alpha})$, $r(\alpha) > 0$, or $\mu(\alpha) > 0$. It is easy to check that if $\theta = 0.19$, $\sin(\bar{\alpha}) = 0.6158$ is the solution of $r(\alpha) = 0$. ■

Remark 10.1 Corollary 10.1 indicates that for any $\theta \in (0,1)$, there is a positive $\bar{\alpha}$ such that for $\alpha \leq \bar{\alpha}$, $\mu(\alpha) > 0$. Intuitively, to search in a wider region will generate a longer step. Therefore, the larger the θ is, the better. To derive the convergence result, $\theta \leq 0.148$ is imposed in Lemma 9.12 and $\theta \leq 0.19$ is imposed in Lemma 10.13. This is an indication that an algorithm specifically derived for convex QP with box constraint will be more efficient than the general algorithm for convex QP proposed in Chapter 9. ■

To reduce the duality measure in an iteration, it must have $\mu(\alpha) \leq \mu$. For linear programming, it has been shown in Chapter 5 that $\mu(\alpha) \leq \mu$ for $\alpha \in [0, \hat{\alpha}]$ with $\hat{\alpha} = \frac{\pi}{2}$, and the larger the α in the interval is, the smaller the $\mu(\alpha)$ will be. This claim is not true for the convex quadratic programming with box constraints and it needs to be modified as follows.

Lemma 10.6

Let $(\mathbf{x}, \mathbf{p}, \omega) = (\mathbf{x}, \mathbf{y}, \mathbf{z}, \lambda, \gamma) \in \mathcal{N}_2(\theta)$, $(\dot{\mathbf{x}}, \dot{\mathbf{y}}, \dot{\mathbf{z}}, \dot{\lambda}, \dot{\gamma})$ and $(\ddot{\mathbf{x}}, \ddot{\mathbf{y}}, \ddot{\mathbf{z}}, \ddot{\lambda}, \ddot{\gamma})$ meet equations (10.12) and (10.13). Let $\mathbf{x}(\alpha)$, $\mathbf{y}(\alpha)$, $\mathbf{z}(\alpha)$, $\lambda(\alpha)$, and $\gamma(\alpha)$ be defined by (10.18), (10.19), (10.20), (10.21), and (10.22). Then, there exists

$$\hat{\alpha} = \begin{cases} \frac{\pi}{2}, & \textit{if } \frac{\dot{\mathbf{x}}^{\mathrm{T}}\mathbf{H}\dot{\mathbf{x}}}{n\mu} \leq \frac{1}{2} \\ \sin^{-1}(g), & \textit{if } \frac{\dot{\mathbf{x}}^{\mathrm{T}}\mathbf{H}\dot{\mathbf{x}}}{n\mu} > \frac{1}{2} \end{cases} \tag{10.49}$$

where

$$g = \sqrt[3]{\frac{n\mu}{\ddot{\mathbf{x}}^{\mathrm{T}}\mathbf{H}\ddot{\mathbf{x}}} + \sqrt{\left(\frac{n\mu}{\ddot{\mathbf{x}}^{\mathrm{T}}\mathbf{H}\ddot{\mathbf{x}}}\right)^2 + \left(\frac{1}{3}\right)^3}} + \sqrt[3]{\frac{n\mu}{\ddot{\mathbf{x}}^{\mathrm{T}}\mathbf{H}\ddot{\mathbf{x}}} - \sqrt{\left(\frac{n\mu}{\ddot{\mathbf{x}}^{\mathrm{T}}\mathbf{H}\ddot{\mathbf{x}}}\right)^2 + \left(\frac{1}{3}\right)^3}},$$

such that for every $\alpha \in [0, \hat{\alpha}]$, $\mu(\alpha) \le \mu$.

Proof 10.7 From the second inequality of (10.45), we have

$$\mu(\alpha) - \mu \le \mu \sin(\alpha) \left(-1 + \frac{\ddot{\mathbf{x}}^{\mathrm{T}}\mathbf{H}\ddot{\mathbf{x}}}{2n\mu} \sin(\alpha) + \frac{\ddot{\mathbf{x}}^{\mathrm{T}}\mathbf{H}\ddot{\mathbf{x}}}{2n\mu} \sin^3(\alpha) \right). \tag{10.50}$$

Clearly, if $\frac{\ddot{\mathbf{x}}^{\mathrm{T}}\mathbf{H}\ddot{\mathbf{x}}}{2n\mu} \le \frac{1}{2}$, for any $\alpha \in [0, \frac{\pi}{2}]$, the function

$$f(\alpha) := \left(-1 + \frac{\ddot{\mathbf{x}}^{\mathrm{T}}\mathbf{H}\ddot{\mathbf{x}}}{2n\mu} \sin(\alpha) + \frac{\ddot{\mathbf{x}}^{\mathrm{T}}\mathbf{H}\ddot{\mathbf{x}}}{2n\mu} \sin^3(\alpha) \right) \le 0, \tag{10.51}$$

and $\mu(\alpha) \le \mu$. If $\frac{\ddot{\mathbf{x}}^{\mathrm{T}}\mathbf{H}\ddot{\mathbf{x}}}{2n\mu} > \frac{1}{2}$, noticing that $f(\alpha)$ is a monotonic increasing function of α with $f(0) < 0$ and $f(\frac{\pi}{2}) > 0$, $f(\alpha)$ has a unique positive root in $(0, \frac{\pi}{2})$. The solution is given by Cardano's formula as follows.

$$\sin(\hat{\alpha}) = \sqrt[3]{\frac{n\mu}{\ddot{\mathbf{x}}^{\mathrm{T}}\mathbf{H}\ddot{\mathbf{x}}} + \sqrt{\left(\frac{n\mu}{\ddot{\mathbf{x}}^{\mathrm{T}}\mathbf{H}\ddot{\mathbf{x}}}\right)^2 + \left(\frac{1}{3}\right)^3}} + \sqrt[3]{\frac{n\mu}{\ddot{\mathbf{x}}^{\mathrm{T}}\mathbf{H}\ddot{\mathbf{x}}} - \sqrt{\left(\frac{n\mu}{\ddot{\mathbf{x}}^{\mathrm{T}}\mathbf{H}\ddot{\mathbf{x}}}\right)^2 + \left(\frac{1}{3}\right)^3}}. \tag{10.52}$$

This proves the Lemma. ■

According to Theorem 10.1, Lemmas 10.1, 10.3, 10.4, and 10.6, if α is small enough, then $(\mathbf{p}(\alpha), \omega(\alpha)) > \mathbf{0}$, and $\mu(\alpha) < \mu$, i.e., the search along the ellipse defined by Theorem 10.1 will generate a strictly feasible point with a smaller duality measure. Since $(\mathbf{p}, \omega) > \mathbf{0}$ holds in all iterations, reducing the duality measure to zero means approaching the solution of the convex quadratic programming. This can be achieved by applying a similar idea to the one that was used in Chapters 3 and 5, i.e., starting with an iterate in $\mathcal{N}_2(\theta)$, searching along the approximated central path to reduce the duality measure and to keep the iterate in $\mathcal{N}_2(2\theta)$, and then making a correction to move the iterate back to $\mathcal{N}_2(\theta)$. The following notations will be used.

$$a_0 = -\theta\mu < 0,$$

$$a_1 = \theta\mu > 0,$$

$$a_2 = 2\theta \frac{\dot{\mathbf{p}}^{\mathrm{T}}\dot{\omega}}{2n} = 2\theta \frac{\dot{\mathbf{x}}^{\mathrm{T}}(\dot{\gamma} - \dot{\lambda})}{2n} = 2\theta \frac{\dot{\mathbf{x}}^{\mathrm{T}}\mathbf{H}\dot{\mathbf{x}}}{2n} \ge 0,$$

$$a_3 = \left\|\dot{\mathbf{p}} \circ \ddot{\omega} + \dot{\omega} \circ \ddot{\mathbf{p}} - \frac{1}{2n}(\dot{\mathbf{p}}^{\mathrm{T}}\ddot{\omega} + \dot{\omega}^{\mathrm{T}}\ddot{\mathbf{p}})\mathbf{e}\right\| \geq 0,$$

and

$$\begin{aligned} a_4 &= \left\|\ddot{\mathbf{p}} \circ \ddot{\omega} - \dot{\omega} \circ \dot{\mathbf{p}} - \frac{1}{2n}(\ddot{\mathbf{p}}^{\mathrm{T}}\ddot{\omega} - \dot{\omega}^{\mathrm{T}}\dot{\mathbf{p}})\mathbf{e}\right\| + 2\theta\frac{\dot{\mathbf{p}}^{\mathrm{T}}\dot{\omega}}{2n} \\ &= \left\|\ddot{\mathbf{p}} \circ \ddot{\omega} - \dot{\omega} \circ \dot{\mathbf{p}} - \frac{1}{2n}(\ddot{\mathbf{p}}^{\mathrm{T}}\ddot{\omega} - \dot{\omega}^{\mathrm{T}}\dot{\mathbf{p}})\mathbf{e}\right\| + 2\theta\frac{\dot{\mathbf{x}}^{\mathrm{T}}\mathbf{H}\dot{\mathbf{x}}}{2n} \geq 0. \end{aligned}$$

Denote a quartic polynomial in terms of $\sin(\alpha)$ as follows:

$$q(\alpha) = a_4\sin^4(\alpha) + a_3\sin^3(\alpha) + a_2\sin^2(\alpha) + a_1\sin(\alpha) + a_0 = 0. \quad (10.53)$$

Since $q(\alpha)$ is a monotonic increasing function of $\alpha \in [0, \frac{\pi}{2}]$, $q(0) = -\theta\mu < 0$ and $q(\frac{\pi}{2}) = a_2 + a_3 + a_4 \geq 0$, the polynomial has exactly one positive root in $[0, \frac{\pi}{2}]$. Moreover, since (10.53) is a quartic equation, all the solutions are analytical and the computational cost is independent of m and n, and negligible [113].

Lemma 10.7

Let $(\mathbf{x},\mathbf{y},\mathbf{z},\lambda,\omega) \in \mathcal{N}_2(\theta)$, $(\dot{\mathbf{x}},\dot{\mathbf{y}},\dot{\mathbf{z}},\dot{\lambda},\dot{\omega})$ *and* $(\ddot{\mathbf{x}},\ddot{\mathbf{y}},\ddot{\mathbf{z}},\ddot{\lambda},\ddot{\omega})$ *be calculated from (10.12) and (10.13). Denote by* $\sin(\tilde{\alpha}) \in [0,1]$ *the only positive real solution of (10.53). Assume* $\sin(\alpha) < \min\{\sin(\tilde{\alpha}), \sin(\bar{\alpha})\}$, *let* $(\mathbf{x}(\alpha),\mathbf{y}(\alpha),\mathbf{z}(\alpha),\lambda(\alpha),\gamma(\alpha))$ *and* $\mu(\alpha)$ *be updated as follows:*

$$\begin{aligned} &(\mathbf{x}(\alpha),\mathbf{y}(\alpha),\mathbf{z}(\alpha),\lambda(\alpha),\gamma(\alpha)) \\ = \; &(\mathbf{x},\mathbf{y},\mathbf{z},\lambda,\gamma) - (\dot{\mathbf{x}},\dot{\mathbf{y}},\dot{\mathbf{z}},\dot{\lambda},\dot{\gamma})\sin(\alpha) + (\ddot{\mathbf{x}},\ddot{\mathbf{y}},\ddot{\mathbf{z}},\ddot{\lambda},\ddot{\gamma})(1-\cos(\alpha)), \end{aligned} \quad (10.54)$$

$$\begin{aligned} \mu(\alpha) &= \mu(1-\sin(\alpha)) \\ &+ \frac{1}{2n}\Big((\ddot{\mathbf{p}}^{\mathrm{T}}\ddot{\omega} - \dot{\mathbf{p}}^{\mathrm{T}}\dot{\omega})(1-\cos(\alpha))^2 - (\dot{\mathbf{p}}^{\mathrm{T}}\ddot{\omega} + \ddot{\mathbf{p}}^{\mathrm{T}}\dot{\omega})\sin(\alpha)(1-\cos(\alpha))\Big). \end{aligned} \quad (10.55)$$

Then, $(\mathbf{x}(\alpha),\mathbf{y}(\alpha),\mathbf{z}(\alpha),\lambda(\alpha),\gamma(\alpha)) \in \mathcal{N}_2(2\theta)$.

Proof 10.8 Since $\sin(\tilde{\alpha})$ is the only positive real solution of (10.53) in $[0,1]$ and $q(0) < 0$, substituting a_0, a_1, a_2, a_3 and a_4 into (10.53) yields, for all $\sin(\alpha) \leq \sin(\tilde{\alpha})$,

$$\begin{aligned} &\left(\left\|\ddot{\mathbf{p}} \circ \ddot{\omega} - \dot{\omega} \circ \dot{\mathbf{p}} - \frac{1}{2n}(\ddot{\mathbf{p}}^{\mathrm{T}}\ddot{\omega} - \dot{\omega}^{\mathrm{T}}\dot{\mathbf{p}})\mathbf{e}\right\|\right)\sin^4(\alpha) \\ &+ \left(\left\|\dot{\mathbf{p}} \circ \ddot{\omega} + \dot{\omega} \circ \ddot{\mathbf{p}} - \frac{1}{2n}(\dot{\mathbf{p}}^{\mathrm{T}}\ddot{\omega} + \dot{\omega}^{\mathrm{T}}\ddot{\mathbf{p}})\mathbf{e}\right\|\right)\sin^3(\alpha) \\ &\leq -\left(2\theta\frac{\dot{\mathbf{p}}^{\mathrm{T}}\dot{\omega}}{2n}\right)\sin^4(\alpha) - \left(2\theta\frac{\dot{\mathbf{p}}^{\mathrm{T}}\dot{\omega}}{2n}\right)\sin^2(\alpha) + \theta\mu(1-\sin(\alpha)). \end{aligned} \quad (10.56)$$

Using (10.23), (10.24), (10.16), (10.17), (10.55), Lemma 5.4, (10.56), and the first inequality of (10.45) yields

$$
\begin{aligned}
&\left\|\mathbf{p}(\alpha)\circ\omega(\alpha)-\mu(\alpha)\mathbf{e}\right\| \\
=&\left\|\left(\mathbf{p}-\dot{\mathbf{p}}\sin(\alpha)+\ddot{\mathbf{p}}(1-\cos(\alpha))\right)\circ\left(\omega-\dot{\omega}\sin(\alpha)+\ddot{\omega}(1-\cos(\alpha))\right)-\mu(\alpha)\mathbf{e}\right\| \\
=&\left\|(\mathbf{p}\circ\omega-\mu\mathbf{e})(1-\sin(\alpha))+\left(\ddot{\mathbf{p}}\circ\ddot{\omega}-\dot{\mathbf{p}}\circ\dot{\omega}-\frac{1}{2n}(\ddot{\mathbf{p}}^{\mathrm{T}}\ddot{\omega}-\dot{\mathbf{p}}^{\mathrm{T}}\dot{\omega})\mathbf{e}\right)(1-\cos(\alpha))^2\right. \\
&\left.-\left(\dot{\mathbf{p}}\circ\ddot{\omega}+\dot{\omega}\circ\ddot{\mathbf{p}}-\frac{1}{2n}(\dot{\mathbf{p}}^{\mathrm{T}}\ddot{\omega}+\ddot{\mathbf{p}}^{\mathrm{T}}\dot{\omega})\mathbf{e}\right)\sin(\alpha)(1-\cos(\alpha))\right\| \\
\le&(1-\sin(\alpha))\left\|\mathbf{p}\circ\omega-\mu\mathbf{e}\right\|+\left\|(\ddot{\mathbf{p}}\circ\ddot{\omega}-\dot{\mathbf{p}}\circ\dot{\omega}-\frac{1}{2n}(\ddot{\mathbf{p}}^{\mathrm{T}}\ddot{\omega}-\dot{\mathbf{p}}^{\mathrm{T}}\dot{\omega}))\mathbf{e}\right\|(1-\cos(\alpha))^2 \\
&+\left\|(\dot{\mathbf{p}}\circ\ddot{\omega}+\dot{\omega}\circ\ddot{\mathbf{p}}-\frac{1}{2n}(\dot{\mathbf{p}}^{\mathrm{T}}\ddot{\omega}+\ddot{\mathbf{p}}^{\mathrm{T}}\dot{\omega})\mathbf{e}\right\|\sin(\alpha)(1-\cos(\alpha)) \qquad (10.57) \\
\le&\theta\mu(1-\sin(\alpha))+\left\|(\ddot{\mathbf{p}}\circ\ddot{\omega}-\dot{\mathbf{p}}\circ\dot{\omega}-\frac{1}{2n}(\ddot{\mathbf{p}}^{\mathrm{T}}\ddot{\omega}-\dot{\mathbf{p}}^{\mathrm{T}}\dot{\omega}))\mathbf{e}\right\|\sin^4(\alpha)+a_3\sin^3(\alpha) \\
\le&2\theta\mu(1-\sin(\alpha))-\left(2\theta\frac{\dot{\mathbf{p}}^{\mathrm{T}}\dot{\omega}}{2n}\right)(\sin^4(\alpha)+\sin^2(\alpha)) \\
\le&2\theta\left(\mu(1-\sin(\alpha))-\left(\frac{\dot{\mathbf{x}}^{\mathrm{T}}\mathbf{H}\dot{\mathbf{x}}}{2n}\right)\left((1-\cos(\alpha))^2+\sin^2(\alpha)\right)\right) \\
\le&2\theta\mu(\alpha). \qquad (10.58)
\end{aligned}
$$

Hence, the point $(\mathbf{x}(\alpha),\mathbf{p}(\alpha),\omega(\alpha))$ satisfies the proximity condition for $\mathcal{N}_2(2\theta)$. To check the positivity condition $(\mathbf{p}(\alpha),\omega(\alpha))>\mathbf{0}$, in view of the initial condition $(\mathbf{p},\omega)>\mathbf{0}$, it follows from (10.58) and Corollary 10.1 that, for $\sin(\alpha)<\sin(\bar{\alpha})$ and $\theta<0.5$,

$$
p_i(\alpha)\omega_i(\alpha)\ge(1-2\theta)\mu(\alpha)>0. \qquad (10.59)
$$

Therefore, it cannot have $p_i(\alpha)=0$ or $\omega_i(\alpha)=0$ for any index i when $\alpha\in[0,\sin^{-1}(\bar{\alpha}))$. This proves $(\mathbf{p}(\alpha),\omega(\alpha))>\mathbf{0}$. ■

Remark 10.2 It is worthwhile to note, by examining the proof of Lemma 10.7, that $\sin(\tilde{\alpha})$ is selected for the proximity condition (10.58) to hold, and $\sin(\bar{\alpha})$ is selected for $\mu(\alpha)>0$, thereby insuring that the positivity condition (10.59) holds. ■

The lower bound of $\sin(\bar{\alpha})$ is estimated in Corollary 10.1. To estimate the lower bound of $\sin(\tilde{\alpha})$, the following lemma is needed.

Lemma 10.8

Let $(\mathbf{x},\mathbf{p},\omega)\in\mathcal{N}_2(\theta)$, $(\dot{\mathbf{x}},\dot{\mathbf{p}},\dot{\omega})$ and $(\ddot{\mathbf{x}},\ddot{\mathbf{p}},\ddot{\omega})$ meet equations (10.12) and (10.13). Then,

$$
\left\|\dot{\mathbf{p}}\circ\dot{\omega}\right\|\le\frac{(1+\theta)}{(1-\theta)}n\mu, \qquad (10.60)
$$

$$\left\|\ddot{\mathbf{p}} \circ \ddot{\omega}\right\| \leq \frac{2(1+\theta)^2}{(1-\theta)^3} n^2 \mu, \tag{10.61}$$

$$\left\|\ddot{\mathbf{p}} \circ \dot{\omega}\right\| \leq \frac{2\sqrt{2}(1+\theta)^{\frac{3}{2}}}{(1-\theta)^2} n^{\frac{3}{2}} \mu, \tag{10.62}$$

$$\left\|\dot{\mathbf{p}} \circ \ddot{\omega}\right\| \leq \frac{2\sqrt{2}(1+\theta)^{\frac{3}{2}}}{(1-\theta)^2} n^{\frac{3}{2}} \mu. \tag{10.63}$$

Proof 10.9 Since

$$\left\|\frac{\dot{\mathbf{p}}}{\mathbf{p}}\right\|^2 = \sum_{i=1}^{2n} \left(\frac{\dot{p}_i}{p_i}\right)^2, \quad \left\|\frac{\dot{\omega}}{\omega}\right\|^2 = \sum_{i=1}^{2n} \left(\frac{\dot{\omega}_i}{\omega_i}\right)^2,$$

from Lemma 10.3 and (10.10), it must have

$$\begin{aligned}
& \left(\frac{n}{1-\theta}\right)^2 \\
\geq\ & \left\|\frac{\dot{\mathbf{p}}}{\mathbf{p}}\right\|^2 \left\|\frac{\dot{\omega}}{\omega}\right\|^2 = \left(\sum_{i=1}^{2n} \left(\frac{\dot{p}_i}{p_i}\right)^2\right) \left(\sum_{i=1}^{2n} \left(\frac{\dot{\omega}_i}{\omega_i}\right)^2\right) \\
\geq\ & \sum_{i=1}^{2n} \left(\frac{\dot{p}_i}{p_i} \frac{\dot{\omega}_i}{\omega_i}\right)^2 = \left\|\frac{\dot{\mathbf{p}}}{\mathbf{p}} \circ \frac{\dot{\omega}}{\omega}\right\|^2 \\
\geq\ & \sum_{i=1}^{2n} \left(\frac{\dot{p}_i \dot{\omega}_i}{(1+\theta)\mu}\right)^2 = \frac{1}{(1+\theta)^2 \mu^2} \left\|\dot{\mathbf{p}} \circ \dot{\omega}\right\|^2,
\end{aligned}$$

i.e.,

$$\left\|\dot{\mathbf{p}} \circ \dot{\omega}\right\|^2 \leq \left(\frac{1+\theta}{1-\theta} n\mu\right)^2.$$

This proves (10.60). Using

$$\left\|\frac{\ddot{\mathbf{p}}}{\mathbf{p}}\right\|^2 = \sum_{i=1}^{2n} \left(\frac{\ddot{p}_i}{p_i}\right)^2, \quad \left\|\frac{\ddot{\omega}}{\omega}\right\|^2 = \sum_{i=1}^{2n} \left(\frac{\ddot{\omega}_i}{\omega_i}\right)^2,$$

and Lemma 10.4, then following the same procedure, it is easy to verify (10.61). From (10.32) and (10.40), one obtains

$$\begin{aligned}
& \left(\frac{2n}{(1-\theta)}\right) \left(\frac{4(1+\theta)n^2}{(1-\theta)^3}\right) \geq \left(\left\|\frac{\dot{\mathbf{p}}}{\mathbf{p}}\right\|^2 + \left\|\frac{\dot{\omega}}{\omega}\right\|^2\right) \left(\left\|\frac{\ddot{\mathbf{p}}}{\mathbf{p}}\right\|^2 + \left\|\frac{\ddot{\omega}}{\omega}\right\|^2\right) \\
\geq\ & \left\|\frac{\ddot{\mathbf{p}}}{\mathbf{p}}\right\|^2 \left\|\frac{\dot{\omega}}{\omega}\right\|^2 + \left\|\frac{\dot{\mathbf{p}}}{\mathbf{p}}\right\|^2 \left\|\frac{\ddot{\omega}}{\omega}\right\|^2 \\
=\ & \left(\sum_{i=1}^{2n} \left(\frac{\ddot{p}_i}{p_i}\right)^2\right) \left(\sum_{i=1}^{2n} \left(\frac{\dot{\omega}_i}{\omega_i}\right)^2\right) + \left(\sum_{i=1}^{2n} \left(\frac{\dot{p}_i}{p_i}\right)^2\right) \left(\sum_{i=1}^{2n} \left(\frac{\ddot{\omega}_i}{\omega_i}\right)^2\right)
\end{aligned}$$

$$
\begin{aligned}
&\geq \sum_{i=1}^{2n}\left(\frac{\ddot{p}_i\dot{\omega}_i}{p_i\omega_i}\right)^2+\sum_{i=1}^{2n}\left(\frac{\dot{p}_i\ddot{\omega}_i}{p_i\omega_i}\right)^2\\
&\geq \sum_{i=1}^{2n}\left(\frac{\ddot{p}_i\dot{\omega}_i}{(1+\theta)\mu}\right)^2+\sum_{i=1}^{2n}\left(\frac{\dot{p}_i\ddot{\omega}_i}{(1+\theta)\mu}\right)^2\\
&= \frac{1}{(1+\theta)^2\mu^2}\left(\left\|\ddot{\mathbf{p}}\circ\dot{\omega}\right\|^2+\left\|\dot{\mathbf{p}}\circ\ddot{\omega}\right\|^2\right),
\end{aligned}
\tag{10.64}
$$

i.e.,

$$
\left\|\ddot{\mathbf{p}}\circ\dot{\omega}\right\|^2+\left\|\dot{\mathbf{p}}\circ\ddot{\omega}\right\|^2 \leq \frac{(2n)^3(1+\theta)^3}{(1-\theta)^4}\mu^2.
$$

This proves the lemma. ■

Lemma 10.7 established a relation between the solution of (10.53) and the proximity condition of $(\mathbf{x}(\alpha),\mathbf{p}(\alpha),\omega(\alpha))\in\mathcal{N}_2(2\theta)$. By estimating the solution of (10.53), we will find an estimation of $\sin(\tilde{\alpha})$ such that, for all $\alpha\leq\tilde{\alpha}$, the proximity condition of $(\mathbf{x}(\alpha),\mathbf{p}(\alpha),\omega(\alpha))\in\mathcal{N}_2(2\theta)$ holds. The following lemma gives the estimation of $\sin(\tilde{\alpha})$.

Lemma 10.9
Let $\theta\leq 0.22$. Then, $\sin(\tilde{\alpha})\geq\frac{\theta}{\sqrt{n}}$.

Proof 10.10 First notice from (10.53) that $q(\sin(\alpha))$ is a monotonic increasing function of $\sin(\alpha)$ for $\alpha\in[0,\frac{\pi}{2}]$. Also, $q(\sin(0))<0$ and $q(\sin(\frac{\pi}{2}))\geq 0$ hold, therefore, one needs only to show that $q(\frac{\theta}{\sqrt{n}})<0$ for $\theta\leq 0.22$. Using Lemma 9.11 yields

$$
\left\|\dot{\mathbf{p}}\circ\ddot{\omega}+\dot{\omega}\circ\ddot{\mathbf{p}}-\frac{1}{2n}(\dot{\mathbf{p}}^{\mathrm{T}}\ddot{\omega}+\dot{\omega}^{\mathrm{T}}\ddot{\mathbf{p}})\mathbf{e}\right\|\leq\left\|\dot{\mathbf{p}}\circ\ddot{\omega}\right\|+\left\|\dot{\omega}\circ\ddot{\mathbf{p}}\right\|,
$$

$$
\left\|\ddot{\mathbf{p}}\circ\ddot{\omega}-\dot{\omega}\circ\dot{\mathbf{p}}-\frac{1}{2n}(\ddot{\mathbf{p}}^{\mathrm{T}}\ddot{\omega}-\dot{\omega}^{\mathrm{T}}\dot{\mathbf{p}})\mathbf{e}\right\|\leq\left\|\ddot{\mathbf{p}}\circ\ddot{\omega}\right\|+\left\|\dot{\omega}\circ\dot{\mathbf{p}}\right\|.
$$

Substituting these relations into (10.53) and using Lemmas 10.8, 10.3, and 10.4, we have, for $\alpha\in[0,\frac{\pi}{2}]$,

$$
\begin{aligned}
q(\sin(\alpha))\leq&\left(\left\|\ddot{\mathbf{p}}\circ\ddot{\omega}\right\|+\left\|\dot{\omega}\circ\dot{\mathbf{p}}\right\|+2\theta\frac{\dot{\mathbf{p}}^{\mathrm{T}}\dot{\omega}}{2n}\right)\sin^4(\alpha)\\
&+\left(\left\|\dot{\mathbf{p}}\circ\ddot{\omega}\right\|+\left\|\dot{\omega}\circ\ddot{\mathbf{p}}\right\|\right)\sin^3(\alpha)\\
&+2\theta\frac{\dot{\mathbf{p}}^{\mathrm{T}}\dot{\omega}}{2n}\sin^2(\alpha)+\theta\mu\sin(\alpha)-\theta\mu\\
\leq&\mu\left[\left(\frac{2(1+\theta)^2}{(1-\theta)^3}n^2+\frac{n(1+\theta)}{(1-\theta)}+\frac{\theta(1+\theta)}{(1-\theta)}\right)\sin^4(\alpha)\right.
\end{aligned}
$$

$$+4\sqrt{2}\frac{(1+\theta)^{\frac{3}{2}}}{(1-\theta)^2}n^{\frac{3}{2}}\sin^3(\alpha)$$
$$+\frac{\theta(1+\theta)}{(1-\theta)}\sin^2(\alpha)+\theta\sin(\alpha)-\theta\Bigg].$$

Since $n \geq 1$ and $\theta > 0$, substituting $\sin(\alpha) = \frac{\theta}{\sqrt{n}}$ gives

$$\begin{aligned}
q\left(\frac{\theta}{\sqrt{n}}\right) \leq & \mu\Bigg(\left(\frac{2(1+\theta)^2}{(1-\theta)^3}n^2+\frac{n(1+\theta)}{(1-\theta)}+\frac{\theta(1+\theta)}{(1-\theta)}\right)\frac{\theta^4}{n^2} \\
& +4\sqrt{2}\frac{(1+\theta)^{\frac{3}{2}}n^{\frac{3}{2}}}{(1-\theta)^2}\frac{\theta^3}{n^{\frac{3}{2}}}+\frac{\theta(1+\theta)}{(1-\theta)}\frac{\theta^2}{n}+\theta\frac{\theta}{\sqrt{n}}-\theta\Bigg) \\
= & \theta\mu\Bigg(\frac{2\theta^3(1+\theta)^2}{(1-\theta)^3}+\frac{\theta^3(1+\theta)}{n(1-\theta)}+\frac{\theta^4(1+\theta)}{(1-\theta)n^2} \\
& +\frac{4\sqrt{2}\theta^2(1+\theta)^{\frac{3}{2}}}{(1-\theta)^2}+\frac{\theta^2(1+\theta)}{n(1-\theta)}+\frac{\theta}{\sqrt{n}}-1\Bigg) \\
\leq & \theta\mu\Bigg(\frac{2\theta^3(1+\theta)^2}{(1-\theta)^3}+\frac{\theta^3(1+\theta)}{(1-\theta)}+\frac{\theta^4(1+\theta)}{(1-\theta)} \\
& +\frac{4\sqrt{2}\theta^2(1+\theta)^{\frac{3}{2}}}{(1-\theta)^2}+\frac{\theta^2(1+\theta)}{(1-\theta)}+\theta-1\Bigg) := \theta\mu p(\theta).
\end{aligned} \tag{10.65}$$

Since $p(\theta)$ is monotonic increasing function of $\theta \in [0,1)$, $p(0) < 0$, and it is easy to verify that $p(0.22) < 0$, this proves the lemma. ■

Corollary 10.1, Lemmas 10.7, and 10.9 prove the feasibility of searching optimizer along the ellipse. To move the iterate back to $\mathcal{N}_2(\theta)$, one can use the direction $(\Delta\mathbf{x},\Delta\mathbf{y},\Delta\mathbf{z},\Delta\lambda,\Delta\gamma)$, defined by

$$\begin{bmatrix} \mathbf{H} & \mathbf{0} & \mathbf{0} & \mathbf{I} & -\mathbf{I} \\ \mathbf{I} & \mathbf{I} & \mathbf{0} & \mathbf{0} & \mathbf{0} \\ \mathbf{I} & \mathbf{0} & -\mathbf{I} & \mathbf{0} & \mathbf{0} \\ \mathbf{0} & \Lambda(\alpha) & \mathbf{0} & \mathbf{Y}(\alpha) & \mathbf{0} \\ \mathbf{0} & \mathbf{0} & \Gamma(\alpha) & \mathbf{0} & \mathbf{Z}(\alpha) \end{bmatrix} \begin{bmatrix} \Delta\mathbf{x} \\ \Delta\mathbf{y} \\ \Delta\mathbf{z} \\ \Delta\lambda \\ \Delta\gamma \end{bmatrix} = \begin{bmatrix} \mathbf{0} \\ \mathbf{0} \\ \mathbf{0} \\ \mu(\alpha)\mathbf{e}-\lambda(\alpha)\circ\mathbf{y}(\alpha) \\ \mu(\alpha)\mathbf{e}-\gamma(\alpha)\circ\mathbf{z}(\alpha) \end{bmatrix}. \tag{10.66}$$

and update $(\mathbf{x}^{k+1},\mathbf{p}^{k+1},\omega^{k+1})$ and μ_{k+1} by

$$(\mathbf{x}^{k+1},\mathbf{p}^{k+1},\omega^{k+1}) = (\mathbf{x}(\alpha),\mathbf{p}(\alpha),\omega(\alpha))+(\Delta\mathbf{x},\Delta\mathbf{p},\Delta\omega), \tag{10.67}$$

$$\mu_{k+1} = \frac{\mathbf{p}^{k+1^{\mathrm{T}}}\omega^{k+1}}{2n}, \tag{10.68}$$

where

$$\Delta\mathbf{p} = (\Delta\mathbf{y},\Delta\mathbf{z})$$

and

$$\Delta\omega = (\Delta\lambda, \Delta\gamma).$$

Denote $\mathbf{P}(\alpha) = \begin{bmatrix} \mathbf{Y}(\alpha) & \mathbf{0} \\ \mathbf{0} & \mathbf{Z}(\alpha) \end{bmatrix}$, $\Omega(\alpha) = \begin{bmatrix} \Lambda(\alpha) & \mathbf{0} \\ \mathbf{0} & \Gamma(\alpha) \end{bmatrix}$, and $\mathbf{D} = \mathbf{P}^{\frac{1}{2}}(\alpha)\Omega^{-\frac{1}{2}}(\alpha)$. Then, the last 2 rows of (10.66) can be rewritten as

$$\mathbf{P}\Delta\omega + \Omega\Delta\mathbf{p} = \mu(\alpha)\mathbf{e} - \mathbf{P}(\alpha)\Omega(\alpha)\mathbf{e}. \tag{10.69}$$

Now, we are ready to show that the correction step brings the iterate from $\mathcal{N}_2(2\theta)$ back to $\mathcal{N}_2(\theta)$.

Lemma 10.10
Let $(\mathbf{x}(\alpha), \mathbf{p}(\alpha), \omega(\alpha)) \in \mathcal{N}_2(2\theta)$ and $(\Delta\mathbf{x}, \Delta\mathbf{p}, \Delta\omega)$ be defined as in (10.66). Let $(\mathbf{x}^{k+1}, \mathbf{p}^{k+1}, \omega^{k+1})$ be updated by using (10.67). Then, for $\theta \le 0.29$ and $\sin(\alpha) \le \sin(\bar{\alpha})$, $(\mathbf{x}^{k+1}, \mathbf{p}^{k+1}, \omega^{k+1}) \in \mathcal{N}_2(\theta)$.

Proof 10.11 Using Lemma 9.11 yields

$$0 \le \left\| \Delta\mathbf{p} \circ \Delta\omega - \frac{1}{2n}(\Delta\mathbf{p}^{\mathrm{T}}\Delta\omega)\mathbf{e} \right\|^2 \le \|\Delta\mathbf{p} \circ \Delta\omega\|^2. \tag{10.70}$$

Pre-multiplying $\left(\mathbf{P}(\alpha)\Omega(\alpha)\right)^{-\frac{1}{2}}$ on the both sides of (10.69) yields

$$\mathbf{D}\Delta\omega + \mathbf{D}^{-1}\Delta\mathbf{p} = \left(\mathbf{P}(\alpha)\Omega(\alpha)\right)^{-\frac{1}{2}} \left(\mu(\alpha)\mathbf{e} - \mathbf{P}(\alpha)\Omega(\alpha)\mathbf{e}\right).$$

Let $\mathbf{u} = \mathbf{D}\Delta\omega$, $\mathbf{v} = \mathbf{D}^{-1}\Delta\mathbf{p}$, from the first three rows of (10.66), it must have

$$\mathbf{u}^{\mathrm{T}}\mathbf{v} = \Delta\mathbf{p}^{\mathrm{T}}\Delta\omega = \Delta\mathbf{y}^{\mathrm{T}}\Delta\lambda + \Delta\mathbf{z}^{\mathrm{T}}\Delta\gamma = \Delta\mathbf{x}^{\mathrm{T}}(\Delta\gamma - \Delta\lambda) = \Delta\mathbf{x}^{\mathrm{T}}\mathbf{H}\Delta\mathbf{x} \ge 0. \tag{10.71}$$

Using Lemma 3.2, the last two rows of (10.66), and the assumption of $(\mathbf{x}(\alpha), \mathbf{p}(\alpha), \omega(\alpha)) \in \mathcal{N}_2(2\theta)$ yields

$$\begin{aligned}
\left\|\Delta\mathbf{p} \circ \Delta\omega\right\| &= \left\|\mathbf{u} \circ \mathbf{v}\right\| \le 2^{-\frac{3}{2}} \left\| \left(\mathbf{P}(\alpha)\Omega(\alpha)\right)^{-\frac{1}{2}} \left(\mu(\alpha)\mathbf{e} - \mathbf{P}(\alpha)\Omega(\alpha)\mathbf{e}\right) \right\|^2 \\
&= 2^{-\frac{3}{2}} \sum_{i=1}^{2n} \frac{(\mu(\alpha) - p_i(\alpha)\omega_i(\alpha))^2}{p_i(\alpha)\omega_i(\alpha)} \\
&\le 2^{-\frac{3}{2}} \frac{\|\mu(\alpha)\mathbf{e} - \mathbf{p}(\alpha) \circ \omega(\alpha)\|^2}{\min_i p_i(\alpha)\omega_i(\alpha)} \\
&\le 2^{-\frac{3}{2}} \frac{(2\theta)^2\mu(\alpha)^2}{(1-2\theta)\mu(\alpha)} = 2^{\frac{1}{2}} \frac{\theta^2\mu(\alpha)}{(1-2\theta)}.
\end{aligned} \tag{10.72}$$

Define $(\mathbf{p}^{k+1}(t), \omega^{k+1}(t)) = (\mathbf{p}(\alpha), \omega(\alpha)) + t(\Delta\mathbf{p}, \Delta\omega)$. From (10.69) and (10.25), one gets

$$\mathbf{p}(\alpha)^{\mathrm{T}}\Delta\omega + \omega(\alpha)^{\mathrm{T}}\Delta\mathbf{p} = 2n\mu(\alpha) - \sum_{i=1}^{2n} p_i(\alpha)\omega_i(\alpha) = 0. \tag{10.73}$$

Therefore,

$$\begin{aligned}
\mu_{k+1}(t) &= \frac{\left(\mathbf{p}(\alpha) + t\Delta\mathbf{p}\right)^{\mathrm{T}}\left(\omega(\alpha) + t\Delta\omega\right)}{2n} \\
&= \frac{\mathbf{p}(\alpha)^{\mathrm{T}}\omega(\alpha) + t^2\Delta\mathbf{p}^{\mathrm{T}}\Delta\omega}{2n} = \mu(\alpha) + t^2\frac{\Delta\mathbf{p}^{\mathrm{T}}\Delta\omega}{2n}.
\end{aligned} \tag{10.74}$$

In view of (10.71), we have $\Delta\mathbf{p}^{\mathrm{T}}\Delta\omega = \Delta\mathbf{x}^{\mathrm{T}}\mathbf{H}\Delta\mathbf{x} \geq 0$, therefore, it must have $\mu_{k+1}(t) \geq \mu(\alpha)$. Using (10.74), (10.69), (10.70), and (10.72) yields

$$\begin{aligned}
&\left\|\mathbf{p}^{k+1}(t) \circ \omega^{k+1}(t) - \mu_{k+1}(t)\mathbf{e}\right\| \\
= &\left\|(\mathbf{p}(\alpha) + t\Delta\mathbf{p}) \circ (\omega(\alpha) + t\Delta\omega) - \mu(\alpha)\mathbf{e} - \frac{t^2}{2n}\left(\Delta\mathbf{p}^{\mathrm{T}}\Delta\omega\right)\mathbf{e}\right\| \\
= &\left\|\mathbf{p}(\alpha) \circ \omega(\alpha) + t[\omega(\alpha) \circ \Delta\mathbf{p} + \mathbf{p}(\alpha) \circ \Delta\omega]\right. \\
&\left. + t^2\Delta\mathbf{p} \circ \Delta\omega - \mu(\alpha)\mathbf{e} - \frac{t^2}{2n}\left(\Delta\mathbf{p}^{\mathrm{T}}\Delta\omega\right)\mathbf{e}\right\| \\
= &\left\|\mathbf{p}(\alpha) \circ \omega(\alpha) + t[\mu(\alpha)\mathbf{e} - \mathbf{p}(\alpha) \circ \omega(\alpha)]\right. \\
&\left. + t^2\Delta\mathbf{p} \circ \Delta\omega - \mu(\alpha)\mathbf{e} - \frac{t^2}{2n}\left(\Delta\mathbf{p}^{\mathrm{T}}\Delta\omega\right)\mathbf{e}\right\| \\
= &\left\|(1-t)\left[\mathbf{p}(\alpha) \circ \omega(\alpha) - \mu(\alpha)\mathbf{e}\right] + t^2\left(\Delta\mathbf{p} \circ \Delta\omega - \frac{1}{2n}\left(\Delta\mathbf{p}^{\mathrm{T}}\Delta\omega\right)\mathbf{e}\right)\right\| \\
\leq &\ (1-t)(2\theta)\mu(\alpha) + t^2\frac{2^{\frac{1}{2}}\theta^2}{(1-2\theta)}\mu(\alpha) \\
\leq &\left((1-t)(2\theta) + t^2\frac{2^{\frac{1}{2}}\theta^2}{(1-2\theta)}\right)\mu_{k+1} := f(t,\theta)\mu_{k+1}.
\end{aligned} \tag{10.75}$$

Therefore, taking $t = 1$ gives $\left\|\mathbf{p}^{k+1} \circ \omega^{k+1} - \mu_{k+1}\mathbf{e}\right\| \leq \frac{2^{\frac{1}{2}}\theta^2}{(1-2\theta)}\mu_{k+1}$. It is easy to see that, for $\theta \leq 0.29$,

$$\frac{2^{\frac{1}{2}}\theta^2}{(1-2\theta)} = 0.2832 < \theta.$$

For $\theta \leq 0.29$ and $t \in [0,1]$, noticing

$$0 \leq f(t,\theta) \leq f(t,0.29) \leq 0.58(1-t) + 0.2832t^2 < 1,$$

and using (10.74), (10.71), and Corollary 10.1, one gets, for an additional condition $\sin(\alpha) \le \sin^{-1}(\bar{\alpha})$,

$$\begin{aligned} p_i^{k+1}(t)\omega_i^{k+1}(t) &\ge (1-f(t,\theta))\,\mu_{k+1}(t) \\ &= (1-f(t,\theta))\left(\mu(\alpha)+\frac{t^2}{n}\Delta\mathbf{p}^{\mathrm{T}}\Delta\omega\right) \\ &\ge (1-f(t,\theta))\,\mu(\alpha) \\ &> 0, \end{aligned} \tag{10.76}$$

Therefore, $(\mathbf{p}^{k+1}(t), \omega^{k+1}(t)) > \mathbf{0}$ for $t \in [0,1]$, i.e., $(\mathbf{p}^{k+1}, \omega^{k+1}) > \mathbf{0}$. This finishes the proof. ∎

The next step is to show that the combined step (searching along the arc in $\mathcal{N}_2(2\theta)$ and moving back to $\mathcal{N}_2(\theta)$) will reduce the duality measure of the iterate, i.e., $\mu_{k+1} < \mu_k$, if some appropriate θ and α are selected. The following two Lemmas are introduced for this purpose.

Lemma 10.11
Let $(\mathbf{x}(\alpha), \mathbf{p}(\alpha), \omega(\alpha)) \in \mathcal{N}_2(2\theta)$ and $(\Delta\mathbf{x}, \Delta\mathbf{p}, \Delta\omega)$ be defined as in (10.66). Then,

$$0 \le \frac{\Delta\mathbf{p}^{\mathrm{T}}\Delta\omega}{2n} \le \frac{\theta^2(1+2\theta)}{n(1-2\theta)^2}\mu(\alpha) := \frac{\delta_0}{n}\mu(\alpha). \tag{10.77}$$

Proof 10.12 The first inequality of (10.77) follows from (10.71). Pre-multiplying both sides of (10.69) by $\mathbf{P}^{-\frac{1}{2}}(\alpha)\Omega^{-\frac{1}{2}}(\alpha)$ gives

$$\mathbf{P}^{-\frac{1}{2}}(\alpha)\Omega^{\frac{1}{2}}(\alpha)\Delta\mathbf{p}+\mathbf{P}^{\frac{1}{2}}(\alpha)\Omega^{-\frac{1}{2}}(\alpha)\Delta\omega=\mathbf{P}^{-\frac{1}{2}}(\alpha)\Omega^{-\frac{1}{2}}(\alpha)\Big(\mu(\alpha)\mathbf{e}-\mathbf{P}(\alpha)\Omega(\alpha)\mathbf{e}\Big).$$

Let

$$\begin{aligned} \mathbf{u} &= \mathbf{P}^{-\frac{1}{2}}(\alpha)\Omega^{\frac{1}{2}}(\alpha)\Delta\mathbf{p}, \\ \mathbf{v} &= \mathbf{P}^{\frac{1}{2}}(\alpha)\Omega^{-\frac{1}{2}}(\alpha)\Delta\omega, \end{aligned}$$

and

$$\mathbf{w} = \mathbf{P}^{-\frac{1}{2}}(\alpha)\Omega^{-\frac{1}{2}}(\alpha)\Big(\mu(\alpha)\mathbf{e}-\mathbf{P}(\alpha)\Omega(\alpha)\mathbf{e}\Big),$$

in view of (10.71), it must have

$$\mathbf{u}^{\mathrm{T}}\mathbf{v} = \Delta\mathbf{p}^{\mathrm{T}}\Delta\omega \ge 0.$$

Using Lemma 9.3 and the assumption of $(\mathbf{x}(\alpha), \mathbf{p}(\alpha), \omega(\alpha)) \in \mathcal{N}_2(2\theta)$ yields

$$\|\mathbf{u}\|^2+\|\mathbf{v}\|^2 = \sum_{i=1}^{2n}\left(\frac{(\Delta p_i)^2\omega_i(\alpha)}{p_i(\alpha)}+\frac{(\Delta\omega_i)^2 p_i(\alpha)}{\omega_i(\alpha)}\right)$$

$$
\begin{aligned}
&\le \|\mathbf{w}\|^2 = \sum_{i=1}^{2n} \frac{(\mu(\alpha) - p_i(\alpha)\omega_i(\alpha))^2}{p_i(\alpha)\omega_i(\alpha)} \\
&\le \frac{\sum_{i=1}^{2n} (\mu(\alpha) - p_i(\alpha)\omega_i(\alpha))^2}{\min_i p_i(\alpha)\omega_i(\alpha)} \\
&\le \frac{(2\theta)^2\mu^2(\alpha)}{(1-2\theta)\mu(\alpha)} = \frac{(2\theta)^2\mu(\alpha)}{(1-2\theta)}.
\end{aligned}
\tag{10.78}
$$

Dividing both sides by $\mu(\alpha)$ and using $p_i(\alpha)\omega_i(\alpha) \ge \mu(\alpha)(1-2\theta)$ yields

$$
\begin{aligned}
&\sum_{i=1}^{2n}(1-2\theta)\left(\frac{(\Delta p_i)^2}{p_i^2(\alpha)} + \frac{(\Delta\omega_i)^2}{\omega_i^2(\alpha)}\right) \\
=&(1-2\theta)\left(\left\|\frac{\Delta\mathbf{p}}{\mathbf{p}(\alpha)}\right\|^2 + \left\|\frac{\Delta\omega}{\omega(\alpha)}\right\|^2\right) \\
\le&\frac{(2\theta)^2}{(1-2\theta)},
\end{aligned}
\tag{10.79}
$$

i.e.,

$$
\left\|\frac{\Delta\mathbf{p}}{\mathbf{p}(\alpha)}\right\|^2 + \left\|\frac{\Delta\omega}{\omega(\alpha)}\right\|^2 \le \left(\frac{2\theta}{1-2\theta}\right)^2. \tag{10.80}
$$

Invoking Lemma 5.3, one gets

$$
\left\|\frac{\Delta\mathbf{p}}{\mathbf{p}(\alpha)}\right\|^2 \cdot \left\|\frac{\Delta\omega}{\omega(\alpha)}\right\|^2 \le \frac{1}{4}\left(\frac{2\theta}{1-2\theta}\right)^4. \tag{10.81}
$$

This gives

$$
\left\|\frac{\Delta\mathbf{p}}{\mathbf{p}(\alpha)}\right\| \cdot \left\|\frac{\Delta\omega}{\omega(\alpha)}\right\| \le \frac{2\theta^2}{(1-2\theta)^2}. \tag{10.82}
$$

Using Cauchy-Schwarz inequality leads to

$$
\begin{aligned}
&\frac{(\Delta\mathbf{p})^{\mathrm{T}}(\Delta\omega)}{\mu(\alpha)} \\
\le&\sum_{i=1}^{2n} \frac{|\Delta p_i||\Delta\omega_i|}{\mu(\alpha)} \\
\le&(1+2\theta)\sum_{i=1}^{2n} \frac{|\Delta p_i|}{p_i(\alpha)}\frac{|\Delta\omega_i|}{\omega_i(\alpha)} \\
=&(1+2\theta)\left|\frac{\Delta\mathbf{p}}{\mathbf{p}(\alpha)}\right|^{\mathrm{T}}\left|\frac{\Delta\omega}{\omega(\alpha)}\right|
\end{aligned}
$$

$$\leq (1+2\theta)\left\|\frac{\Delta\mathbf{p}}{\mathbf{p}(\alpha)}\right\|\cdot\left\|\frac{\Delta\omega}{\omega(\alpha)}\right\|$$
$$\leq \frac{2\theta^2(1+2\theta)}{(1-2\theta)^2}. \tag{10.83}$$

Therefore,

$$\frac{(\Delta\mathbf{p})^{\mathrm{T}}(\Delta\omega)}{2n} \leq \frac{\theta^2(1+2\theta)}{n(1-2\theta)^2}\mu(\alpha). \tag{10.84}$$

This proves the lemma. ■

Lemma 10.12
Let $(\mathbf{x}(\alpha),\mathbf{p}(\alpha),\omega(\alpha)) \in \mathcal{N}_2(2\theta)$ *and* $(\Delta\mathbf{x},\Delta\mathbf{p},\Delta\omega)$ *be defined as in (10.66). Let* $(\mathbf{x}^{k+1},\mathbf{p}^{k+1},\omega^{k+1})$ *be defined as in (10.67). Then,*

$$\mu(\alpha) \leq \mu_{k+1} := \frac{\mathbf{p}^{k+1^{\mathrm{T}}}\omega^{k+1}}{2n} \leq \mu(\alpha)\left(1+\frac{\theta^2(1+2\theta)}{n(1-2\theta)^2}\right) = \mu(\alpha)\left(1+\frac{\delta_0}{n}\right).$$

Proof 10.13 Using the fact that $\mathbf{p}(\alpha)^{\mathrm{T}}\Delta\omega + \omega(\alpha)^{\mathrm{T}}\Delta\mathbf{p} = 0$ established in (10.73) in the proof of Lemma 10.10, and Lemma 10.11, it is straightforward to obtain

$$\begin{aligned}
\mu(\alpha) &\leq \frac{\mathbf{p}(\alpha)^{\mathrm{T}}\omega(\alpha)}{2n} + \frac{1}{2n}\Delta\mathbf{p}^{\mathrm{T}}\Delta\omega \\
&= \frac{(\mathbf{p}(\alpha)+\Delta\mathbf{p})^{\mathrm{T}}(\omega(\alpha)+\Delta\omega)}{2n} = \mu_{k+1} \\
&\leq \mu(\alpha) + \frac{\theta^2(1+2\theta)}{n(1-2\theta)^2}\mu(\alpha).
\end{aligned} \tag{10.85}$$

This proves the lemma. ■

For linear programming, it is shown in Chapter 5 that $\mu_{k+1} = \mu(\alpha)$. This claim is not always true for the convex quadratic programming, as pointed out in Lemma 10.12. Therefore, some extra work is needed to make sure that the μ_k will be reduced in every iteration.

Lemma 10.13
For $\theta \leq 0.19$, *if*

$$\sin(\alpha) = \frac{\theta}{\sqrt{n}}, \tag{10.86}$$

then $\mu_{k+1} < \mu_k$. *Moreover, for* $\sin(\alpha) = \frac{\theta}{\sqrt{n}} = \frac{0.19}{\sqrt{n}}$,

$$\mu_{k+1} \leq \mu_k\left(1-\frac{0.0185}{\sqrt{n}}\right). \tag{10.87}$$

Proof 10.14 Using Lemmas 10.12, 10.5, 5.4, 10.2, 10.3, and 10.4, and noticing $\ddot{\mathbf{p}}^{\mathrm{T}}\ddot{\omega} \geq 0$ and $\dot{\mathbf{p}}^{\mathrm{T}}\dot{\omega} \geq 0$ yields

$$\mu_{k+1} \leq \mu(\alpha)\left(1+\frac{\theta^2(1+2\theta)}{n(1-2\theta)^2}\right) = \mu(\alpha)\left(1+\frac{\delta_0}{n}\right) \tag{10.88a}$$

$$
\begin{aligned}
&=\mu_k\left[1-\sin(\alpha)+\left(\frac{\ddot{\mathbf{p}}^{\mathrm{T}}\ddot{\omega}}{2n\mu_k}-\frac{\dot{\mathbf{p}}^{\mathrm{T}}\dot{\omega}}{2n\mu_k}\right)(1-\cos(\alpha))^2\right.\\
&\quad\left.-\left(\frac{\dot{\mathbf{p}}^{\mathrm{T}}\ddot{\omega}}{2n\mu_k}+\frac{\dot{\omega}^{\mathrm{T}}\ddot{\mathbf{p}}}{2n\mu_k}\right)\sin(\alpha)(1-\cos(\alpha))\right]\left(1+\frac{\delta_0}{n}\right)\\
&\leq\mu_k\left(1-\sin(\alpha)+\frac{\ddot{\mathbf{p}}^{\mathrm{T}}\ddot{\omega}}{2n\mu_k}\sin^4(\alpha)+\left(\left|\frac{\dot{\mathbf{p}}^{\mathrm{T}}\ddot{\omega}}{2n\mu_k}\right|+\left|\frac{\dot{\omega}^{\mathrm{T}}\ddot{\mathbf{p}}}{2n\mu_k}\right|\right)\sin^3(\alpha)\right)\left(1+\frac{\delta_0}{n}\right)\\
&\leq\mu_k\left(1-\sin(\alpha)+\frac{n(1+\theta)^2}{(1-\theta)^3}\sin^4(\alpha)+\frac{2(2n)^{\frac{1}{2}}(1+\theta)^{\frac{3}{2}}}{(1-\theta)^2}\sin^3(\alpha)\right)\left(1+\frac{\delta_0}{n}\right).
\end{aligned} \tag{10.88b}
$$

Substituting $\sin(\alpha)=\frac{\theta}{\sqrt{n}}$ into (10.88b) gives

$$
\begin{aligned}
\mu_{k+1} \leq&\mu_k\left(1-\frac{\theta}{\sqrt{n}}+\frac{n(1+\theta)^2}{(1-\theta)^3}\frac{\theta^4}{n^2}+\frac{2(2n)^{\frac{1}{2}}(1+\theta)^{\frac{3}{2}}}{(1-\theta)^2}\frac{\theta^3}{n^{\frac{3}{2}}}\right)\left(1+\frac{\delta_0}{n}\right)\\
=&\mu_k\left(1-\frac{\theta}{\sqrt{n}}+\frac{\theta^4(1+\theta)^2}{n(1-\theta)^3}+\frac{2^{\frac{3}{2}}\theta^3(1+\theta)^{\frac{3}{2}}}{n(1-\theta)^2}\right)\left(1+\frac{\delta_0}{n}\right)\\
=&\mu_k\left(1-\frac{\theta}{\sqrt{n}}+\frac{\delta_0}{n}+\frac{\theta^4(1+\theta)^2}{n(1-\theta)^3}+\frac{2^{\frac{3}{2}}\theta^3(1+\theta)^{\frac{3}{2}}}{n(1-\theta)^2}-\frac{\theta\delta_0}{n^{\frac{3}{2}}}\right.\\
&\left.+\frac{\delta_0}{n}\left[\frac{\theta^4(1+\theta)^2}{n(1-\theta)^3}+\frac{2^{\frac{3}{2}}\theta^3(1+\theta)^{\frac{3}{2}}}{n(1-\theta)^2}\right]\right)\\
=&\mu_k\left(1-\frac{\theta}{\sqrt{n}}\left[1-\frac{\delta_0}{\sqrt{n}\theta}-\frac{\theta^3(1+\theta)^2}{\sqrt{n}(1-\theta)^3}-\frac{2^{\frac{3}{2}}\theta^2(1+\theta)^{\frac{3}{2}}}{\sqrt{n}(1-\theta)^2}\right]\right.\\
&\left.-\frac{\theta\delta_0}{n^{\frac{3}{2}}}\left[1-\frac{\theta^3(1+\theta)^2}{\sqrt{n}(1-\theta)^3}-\frac{2^{\frac{3}{2}}\theta^2(1+\theta)^{\frac{3}{2}}}{\sqrt{n}(1-\theta)^2}\right]\right).
\end{aligned}
$$

Since

$$
\begin{aligned}
&1-\frac{\theta^3(1+\theta)^2}{\sqrt{n}(1-\theta)^3}-\frac{2^{\frac{3}{2}}\theta^2(1+\theta)^{\frac{3}{2}}}{\sqrt{n}(1-\theta)^2}\\
\geq\ &1-\frac{\theta^3(1+\theta)^2}{(1-\theta)^3}-\frac{2^{\frac{3}{2}}\theta^2(1+\theta)^{\frac{3}{2}}}{(1-\theta)^2}:=f(\theta),
\end{aligned}
$$

where $f(\theta)$ is a monotonic decreasing function of θ, and for $\theta \leq 0.37$, $f(\theta)>0$.

Therefore, for $\theta \le 0.37$, the following relation holds.

$$\begin{aligned}\mu_{k+1} \le & \mu_k \left(1 - \frac{\theta}{\sqrt{n}}\left[1 - \frac{\delta_0}{\sqrt{n}\theta} - \frac{\theta^3(1+\theta)^2}{\sqrt{n}(1-\theta)^3} - \frac{2^{\frac{3}{2}}\theta^2(1+\theta)^{\frac{3}{2}}}{\sqrt{n}(1-\theta)^2}\right]\right) \\ = & \mu_k \left(1 - \frac{\theta}{\sqrt{n}}\left[1 - \frac{\theta(1+2\theta)}{\sqrt{n}(1-2\theta)^2} - \frac{\theta^3(1+\theta)^2}{\sqrt{n}(1-\theta)^3} - \frac{2^{\frac{3}{2}}\theta^2(1+\theta)^{\frac{3}{2}}}{\sqrt{n}(1-\theta)^2}\right]\right).\end{aligned} \tag{10.89}$$

Since

$$\begin{aligned}& 1 - \frac{\theta(1+2\theta)}{\sqrt{n}(1-2\theta)^2} - \frac{\theta^3(1+\theta)^2}{\sqrt{n}(1-\theta)^3} - \frac{2^{\frac{3}{2}}\theta^2(1+\theta)^{\frac{3}{2}}}{\sqrt{n}(1-\theta)^2} \\ \ge \; & 1 - \frac{\theta(1+2\theta)}{(1-2\theta)^2} - \frac{\theta^3(1+\theta)^2}{(1-\theta)^3} - \frac{2^{\frac{3}{2}}\theta^2(1+\theta)^{\frac{3}{2}}}{(1-\theta)^2} := g(\theta),\end{aligned} \tag{10.90}$$

where $g(\theta)$ is a monotonic decreasing function of θ, one can conclude, for $\theta \le 0.19$, $g(\theta) > 0.0976 > 0$. For $\theta = 0.19$, it must have $\theta g(\theta) > 0.0185$ and

$$\mu_{k+1} \le \mu_k \left(1 - \frac{0.0185}{\sqrt{n}}\right).$$

This proves (10.87). ∎

Remark 10.3 As seen in this section, that starting with $(\mathbf{x}^0, \mathbf{p}^0, \omega^0)$, the arc-search interior-point algorithm is carried out to find $(\mathbf{x}(\alpha), \mathbf{p}(\alpha), \omega(\alpha)) \in \mathcal{N}_2(2\theta)$ and $(\mathbf{x}^{k+1}, \mathbf{p}^{k+1}, \omega^{k+1}) \in \mathcal{N}_2(\theta)$ such that $\mu_{k+1} < \mu_k$. In view of the proofs of Lemmas 10.7, 10.10, and 10.13, the positivity conditions of $(\mathbf{x}(\alpha), \mathbf{p}(\alpha), \omega(\alpha)) > \mathbf{0}$ and $(\mathbf{x}^{k+1}, \mathbf{p}^{k+1}, \omega^{k+1}) > \mathbf{0}$ relies on $\mu(\alpha) > 0$ which, according to Corollary 10.1, is achievable for any $\theta \le 0.19$ and is given by a bound in terms of $\bar{\alpha}$. The proximity condition for $(\mathbf{x}(\alpha), \mathbf{p}(\alpha), \omega(\alpha))$ relies on the real positive root of $q(\sin(\alpha)) = 0$, denoted by $\sin(\tilde{\alpha})$, which is conservatively estimated in Lemma 10.9 under the condition that $\theta \le 0.22$; the proximity condition for $(\mathbf{x}^{k+1}, \mathbf{p}^{k+1}, \omega^{k+1})$ is established in Lemma 10.10 under the condition that $\theta \le 0.29$. Finally, duality measure reduction $\mu_{k+1} < \mu_k$ is established in Lemma 10.13 under the condition that $\theta \le 0.19$. For all these results to hold, it just needs to take the smallest bound $\theta = 0.19$. ∎

Summarizing all the results in this section leads to the following theorem.

Theorem 10.2
Let $\theta = 0.19$ and $(\mathbf{x}^k, \mathbf{p}^k, \omega^k) \in \mathcal{N}_2(\theta)$. Then, $(\mathbf{x}(\alpha), \mathbf{p}(\alpha), \omega(\alpha)) \in \mathcal{N}_2(2\theta)$; $(\mathbf{x}^{k+1}, \mathbf{p}^{k+1}, \omega^{k+1}) \in \mathcal{N}_2(\theta)$; and $\mu_{k+1} \le \mu_k \left(1 - \frac{0.0185}{\sqrt{n}}\right)$.

Proof 10.15 By examining Corollary 10.1 and Lemma 10.9, one can select $\sin(\alpha) \le \min\{\sin(\tilde{\alpha}), \sin(\bar{\alpha})\}$. Therefore, Lemma 10.7 holds. This means that the relation $(\mathbf{x}(\alpha), \mathbf{p}(\alpha), \omega(\alpha)) \in \mathcal{N}_2(2\theta)$ holds. Since $\sin(\alpha) \leq \sin(\bar{\alpha})$ and $(\mathbf{x}(\alpha), \mathbf{p}(\alpha), \omega(\alpha)) \in \mathcal{N}_2(2\theta)$, Lemma 10.10 states $(\mathbf{x}^{k+1}, \mathbf{p}^{k+1}, \omega^{k+1}) \in \mathcal{N}_2(\theta)$. For $\theta = 0.19$ and $\sin(\alpha) = \frac{\theta}{\sqrt{n}}$, Lemma 10.13 states $\mu_{k+1} \leq \mu_k \left(1 - \frac{0.0185}{\sqrt{n}}\right)$. This finishes the proof. ∎

Remark 10.4 It is worthwhile to point out that $\theta = 0.19$ for the box constrained quadratic optimization problem is larger than the $\theta = 0.148$ for linearly constrained quadratic optimization problem. This makes the searching neighborhood larger and the following algorithm more efficient than the algorithm in Chapter 9. ∎

The proposed algorithm is presented as below:

Algorithm 10.1
(Arc-search path-following)
Data: $\mathbf{H} \succeq 0$, $\mathbf{c}$, n, $\theta = 0.19$, $\varepsilon > 0$.
Initial point $(\mathbf{x}^0, \mathbf{p}^0, \omega^0) \in \mathcal{N}_2(\theta)$, *and* $\mu_0 = \frac{\mathbf{p}^{0^{\mathrm{T}}} \omega^0}{2n}$.
for *iteration* $k = 0, 1, 2, \ldots$

Step 1: Solve the linear systems of equations (10.12) and (10.13) to get $(\dot{\mathbf{x}}, \dot{\mathbf{p}}, \dot{\omega})$ *and* $(\ddot{\mathbf{x}}, \ddot{\mathbf{p}}, \ddot{\omega})$.

Step 2: Let $\sin(\alpha) = \frac{\theta}{\sqrt{n}}$. *Update* $(\mathbf{x}(\alpha), \mathbf{p}(\alpha), \omega(\alpha))$ *and* $\mu(\alpha)$ *by (10.54) and (10.55).*

Step 3: Solve (10.66) to get $(\Delta\mathbf{x}, \Delta\mathbf{p}, \Delta\omega)$, *update* $(\mathbf{x}^{k+1}, \mathbf{p}^{k+1}, \omega^{k+1})$ *and* μ_{k+1} *by using (10.67) and (10.68).*

Step 4: Set $k+1 \rightarrow k$. *Go back to Step 1.*

end (for)

10.3 Convergence Analysis

The first result in this section extends a result of Chapter 9 to convex quadratic programming subject to box constraints.

Lemma 10.14
Suppose $\mathcal{F}^o \neq \emptyset$. *Then, for each* $K \geq 0$, *the set*

$$\{(\mathbf{x}, \mathbf{p}, \omega) \mid (\mathbf{x}, \mathbf{p}, \omega) \in \mathcal{F}, \quad \mathbf{p}^{\mathrm{T}} \omega \leq K\}$$

is bounded.

Proof 10.16 The proof is similar to the proof in Chapter 9. It is given here for completeness. First, $\mathbf{x}$ is bounded because $-\mathbf{e} \leq \mathbf{x} \leq \mathbf{e}$. Since $\mathbf{x}+\mathbf{y}=\mathbf{e}$ and $-\mathbf{e} \leq \mathbf{x} \leq \mathbf{e}$, it is easy to see $\mathbf{0} \leq \mathbf{y} = \mathbf{e}-\mathbf{x} \leq 2\mathbf{e}$. Since $\mathbf{x}-\mathbf{z}=-\mathbf{e}$, it is easy to see $\mathbf{0} \leq \mathbf{z} = \mathbf{x}+\mathbf{e} \leq 2\mathbf{e}$. Therefore, $\mathbf{y}$ and $\mathbf{z}$ are also bounded. Let $(\bar{\mathbf{x}}, \bar{\mathbf{y}}, \bar{\mathbf{z}}, \bar{\lambda}, \bar{\gamma})$ be any fixed point in $\mathcal{F}^o$, and $(\mathbf{x}, \mathbf{y}, \mathbf{z}, \lambda, \gamma)$ be any point in $\mathcal{F}$ with $\mathbf{y}^{\mathrm{T}}\lambda + \mathbf{z}^{\mathrm{T}}\gamma \leq K$. Using the definition of $\mathcal{F}^o$ and $\mathcal{F}$ yields

$$\mathbf{H}(\bar{\mathbf{x}}-\mathbf{x}) + (\bar{\lambda}-\lambda) - (\bar{\gamma}-\gamma) = \mathbf{0}.$$

Therefore,

$$(\bar{\mathbf{x}}-\mathbf{x})^{\mathrm{T}}\mathbf{H}(\bar{\mathbf{x}}-\mathbf{x}) + (\bar{\mathbf{x}}-\mathbf{x})^{\mathrm{T}}(\bar{\lambda}-\lambda) - (\bar{\mathbf{x}}-\mathbf{x})^{\mathrm{T}}(\bar{\gamma}-\gamma) = 0,$$

or equivalently,

$$(\bar{\mathbf{x}}-\mathbf{x})^{\mathrm{T}}(\bar{\gamma}-\gamma) - (\bar{\mathbf{x}}-\mathbf{x})^{\mathrm{T}}(\bar{\lambda}-\lambda) = (\bar{\mathbf{x}}-\mathbf{x})^{\mathrm{T}}\mathbf{H}(\bar{\mathbf{x}}-\mathbf{x}) \geq 0.$$

This gives

$$((\bar{\mathbf{x}}+\mathbf{e}) - (\mathbf{x}+\mathbf{e}))^{\mathrm{T}}(\bar{\gamma}-\gamma) - ((\bar{\mathbf{x}}-\mathbf{e}) - (\mathbf{x}-\mathbf{e}))^{\mathrm{T}}(\bar{\lambda}-\lambda) \geq 0.$$

Substituting $\mathbf{x}-\mathbf{e} = -\mathbf{y}$ and $\mathbf{x}+\mathbf{e} = \mathbf{z}$ yields

$$(\bar{\mathbf{z}}-\mathbf{z})^{\mathrm{T}}(\bar{\gamma}-\gamma) + (\bar{\mathbf{y}}-\mathbf{y})^{\mathrm{T}}(\bar{\lambda}-\lambda) \geq 0.$$

This leads to

$$\bar{\mathbf{z}}^{\mathrm{T}}\bar{\gamma} + \mathbf{z}^{\mathrm{T}}\gamma - \mathbf{z}^{\mathrm{T}}\bar{\gamma} - \bar{\mathbf{z}}^{\mathrm{T}}\gamma + \bar{\mathbf{y}}^{\mathrm{T}}\bar{\lambda} + \mathbf{y}^{\mathrm{T}}\lambda - \mathbf{y}^{\mathrm{T}}\bar{\lambda} - \bar{\mathbf{y}}^{\mathrm{T}}\lambda \geq 0,$$

or in a compact form

$$\bar{\mathbf{p}}^{\mathrm{T}}\bar{\omega} + \mathbf{p}^{\mathrm{T}}\omega - \mathbf{p}^{\mathrm{T}}\bar{\omega} - \bar{\mathbf{p}}^{\mathrm{T}}\omega \geq 0.$$

Since $(\bar{\mathbf{p}}, \bar{\omega}) > \mathbf{0}$ is fixed, let

$$\xi = \min_{i=1,\cdots,n} \min\{\bar{p}_i, \bar{\omega}_i\},$$

then, using $\mathbf{p}^{\mathrm{T}}\omega \leq K$,

$$\bar{\mathbf{p}}^{\mathrm{T}}\bar{\omega} + K \geq \xi \mathbf{e}^{\mathrm{T}}(\mathbf{p}+\omega) \geq \max_{i=1,\cdots,n} \max\{\xi p_i, \xi \omega_i\},$$

i.e., for $i \in \{1, \cdots, n\}$,

$$0 \leq p_i \leq \frac{1}{\xi}(K + \bar{\mathbf{p}}^{\mathrm{T}}\bar{\omega}), \qquad 0 \leq \omega_i \leq \frac{1}{\xi}(K + \bar{\mathbf{p}}^{\mathrm{T}}\bar{\omega}).$$

This proves the lemma. ■

The following theorem is a direct result of Lemmas 10.14, 10.1, Theorem 10.2, KKT conditions, Theorem 1.5.

Theorem 10.3
Suppose that Assumption 1 holds, then the sequence generated by Algorithm 10.1 converges to a set of accumulation points, and all these accumulation points are global optimal solutions of the convex quadratic programming subject to box constraints.

Let $(\mathbf{x}^*, \mathbf{p}^*, \omega^*)$ be any solution of (10.2), denote index sets $\mathcal{B}$, $\mathcal{S}$, and $\mathcal{T}$ as

$$\mathcal{B} = \{j \in \{1, \ldots, 2n\} \mid p_j^* \neq 0\}. \tag{10.91}$$

$$\mathcal{S} = \{j \in \{1, \ldots, 2n\} \mid \omega_j^* \neq 0\}. \tag{10.92}$$

$$\mathcal{T} = \{j \in \{1, \ldots, 2n\} \mid p_j^* = \omega_j^* = 0\}. \tag{10.93}$$

According to Goldman-Tucker theorem [40], for the linear programming, $\mathcal{B} \cap \mathcal{S} = \emptyset = \mathcal{T}$ and $\mathcal{B} \cup \mathcal{S} = \{1, \ldots, 2n\}$. A solution with this property is called strictly complementary. An example in Chapter 9 shows that there may have quadratic programming problems that do not have strictly complementary solution. But if a convex quadratic programming subject to box constraints has strictly complementary solution(s), an interior-point algorithm will generate a sequence to approach strict complementary solution(s). As a matter of fact, from Lemma 10.14, we can show that Lemma 9.18 has a parallel result for convex quadratic programming subject to box constraints given as follows:

Lemma 10.15
Let $\mu_0 > 0$, *and* $\rho \in (0, 1)$. *Assume that the convex BQP (10.1) has strictly complementary solution(s). Then, for all points* $(\mathbf{x}, \mathbf{p}, \omega) \in \mathcal{F}^o$, $p_i \omega_i > \rho \mu$, *and* $\mu < \mu_0$, *there are constants M,* C_1, *and* C_2 *such that*

$$\|(\mathbf{p}, \omega)\| \leq M, \tag{10.94}$$

$$0 < p_i \leq \mu / C_1 \quad (i \in \mathcal{S}), \qquad 0 < \omega_i \leq \mu / C_1 \quad (i \in \mathcal{B}). \tag{10.95}$$

$$\omega_i \geq C_2 \rho \quad (i \in \mathcal{S}), \qquad p_i \geq C_2 \rho \quad (i \in \mathcal{B}). \tag{10.96}$$

Proof 10.17 The first result (10.94) follows immediately from Lemma 10.14. Let $(\mathbf{x}^*, \mathbf{p}^*, \omega^*)$ be any strictly complementary solution. Since $(\mathbf{x}^*, \mathbf{p}^*, \omega^*)$ and $(\mathbf{x}, \mathbf{p}, \omega)$ are both feasible, it must have

$$(\mathbf{y} - \mathbf{y}^*) = -(\mathbf{x} - \mathbf{x}^*) = -(\mathbf{z} - \mathbf{z}^*), \qquad \mathbf{H}(\mathbf{x} - \mathbf{x}^*) + (\lambda - \lambda^*) - (\gamma - \gamma^*) = \mathbf{0}.$$

Therefore,

$$(\mathbf{y} - \mathbf{y}^*)^{\mathrm{T}} (\lambda - \lambda^*) + (\mathbf{z} - \mathbf{z}^*)^{\mathrm{T}} (\gamma - \gamma^*) = (\mathbf{x} - \mathbf{x}^*)^{\mathrm{T}} \mathbf{H} (\mathbf{x} - \mathbf{x}^*) \geq 0,$$

or

$$\mathbf{y}^{\mathrm{T}} \lambda + \mathbf{z}^{\mathrm{T}} \gamma \geq \mathbf{y}^{\mathrm{T}} \lambda^* + \mathbf{z}^{\mathrm{T}} \gamma^* + ((\mathbf{y}^*)^{\mathrm{T}} \lambda + (\mathbf{z}^*)^{\mathrm{T}} \gamma) \tag{10.97}$$

as $(\mathbf{p}^*)^{\mathrm{T}}\omega^* = 0$. Since $(\mathbf{x}^*, \mathbf{y}^*, \mathbf{z}^*, \lambda^*, \gamma^*) = (\mathbf{x}^*, \mathbf{p}^*, \omega^*)$ is strictly complementary solution, it must have $\mathcal{T} = \emptyset$, $p_i^* = 0$ for $i \in \mathcal{S}$, and $\omega_i^* = 0$ for $i \in \mathcal{B}$. Since $\mathbf{p}^{\mathrm{T}}\omega = 2n\mu$, from (10.97), it must have

$$\begin{aligned}
\mathbf{p}^{\mathrm{T}}\omega &= \mathbf{y}^{\mathrm{T}}\lambda + \mathbf{z}^{\mathrm{T}}\gamma \\
&\geq \mathbf{y}^{\mathrm{T}}\lambda^* + \mathbf{z}^{\mathrm{T}}\gamma^* + \left((\mathbf{y}^*)^{\mathrm{T}}\lambda + (\mathbf{z}^*)^{\mathrm{T}}\gamma\right) \\
&= \mathbf{p}^{\mathrm{T}}\omega^* + \omega^{\mathrm{T}}\mathbf{p}^* \\
\iff & \quad 2n\mu \geq \mathbf{p}^{\mathrm{T}}\omega^* + \omega^{\mathrm{T}}\mathbf{p}^* = \sum_{i\in\mathcal{S}} p_i\omega_i^* + \sum_{i\in\mathcal{B}} p_i^*\omega_i. \qquad (10.98)
\end{aligned}$$

Since each term in the summations is positive and bounded above by $2n\mu$, it must have $\omega_i^* > 0$ for any $i \in \mathcal{S}$; therefore,

$$0 < p_i \leq \frac{2n\mu}{\omega_i^*}.$$

Denote $D^* = \{(\mathbf{p}^*, \omega^*) \mid \omega_i^* > 0\}$ and $P^* = \{(\mathbf{p}^*, \omega^*) \mid p_i^* > 0\}$, it must have

$$0 < p_i \leq \frac{2n\mu}{\sup_{(\mathbf{p}^*,\omega^*)\in D^*} \omega_i^*}.$$

This leads to

$$\max_{i\in\mathcal{S}} p_i \leq \frac{2n\mu}{\min_{i\in\mathcal{S}} \sup_{(\mathbf{p}^*,\omega^*)\in D^*} \omega_i^*}.$$

Similarly,

$$\max_{i\in\mathcal{B}} \omega_i \leq \frac{2n\mu}{\min_{i\in\mathcal{B}} \sup_{(\mathbf{p}^*,\omega^*)\in P^*} p_i^*}.$$

Combining these two inequalities gives

$$\begin{aligned}
&\max\{\max_{i\in\mathcal{S}} p_i, \max_{i\in\mathcal{B}} \omega_i\} \\
\leq\; & \frac{2n\mu}{\min\{\min_{i\in\mathcal{S}} \sup_{(\mathbf{p}^*,\omega^*)\in D^*} \omega_i^*, \min_{i\in\mathcal{B}} \sup_{(\mathbf{p}^*,\omega^*)\in P^*} p_i^*\}} \\
=\; & \frac{\mu}{C_1}. \qquad (10.99)
\end{aligned}$$

This proves (10.95). Finally, since $p_i\omega_i \geq \rho\mu$, we have, for any $i \in \mathcal{S}$,

$$\omega_i \geq \frac{\rho\mu}{p_i} \geq \frac{\rho\mu}{\mu/C_1} = C_2\rho.$$

Similarly, for any $i \in \mathcal{B}$,

$$p_i \geq \frac{\rho\mu}{\omega_i} \geq \frac{\rho\mu}{\mu/C_1} = C_2\rho.$$

This completes the proof. ∎

Lemma 10.15 leads to the following

Theorem 10.4
Let $(\mathbf{x}^k, \mathbf{p}^k, \omega^k) \in \mathcal{N}_2(\theta)$ *be generated by Algorithm 10.1. Assume that the convex QP with box constraints has strictly complementary solution(s). Then, every limit point of the sequence is a strictly complementary solution of the convex quadratic programming with box constraints, i.e.,*

$$\omega_i^* \geq C_2\rho \quad (i \in \mathcal{S}), \qquad p_i^* \geq C_2\rho \quad (i \in \mathcal{B}). \tag{10.100}$$

Proof 10.18 From Lemma 10.15, $(\mathbf{p}^k, \omega^k)$ is bounded; therefore, there is at least one limit point $(\mathbf{p}^*, \omega^*)$. Since (p_i^k, ω_i^k) is in the neighborhood of the central path, i.e., $p_i^k \omega_i^k > \rho\mu_k := (1-\theta)\mu_k$,

$$\omega_i^k \geq C_2\rho \quad (i \in \mathcal{S}), \qquad p_i^k \geq C_2\rho \quad (i \in \mathcal{B}),$$

every limit point will meet (10.100) due to the fact that $C_2\rho$ is a constant. ∎

We are now ready to prove the complexity bound of Algorithm 10.1. Combining Lemma 10.13 and Theorem 1.4 gives

Theorem 10.5
The complexity of Algorithm 10.1 is bounded by $O(\sqrt{n}\log(1/\varepsilon))$.

10.4 Implementation Issues

Algorithm 10.1 is presented in a form that is convenient for the convergence analysis. Some implementation details that make the algorithm more efficient are discussed in this section.

10.4.1 Termination criterion

Algorithm 10.1 needs a termination criterion in real implementation. One can use

$$\mu_k \leq \varepsilon, \tag{10.101a}$$

$$\|\mathbf{r}_X\| = \|\mathbf{H}\mathbf{x}^k + \lambda^k - \gamma^k + \mathbf{c}\| \leq \varepsilon, \tag{10.101b}$$

$$\|\mathbf{r}_Y\| = \|\mathbf{x}^k + \mathbf{y}^k - \mathbf{e}\| \leq \varepsilon, \tag{10.101c}$$

$$\|\mathbf{r}_Z\| = \|\mathbf{x}^k - \mathbf{z}^k + \mathbf{e}\| \leq \varepsilon, \tag{10.101d}$$

$$(\mathbf{p}^k, \omega^k) > \mathbf{0}. \tag{10.101e}$$

An alternate criterion is similar to the one used in `linprog` [172]

$$\kappa := \frac{\|\mathbf{r}_Y\| + \|\mathbf{r}_Z\|}{2n} + \frac{\|\mathbf{r}_X\|}{\max\{1, \|\mathbf{c}\|\}} + \frac{\mu_k}{\max\{1, \|\mathbf{x}^{k^{\mathrm{T}}}\mathbf{H}\mathbf{x}^k + \mathbf{c}^{\mathrm{T}}\mathbf{x}^k\|\}} \le \varepsilon. \quad (10.102)$$

10.4.2 A feasible initial point

For feasible interior-point algorithms, an important prerequisite is to start with a feasible interior point. While finding an *initial feasible point* may not be a simple and trivial task for even linear programming with equality constraints [144], for quadratic programming subject to box constraints, finding the initial point is not an issue. As a matter of fact, the following initial point $(\mathbf{x}^0, \mathbf{y}^0, \mathbf{z}^0, \lambda^0, \gamma^0)$ is an interior point, moreover, $(\mathbf{x}^0, \mathbf{y}^0, \mathbf{z}^0, \lambda^0, \gamma^0) \in \mathcal{N}_2(\theta)$.

$$\mathbf{x}^0 = 0, \;\; \mathbf{y}^0 = \mathbf{z}^0 = \mathbf{e} > \mathbf{0}, \quad (10.103a)$$

$$\lambda_i^0 = 4(1 + \|\mathbf{c}\|^2) - \frac{c_i}{2} > \mathbf{0}, \quad (10.103b)$$

$$\gamma_i^0 = 4(1 + \|\mathbf{c}\|^2) + \frac{c_i}{2} > \mathbf{0}. \quad (10.103c)$$

It is easy to see that this selected point meets (10.5). First, the following relations, $\mathbf{H}\mathbf{x}^0 + \mathbf{c} + \lambda^0 - \gamma^0 = \mathbf{0}$, $(\mathbf{y}^0, \mathbf{z}^0, \lambda^0, \gamma^0) > \mathbf{0}$, $\mathbf{x}^0 + \mathbf{y}^0 = \mathbf{e}$, and $\mathbf{x}^0 - \mathbf{z}^0 = -\mathbf{e}$ are easy to verify. Since

$$\mu_0 = \frac{\sum_{i=1}^n (\lambda_i^0 + \gamma_i^0)}{2n} = \frac{\sum_{i=1}^n (8(1 + \|\mathbf{c}\|^2))}{2n} = 4(1 + \|\mathbf{c}\|^2), \quad (10.104)$$

for $\theta = 0.19$, it must have

$$\begin{aligned} \left\| \mathbf{p}^0 \circ \omega^0 - \mu_0 \mathbf{e} \right\|^2 &= \sum_{i=1}^n (\lambda_i^0 - \mu_0)^2 + \sum_{i=1}^n (\gamma_i^0 - \mu_0)^2 \\ &= \frac{\|\mathbf{c}\|^2}{2} \le 16\theta^2 (1 + \|\mathbf{c}\|^2)^2 = \theta^2 (\mu_0)^2. \end{aligned}$$

This shows that $(\mathbf{x}^0, \mathbf{y}^0, \mathbf{z}^0, \lambda^0, \gamma^0) \in \mathcal{N}_2(\theta)$.

10.4.3 Step size

Directly using $\sin(\alpha) = \frac{\theta}{\sqrt{n}}$ in Algorithm 10.1 provides an effective formula to prove the polynomiality. However, this choice of $\sin(\alpha)$ is too conservative in practice because this search step in $\mathcal{N}_2(2\theta)$ is too small and the speed of duality measure reduction is slow. A better choice of $\sin(\alpha)$ should have a larger step in every iteration so that the polynomiality is reserved and fast convergence is achieved. In view of Remark 10.3, *conditions that restrict step size are positivity conditions, proximity conditions, and duality reduction condition*. This subsection examines how to enlarge the step size under these restrictions.

First, from (10.59) and (10.76), $\mu(\alpha) > 0$ is required for positivity conditions $(\mathbf{p}(\alpha), \omega(\alpha)) > \mathbf{0}$ and $(\mathbf{p}^{k+1}, \omega^{k+1}) > \mathbf{0}$ to hold. Since $\sin(\bar{\alpha})$, estimated in Corollary 10.1, is conservative, a better selection of $\bar{\alpha}$ is to use (10.45), Lemmas 5.4 and 10.2:

$$\begin{aligned}
\mu(\alpha) &\geq \mu(1-\sin(\alpha)) - \frac{1}{2n}\dot{\mathbf{x}}^{\mathrm{T}}\mathbf{H}\dot{\mathbf{x}}\left((1-\cos(\alpha))^2+\sin^2(\alpha)\right) \\
&\geq \mu(1-\sin(\alpha)) - \frac{1}{2n}(\dot{\mathbf{p}}^{\mathrm{T}}\dot{\omega})\Big(\sin^4(\alpha)+\sin^2(\alpha)\Big) \\
&:= f(\sin(\alpha)) = \sigma, \qquad (10.105)
\end{aligned}$$

where $\sigma > 0$ is a small number, and $f(\sin(\alpha))$ is a monotonic decreasing function of $\sin(\alpha)$ with $f(\sin(0)) = \mu > 0$ and $f(\sin(\frac{\pi}{2})) < 0$. Therefore, equation (10.105) has a unique positive real solution for $\alpha \in [0, \frac{\pi}{2}]$. Since (10.105) is a quartic function of $\sin(\alpha)$, the cost of finding the smallest positive solution is negligible [113].

Second, in view of (10.75), the *proximity condition* for

$$(\mathbf{x}^{k+1}, \mathbf{y}^{k+1}, \mathbf{z}^{k+1}, \lambda^{k+1}, \gamma^{k+1}) \in \mathcal{N}_2(\theta)$$

holds for $\theta \leq 0.19$ without further restriction. The proximity condition (10.58) is met for $\sin(\alpha) \in [0, \sin(\tilde{\alpha})]$, where $\sin(\tilde{\alpha})$ is the smallest positive solution of (10.53) and it is estimated very conservatively in Lemma 10.9. An efficient implementation should use $\sin(\tilde{\alpha})$, the smallest positive solution of (10.53). Actually, there exists a $\acute{\alpha}$ which is normally larger than $\tilde{\alpha}$ such that the proximity condition (10.58) is met for $\sin(\alpha) \in [0, \sin(\acute{\alpha})]$. Let

$$b_0 = -\theta\mu < 0,$$

$$b_1 = \theta\mu > 0,$$

$$\begin{aligned}
b_3 &= \left\|\dot{\mathbf{p}}\circ\ddot{\omega} + \dot{\omega}\circ\ddot{\mathbf{p}} - \tfrac{1}{2n}(\dot{\mathbf{p}}^{\mathrm{T}}\ddot{\omega} + \dot{\omega}^{\mathrm{T}}\ddot{\mathbf{p}})\mathbf{e}\right\| + \tfrac{\theta}{n}\left(\dot{\mathbf{p}}^{\mathrm{T}}\ddot{\omega} + \ddot{\mathbf{p}}^{\mathrm{T}}\dot{\omega}\right) \\
&= a_3 + \tfrac{\theta}{n}\left(\dot{\mathbf{p}}^{\mathrm{T}}\ddot{\omega} + \ddot{\mathbf{p}}^{\mathrm{T}}\dot{\omega}\right), \\
b_4 &= \left\|\ddot{\mathbf{p}}\circ\ddot{\omega} - \dot{\omega}\circ\dot{\mathbf{p}} - \tfrac{1}{2n}(\ddot{\mathbf{p}}^{\mathrm{T}}\ddot{\omega} - \dot{\omega}^{\mathrm{T}}\dot{\mathbf{p}})\mathbf{e}\right\| - \tfrac{\theta}{n}\left(\ddot{\mathbf{p}}^{\mathrm{T}}\ddot{\omega} - \dot{\mathbf{p}}^{\mathrm{T}}\dot{\omega}\right) \\
&= a_4 - \tfrac{\theta}{n}\ddot{\mathbf{p}}^{\mathrm{T}}\ddot{\omega},
\end{aligned}$$

and

$$p(\alpha) := b_4(1-\cos(\alpha))^2 + b_3\sin(\alpha)(1-\cos(\alpha)) + b_1\sin(\alpha) + b_0. \qquad (10.106)$$

Applying the second inequality of (10.30) to $\frac{\theta}{n}\left(\dot{\mathbf{p}}^{\mathrm{T}}\ddot{\omega} + \ddot{\mathbf{p}}^{\mathrm{T}}\dot{\omega}\right)\sin(\alpha)(1 - \cos(\alpha))$, we have

$$\begin{aligned}
& b_3\sin(\alpha)(1-\cos(\alpha)) \\
=\ & \left[a_3 + \tfrac{\theta}{n}\left(\dot{\mathbf{p}}^{\mathrm{T}}\ddot{\omega} + \ddot{\mathbf{p}}^{\mathrm{T}}\dot{\omega}\right)\right]\sin(\alpha)(1-\cos(\alpha)) \\
\leq\ & a_3\sin^3(\alpha) + \tfrac{\theta}{n}\left(\dot{\mathbf{x}}^{\mathrm{T}}\mathbf{H}\dot{\mathbf{x}}\sin^2(\alpha) + \ddot{\mathbf{x}}^{\mathrm{T}}\mathbf{H}\ddot{\mathbf{x}}(1-\cos(\alpha))^2\right).
\end{aligned}$$

Therefore,

$$
\begin{array}{rl}
& b_4(1-\cos(\alpha))^2 + b_3\sin(\alpha)(1-\cos(\alpha)) \\
\leq & \left[a_4 - \frac{\theta}{n}\ddot{\mathbf{x}}^{\mathrm{T}}\mathbf{H}\ddot{\mathbf{x}}\right](1-\cos(\alpha))^2 \\
& + a_3\sin^3(\alpha) + \frac{\theta}{n}\left(\dot{\mathbf{x}}^{\mathrm{T}}\mathbf{H}\dot{\mathbf{x}}\sin^2(\alpha) + \ddot{\mathbf{x}}^{\mathrm{T}}\mathbf{H}\ddot{\mathbf{x}}(1-\cos(\alpha))^2\right) \\
= & a_4(1-\cos(\alpha))^2 + a_3\sin^3(\alpha) + \frac{\theta}{n}\dot{\mathbf{x}}^{\mathrm{T}}\mathbf{H}\dot{\mathbf{x}}\sin^2(\alpha) \\
\leq & a_4\sin^4(\alpha) + a_3\sin^3(\alpha) + a_2\sin^2(\alpha).
\end{array}
$$

This shows that

$$
\begin{array}{rcl}
p(\alpha) & = & b_4(1-\cos(\alpha))^2 + b_3\sin(\alpha)(1-\cos(\alpha)) + b_1\sin(\alpha) + b_0 \\
& \leq & a_4\sin^4(\alpha) + a_3\sin^3(\alpha) + a_2\sin^2(\alpha) + a_1\sin(\alpha) + a_0 \\
& = & q(\alpha)
\end{array}
$$

where $q(\alpha)$ is defined in (10.53). Therefore, the smallest positive solution $\grave{\alpha}$ of $p(\alpha)$ is larger than the smallest positive solution $\tilde{\alpha}$ of $q(\alpha)$. Hence, the goal is to show that for $\sin(\alpha) \in [0, \sin(\grave{\alpha})]$, the proximity condition (10.58) holds. Since for $\sin(\alpha) \in [0, \sin(\grave{\alpha})]$, $p(\alpha) \leq 0$ is equivalent to

$$
\begin{array}{rl}
& \left\|\ddot{\mathbf{p}}\circ\ddot{\omega} - \dot{\omega}\circ\dot{\mathbf{p}} - \frac{1}{2n}(\ddot{\mathbf{p}}^{\mathrm{T}}\ddot{\omega} - \dot{\omega}^{\mathrm{T}}\dot{\mathbf{p}})\mathbf{e}\right\|(1-\cos(\alpha))^2 \\
& + \left\|\dot{\mathbf{p}}\circ\ddot{\omega} + \dot{\omega}\circ\ddot{\mathbf{p}} - \frac{1}{2n}(\dot{\mathbf{p}}^{\mathrm{T}}\ddot{\omega} + \dot{\omega}^{\mathrm{T}}\ddot{\mathbf{p}})\mathbf{e}\right\|\sin(\alpha)(1-\cos(\alpha)) \\
\leq & (2\theta)\left(\frac{1}{2n}\left(\ddot{\mathbf{p}}^{\mathrm{T}}\ddot{\omega} - \dot{\mathbf{p}}^{\mathrm{T}}\dot{\omega}\right)(1-\cos(\alpha))^2 - \frac{1}{2n}\left(\dot{\mathbf{p}}^{\mathrm{T}}\ddot{\omega} + \ddot{\mathbf{p}}^{\mathrm{T}}\dot{\omega}\right)\sin(\alpha)(1-\cos(\alpha))\right) \\
& + \theta\mu(1-\sin(\alpha)).
\end{array}
\tag{10.107}
$$

Substituting this inequality into (10.57) and using the formula for $\mu(\alpha)$ in Lemma 10.45, we have

$$
\begin{array}{rl}
& \left\|\mathbf{p}(\alpha)\circ\omega(\alpha) - \mu(\alpha)\mathbf{e}\right\| \\
\leq & (1-\sin(\alpha))\left\|\mathbf{p}\circ\omega - \mu\mathbf{e}\right\| \\
& + \left\|(\ddot{\mathbf{p}}\circ\ddot{\omega} - \dot{\mathbf{p}}\circ\dot{\omega} - \frac{1}{2n}(\ddot{\mathbf{p}}^{\mathrm{T}}\ddot{\omega} - \dot{\mathbf{p}}^{\mathrm{T}}\dot{\omega}))\mathbf{e}\right\|(1-\cos(\alpha))^2 \\
& + \left\|(\dot{\mathbf{p}}\circ\ddot{\omega} + \dot{\omega}\circ\ddot{\mathbf{p}} - \frac{1}{2n}(\dot{\mathbf{p}}^{\mathrm{T}}\ddot{\omega} + \ddot{\mathbf{p}}^{\mathrm{T}}\dot{\omega})\mathbf{e}\right\|\sin(\alpha)(1-\cos(\alpha)) \\
\leq & \theta\mu(1-\sin(\alpha)) \\
& + \frac{\theta}{n}\left[\left(\ddot{\mathbf{p}}^{\mathrm{T}}\ddot{\omega} - \dot{\mathbf{p}}^{\mathrm{T}}\dot{\omega}\right)(1-\cos(\alpha))^2 - \left(\dot{\mathbf{p}}^{\mathrm{T}}\ddot{\omega} + \ddot{\mathbf{p}}^{\mathrm{T}}\dot{\omega}\right)\sin(\alpha)(1-\cos(\alpha))\right] \\
& + \theta\mu(1-\sin(\alpha)) \\
\leq & 2\theta\left[\mu(1-\sin(\alpha)) + \frac{1}{2n}\left(\ddot{\mathbf{x}}^{\mathrm{T}}(\ddot{\gamma} - \ddot{\lambda}) - \dot{\mathbf{x}}^{\mathrm{T}}(\dot{\gamma} - \dot{\lambda})\right)(1-\cos(\alpha))^2\right. \\
& \left. - \frac{1}{2n}\left(\dot{\mathbf{x}}^{\mathrm{T}}(\ddot{\gamma} - \ddot{\lambda}) + \ddot{\mathbf{x}}^{\mathrm{T}}(\dot{\gamma} - \dot{\lambda})\right)\sin(\alpha)(1-\cos(\alpha))\right] \\
= & 2\theta\mu(\alpha). \qquad \text{[use Lemma 10.45]}
\end{array}
\tag{10.108}
$$

This is the proximity condition for $(\mathbf{x}(\alpha),\mathbf{y}(\alpha),\mathbf{z}(\alpha),\lambda(\alpha),\gamma(\alpha))$. Therefore, the smallest positive solution $\grave{\alpha}$ of $p(\alpha)$ gives a better condition for the proximity condition (10.108) to hold. However, $p(\alpha)$ may not be a monotonic increasing function, therefore, it may not be solved efficiently. Denote $\hat{b}_0 = b_0$, $\hat{b}_1 = b_1$,

$$\hat{b}_3 = \begin{cases} b_3 & \text{if } b_3 \geq 0, \\ 0 & \text{if } b_3 < 0, \end{cases} \qquad \hat{b}_4 = \begin{cases} b_4 & \text{if } b_4 \geq 0, \\ 0 & \text{if } b_4 < 0, \end{cases}$$

and

$$\hat{p}(\alpha) := \hat{b}_4(1-\cos(\alpha))^2 + \hat{b}_3\sin(\alpha)(1-\cos(\alpha)) + \hat{b}_1\sin(\alpha) + \hat{b}_0. \quad (10.109)$$

Since $\hat{p}(\alpha) \geq p(\alpha)$, the smallest positive solution $\acute{\alpha}$ of $\hat{p}(\alpha)$ is smaller than smallest positive solution $\grave{\alpha}$ of $p(\alpha)$. To estimate the smallest solution of $\acute{\alpha}$, by noticing that $\hat{p}(\alpha)$ is a monotonic increasing function of α and $\hat{p}(0) = -\theta\mu < 0$, one can simply use the bisection method. The computational cost is independent of the problem size n and is negligible. Since both estimated step sizes $\acute{\alpha}$ and $\tilde{\alpha}$ guarantee the proximity condition for $(\mathbf{x}(\alpha),\mathbf{y}(\alpha),\mathbf{z}(\alpha),\lambda(\alpha),\gamma(\alpha))$ to hold, one can select $\breve{\alpha} = \max\{\acute{\alpha},\tilde{\alpha}\} \geq \tilde{\alpha}$, which guarantees that the polynomiality claim will hold.

Third, for duality measure reduction, from the first inequality of (10.29), we have

$$\begin{aligned} & -\left[\ddot{\mathbf{x}}^{\mathrm{T}}(\dot{\gamma}-\dot{\lambda}) + \dot{\mathbf{x}}^{\mathrm{T}}(\ddot{\gamma}-\ddot{\lambda})\right]\sin(\alpha)(1-\cos(\alpha)) \\ \leq\ & (\dot{\mathbf{x}}^{\mathrm{T}}\mathbf{H}\dot{\mathbf{x}})(1-\cos(\alpha))^2 + (\ddot{\mathbf{x}}^{\mathrm{T}}\mathbf{H}\ddot{\mathbf{x}})\sin^2(\alpha). \end{aligned}$$

Substituting this relation into $\mu(\alpha)$ in Lemma 10.5 and using Lemmas 10.2 and 5.4 yields

$$\begin{aligned} \mu(\alpha) &= \mu(1-\sin(\alpha)) + \tfrac{1}{2n}\left(\ddot{\mathbf{x}}^{\mathrm{T}}(\ddot{\gamma}-\ddot{\lambda}) - \dot{\mathbf{x}}^{\mathrm{T}}(\dot{\gamma}-\dot{\lambda})\right)(1-\cos(\alpha))^2 \\ &\quad -\tfrac{1}{2n}\left(\dot{\mathbf{x}}^{\mathrm{T}}(\ddot{\gamma}-\ddot{\lambda}) + \ddot{\mathbf{x}}^{\mathrm{T}}(\dot{\gamma}-\dot{\lambda})\right)\sin(\alpha)(1-\cos(\alpha)) \\ &\leq \mu(1-\sin(\alpha)) + \tfrac{1}{2n}\left(\ddot{\mathbf{x}}^{\mathrm{T}}(\ddot{\gamma}-\ddot{\lambda}) - \dot{\mathbf{x}}^{\mathrm{T}}(\dot{\gamma}-\dot{\lambda})\right)(1-\cos(\alpha))^2 \\ &\quad +\tfrac{1}{2n}\left[(\dot{\mathbf{x}}^{\mathrm{T}}\mathbf{H}\dot{\mathbf{x}})(1-\cos(\alpha))^2 + (\ddot{\mathbf{x}}^{\mathrm{T}}\mathbf{H}\ddot{\mathbf{x}})\sin^2(\alpha)\right] \\ &\leq \mu(1-\sin(\alpha)) + \tfrac{1}{2n}\ddot{\mathbf{x}}^{\mathrm{T}}(\ddot{\gamma}-\ddot{\lambda})(1-\cos(\alpha))^2 + \tfrac{1}{2n}(\ddot{\mathbf{x}}^{\mathrm{T}}\mathbf{H}\ddot{\mathbf{x}})\sin^2(\alpha) \\ &\leq \mu(1-\sin(\alpha)) + \tfrac{1}{2n}\ddot{\mathbf{x}}^{\mathrm{T}}\mathbf{H}\ddot{\mathbf{x}}\left[\sin^4(\alpha) + \sin^2(\alpha)\right] \end{aligned}$$

Substituting this inequality into (10.88a) yields

$$\begin{aligned} \mu_{k+1} &\leq \mu(\alpha)\left(1 + \frac{\theta^2(1+2\theta)}{n(1-2\theta)^2}\right) \\ &\leq \mu_k\left[1 + \frac{\theta^2(1+2\theta)}{n(1-2\theta)^2} - \left(1 + \frac{\theta^2(1+2\theta)}{n(1-2\theta)^2}\right)\sin(\alpha)\right. \\ &\quad \left. + \left(1 + \frac{\theta^2(1+2\theta)}{n(1-2\theta)^2}\right)\frac{\ddot{\mathbf{p}}^{\mathrm{T}}\ddot{\omega}}{2n\mu}\left(\sin^2(\alpha) + \sin^4(\alpha)\right)\right]. \quad (10.110) \end{aligned}$$

For $\mu_{k+1} \le \mu_k$ to hold, one needs

$$\frac{\theta^2(1+2\theta)}{n(1-2\theta)^2} - \left(1+\frac{\theta^2(1+2\theta)}{n(1-2\theta)^2}\right)\sin(\alpha) + \left(1+\frac{\theta^2(1+2\theta)}{n(1-2\theta)^2}\right)\frac{\ddot{\mathbf{p}}^{\mathrm{T}}\ddot{\omega}}{2n\mu}\left(\sin^2(\alpha)+\sin^4(\alpha)\right) \le 0.$$

For the sake of convenience in convergence analysis, a conservative estimate is used in Lemma 10.13. For an efficient implementation, the following solution should be adopted. Denote $u = \frac{\theta^2(1+2\theta)}{n(1-2\theta)^2} > 0$, $v = \frac{\ddot{\mathbf{p}}^{\mathrm{T}}\ddot{\omega}}{2n\mu} > 0$, $z = \sin(\alpha) \in [0,1]$, and

$$F(z) = (1+u)vz^4 + (1+u)vz^2 - (1+u)z + u.$$

For $z \in [0,1]$ and $v \le \frac{1}{6}$, $F'(z) = (1+u)(4vz^3+2vz-1) \le 0$; therefore, the upper bound of the duality measure μ_{k+1} is a monotonic decreasing function of $\sin(\alpha)$ for $\alpha \in [0, \frac{\pi}{2}]$. The larger α is, the smaller the upper bound of the duality measure will be. For $v > \frac{1}{6}$, in order to minimize the upper bound of the duality measure, one can find the solution of $F'(z) = 0$. It is easy to check from discriminator in Lemma 9.7 that the cubic polynomial $F'(z)$ has only one real solution which is given by

$$\begin{aligned} \sin(\breve{\alpha}) &= \sqrt[3]{\frac{n\mu}{4\ddot{\mathbf{p}}^{\mathrm{T}}\ddot{\omega}} + \sqrt{\left(\frac{n\mu}{4\ddot{\mathbf{p}}^{\mathrm{T}}\ddot{\omega}}\right)^2 + \left(\frac{1}{6}\right)^3}} + \sqrt[3]{\frac{n\mu}{4\ddot{\mathbf{p}}^{\mathrm{T}}\ddot{\omega}} - \sqrt{\left(\frac{n\mu}{4\ddot{\mathbf{p}}^{\mathrm{T}}\ddot{\omega}}\right)^2 + \left(\frac{1}{6}\right)^3}} \\ &:= \chi. \end{aligned} \tag{10.111}$$

Since $F''(\sin(\breve{\alpha})) = (1+u)(12v\sin^2(\breve{\alpha})+2v) > 0$ at $\sin(\breve{\alpha}) \in [0,1)$, the upper bound of the duality measure is minimized. Therefore, one can define

$$\breve{\alpha} = \begin{cases} \frac{\pi}{2}, & \text{if } \frac{\ddot{\mathbf{p}}^{\mathrm{T}}\ddot{\omega}}{2n\mu} \le \frac{1}{6} \\ \sin^{-1}(\chi), & \text{if } \frac{\ddot{\mathbf{p}}^{\mathrm{T}}\ddot{\omega}}{2n\mu} > \frac{1}{6}. \end{cases} \tag{10.112}$$

It is worthwhile to note that for $\alpha < \breve{\alpha}$, $F'(\sin(\alpha)) < 0$, i.e., $F(\sin(\alpha))$ is a monotonic decreasing function of $\alpha \in [0, \breve{\alpha}]$.

The step size selection process, is therefore, a simple algorithm, given below.

Algorithm 10.2
(Step Size Selection)
Data: $\sigma > 0$.
Step 1: Find the positive real solution of (10.105) to get $\sin(\bar{\alpha})$.
Step 2: Find the smallest positive real solution of (10.109) to get $\sin(\acute{\alpha})$, *the smallest positive real solution of (10.53) to get* $\sin(\tilde{\alpha})$, *and set* $\sin(\check{\alpha}) = \max\{\sin(\tilde{\alpha}), \sin(\acute{\alpha})\}$.
Step 3: Calculate $\breve{\alpha}$ *given by (10.112).*
Step 4: The step size is obtained as $\sin(\alpha) = \min\{\sin(\bar{\alpha}), \sin(\check{\alpha}), \sin(\breve{\alpha})\}$.

10.4.4 The practical implementation

Therefore, a more efficient implementation than Algorithm 10.1 is the following algorithm:

Algorithm 10.3
(Arc-search path-following)
Data: $\mathbf{H} \succeq \mathbf{0}$, $\mathbf{c}$, n, $\theta = 0.19$, $\varepsilon > \sigma > 0$.
Step 0: Find initial point $(\mathbf{x}^0, \mathbf{p}^0, \omega^0) \in \mathcal{N}_2(\theta)$ *using (10.103),* κ *using (10.102), and* μ_0 *using (10.104).*
while $\kappa > \varepsilon$

Step 1: Compute $(\dot{\mathbf{x}}, \dot{\mathbf{p}}, \dot{\omega})$ *and* $(\ddot{\mathbf{x}}, \ddot{\mathbf{p}}, \ddot{\omega})$ *using (10.12) and (10.13).*

Step 2: Select $\sin(\alpha)$ *using Algorithm 10.2. Update* $(\mathbf{x}(\alpha), \mathbf{p}(\alpha), \omega(\alpha))$ *and* $\mu(\alpha)$ *using (10.54) and (10.55).*

Step 3: Compute $(\Delta\mathbf{x}, \Delta\mathbf{p}, \Delta\omega)$ *using (10.66), update* $(\mathbf{x}^{k+1}, \mathbf{p}^{k+1}, \omega^{k+1})$ *and* μ_{k+1} *using (10.67) and (10.68).*

Step 4: Computer κ *using (10.102).*

Step 5: If $\kappa \le \varepsilon$*, quit. Otherwise, set* $k+1 \to k$*. Go back to Step 1.*

end (while)

Remark 10.5 The condition $\mu > \sigma$ guarantees that the equation (10.105) has a positive solution before termination criterion is met. ■

10.5 A Design Example

We claim that the convex quadratic programming problem with box constraints has a lot of applications. One example is the widely-used linear quadratic regulator with control saturation. In this section, the OrbView-2 spacecraft orbit-raising design example discussed in [160] is used to demonstrate the effectiveness and efficiency of the proposed algorithm[1]. Let $\mathbf{w} = (w_x, w_y, w_z)$ be the spacecraft body rate with respect to the reference frame expressed in the body frame, $\bar{\mathbf{q}} = (q_0, q_1, q_2, q_3)$ be the quaternion of the spacecraft attitude with respect to the reference frame represented in the body frame and $\mathbf{q} = (q_1, q_2, q_3)$ be the reduced quaternion, $\mathbf{J} = \mathrm{diag}(J_x, J_y, J_z)$ be the spacecraft inertia matrix, and h_w be the angular momentum produced by a momentum wheel. Orbit-raising is performed by 4 fixed thrusters (1 Newton) with on/off switches which are mounted on the anti-nadir face of the spacecraft in each corner of a square with a side length of $2d$

[1] There are algorithms to solve this type of problem [9], but the new method is more efficient.

meter. The thrusters point to $+z$ direction and canted 5 degree from z-axis. (more details were provided in [160]). The matrices of the thruster force direction $\mathbf{F}$ and moment arms $\mathbf{R}$ in the body frame are given as

$$\mathbf{F} = [\mathbf{f}_1, \mathbf{f}_2, \mathbf{f}_3, \mathbf{f}_4] = \begin{bmatrix} -a & -a & a & a \\ a & -a & -a & a \\ 1 & 1 & 1 & 1 \end{bmatrix}$$

$$\mathbf{R} = [\mathbf{r}_1, \mathbf{r}_2, \mathbf{r}_3, \mathbf{r}_4] = \begin{bmatrix} -d & -d & d & d \\ -d & d & d & -d \\ -\ell & -\ell & -\ell & -\ell \end{bmatrix}.$$

Let $\mathbf{x} = (w_x, w_y, w_z, q_1, q_2, q_3)$ be the states of the attitude and $\mathbf{u} = (T_1, T_2, T_3, T_4)$ be the control variable with T_1, T_2, T_3, T_4 the thrust level of the four thrusters. The linear time-invariant system under consideration is represented in a reduced quaternion model (see [160]).

$$\begin{aligned} \dot{\mathbf{x}} &= \begin{bmatrix} 0 & 0 & \frac{h_w}{J_x} & 0 & 0 & 0 \\ 0 & 0 & 0 & 0 & 0 & 0 \\ \frac{h_w}{J_z} & 0 & 0 & 0 & 0 & 0 \\ 0.5 & 0 & 0 & 0 & 0 & 0 \\ 0 & 0.5 & 0 & 0 & 0 & 0 \\ 0 & 0 & 0.5 & 0 & 0 & 0 \end{bmatrix} \mathbf{x} \\ &+ \begin{bmatrix} \frac{1}{J_x} 0 & 0 & \\ 0 & \frac{1}{J_y} & 0 \\ 0 & 0 & \frac{1}{J_z} \\ 0 & 0 & 0 \\ 0 & 0 & 0 \\ 0 & 0 & 0 \end{bmatrix} \begin{bmatrix} \mathbf{r}_1 \times \mathbf{f}_1 \\ \mathbf{r}_2 \times \mathbf{f}_2 \\ \mathbf{r}_3 \times \mathbf{f}_3 \\ \mathbf{r}_4 \times \mathbf{f}_4 \end{bmatrix}^{\mathrm{T}} \begin{bmatrix} T_1 \\ T_2 \\ T_3 \\ T_4 \end{bmatrix} \\ &= \mathbf{A}\mathbf{x} + \mathbf{B}\mathbf{u}, \end{aligned} \tag{10.113}$$

with the control constraints

$$-\mathbf{e} \le \mathbf{u} = (T_1, T_2, T_3, T_4) \le \mathbf{e}. \tag{10.114}$$

The problem is converted to discrete model using MATLAB function `c2d` with sampling time 1 second. The design is to minimize

$$J = \min_{\mathbf{u}_0, \mathbf{u}_1, \cdots, \mathbf{u}_{N-1}} \frac{1}{2} \mathbf{x}_N^{\mathrm{T}} \mathbf{P} \mathbf{x}_N + \frac{1}{2} \sum_{k=0}^{N-1} \left[\mathbf{x}_k^{\mathrm{T}} \mathbf{Q} \mathbf{x}_k + \mathbf{u}_k^{\mathrm{T}} \mathbf{R} \mathbf{u}_k, \right] \tag{10.115}$$

where the horizon number $N = 30$, the matrices $\mathbf{P}$, $\mathbf{Q}$, and $\mathbf{R}$ are given by

$$\mathbf{P} = \mathbf{Q} = \begin{bmatrix} \frac{1}{2.5}\mathbf{I}_3 & \mathbf{0} \\ \mathbf{0} & 10000\mathbf{I}_3 \end{bmatrix}, \quad \mathbf{R} = \mathbf{I}_6.$$

Other spacecraft parameters ($d = 0.248$m, $\ell = 0.815$m, $I_x = 189kg.m^2$, $I_y = 159kg.m^2$, and $I_z = 114kg.m^2$, and $h_w = -2.8N.m.s$) are the same as the ones of [128]. The algorithm is implemented in MATLAB. In our implementation of Algorithm 10.3, $\varepsilon = 10^{-6}$ and $\sigma = 10^{-10}$ are selected. After 20 iterations, the algorithm converges. The optimal thrust controller design is obtained.

10.6 Concluding Remarks

As explained at the beginning of the chapter, there is no need to work on infeasible interior-point algorithms for convex quadratic programming problem with box constraints because a feasible initial point is always available. There is, however, a need to develop feasible interior-point algorithms that search for the optimizer in a wider neighborhood as this strategy may lead to better algorithms than the one proposed in this chapter.

Chapter 11

An Arc-Search Algorithm for LCP

Linear complementarity problem (LCP) [21] is a class of optimization problems that includes linear programming, quadratic programming, and linear complementarity problem. Because of the great success of interior-point method in linear programming [82, 83, 89], researchers quickly turned their strike direction to solve LCP using interior-point method and achieved the expected success [72, 171].

After arc-search method demonstrated its superiority in interior-point method [152, 153], quite a few papers on arc-search interior-point algorithms for LCP have been published since 2015. The first one is probably due to Kheirfam and Chitsaz [64], which consider a special LCP problem, the so called $P_*(\kappa)$-linear complementarity problem. Pirhaji, Zangiabadi, and Mansouri [110] discussed a ℓ_2-neighborhood infeasible interior-point algorithm for LCP. Pirhaji and Kheirfam developed an arc-search infeasible interior-point algorithm for horizontal linear complementarity problem in the $N_{-\infty}$ neighborhood of the central path [61]. Zangiabadi, Mansouri, and Amin [112] proposed an arc-search interior-point algorithm for monotone linear complementary problem over symmetric cones. Shahraki et al. [123] discussed a wide neighborhood infeasible-interior-point method with arc-search for linear complementarity problem over symmetric cones (SCLCPs). Yuan, Zhang, and Huang [168] devised a wide neighborhood interior-point algorithm with arc-search for $P_*(k)$ LCP. Shahraki and Delavarkhalafi [122] proposed an arc-search predictor-corrector infeasible-interior-point algorithm for $P_*(k)$-LCPs. This chapter presents an arc-search feasible interior-point algorithm for LCP [167], which is tied closely to the method discussed in Chapter 9.

11.1 Linear Complementarity Programming

In Section 1.3.3, we introduced a linear complementarity problem and showed that it includes linear programming and quadratic programming. Another form of general LCP is the so-called horizontal linear complementarity problem [171]. By the horizontal linear complementarity problem, we mean the following non-linear system with non-negativity constraints:

$$\mathbf{M}\mathbf{x}+\mathbf{N}\mathbf{s}=\mathbf{q}, \quad \mathbf{x}\circ\mathbf{s}=\mathbf{0}, \quad (\mathbf{x},\mathbf{s})\geq\mathbf{0}, \tag{11.1}$$

where, given vector $\mathbf{q}\in\mathbb{R}^n$ and matrices $\mathbf{M},\mathbf{N}\in\mathbb{R}^{n\times n}$, $\mathbf{x},\mathbf{s}\in\mathbb{R}^n$ are variables to be determined. This problem is general enough to include the standard (monotone) linear complementarity problem (monotone LCP), linear, and quadratic programs. It is easy to see that when $\mathbf{N}=-\mathbf{I}$, problem (11.1) is equivalent to the monotone linear complementarity problem given in (1.38). Moreover, consider the quadratic programming:

$$\min \quad \mathbf{c}^{\mathrm{T}}\mathbf{x}+\frac{1}{2}\mathbf{x}^{\mathrm{T}}\mathbf{H}\mathbf{x}, \qquad \text{s.t.} \quad \mathbf{A}\mathbf{x}=\mathbf{b}, \quad \mathbf{x}\geq\mathbf{0}, \tag{11.2}$$

where given $\mathbf{c}\in\mathbb{R}^n$, $\mathbf{b}\in\mathbb{R}^m$, $\mathbf{A}\in\mathbb{R}^{m\times n}$, and symmetric $\mathbf{H}\in\mathbb{R}^{n\times n}$, $\mathbf{x}\in\mathbb{R}^n$ is the variable vector to be optimized. The KKT conditions of (1.37) can be rewritten as horizontal LCP with:

$$\mathbf{M}=\begin{bmatrix}\mathbf{A}\\ -\hat{\mathbf{A}}\mathbf{H}\end{bmatrix}, \quad \mathbf{N}=\begin{bmatrix}\mathbf{0}\\ \hat{\mathbf{A}}\end{bmatrix}, \quad \mathbf{q}=\begin{bmatrix}\mathbf{b}\\ \hat{\mathbf{A}}\mathbf{c}\end{bmatrix}, \tag{11.3}$$

where $\hat{\mathbf{A}}\in\mathbb{R}^{(n-m)\times n}$ is the full rank matrix that meets $\hat{\mathbf{A}}\mathbf{A}^{\mathrm{T}}=\mathbf{0}$, i.e., the rows of $\hat{\mathbf{A}}$ span the null space of $\mathbf{A}$. When $\mathbf{H}=\mathbf{0}$, the problem (11.2) is reduced to a linear programming problem.

On the other hand, horizontal linear complementarity problem can also be represented as a standard LCP. Assuming $\mathbf{N}$ is invertible, problem (11.1) can be rewritten as:

$$-\bar{\mathbf{M}}\mathbf{x}+\mathbf{s}=\bar{\mathbf{q}}, \quad \mathbf{x}\circ\mathbf{s}=\mathbf{0}, \quad (\mathbf{x},\mathbf{s})\geq\mathbf{0}, \tag{11.4}$$

which is a standard LCP with $-\bar{\mathbf{M}}=\mathbf{N}^{-1}\mathbf{M}$ and $\bar{\mathbf{q}}=\mathbf{N}^{-1}\mathbf{q}$. For this reason, in this chapter, we will discuss the standard LCP problem (1.38) which is rewritten below:

$$\mathbf{s}=\mathbf{M}\mathbf{x}+\mathbf{q}, \quad (\mathbf{x},\mathbf{s})\geq\mathbf{0}, \quad \mathbf{x}\circ\mathbf{s}=\mathbf{0}, \tag{11.5}$$

We assume, as in [69], that $\mathbf{0}\preceq\mathbf{M}\in\mathbb{R}^{n\times n}$ is positive semi-definite, and $n\geq 2$ (because $n=1$ is trivial).

11.2 An Arc-search Interior-point Algorithm

For convenience of reference, we define the feasible set of problem (11.5) by

$$\mathcal{F}=\{(\mathbf{x},\mathbf{s})\in\mathbb{R}^n\times\mathbb{R}^n \mid \mathbf{s}=\mathbf{M}\mathbf{x}+\mathbf{q}, \quad (\mathbf{x},\mathbf{s})\geq\mathbf{0}\}, \tag{11.6}$$

and the strict feasible set

$$\mathcal{F}^0 = \{(\mathbf{x},\mathbf{s}) \in \mathbb{R}^n \times \mathbb{R}^n \mid \mathbf{s} = \mathbf{M}\mathbf{x} + \mathbf{q},\ (\mathbf{x},\mathbf{s}) > \mathbf{0}\}. \tag{11.7}$$

We assume that the interior-point condition holds [118], i.e., $\mathcal{F}^0$ is not empty. The basic idea of the feasible interior-point method is to replace the $\mathbf{x} \circ \mathbf{s} = \mathbf{0}$ in (11.5) by the perturbed equation $\mathbf{x} \circ \mathbf{s} = \mu \mathbf{e}$ in order to get the following parametrized system

$$\mathbf{s} = \mathbf{M}\mathbf{x} + \mathbf{q}, \tag{11.8a}$$

$$\mathbf{x} \circ \mathbf{s} = \mu \mathbf{e}, \tag{11.8b}$$

$$\mathbf{x}, \mathbf{s} \geq \mathbf{0}. \tag{11.8c}$$

If the LCP satisfies the interior-point condition, then system (11.8) has a unique solution for every $\mu > 0$, denoted by $(\mathbf{x}(\mu), \mathbf{s}(\mu))$. The set of all such solutions forms a homotopy path, which is called the central path of the LCP and is used as a guideline to the solution of LCP [174]. Thus, the central path is an arc in $\mathbb{R}^{2n}$ parametrized as a function of μ, defined as

$$\mathcal{C} = \{(\mathbf{x}(\mu), \mathbf{s}(\mu)) \mid \mu > 0\}. \tag{11.9}$$

If $\mu \to 0$, then the limit of the central path exists and is an optimal solution for LCP (11.5).

Similar to the definition in Chapter 9, we define an ellipse $\mathcal{E}(\alpha)$ in $2n$-dimensional space in order to approximate the central path $\mathcal{C}$ as follows

$$\mathcal{E}(\alpha) = \{(\mathbf{x}(\alpha), \mathbf{s}(\alpha)) \mid (\mathbf{x}(\alpha), \mathbf{s}(\alpha)) = \vec{\mathbf{a}}\cos(\alpha) + \vec{\mathbf{b}}\sin(\alpha) + \vec{\mathbf{c}}\}, \tag{11.10}$$

where $\vec{\mathbf{a}}, \vec{\mathbf{b}} \in \mathbf{R}^{2n}$ are the axes of the ellipse, and $\vec{\mathbf{c}} \in \mathbf{R}^{2n}$ is the center of the ellipse.

Let $\mathbf{z} = (\mathbf{x}(\mu), \mathbf{s}(\mu)) = (\mathbf{x}(\alpha_0), \mathbf{s}(\alpha_0)) \in \mathcal{E}(\alpha)$ be close to or on the central path. Using the approach of Chapter 9, we define the first and second derivatives at $(\mathbf{x}(\alpha_0), \mathbf{s}(\alpha_0))$ to have the form as if they were on the central path, satisfying

$$\begin{bmatrix} \mathbf{M} & -\mathbf{I} \\ \mathbf{S} & \mathbf{X} \end{bmatrix} \begin{bmatrix} \dot{\mathbf{x}} \\ \dot{\mathbf{s}} \end{bmatrix} = \begin{bmatrix} \mathbf{0} \\ \mathbf{x} \circ \mathbf{s} - \sigma \mu \mathbf{e} \end{bmatrix}, \tag{11.11}$$

$$\begin{bmatrix} \mathbf{M} & -\mathbf{I} \\ \mathbf{S} & \mathbf{X} \end{bmatrix} \begin{bmatrix} \ddot{\mathbf{x}} \\ \ddot{\mathbf{s}} \end{bmatrix} = \begin{bmatrix} \mathbf{0} \\ -2\dot{\mathbf{x}}\dot{\mathbf{s}} \end{bmatrix}, \tag{11.12}$$

where $\mu = \frac{\mathbf{x}^{\mathrm{T}}\mathbf{s}}{n}$, $\sigma \in (0, \frac{1}{4})$ is the centering parameter.

Let $\alpha \in [0, \frac{\pi}{2}]$ and $(\mathbf{x}(\alpha), \mathbf{s}(\alpha))$ be updated from $(\mathbf{x}, \mathbf{s})$ after the searching along the ellipse. Similar to the previous chapters, we have the following theorem.

Theorem 11.1
Let $(\mathbf{x}(\alpha),\mathbf{s}(\alpha))$ *be an arc defined by* (11.10) *passing through a point* $(\mathbf{x},\mathbf{s}) \in \mathcal{E}(\alpha)$, *and its first and second derivatives at* $(\mathbf{x},\mathbf{s})$ *be* $(\dot{\mathbf{x}},\dot{\mathbf{s}})$ *and* $(\ddot{\mathbf{x}},\ddot{\mathbf{s}})$, *which are defined by (11.11) and (11.12). Then, an ellipse approximation of the central path is given by*

$$\mathbf{x}(\alpha) = \mathbf{x} - \sin(\alpha)\dot{\mathbf{x}} + (1-\cos(\alpha))\ddot{\mathbf{x}}, \tag{11.13}$$

$$\mathbf{s}(\alpha) = \mathbf{s} - \sin(\alpha)\dot{\mathbf{s}} + (1-\cos(\alpha))\ddot{\mathbf{s}}. \tag{11.14}$$

To simplify the notation, let $g(\alpha) := 1-\cos(\alpha)$. Using Theorem 11.1, Equations (11.11) and (11.12), we get

$$\mathbf{x}(\alpha) \circ \mathbf{s}(\alpha) = \mathbf{x} \circ \mathbf{s} - \sin(\alpha)(\mathbf{x} \circ \mathbf{s} - \sigma\mu\mathbf{e}) + \chi(\alpha), \tag{11.15}$$

where

$$\chi(\alpha) := -g^2(\alpha)\dot{\mathbf{x}} \circ \dot{\mathbf{s}} - \sin(\alpha)g(\alpha)(\dot{\mathbf{x}} \circ \ddot{\mathbf{s}} + \dot{\mathbf{s}} \circ \ddot{\mathbf{x}}) + g^2(\alpha)\ddot{\mathbf{x}} \circ \ddot{\mathbf{s}}. \tag{11.16}$$

The search for the optimizer is carried out in the wide neighborhood of the central path, defined as

$$\mathcal{N}_{-\infty}(\gamma) = \{(\mathbf{x},\mathbf{s}) \in \mathcal{F}^0 \mid \min(\mathbf{x} \circ \mathbf{s}) \geq \gamma\mu\}, \tag{11.17}$$

where $\gamma \in (0, \frac{1}{2})$ is a constant independent of n. In order to facilitate the analysis of the algorithm, we denote

$$\sin(\hat{\alpha}) := \max\left\{ \sin(\alpha) \mid (\mathbf{x}(\alpha),\mathbf{s}(\alpha)) \in \mathcal{N}_{-\infty}(\gamma),\ \alpha \in \left[0, \frac{\pi}{2}\right] \right\}. \tag{11.18}$$

The proposed algorithm starts from an initial point $(\mathbf{x}^0,\mathbf{s}^0) \in N_{-\infty}(\gamma)$, which is close to or on the central path $\mathcal{C}$. Using the ellipse that passes through the point $(\mathbf{x}^0,\mathbf{s}^0)$, the algorithm searches along the ellipse and gets a new iterate $(\mathbf{x}(\alpha),\mathbf{s}(\alpha))$, which reduces the value of the duality gap. The procedure is repeated until an ε-approximate solution of the LCP is found. It is worthwhile to indicate that this algorithm is a feasible interior-point algorithm. A formal description of the algorithm is presented below:

Algorithm 11.1
(Arc-search IPM for LCP)
data: $\varepsilon > 0$, $\gamma \in (0, \frac{1}{2})$, *and* $\sigma \in (0, \frac{1}{4})$.
initial point: $(\mathbf{x}^0,\mathbf{s}^0) \in N_{-\infty}(\gamma)$ *and* $(\mathbf{x}^0)^{\mathrm{T}}\mathbf{s}^0$.
for *iteration* $k = 0, 1, 2, \ldots$

Step 1: If $(\mathbf{x}^k)^{\mathrm{T}}\mathbf{s}^k \leq \varepsilon$, *then stop.*

Step 2: Solve the systems (11.11) and (11.12) to get $(\dot{\mathbf{x}},\dot{\mathbf{s}})$ *and* $(\ddot{\mathbf{x}},\ddot{\mathbf{s}})$.

Step 3: Compute $\sin(\hat{\alpha}^k)$ *by (11.18), and let* $(\mathbf{x}^{k+1},\mathbf{s}^{k+1}) = (\mathbf{x}(\hat{\alpha}^k),\mathbf{s}(\hat{\alpha}^k))$.

Step 4: Calculate $\mu_{k+1} = \frac{(\mathbf{x}^{k+1})^{\mathrm{T}}\mathbf{s}^{k+1}}{n}$, *and set* $k := k+1$.

end(for)

11.3 Convergence Analysis

We first establish several lemmas which are useful for the convergence analysis of the algorithm. First, starting from a feasible solution, searching for the optimizer along the ellipse, the iterate always meets the equality constraints. For the sake of simplicity, in the rest of the analysis, we omit the index k if confusion will not be introduced.

Lemma 11.1
Let $(\mathbf{x},\mathbf{s})$ *be a strictly feasible point of LCP,* $(\dot{\mathbf{x}},\dot{\mathbf{s}})$ *and* $(\ddot{\mathbf{x}},\ddot{\mathbf{s}})$ *meet (11.11) and (11.12), respectively,* $(\mathbf{x}(\alpha),\mathbf{s}(\alpha))$ *be calculated using* (11.13) *and* 11.14)*, then*

$$\mathbf{M}\mathbf{x}(\alpha)-\mathbf{s}(\alpha)=-\mathbf{q}.$$

Proof 11.1 Since $(\mathbf{x},\mathbf{s})$ is a strict feasible point, by Theorem 11.1, equations (11.11) and (11.12), we have

$$\begin{aligned}
& \mathbf{M}\mathbf{x}(\alpha)-\mathbf{s}(\alpha) \\
= \ & \mathbf{M}[\mathbf{x}-\sin(\alpha)\dot{\mathbf{x}}+(1-\cos(\alpha))\ddot{\mathbf{x}}]-[\mathbf{s}-\sin(\alpha)\dot{\mathbf{s}}+(1-\cos(\alpha))\ddot{\mathbf{s}}] \\
= \ & \mathbf{M}\mathbf{x}-\sin(\alpha)\mathbf{M}\dot{\mathbf{x}}+(1-\cos(\alpha))\mathbf{M}\ddot{\mathbf{x}}-\mathbf{s}+\sin(\alpha)\dot{\mathbf{s}}-(1-\cos(\alpha))\ddot{\mathbf{s}} \\
= \ & \mathbf{M}\mathbf{x}-\mathbf{s}-\sin(\alpha)(\mathbf{M}\dot{\mathbf{x}}-\dot{\mathbf{s}})+(1-\cos(\alpha))(\mathbf{M}\ddot{\mathbf{x}}-\ddot{\mathbf{s}}) \\
= \ & -\mathbf{q}.
\end{aligned}$$

This completes the proof. ∎

The next lemma provides several useful relations.

Lemma 11.2
Let $\dot{\mathbf{x}}$, $\dot{\mathbf{s}}$, $\ddot{\mathbf{x}}$ *and* $\ddot{\mathbf{s}}$ *be defined in (11.11) and (11.12), and* $\mathbf{M}$ *be a positive semi-define matrix. Then, the following relations hold:*

$$\dot{\mathbf{x}}^{\mathrm{T}}\dot{\mathbf{s}}=\dot{\mathbf{x}}^{\mathrm{T}}\mathbf{M}\dot{\mathbf{x}}\geq 0, \tag{11.19}$$

$$\ddot{\mathbf{x}}^{\mathrm{T}}\ddot{\mathbf{s}}=\ddot{\mathbf{x}}^{\mathrm{T}}\mathbf{M}\ddot{\mathbf{x}}\geq 0, \tag{11.20}$$

$$\ddot{\mathbf{x}}^{\mathrm{T}}\dot{\mathbf{s}}=\dot{\mathbf{x}}^{\mathrm{T}}\ddot{\mathbf{s}}=\dot{\mathbf{x}}^{\mathrm{T}}\mathbf{M}\ddot{\mathbf{x}}, \tag{11.21}$$

$$\begin{aligned}
& -(\dot{\mathbf{x}}^{\mathrm{T}}\dot{\mathbf{s}})(1-\cos(\alpha))^2-(\ddot{\mathbf{x}}^{\mathrm{T}}\ddot{\mathbf{s}})\sin^2(\alpha) \\
\leq \ & [\ddot{\mathbf{x}}^{\mathrm{T}}\dot{\mathbf{s}}+\dot{\mathbf{x}}^{\mathrm{T}}\ddot{\mathbf{s}}]\sin(\alpha)(1-\cos(\alpha)) \\
\leq \ & (\dot{\mathbf{x}}^{\mathrm{T}}\dot{\mathbf{s}})(1-\cos(\alpha))^2+(\ddot{\mathbf{x}}^{\mathrm{T}}\ddot{\mathbf{s}})\sin^2(\alpha),
\end{aligned} \tag{11.22}$$

$$\begin{aligned}
& -(\dot{\mathbf{x}}^{\mathrm{T}}\dot{\mathbf{s}})\sin^2(\alpha)-(\ddot{\mathbf{x}}^{\mathrm{T}}\ddot{\mathbf{s}})(1-\cos(\alpha))^2 \\
\leq \ & [\ddot{\mathbf{x}}^{\mathrm{T}}\dot{\mathbf{s}}+\dot{\mathbf{x}}^{\mathrm{T}}\ddot{\mathbf{s}}]\sin(\alpha)(1-\cos(\alpha)) \\
\leq \ & (\dot{\mathbf{x}}^{\mathrm{T}}\dot{\mathbf{s}})\sin^2(\alpha)+(\ddot{\mathbf{x}}^{\mathrm{T}}\ddot{\mathbf{s}})(1-\cos(\alpha))^2.
\end{aligned} \tag{11.23}$$

Proof 11.2 Pre-multiplying $\dot{\mathbf{x}}^{\mathrm{T}}$ and $\ddot{\mathbf{x}}^{\mathrm{T}}$ to the first rows of (11.11) and (11.12), respectively, we have $\dot{\mathbf{x}}^{\mathrm{T}}\dot{\mathbf{s}} = \dot{\mathbf{x}}^{\mathrm{T}}\mathbf{M}\dot{\mathbf{x}}$ and $\ddot{\mathbf{x}}^{\mathrm{T}}\ddot{\mathbf{s}} = \ddot{\mathbf{x}}^{\mathrm{T}}\mathbf{M}\ddot{\mathbf{x}}$. The second two inequalities of (11.19) and (11.20) follow from the fact that $\mathbf{M}$ is positive semi-definite. Similarly, we also have $\ddot{\mathbf{x}}^{\mathrm{T}}\mathbf{M}\dot{\mathbf{x}} = \ddot{\mathbf{x}}^{\mathrm{T}}\dot{\mathbf{s}}$ and $\dot{\mathbf{x}}^{\mathrm{T}}\mathbf{M}\ddot{\mathbf{x}} = \dot{\mathbf{x}}^{\mathrm{T}}\ddot{\mathbf{s}}$, which means $\ddot{\mathbf{x}}^{\mathrm{T}}\dot{\mathbf{s}} = \dot{\mathbf{x}}^{\mathrm{T}}\ddot{\mathbf{s}} = \dot{\mathbf{x}}^{\mathrm{T}}\mathbf{M}\ddot{\mathbf{x}}$. Since $\mathbf{M}$ is a positive semi-definite matrix, we have

$$\begin{aligned}
& [(1-\cos(\alpha))\dot{\mathbf{x}} + \sin(\alpha)\ddot{\mathbf{x}}]^{\mathrm{T}}\mathbf{M}[(1-\cos(\alpha))\dot{\mathbf{x}} + \sin(\alpha)\ddot{\mathbf{x}}] \\
= \ & (\dot{\mathbf{x}}^{\mathrm{T}}\mathbf{M}\dot{\mathbf{x}})(1-\cos(\alpha))^2 + 2(\dot{\mathbf{x}}^{\mathrm{T}}\mathbf{M}\ddot{\mathbf{x}})\sin(\alpha)(1-\cos(\alpha)) + (\ddot{\mathbf{x}}^{\mathrm{T}}\mathbf{M}\ddot{\mathbf{x}})\sin^2(\alpha) \\
= \ & (\dot{\mathbf{x}}^{\mathrm{T}}\dot{\mathbf{s}})(1-\cos(\alpha))^2 + (\ddot{\mathbf{x}}^{\mathrm{T}}\ddot{\mathbf{s}})\sin^2(\alpha) + (\ddot{\mathbf{x}}^{\mathrm{T}}\dot{\mathbf{s}} + \dot{\mathbf{x}}^{\mathrm{T}}\ddot{\mathbf{s}})\sin(\alpha)(1-\cos(\alpha)) \\
\geq \ & 0. \qquad (11.24)
\end{aligned}$$

From the above result, we can get the first inequality of (11.22). Similarly, we also have

$$\begin{aligned}
& [(1-\cos(\alpha))\dot{\mathbf{x}} - \sin(\alpha)\ddot{\mathbf{x}}]^{\mathrm{T}}\mathbf{M}[(1-\cos(\alpha))\dot{\mathbf{x}} - \sin(\alpha)\ddot{\mathbf{x}}] \\
= \ & (\dot{\mathbf{x}}^{\mathrm{T}}\mathbf{M}\dot{\mathbf{x}})(1-\cos(\alpha))^2 - 2(\dot{\mathbf{x}}^{\mathrm{T}}\mathbf{M}\ddot{\mathbf{x}})\sin(\alpha)(1-\cos(\alpha)) + (\ddot{\mathbf{x}}^{\mathrm{T}}\mathbf{M}\ddot{\mathbf{x}})\sin^2(\alpha) \\
= \ & (\dot{\mathbf{x}}^{\mathrm{T}}\dot{\mathbf{s}})(1-\cos(\alpha))^2 + (\ddot{\mathbf{x}}^{\mathrm{T}}\ddot{\mathbf{s}})\sin^2(\alpha) - (\ddot{\mathbf{x}}^{\mathrm{T}}\dot{\mathbf{s}} + \dot{\mathbf{x}}^{\mathrm{T}}\ddot{\mathbf{s}})\sin(\alpha)(1-\cos(\alpha)) \\
\geq \ & 0, \qquad (11.25)
\end{aligned}$$

which implies the second inequality of (11.22). Substituting $(1-\cos(\alpha))\dot{\mathbf{x}}$ and $\sin(\alpha)\ddot{\mathbf{x}}$ by $\sin(\alpha)\dot{\mathbf{x}}$ and $(1-\cos(\alpha))\ddot{\mathbf{x}}$, respectively, following the same way, we can obtain the inequalities of (11.23). ∎

The next lemma was used in previous chapters. We repeat it here for easy reference.

Lemma 11.3
For $\alpha \in [0, \frac{\pi}{2}]$, the following inequalities hold.

$$\sin(\alpha) \geq \sin^2(\alpha) = 1 - \cos^2(\alpha) \geq 1 - \cos(\alpha). \qquad (11.26)$$

The following lemma is useful for the estimation of the upper bound of $\mu(\alpha)$.

Lemma 11.4
Let $(\mathbf{x}, \mathbf{s}) \in \mathcal{N}_{-\infty}(\gamma)$, $(\dot{\mathbf{x}}, \dot{\mathbf{s}})$ and $(\ddot{\mathbf{x}}, \ddot{\mathbf{s}})$ be calculated from (11.11) and (11.12). Let $\mathbf{x}(\alpha)$ and $\mathbf{s}(\alpha)$ be defined as (11.13) and (11.14), respectively. Then,

$$\begin{aligned}
\mu(\alpha) \ &= \ \mu[1-(1-\sigma)\sin(\alpha)] + \frac{1}{n}\Big[(1-\cos(\alpha))^2(\ddot{\mathbf{x}}^{\mathrm{T}}\ddot{\mathbf{s}} - \dot{\mathbf{x}}^{\mathrm{T}}\dot{\mathbf{s}}) \\
& \quad - \sin(\alpha)(1-\cos(\alpha))(\dot{\mathbf{x}}^{\mathrm{T}}\ddot{\mathbf{s}} + \dot{\mathbf{s}}^{\mathrm{T}}\ddot{\mathbf{x}})\Big] \\
&\leq \ \mu[1-(1-\sigma)\sin(\alpha)] + \frac{1}{n}\Big[\ddot{\mathbf{x}}^{\mathrm{T}}\ddot{\mathbf{s}}(\sin^4(\alpha) + \sin^2(\alpha))\Big]. \qquad (11.27)
\end{aligned}$$

Proof 11.3 Using (11.13), (11.14), (11.11) and (11.12), we get

$$
\begin{aligned}
n\mu(\alpha) &= \mathbf{x}(\alpha)^{\mathrm{T}}\mathbf{s}(\alpha) \\
&= \left[\mathbf{x}^{\mathrm{T}} - \dot{\mathbf{x}}^{\mathrm{T}}\sin(\alpha) + \ddot{\mathbf{x}}^{\mathrm{T}}(1-\cos(\alpha))\right]\left[\mathbf{s} - \dot{\mathbf{s}}\sin(\alpha) + \ddot{\mathbf{s}}(1-\cos(\alpha))\right] \\
&= \mathbf{x}^{\mathrm{T}}\mathbf{s} - \mathbf{x}^{\mathrm{T}}\dot{\mathbf{s}}\sin(\alpha) + \mathbf{x}^{\mathrm{T}}\ddot{\mathbf{s}}(1-\cos(\alpha)) \\
&\quad -\dot{\mathbf{x}}^{\mathrm{T}}\mathbf{s}\sin(\alpha) + \dot{\mathbf{x}}^{\mathrm{T}}\dot{\mathbf{s}}\sin^2(\alpha) - \dot{\mathbf{x}}^{\mathrm{T}}\ddot{\mathbf{s}}\sin(\alpha)(1-\cos(\alpha)) \\
&\quad +\ddot{\mathbf{x}}^{\mathrm{T}}\mathbf{s}(1-\cos(\alpha)) - \ddot{\mathbf{x}}^{\mathrm{T}}\dot{\mathbf{s}}\sin(\alpha)(1-\cos(\alpha)) + \ddot{\mathbf{x}}^{\mathrm{T}}\ddot{\mathbf{s}}(1-\cos(\alpha))^2 \\
&= \mathbf{x}^{\mathrm{T}}\mathbf{s} - (\mathbf{x}^{\mathrm{T}}\dot{\mathbf{s}} + \mathbf{s}^{\mathrm{T}}\dot{\mathbf{x}})\sin(\alpha) + (\mathbf{x}^{\mathrm{T}}\ddot{\mathbf{s}} + \mathbf{s}^{\mathrm{T}}\ddot{\mathbf{x}})(1-\cos(\alpha)) \\
&\quad -(\dot{\mathbf{x}}^{\mathrm{T}}\ddot{\mathbf{s}} + \dot{\mathbf{s}}^{\mathrm{T}}\ddot{\mathbf{x}})\sin(\alpha)(1-\cos(\alpha)) + \dot{\mathbf{x}}^{\mathrm{T}}\dot{\mathbf{s}}\sin^2(\alpha) + \ddot{\mathbf{x}}^{\mathrm{T}}\ddot{\mathbf{s}}(1-\cos(\alpha))^2 \\
&= n\mu[1-(1-\sigma)\sin(\alpha)] + (1-\cos(\alpha))^2(\ddot{\mathbf{x}}^{\mathrm{T}}\ddot{\mathbf{s}} - \dot{\mathbf{x}}^{\mathrm{T}}\dot{\mathbf{s}}) \\
&\quad -(\dot{\mathbf{x}}^{\mathrm{T}}\ddot{\mathbf{s}} + \dot{\mathbf{s}}^{\mathrm{T}}\ddot{\mathbf{x}})\sin(\alpha)(1-\cos(\alpha)) \\
&\leq n\mu[1-(1-\sigma)\sin(\alpha)] + (1-\cos(\alpha))^2(\ddot{\mathbf{x}}^{\mathrm{T}}\ddot{\mathbf{s}} - \dot{\mathbf{x}}^{\mathrm{T}}\dot{\mathbf{s}}) \\
&\quad +(1-\cos(\alpha))^2\dot{\mathbf{x}}^{\mathrm{T}}\dot{\mathbf{s}} + \sin^2(\alpha)\ddot{\mathbf{x}}^{\mathrm{T}}\ddot{\mathbf{s}} \\
&= n\mu[1-(1-\sigma)\sin(\alpha)] + (1-\cos(\alpha))^2\ddot{\mathbf{x}}^{\mathrm{T}}\ddot{\mathbf{s}} + \sin^2(\alpha)\ddot{\mathbf{x}}^{\mathrm{T}}\ddot{\mathbf{s}} \\
&\leq n\mu[1-(1-\sigma)\sin(\alpha)] + (\sin^4(\alpha) + \sin^2(\alpha))(\ddot{\mathbf{x}}^{\mathrm{T}}\ddot{\mathbf{s}}),
\end{aligned}
$$

where the first inequality follows from Lemma 11.2, and the second inequality is from Lemma 11.3. This proves the lemma. ∎

Now, we present some technical results which are used to prove the main results of convergence analysis. Denote $\mathbf{D} = \mathbf{X}^{\frac{1}{2}}\mathbf{S}^{-\frac{1}{2}}$, where $\mathbf{0} < (\mathbf{x},\mathbf{s}) \in \mathbb{R}^{2n}$. The next result was presented in [174].

Lemma 11.5
Let $(\mathbf{u},\mathbf{v}) \geq \mathbf{0}$ *and* $\mathbf{u},\mathbf{v} \in \mathbb{R}^n$*, then,*

$$\|\mathbf{u}\circ\mathbf{v}\|_1 \leq \|\mathbf{D}\mathbf{u}\|\;\|\mathbf{D}^{-1}\mathbf{v}\| \leq \frac{1}{2}\left(\|\mathbf{D}\mathbf{u}\|^2 + \|\mathbf{D}^{-1}\mathbf{v}\|^2\right).$$

Proof 11.4 Since

$$
\begin{aligned}
\|\mathbf{u}\circ\mathbf{v}\|_1 &= \|\mathbf{D}\mathbf{u}\circ\mathbf{D}^{-1}\mathbf{v}\|_1 = |(D_1u_1)(D_1^{-1}v_1)| + \ldots + |(D_nu_n)(D_n^{-1}v_n)| \\
&= (D_1u_1)(D_1^{-1}v_1) + \ldots + (D_nu_n)(D_n^{-1}v_n),
\end{aligned}
$$

and

$$\|\mathbf{D}\mathbf{u}\|_2\;\|\mathbf{D}^{-1}\mathbf{v}\|_2 = \sqrt{(D_1u_1)^2 + \ldots + (D_nu_n)^2}\sqrt{(D_1^{-1}v_1)^2 + \ldots + (D_n^{-1}v_n)^2},$$

it follows directly from Cauchy-Schwarz inequality that $\|\mathbf{u}\circ\mathbf{v}\|_1 \leq \|\mathbf{D}\mathbf{u}\|_2\;\|\mathbf{D}^{-1}\mathbf{v}\|_2$. This proves the first inequality. The second inequality follows the fact that $ab \leq \frac{1}{2}(a^2+b^2)$ for any scalars a and b. ∎

In what follows, we give the upper bounds for $\|\mathbf{D}^{-1}\dot{\mathbf{x}}\|$, $\|\mathbf{D}\dot{\mathbf{s}}\|$, $\|\mathbf{D}^{-1}\ddot{\mathbf{x}}\|$ and $\|\mathbf{D}\ddot{\mathbf{s}}\|$, which are essential to the convergence analysis.

Lemma 11.6
For $(\mathbf{x},\mathbf{s}) \in \mathcal{N}_{-\infty}(\gamma)$*, let* $(\dot{\mathbf{x}},\dot{\mathbf{s}})$ *be the solution of (11.11), then, we have*

$$\|\mathbf{D}^{-1}\dot{\mathbf{x}}\|^2 + \|\mathbf{D}\dot{\mathbf{s}}\|^2 \le \beta_1\mu n, \quad \|\dot{\mathbf{x}}\circ\dot{\mathbf{s}}\|_1 \le \frac{\beta_1\mu n}{2},$$

and

$$\|\mathbf{D}^{-1}\dot{\mathbf{x}}\| \le \sqrt{\beta_1\mu n}, \quad \|\mathbf{D}\dot{\mathbf{s}}\| \le \sqrt{\beta_1\mu n},$$

where $\beta_1 \ge 1 \ge 1-2\sigma+\frac{\sigma^2}{\gamma}$.

Proof 11.5 Pre-Multiplying the second equation in (11.11) by $(\mathbf{XS})^{-\frac{1}{2}}$, we get

$$\mathbf{D}^{-1}\dot{\mathbf{x}}+\mathbf{D}\dot{\mathbf{s}} = (\mathbf{XS})^{-\frac{1}{2}}(\mathbf{x}\circ\mathbf{s}-\sigma\mu\mathbf{e}).$$

Then, taking the squared norm on both sides and using $x_i s_i \ge \gamma\mu$ yields

$$\begin{aligned}
\|\mathbf{D}^{-1}\dot{\mathbf{x}}\|^2+2\dot{\mathbf{x}}^{\mathrm{T}}\dot{\mathbf{s}}+\|\mathbf{D}\dot{\mathbf{s}}\|^2 &= \|(\mathbf{x}\circ\mathbf{s})^{\frac{1}{2}}\|^2-2\sigma\mu n+\|(\mathbf{XS})^{-\frac{1}{2}}\sigma\mu\mathbf{e}\|^2\\
&\le \mathbf{x}^{\mathrm{T}}\mathbf{s}-2\sigma\mu n+\frac{\|\sigma\mu\mathbf{e}\|^2}{\min_i(x_i s_i)}\\
&\le \mu n-2\sigma\mu n+\frac{\sigma^2\mu n}{\gamma}\\
&\le \beta_1\mu n. \qquad (11.28)
\end{aligned}$$

Since $\dot{\mathbf{x}}^{\mathrm{T}}\dot{\mathbf{s}} = \dot{\mathbf{x}}^{\mathrm{T}}\mathbf{M}\dot{\mathbf{x}} \ge 0$, we have

$$\|\mathbf{D}^{-1}\dot{\mathbf{x}}\|^2+\|\mathbf{D}\dot{\mathbf{s}}\|^2 \le \beta_1\mu n.$$

Using this result and Lemma 11.5, we obtain

$$\|\mathbf{D}^{-1}\dot{\mathbf{x}}\| \le \sqrt{\beta_1\mu n}, \quad \|\mathbf{D}\dot{\mathbf{s}}\| \le \sqrt{\beta_1\mu n},$$

and

$$\|\dot{\mathbf{x}}\circ\dot{\mathbf{s}}\|_1 \le \frac{\beta_1\mu n}{2}.$$

This proves the lemma. ■

Lemma 11.7
If $(\mathbf{x},\mathbf{s}) \in \mathcal{N}_{-\infty}(\gamma)$*, let* $(\ddot{\mathbf{x}},\ddot{\mathbf{s}})$ *be the solution of (11.12), then we have*

$$\|\mathbf{D}^{-1}\ddot{\mathbf{x}}\|^2+\|\mathbf{D}\ddot{\mathbf{s}}\|^2 \le \beta_2\mu n^2, \quad \|\ddot{\mathbf{x}}\circ\ddot{\mathbf{s}}\|_1 \le \frac{\beta_2\mu n^2}{2},$$

and

$$\|\mathbf{D}^{-1}\ddot{\mathbf{x}}\| \le \sqrt{\beta_2\mu}n, \quad \|\mathbf{D}\ddot{\mathbf{s}}\| \le \sqrt{\beta_2\mu}n,$$

where $\beta_2 = \frac{\beta_1^2}{\gamma} \ge 1$.

Proof 11.6 Pre-multiplying $(\mathbf{XS})^{-\frac{1}{2}}$ to the second equation in (11.12), taking the squared norm on both sides, and using the fact that $\ddot{\mathbf{x}}^{\mathrm{T}}\ddot{\mathbf{s}} \geq \mathbf{0}$, we obtain

$$\begin{aligned} \|\mathbf{D}^{-1}\ddot{\mathbf{x}}\|^2 + \|\mathbf{D}\ddot{\mathbf{s}}\|^2 &\leq \|(\mathbf{XS})^{-\frac{1}{2}}(-2\dot{\mathbf{x}}\circ\dot{\mathbf{s}})\|^2 \\ &\leq \frac{\|-2\dot{\mathbf{x}}\circ\dot{\mathbf{s}}\|^2}{\min_i(x_i s_i)} \\ &\leq \frac{4\|\dot{\mathbf{x}}\circ\dot{\mathbf{s}}\|_1^2}{\gamma\mu} \\ &\leq \beta_2\mu n^2, \end{aligned} \tag{11.29}$$

From the above inequality and Lemma 11.5, we get

$$\|\mathbf{D}^{-1}\ddot{\mathbf{x}}\| \leq \sqrt{\beta_2\mu}n, \quad \|\mathbf{D}\ddot{\mathbf{s}}\| \leq \sqrt{\beta_2\mu}n,$$

and

$$\|\ddot{\mathbf{x}}\circ\ddot{\mathbf{s}}\|_1 \leq \frac{\beta_2\mu n^2}{2}.$$

This completes the proof. ∎

The next result follows directly from Lemmas 11.6 and 11.7.

Lemma 11.8
For $\beta_3 = \sqrt{\beta_1\beta_2} \geq 1$, we have

$$\|\dot{\mathbf{x}}\circ\ddot{\mathbf{s}}\|_1 \leq \|\mathbf{D}^{-1}\dot{\mathbf{x}}\|\|\mathbf{D}\ddot{\mathbf{s}}\| \leq \beta_3\mu n^{\frac{3}{2}}, \quad \|\ddot{\mathbf{x}}\circ\dot{\mathbf{s}}\|_1 \leq \|\mathbf{D}^{-1}\ddot{\mathbf{x}}\|\|\mathbf{D}\dot{\mathbf{s}}\| \leq \beta_3\mu n^{\frac{3}{2}}.$$

Let $\sin(\hat{\alpha}_0) = \frac{1}{\beta_4 n}$, where $\beta_4 = \frac{\beta_1+\beta_2+4\beta_3}{\sigma(1-\gamma)} \geq 1$, and we have the following lemma.

Lemma 11.9
Let $\chi(\alpha)$ be defined as (11.16), then, for all $\alpha \in (0, \hat{\alpha}_0]$, we have $\chi(\alpha) \geq -\frac{1}{2}\sin(\alpha)\mu(1-\gamma)\sigma\mathbf{e}$.

Proof 11.7 Using Lemmas 11.6, 11.7, 11.8 and $g(\alpha) \leq \sin^2(\alpha)$, we have

$$\begin{aligned} \chi(\alpha) &= -g^2(\alpha)\dot{\mathbf{x}}\circ\dot{\mathbf{s}} - \sin(\alpha)g(\alpha)(\dot{\mathbf{x}}\circ\ddot{\mathbf{s}} + \dot{\mathbf{s}}\circ\ddot{\mathbf{x}}) + g^2(\alpha)\ddot{\mathbf{x}}\circ\ddot{\mathbf{s}} \\ &\geq \left[-\sin^4(\alpha)\|\dot{\mathbf{x}}\circ\dot{\mathbf{s}}\|_1 - \sin^3(\alpha)\|\dot{\mathbf{x}}\circ\ddot{\mathbf{s}} + \dot{\mathbf{s}}\circ\ddot{\mathbf{x}}\|_1 - \sin^4(\alpha)\|\ddot{\mathbf{x}}\circ\ddot{\mathbf{s}}\|_1\right]\mathbf{e} \\ &\geq -\left[\frac{1}{2}\sin^4(\alpha)\beta_1\mu n + 2\sin^3(\alpha)\beta_3\mu n^{\frac{3}{2}} + \frac{1}{2}\sin^4(\alpha)\beta_2\mu n^2\right]\mathbf{e} \end{aligned}$$

$$
\begin{aligned}
&\geq -\frac{1}{2}\sin(\alpha)\mu\left[\sin^3(\hat{\alpha}_0)\beta_1 n + 4\sin^2(\hat{\alpha}_0)\beta_3 n^{\frac{3}{2}} + \sin^3(\hat{\alpha}_0)\beta_2 n^2\right]\mathbf{e} \\
&= -\frac{1}{2}\sin(\alpha)\mu\left[\frac{\beta_1}{\beta_4^3 n^2} + \frac{4\beta_3}{\beta_4^2 n^{\frac{1}{2}}} + \frac{\beta_2}{\beta_4^3 n}\right]\mathbf{e} \\
&\geq -\frac{1}{2}\sin(\alpha)\mu\frac{1}{\beta_4}[\beta_1+\beta_2+4\beta_3]\mathbf{e} \\
&= -\frac{1}{2}\sin(\alpha)\mu(1-\gamma)\sigma\mathbf{e}. \qquad (11.30)
\end{aligned}
$$

This completes the proof. ■

Lemma 11.10

Let $(\mathbf{x},\mathbf{s}) \in \mathcal{N}_{-\infty}(\gamma)$*, and* $\sin(\hat{\alpha})$ *be defined by* (11.18)*. Then, for all* $\sin(\alpha) \in [0, \sin(\hat{\alpha}_0)]$*, we have* $(\mathbf{x}(\alpha),\mathbf{s}(\alpha)) \in \mathcal{N}_{-\infty}(\gamma)$ *and* $\sin(\hat{\alpha}) \geq \sin(\hat{\alpha}_0)$*.*

Proof 11.8 Using (11.15), (11.17), Lemmas 11.9, 11.3, and 11.4, we have

$$
\begin{aligned}
& \min(\mathbf{x}(\alpha)\circ\mathbf{s}(\alpha)) \\
= & \min[\mathbf{x}\circ\mathbf{s} - \sin(\alpha)(\mathbf{x}\circ\mathbf{s}-\sigma\mu\mathbf{e}) + \chi(\alpha)] \\
\geq & \min(\mathbf{x}\circ\mathbf{s})(1-\sin(\alpha)) + \sin(\alpha)\sigma\mu + \min(\chi(\alpha)) \\
\geq & (1-\sin(\alpha))\gamma\mu + \sin(\alpha)\sigma\mu - \frac{1}{2}\sin(\alpha)(1-\gamma)\sigma\mu \\
= & \gamma\mu(\alpha) - \sigma\gamma\mu\sin(\alpha) + \sin(\alpha)\sigma\mu - \frac{1}{2}\sin(\alpha)(1-\gamma)\sigma\mu \\
& -\frac{\gamma}{n}\left[(1-\cos(\alpha))^2(\ddot{\mathbf{x}}^{\mathrm{T}}\ddot{\mathbf{s}} - \dot{\mathbf{x}}^{\mathrm{T}}\dot{\mathbf{s}}) - \sin(\alpha)(1-\cos(\alpha))(\dot{\mathbf{x}}^{\mathrm{T}}\ddot{\mathbf{s}} + \dot{\mathbf{s}}^{\mathrm{T}}\ddot{\mathbf{x}})\right] \\
= & \gamma\mu(\alpha) + \sin(\alpha)(1-\gamma)\sigma\mu - \frac{1}{2}\sin(\alpha)(1-\gamma)\sigma\mu \\
& -\frac{\gamma}{n}(1-\cos(\alpha))^2(\ddot{\mathbf{x}}^{\mathrm{T}}\ddot{\mathbf{s}} - \dot{\mathbf{x}}^{\mathrm{T}}\dot{\mathbf{s}}) \\
& +\frac{\gamma}{n}\sin(\alpha)(1-\cos(\alpha))(\dot{\mathbf{x}}^{\mathrm{T}}\ddot{\mathbf{s}} + \dot{\mathbf{s}}^{\mathrm{T}}\ddot{\mathbf{x}}) \\
\geq & \gamma\mu(\alpha) + \frac{1}{2}\sin(\alpha)(1-\gamma)\sigma\mu - \frac{\gamma}{n}(1-\cos(\alpha))^2(\ddot{\mathbf{x}}^{\mathrm{T}}\ddot{\mathbf{s}} + \dot{\mathbf{x}}^{\mathrm{T}}\dot{\mathbf{s}}) \\
& +\frac{\gamma}{n}\sin(\alpha)(1-\cos(\alpha))(\dot{\mathbf{x}}^{\mathrm{T}}\ddot{\mathbf{s}} + \dot{\mathbf{s}}^{\mathrm{T}}\ddot{\mathbf{x}}) \\
\geq & \gamma\mu(\alpha) + \frac{1}{2}\sin(\alpha)(1-\gamma)\sigma\mu - \frac{\gamma}{n}\sin^2(\alpha)(\dot{\mathbf{x}}^{\mathrm{T}}\dot{\mathbf{s}}) \\
& -\frac{\gamma}{n}(1-\cos(\alpha))^2(\ddot{\mathbf{x}}^{\mathrm{T}}\ddot{\mathbf{s}}) \\
& +\frac{\gamma}{n}\sin(\alpha)(1-\cos(\alpha))(\dot{\mathbf{x}}^{\mathrm{T}}\ddot{\mathbf{s}} + \dot{\mathbf{s}}^{\mathrm{T}}\ddot{\mathbf{x}}) \qquad (11.31)
\end{aligned}
$$

where the last two inequalities follow from the relations $\dot{\mathbf{x}}^{\mathrm{T}}\dot{\mathbf{s}} \geq 0$ and Lemma 11.3, respectively. Using Lemmas 11.6, 11.7 and 11.8, we can verify that the following relation holds.

$$\begin{aligned}
& \frac{1}{2}\sin(\alpha)(1-\gamma)\sigma\mu - \frac{\gamma}{n}\sin^2(\alpha)(\dot{\mathbf{x}}^{\mathrm{T}}\dot{\mathbf{s}}) \\
& -\frac{\gamma}{n}(1-\cos(\alpha))^2(\ddot{\mathbf{x}}^{\mathrm{T}}\ddot{\mathbf{s}}) \\
& +\frac{\gamma}{n}\sin(\alpha)(1-\cos(\alpha))(\dot{\mathbf{x}}^{\mathrm{T}}\ddot{\mathbf{s}}+\dot{\mathbf{s}}^{\mathrm{T}}\ddot{\mathbf{x}}) \\
\geq\ & \frac{\gamma}{n}\mu\sin(\alpha)\left[\frac{n}{2\gamma}(1-\gamma)\sigma - \frac{\beta_1 n}{2}\sin(\alpha) - \frac{\beta_2 n^2}{2}\sin^3(\alpha) - 2\beta_3 n^{3/2}\sin^2(\alpha)\right] \\
\geq\ & \frac{\gamma}{2n}\mu\sin(\alpha)\left[\frac{n}{\gamma}(1-\gamma)\sigma - \frac{\beta_1}{\beta_4} - \frac{\beta_2}{\beta_4^3 n} - \frac{4\beta_3}{\beta_4^2 n^{1/2}}\right] \qquad [\text{use } \sin(\hat{\alpha}_0) = \tfrac{1}{\beta_4 n}] \\
\geq\ & \frac{\gamma}{2n}\mu\sin(\alpha)\left[\frac{n}{\gamma}(1-\gamma)\sigma - \frac{\beta_1}{\beta_4} - \frac{\beta_2}{\beta_4} - \frac{4\beta_3}{\beta_4}\right] \\
\geq\ & \frac{\gamma}{2n}\mu\sin(\alpha)\left[\frac{n}{\gamma}(1-\gamma)\sigma - (1-\gamma)\sigma\right] \qquad [\text{use } \beta_4 = \tfrac{\beta_1+\beta_2+4\beta_3}{\sigma(1-\gamma)}] \\
\geq\ & 0.
\end{aligned} \tag{11.32}$$

Combining (11.31) and (11.32) shows that :

$$\min(\mathbf{x}(\alpha)\circ\mathbf{s}(\alpha)) \geq \gamma\mu(\alpha).$$

This proves that $\mathbf{x}(\alpha)\circ\mathbf{s}(\alpha) > \mathbf{0}$. Since $\mathbf{x}(\alpha)$ and $\mathbf{s}(\alpha)$ are continuous in $[0,\hat{\alpha}_0]$ and $(\mathbf{x},\mathbf{s}) > \mathbf{0}$, hence, $\mathbf{x}(\alpha) > \mathbf{0}$, $\mathbf{s}(\alpha) > \mathbf{0}$. By (11.17) and (11.18), we have $(\mathbf{x}(\alpha),\mathbf{s}(\alpha)) \in \mathcal{N}_{-\infty}(\gamma)$. In view of (11.18), we obtain $\sin(\hat{\alpha}) \geq \sin(\hat{\alpha}_0)$. ∎

Now, we are ready to present the iteration complexity bound of Algorithm 11.1. To this end, we need to obtain a new upper bound of $\mu(\alpha)$. From Lemmas 11.4, 11.3, and 11.7, it follows that for $\alpha \in [0,\hat{\alpha}_0]$,

$$\begin{aligned}
\mu(\alpha) &\leq \mu[1-(1-\sigma)\sin(\alpha)] + \frac{1}{n}\left[\ddot{\mathbf{x}}^{\mathrm{T}}\ddot{\mathbf{s}}(\sin^4(\alpha)+\sin^2(\alpha))\right] \\
&\leq \mu[1-(1-\sigma)\sin(\alpha)] + \frac{1}{n}\left[\|\ddot{\mathbf{x}}\circ\ddot{\mathbf{s}}\|_1(\sin^4(\alpha)+\sin^2(\alpha))\right] \\
&\leq \mu\left[1-(1-\sigma-\frac{\beta_2 n}{2}(\sin^3(\alpha)+\sin(\alpha)))\sin(\alpha)\right] \\
&\leq \mu\left[1-\left(1-\sigma-\frac{\beta_2}{2\beta_4^3 n^2}-\frac{\beta_2}{2\beta_4}\right)\sin(\alpha)\right] \\
&\leq \left(1-\frac{1}{8}\sin(\alpha)\right)\mu,
\end{aligned}$$

where the last two inequalities come from $\sin(\hat{\alpha}_0) = \frac{1}{\beta_4 n}$, $0 < \sigma < \frac{1}{4}$, and $\beta_4 = \frac{\beta_1+\beta_2+4\beta_3}{(1-\gamma)\sigma} \geq \frac{\beta_2}{\sigma} \geq 4\beta_2$, and $n \geq 2$.

Theorem 11.2
Let $(\mathbf{x}^0,\mathbf{s}^0) \in \mathcal{N}_{-\infty}(\gamma)$, $\sin(\hat{\alpha}_0) = \frac{1}{\beta_4 n}$, *then Algorithm 11.1 terminates at most* $O(n\log\frac{(\mathbf{x}^0)^{\mathrm{T}}\mathbf{s}^0}{\varepsilon})$ *iterations.*

Proof 11.9 Using the relation $\sin(\hat{\alpha}) \geq \sin(\hat{\alpha}_0)$, we have

$$\mu(\hat{\alpha}) \leq \left(1-\frac{1}{8}\sin(\hat{\alpha})\right)\mu \leq \left(1-\frac{1}{8}\sin(\hat{\alpha}_0)\right)\mu = \left(1-\frac{1}{8\beta_4 n}\right)\mu.$$

Then, we obtain

$$(\mathbf{x}^k)^{\mathrm{T}}\mathbf{s}^k \leq \left(1-\frac{1}{8\beta_4 n}\right)^k (\mathbf{x}^0)^{\mathrm{T}}\mathbf{s}^0,$$

which implies,

$$\left(1-\frac{1}{8\beta_4 n}\right)^k (\mathbf{x}^0)^{\mathrm{T}}\mathbf{s}^0 \leq \varepsilon,$$

and

$$k\log\left(1-\frac{1}{8\beta_4 n}\right) \leq \log\frac{\varepsilon}{(\mathbf{x}^0)^{\mathrm{T}}\mathbf{s}^0}.$$

Since

$$\log\left(1-\frac{1}{8\beta_4 n}\right) \leq -\frac{1}{8\beta_4 n} \leq 0,$$

Thus, for $k \geq 8\beta_4 n\log\frac{(\mathbf{x}^0)^{\mathrm{T}}\mathbf{s}^0}{\varepsilon}$, we have

$$(\mathbf{x}^k)^{\mathrm{T}}\mathbf{s}^k \leq \varepsilon.$$

This completes the proof. ■

11.4 Concluding Remarks

Although there are quite a few papers that discuss arc-search algorithms for LCP problems, the majority of the research is focused on a very special type of LCP, i.e., $P_*(k)$-LCPs, because of the technical difficulty of using higher-order method for general monotone LCPs. There are many problems which are worth investigating for general LCPs. For example, are there efficient infeasible arc-search interior-point algorithms for general monotone LCPs? Moreover, if such algorithms exist, what is the lowest polynomial bound achievable for infeasible arc-search interior-point algorithms? Computationally, how efficient are the arc-search interior-point algorithms in comparison to traditional interior-point algorithms?

Chapter 12

An Arc-Search Algorithm for Semidefinite Programming

Semidefinite programming (SDP) is a convex optimization problem over the intersecting of affine set and the cone of positive semidefinite matrices [66, 165]. SDP has applications in many areas, such as combinatorial optimization [2], system and control theory [14], and eigenvalue optimization problems [107]. In the last two decades, SDP has become an active research area in mathematical programming as it became clear that the interior-point methods for linear programming (LP) can often be extended to the more general SDP case. Several interior-point methods (IPMs) designed for LP have been successfully generalized to SDP, for example, the ones in [73, 74, 103, 138].

As discussed in Chapters 2, 3 and 4, the majority of IPMs search optimizer along a straight line related to the first-order and higher-order derivatives of the central path [55, 56]. However, most optimization problems are nonlinear, and the ideal search should be along arcs, such as the central path in LP, not straight lines. After this author developed the idea of primal-dual IPMs with arc-search, the idea has been used by different authors to solve various IPM problems. For example, Yang et al. [145, 149, 147, 148, 161], Kheirfam and Chitsaz [64], and Mansouri et al. [85, 111] investigated the infeasible IPMs with arc-search for LP and symmetric conic optimization (SCO). Later, Pirhaji et al. [110] introduced a l_2-neighborhood infeasible IPM with arc-search for linear complementary problem (LCP). Kheirfam [60, 61] proposed infeasible arc-search algorithms in the negative infinity neighborhood for SCO and horizontal LCP. Many of these works improved the theoretical complexity bounds over traditional line search IPMs.

Recently, Kheirfam and Moslemi [62] extended arc-search strategy to the SDP case. The derived complexity bound is of the same order as the one for LP case. This chapter discusses an arc-search algorithm for SDP problem, which is proposed by Zhang et al. [170].

12.1 Preliminary

The following notations are used throughout the chapter. The set of all symmetric $n \times n$ real matrices is denoted by $\mathcal{S}^n$. For $\mathbf{M} \in \mathcal{S}^n$, we write $\mathbf{M} \succ \mathbf{0}(\mathbf{M} \succeq \mathbf{0})$ if $\mathbf{M}$ is positive definite (positive semidefinite). We also denote by $\mathcal{S}^n_{++}(\mathcal{S}^n_{+})$ the set of all matrices in $\mathcal{S}^n$ which are positive definite (positive semidefinite). For a matrix $\mathbf{M}$ with all real eigenvalues, we denote its eigenvalues by $\lambda_i(\mathbf{M})$, $i = 1, 2, ..., n$, in increasing order, and its smallest and largest eigenvalues by $\lambda_{\min}(\mathbf{M}) = \lambda_1(\mathbf{M})$ and $\lambda_{\max}(\mathbf{M}) = \lambda_n(\mathbf{M})$, respectively. Moreover, the spectral condition number of a symmetric matrix $\mathbf{M}$ is denoted by $cond(\mathbf{M}) = \lambda_{\max}(\mathbf{M})/\lambda_{\min}(\mathbf{M})$. The spectral radius of $\mathbf{M}$ is denoted by $\rho(\mathbf{M}) := \max\{|\lambda_i(\mathbf{M})| : i = 1, 2, ..., n\}$. Given $\mathbf{G}$ and $\mathbf{H}$ in $\mathbb{R}^{m\times n}$, the inner product between them is defined as $\mathbf{G} \bullet \mathbf{H} := Tr(\mathbf{G}^{\mathrm{T}}\mathbf{H})$, where $Tr(\mathbf{M})$ is the trace of $\mathbf{M} \in \mathbb{R}^{n\times n}$. For $\mathbf{M} \in \mathbb{R}^{n\times n}$, $\|\mathbf{M}\|_F$, $\|\mathbf{M}\|$ denote the Frobenius norm and matrix 2-norm, respectively.

Some matrix analysis results will play important roles in this chapter. The following formulas involving matrix trace (see [54, 109]) will be used repeatedly.

Lemma 12.1
Let $\mathbf{A}, \mathbf{B}, \mathbf{X} \in \mathbb{R}^{n\times n}$. Then, we have

$$Tr(\mathbf{A}) = Tr(\mathbf{A}^{\mathrm{T}}), \tag{12.1}$$

$$Tr(\mathbf{A}+\mathbf{B}) = Tr(\mathbf{A}) + Tr(\mathbf{B}), \tag{12.2}$$

$$Tr(\mathbf{AB}) = Tr(\mathbf{BA}), \tag{12.3}$$

$$\frac{\partial}{\partial \mathbf{X}} Tr(\mathbf{A}\mathbf{X}^{\mathrm{T}}) = \mathbf{A}, \quad \Rightarrow \quad \textit{if } \mathbf{X} \in \mathbf{S}^n, \textit{ then } \quad \frac{\partial}{\partial \mathbf{X}} Tr(\mathbf{A}\mathbf{X}) = \mathbf{A}. \tag{12.4}$$

The Kronecker product of an $m \times n$ matrix $\mathbf{A}$ and a $p \times q$ matrix $\mathbf{B}$, is an $mp \times nq$ matrix, $\mathbf{A} \otimes \mathbf{B}$, defined as

$$\mathbf{A} \otimes \mathbf{B} = \begin{bmatrix} a_{11}\mathbf{B} & a_{12}\mathbf{B} & \dots & a_{1n}\mathbf{B} \\ a_{21}\mathbf{B} & a_{22}\mathbf{B} & \dots & a_{2n}\mathbf{B} \\ \vdots & \vdots & \vdots & \vdots \\ a_{m1}\mathbf{B} & a_{m2}\mathbf{B} & \dots & a_{mn}\mathbf{B} \end{bmatrix} \tag{12.5}$$

Some Kronecker product identities are useful, these are listed below. The proofs can be found in [54, 109].

Lemma 12.2
Let **A**, **B**, **C**, *and* **D** *be matrices with appropriate dimensions. Then, we have*

$$(\mathbf{A} \otimes \mathbf{B})^{\mathrm{T}} = \mathbf{A}^{\mathrm{T}} \otimes \mathbf{B}^{\mathrm{T}}, \tag{12.6}$$

$$(\mathbf{A} \otimes \mathbf{B})^{-1} = \mathbf{A}^{-1} \otimes \mathbf{B}^{-1}, \tag{12.7}$$

$$(\mathbf{A} \otimes \mathbf{B})(\mathbf{C} \otimes \mathbf{D}) = \mathbf{AC} \otimes \mathbf{BD}, \tag{12.8}$$

$$Tr(\mathbf{A} \otimes \mathbf{B}) = Tr(\mathbf{A})Tr(\mathbf{B}) = Tr(\Lambda_{\mathbf{A}} \otimes \Lambda_{\mathbf{B}}) \tag{12.9}$$

$$\text{eig}(\mathbf{A} \otimes \mathbf{B}) = \text{eig}(\mathbf{A}) \otimes \text{eig}(\mathbf{B}). \tag{12.10}$$

where $\Lambda_{\mathbf{A}}$ *denotes the diagonal matrix with eigenvalues of* **A**.

Corollary 12.1
The Kronecker product $\mathbf{A} \otimes \mathbf{B}$ *is positive semidefinite if and only if both* **A** *and* **B** *are positive semidefinite or negative semidefinite.*

Proof 12.1 In view of the last formula of Lemma 12.2, the claim follows from that, given the eigenvalues $\lambda_1, \ldots, \lambda_n$ for **A** and $\sigma_1, , \ldots, \sigma_n$ for **B**, the eigenvalues of $\mathbf{A} \otimes \mathbf{B}$ are

$$\forall i, j, \lambda_i \sigma_j.$$

This completes the proof. ■

The vector operator applied on a matrix **A**, denoted by $\text{vec}(\mathbf{A})$, stacks the columns one by one from left to the right into a vector. For example, a 2×2 matrix **A** and $\text{vec}(\mathbf{A})$ can be expressed as

$$\mathbf{A} = \begin{bmatrix} a_{11} & a_{12} \\ a_{21} & a_{22} \end{bmatrix}, \quad \text{vec}(\mathbf{A}) = \begin{bmatrix} a_{11} \\ a_{21} \\ a_{12} \\ a_{22} \end{bmatrix}.$$

The Kronecker product and the vector-operator are related by the following lemma, which can be found in [54, 109].

Lemma 12.3
Let **A**, **B**, *and* **X** *be matrices with appropriate dimensions,* **a**, *and* **b** *be vectors with appropriate dimensions, and* α *be a scalar. Then, the following equations hold.*

$$\text{vec}(\mathbf{AXB}) = (\mathbf{B}^{\mathrm{T}} \otimes \mathbf{A})\text{vec}(\mathbf{X}), \tag{12.11}$$

$$Tr(\mathbf{A}^{\mathrm{T}}\mathbf{B}) = \text{vec}(\mathbf{A})^{\mathrm{T}}\text{vec}(\mathbf{B}), \tag{12.12}$$

$$\text{vec}(\mathbf{A}+\mathbf{B}) = \text{vec}(\mathbf{A})+\text{vec}(\mathbf{B}), \tag{12.13}$$

$$\text{vec}(\alpha\mathbf{A}) = \alpha\cdot\text{vec}(\mathbf{A}), \tag{12.14}$$

$$\mathbf{a}^{\mathrm{T}}\mathbf{X}\mathbf{B}\mathbf{X}^{\mathrm{T}}\mathbf{b} = \text{vec}(\mathbf{X})^{\mathrm{T}}(\mathbf{B}\otimes\mathbf{b}\mathbf{a}^{\mathrm{T}})\text{vec}(\mathbf{X}). \tag{12.15}$$

Proof 12.2 We prove only (12.12) and leave the rest for readers. Let $\mathbf{A} = [\mathbf{a}_1,\mathbf{a}_2,\ldots,\mathbf{a}_n] \in \mathbb{R}^{n\times n}$, and $\mathbf{B} = [\mathbf{b}_1,\mathbf{b}_2,\ldots,\mathbf{b}_n] \in \mathbb{R}^{n\times n}$. Then, we have

$$(\text{vec}(\mathbf{A}))^{\mathrm{T}}\text{vec}(\mathbf{B}) = [\mathbf{a}_1^{\mathrm{T}},\mathbf{a}_2^{\mathrm{T}},\ldots,\mathbf{a}_n^{\mathrm{T}}]\begin{bmatrix}\mathbf{b}_1\\ \vdots\\ \mathbf{b}_n\end{bmatrix}$$

$$= \sum_{i=1}^{n}\mathbf{a}_i^{\mathrm{T}}\mathbf{b}_i = Tr\left(\begin{bmatrix}\mathbf{a}_1^{\mathrm{T}}\\ \vdots\\ \mathbf{a}_n^{\mathrm{T}}\end{bmatrix}[\mathbf{b}_1,\mathbf{b}_2,\ldots,\mathbf{b}_n]\right) = Tr(\mathbf{A}^{\mathrm{T}}\mathbf{B}). \tag{12.16}$$

■

Proposition 12.1
Let vectors $\mathbf{u},\mathbf{v},\mathbf{r}\in\mathbb{R}^n$ *and nonsingular matrices* $\mathbf{E},\mathbf{F}\in\mathbb{R}^{n\times n}$ *satisfy*

$$\mathbf{E}\mathbf{u}+\mathbf{F}\mathbf{v} = \mathbf{r}.$$

If $\mathbf{S} = \mathbf{F}\mathbf{E}^{\mathrm{T}}$ *is symmetric positive definite, then*

$$\mathbf{D} = \mathbf{S}^{-1/2}\mathbf{F} = \mathbf{S}^{1/2}\mathbf{E}^{-\mathrm{T}}, \quad \mathbf{D}^{-\mathrm{T}} = \mathbf{S}^{-1/2}\mathbf{E} = \mathbf{S}^{1/2}\mathbf{F}^{-\mathrm{T}}, \tag{12.17}$$

and

$$\|\mathbf{D}^{-\mathrm{T}}\mathbf{u}\|^2 + \|\mathbf{D}\mathbf{v}\|^2 + 2\mathbf{u}^{\mathrm{T}}\mathbf{v} = \|\mathbf{S}^{-1/2}\mathbf{r}\|^2. \tag{12.18}$$

Proof 12.3 First, the following relations are equivalent:

$$\begin{aligned} & \mathbf{S} = \mathbf{F}\mathbf{E}^{\mathrm{T}} \\ \Longleftrightarrow\quad & \mathbf{S}\mathbf{E}^{-\mathrm{T}} = \mathbf{F} \\ \Longleftrightarrow\quad & \mathbf{S}^{\frac{1}{2}}\mathbf{E}^{-\mathrm{T}} = \mathbf{S}^{-\frac{1}{2}}\mathbf{F} := \mathbf{D}. \end{aligned}$$

Similarly, we have the following equivalent relations:

$$\begin{aligned} & \mathbf{S}^{-1} = \mathbf{E}^{-\mathrm{T}}\mathbf{F}^{-1} = \mathbf{F}^{-\mathrm{T}}\mathbf{E}^{-1} \\ \Longleftrightarrow\quad & \mathbf{S}^{-1}\mathbf{E} = \mathbf{F}^{-\mathrm{T}} \\ \Longleftrightarrow\quad & \mathbf{S}^{-\frac{1}{2}}\mathbf{E} = \mathbf{S}^{\frac{1}{2}}\mathbf{F}^{-\mathrm{T}}. \end{aligned}$$

Using $\mathbf{D}^{-\mathrm{T}} = (\mathbf{S}^{-\frac{1}{2}}\mathbf{F})^{-\mathrm{T}} = (\mathbf{F}^{\mathrm{T}}\mathbf{S}^{-\frac{1}{2}})^{-1} = \mathbf{S}^{\frac{1}{2}}\mathbf{F}^{-\mathrm{T}}$, we get (12.17). Pre-multiplying both side of $\mathbf{E}\mathbf{u}+\mathbf{F}\mathbf{v} = \mathbf{r}$ by $\mathbf{S}^{-1/2}$, using (12.17) (noticing that $\mathbf{u}^{\mathrm{T}}\mathbf{E}\mathbf{S}^{-1}\mathbf{F}\mathbf{v} = \mathbf{u}^{\mathrm{T}}\mathbf{v}$), and taking 2-norm squared show (12.18). ■

12.2 A Long Step Feasible SDP Arc-Search Algorithm

Let $\mathcal{S}^n$ be the set of real symmetric $n \times n$ matrices. We consider the SDP in standard form, which is given as :

$$\min_{\mathbf{X} \in \mathcal{S}^n} \mathbf{C} \bullet \mathbf{X}, \quad \text{s.t.} \quad \mathbf{A}_i \bullet \mathbf{X} = b_i, \quad i = 1, \dots, m, \quad \mathbf{X} \succeq \mathbf{0}, \tag{12.19}$$

where $\mathbf{C} \in \mathcal{S}^n$, $\mathbf{A}_i \in \mathcal{S}^n$ for $k = 1, \dots, m$, and $\mathbf{b} = (b_1, \dots, b_m) \in \mathbb{R}^n$ are given, and $\mathbf{0} \preceq \mathbf{X} \in \mathcal{S}^n$ are the primal variables to be optimized. Because of the symmetric restriction, there are $n(n+1)/2$ free variables (not n^2 variables) in $\mathbf{X}$. The dual problem of (12.19) is given by

$$\min_{\mathbf{X} \in \mathcal{S}^n} \mathbf{b}^{\mathrm{T}} \mathbf{y}, \quad \text{s.t.} \quad \sum_{i=1}^{m} y_i \mathbf{A}_i + \mathbf{S} = \mathbf{C}, \quad \mathbf{S} \succeq \mathbf{0}, \tag{12.20}$$

where $\mathbf{y} = [y_1, \dots, y_m]^{\mathrm{T}} \in \mathbb{R}^m$ and $\mathbf{0} \preceq \mathbf{S} \in \mathcal{S}^n$ are the dual variables. Again, there are $n(n+1)/2$ free variables in $\mathbf{S}$ because of the symmetric restriction. The set of primal-dual feasible solutions is denoted by

$$\mathcal{F} := \{(\mathbf{X}, \mathbf{y}, \mathbf{S}) \in \mathcal{S}^n \times \mathbb{R}^m \times \mathcal{S}^n : \mathbf{A}_i \bullet \mathbf{X} = b_i, \ \sum_{i=1}^{m} y_i \mathbf{A}_i + \mathbf{S} = \mathbf{C}, \ \mathbf{X}, \mathbf{S} \succeq \mathbf{0}\}, \tag{12.21}$$

and the set of the primal-dual strict feasible set is

$$\mathcal{F}^0 := \{(\mathbf{X}, \mathbf{y}, \mathbf{S}) \in \mathcal{S}^n \times \mathbb{R}^m \times \mathcal{S}^n : \mathbf{A}_i \bullet \mathbf{X} = b_i, \ \sum_{i=1}^{m} y_i \mathbf{A}_i + \mathbf{S} = \mathbf{C}, \ \mathbf{X}, \mathbf{S} \succ \mathbf{0}\}. \tag{12.22}$$

Throughout this chapter, we also make two standard assumptions:
Assumptions:

1. $\mathcal{F}^0 = \emptyset$.

2. $\mathbf{A}_i$ is linearly independent.

The first assumption makes sure that the SDP problem has an optimal solution. The second assumption guarantees that there are solutions for the symmetric first and second order derivatives (see proposition 12.2).

The KKT conditions for the positive semidefinite problem is given by:

$$\mathbf{A}_i \bullet \mathbf{X} = Tr(\mathbf{A}_i \mathbf{X}) = b_i, \quad i = 1, \dots, m, \quad \mathbf{X} \succeq \mathbf{0} \tag{12.23a}$$

$$\sum_{i=1}^{m} y_i \mathbf{A}_i + \mathbf{S} = \mathbf{C}, \quad \mathbf{S} \succeq \mathbf{0}, \tag{12.23b}$$

$$\mathbf{X}\mathbf{S} = \mathbf{0}. \tag{12.23c}$$

The KKT system of equations has total $n(n+1)+m$ variables.

Denote the duality measure for the SDP as:

$$\mu = \mathbf{X} \bullet \mathbf{S}/n = Tr(\mathbf{XS})/n. \tag{12.24}$$

The perturbed KKT conditions for the positive semidefinite problem is, therefore, given by :

$$\mathbf{A}_i \bullet \mathbf{X} = b_i, \quad i=1,\ldots,m, \quad \mathbf{X} \succ \mathbf{0} \tag{12.25a}$$

$$\sum_{i=1}^{m} y_i \mathbf{A}_i + \mathbf{S} = \mathbf{C}, \quad \mathbf{S} \succ \mathbf{0}, \tag{12.25b}$$

$$\mathbf{XS} = \mu \mathbf{I}. \tag{12.25c}$$

Given a value for μ, there is a solution of (12.25). Therefore, the systems of equations (12.25) define a curve, named the central path in $n(n+1)+m$ dimensional space. The so-called path-following interior-point method approximates the central path using the derivatives and searches for the optimizers along the approximated central path. To approximate the central path, the easiest way is to use a linear approximation, which involves some formulas equal or similar to the following one

$$\mathbf{XS} + \mathbf{X}\dot{\mathbf{S}} + \mathbf{S}\dot{\mathbf{X}} = \mu \mathbf{I} \tag{12.26}$$

where both $\dot{\mathbf{S}}$ and $\dot{\mathbf{X}}$ are required to be symmetric so that $\mathbf{X}+\dot{\mathbf{X}}$ and $\mathbf{S}+\dot{\mathbf{S}}$ are symmetric. Unfortunately, none of the products on left hand side items in (12.26) are guaranteed to be symmetric, neither $\dot{\mathbf{S}}$ nor $\dot{\mathbf{X}}$ will be symmetric for sure. However, it is straightforward (but very tedious) to verify [4] that a modified equation of (12.26)

$$\frac{1}{2}\left[\mathbf{XS} + \mathbf{SX} + \mathbf{X}\dot{\mathbf{S}} + \dot{\mathbf{S}}\mathbf{X} + \mathbf{S}\dot{\mathbf{X}} + \dot{\mathbf{X}}\mathbf{S}\right] = \mu \mathbf{I} \tag{12.27}$$

gives symmetric $\dot{\mathbf{X}}$ and $\dot{\mathbf{S}}$!

Let us consider a little more general case $\mathbf{M} \in \mathbb{R}^{n\times n}$ and $\mathbf{H}(\mathbf{M}) = \frac{1}{2}\left[\mathbf{M}+\mathbf{M}^{\mathrm{T}}\right]$. The following simple observation is useful for the future analysis: for any scalar α and matrices $\mathbf{M}, \mathbf{M}_1, \mathbf{M}_2 \in \mathbb{R}^{n\times n}$, it is easy to verify

$$\mathbf{H}(\mathbf{M}) = (\mathbf{H}(\mathbf{M}))^{\mathrm{T}}, \tag{12.28}$$

$$\mathbf{H}(\alpha\mathbf{M}) = \alpha\mathbf{H}(\mathbf{M}), \tag{12.29}$$

$$\mathbf{H}(\mathbf{M}_1+\mathbf{M}_2) = \mathbf{H}(\mathbf{M}_1) + \mathbf{H}(\mathbf{M}_2). \tag{12.30}$$

Zhang [173] introduced a more general similarly transformed symmetrization operator

$$\mathbf{H}_{\mathbf{P}}(\mathbf{M}) = \frac{1}{2}\left[\mathbf{PMP}^{-1} + (\mathbf{PMP}^{-1})^{\mathrm{T}}\right]. \tag{12.31}$$

For the special case $\mathbf{P}=\mathbf{I}$, we define $\mathbf{H_P}(\mathbf{M})=\mathbf{H_I}(\mathbf{M})=\mathbf{H}(\mathbf{M})$, the last notation is used for this special case.

Lemma 12.4
For any scalar α and matrices $\mathbf{M},\mathbf{M}_1,\mathbf{M}_2 \in \mathbb{R}^{n\times n}$, it is easy to verify

$$\mathbf{H_P}(\mathbf{M}) = (\mathbf{H_P}(\mathbf{M}))^{\mathrm{T}}, \tag{12.32}$$
$$\mathbf{H_P}(\alpha\mathbf{M}) = \alpha\mathbf{H_P}(\mathbf{M}), \tag{12.33}$$
$$\mathbf{H_P}(\mathbf{M}_1+\mathbf{M}_2) = \mathbf{H_P}(\mathbf{M}_1)+\mathbf{H_P}(\mathbf{M}_2), \tag{12.34}$$
$$Tr(\mathbf{H_P}(\mathbf{M})) = Tr(\mathbf{M}). \tag{12.35}$$

Therefore, the modified perturbed KKT conditions, which guarantee to have symmetric $\dot{\mathbf{X}}$ and $\dot{\mathbf{S}}$, can be rewritten as follows:

$$\mathbf{A}_i \bullet \dot{\mathbf{X}} = 0, \quad i=1,\dots,m, \tag{12.36a}$$
$$\sum_{i=1}^{m} \dot{\mathbf{y}}_i\mathbf{A}_i + \dot{\mathbf{S}} = \mathbf{0}, \tag{12.36b}$$
$$\mathbf{H_P}(\mathbf{X}\dot{\mathbf{S}}+\dot{\mathbf{X}}\mathbf{S}) = \mathbf{H_P}(\mathbf{X}\mathbf{S}) - \sigma\mu\mathbf{I}, \tag{12.36c}$$

where $\sigma \in [0,1]$ is the centering parameter. Similarly, we have the second order equations:

$$\mathbf{A}_i \bullet \ddot{\mathbf{X}} = 0, \quad i=1,\dots,m, \tag{12.37a}$$
$$\sum_{i=1}^{m} \ddot{\mathbf{y}}_i\mathbf{A}_i + \ddot{\mathbf{S}} = \mathbf{0}, \tag{12.37b}$$
$$\mathbf{H_P}(\mathbf{X}\ddot{\mathbf{S}}+\ddot{\mathbf{X}}\mathbf{S}) = -2\mathbf{H_P}(\dot{\mathbf{X}}\dot{\mathbf{S}}). \tag{12.37c}$$

It is important to emphasize that (12.36) and (12.37) guarantee that $\dot{\mathbf{X}}$, $\dot{\mathbf{S}}$, $\ddot{\mathbf{X}}$ and $\ddot{\mathbf{S}}$ are all symmetric which in turn guarantee that $\mathbf{X}^k$ and $\mathbf{S}^k$ will be symmetric. By taking extra care of the iteration, we can prove that they are actually positive definite in all iterations.

Let $\mathbf{P}$ be nonsingular matrix, and define

$$\hat{\mathbf{X}} = \mathbf{PXP}, \quad \dot{\hat{\mathbf{X}}} = \mathbf{P}\dot{\mathbf{X}}\mathbf{P}, \quad \hat{\mathbf{S}} = \mathbf{P}^{-1}\mathbf{S}\mathbf{P}^{-1}, \quad \dot{\hat{\mathbf{S}}} = \mathbf{P}^{-1}\dot{\mathbf{S}}\mathbf{P}^{-1}. \tag{12.38}$$

Then, we have

$$\begin{aligned}
& \mathbf{H_P}(\mathbf{X}\dot{\mathbf{S}}+\dot{\mathbf{X}}\mathbf{S}) \\
= \; & \mathbf{P}(\mathbf{X}\dot{\mathbf{S}}+\dot{\mathbf{X}}\mathbf{S})\mathbf{P}^{-1} + \mathbf{P}^{-\mathrm{T}}(\dot{\mathbf{S}}\mathbf{X}+\mathbf{S}\dot{\mathbf{X}})\mathbf{P}^{\mathrm{T}} \\
= \; & \mathbf{PXPP}^{-1}\dot{\mathbf{S}}\mathbf{P}^{-1} + \mathbf{P}\dot{\mathbf{X}}\mathbf{PP}^{-1}\mathbf{SP}^{-1} \\
& +\mathbf{P}^{-\mathrm{T}}\dot{\mathbf{S}}\mathbf{P}^{-\mathrm{T}}\mathbf{P}^{\mathrm{T}}\mathbf{XP}^{\mathrm{T}} + \mathbf{P}^{-\mathrm{T}}\mathbf{SP}^{-\mathrm{T}}\mathbf{P}^{\mathrm{T}}\dot{\mathbf{X}}\mathbf{P}^{\mathrm{T}}
\end{aligned}$$

$$
\begin{aligned}
&= \hat{\mathbf{X}}\dot{\hat{\mathbf{S}}} + \dot{\hat{\mathbf{X}}}\hat{\mathbf{S}} + (\hat{\mathbf{X}}\dot{\hat{\mathbf{S}}} + \dot{\hat{\mathbf{X}}}\hat{\mathbf{S}})^{\mathrm{T}} \\
&= \mathbf{H}(\hat{\mathbf{X}}\dot{\hat{\mathbf{S}}} + \dot{\hat{\mathbf{X}}}\hat{\mathbf{S}}). \qquad (12.39)
\end{aligned}
$$

This means that $\mathbf{H_P}(\mathbf{X}\dot{\mathbf{S}} + \dot{\mathbf{X}}\mathbf{S})$ is equivalent to perform a scaling first and then perform the symmetrization. Therefore, the last equation of (12.36) can be rewritten as

$$\mathbf{H}(\hat{\mathbf{X}}\dot{\hat{\mathbf{S}}} + \dot{\hat{\mathbf{X}}}\hat{\mathbf{S}}) = \mathbf{H}(\hat{\mathbf{X}}\hat{\mathbf{S}}) - \sigma\mu\mathbf{I}, \qquad (12.40)$$

Similarly, the last equation of (12.37) can be rewritten as

$$\mathbf{H}(\hat{\mathbf{X}}\ddot{\hat{\mathbf{S}}} + \ddot{\hat{\mathbf{X}}}\hat{\mathbf{S}}) = -2\mathbf{H}(\dot{\hat{\mathbf{X}}}\dot{\hat{\mathbf{S}}}). \qquad (12.41)$$

Applying vector-operator (see Lemma 12.3) to (12.40) gives

$$\hat{\mathbf{E}}\mathrm{vec}(\dot{\hat{\mathbf{X}}}) + \hat{\mathbf{F}}\mathrm{vec}(\dot{\hat{\mathbf{S}}}) = \mathrm{vec}(\mathbf{H}(\hat{\mathbf{X}}\hat{\mathbf{S}}) - \sigma\mu\mathbf{I}), \qquad (12.42)$$

where

$$\hat{\mathbf{E}} \equiv \frac{1}{2}(\hat{\mathbf{S}}\otimes\mathbf{I} + \mathbf{I}\otimes\hat{\mathbf{S}}), \;\; \hat{\mathbf{F}} \equiv \frac{1}{2}(\hat{\mathbf{X}}\otimes\mathbf{I} + \mathbf{I}\otimes\hat{\mathbf{X}}). \qquad (12.43)$$

In view of Corollary 12.1, it follows that $\hat{\mathbf{E}} \in \mathcal{S}^{n^2}_{++}$ if $\mathbf{S} \succ \mathbf{0}$ ($\hat{\mathbf{E}} \in \mathcal{S}^{n^2}_{+}$ if $\mathbf{S} \succeq \mathbf{0}$) and $\hat{\mathbf{F}} \in \mathcal{S}^{n^2}_{++}$ if $\mathbf{X} \succ \mathbf{0}$ ($\hat{\mathbf{F}} \in \mathcal{S}^{n^2}_{+}$ if $\mathbf{X} \succeq \mathbf{0}$). Applying vector-operator to (12.41) gives

$$\hat{\mathbf{E}}\mathrm{vec}(\ddot{\hat{\mathbf{X}}}) + \hat{\mathbf{F}}\mathrm{vec}(\ddot{\hat{\mathbf{S}}}) = -2\mathrm{vec}(\mathbf{H}(\dot{\hat{\mathbf{X}}}\dot{\hat{\mathbf{S}}})). \qquad (12.44)$$

When $\mathbf{P} \succeq \mathbf{0}$ holds, the following proposition, due to Shida, Shindoh and Kojima [98, 124], is useful.

Proposition 12.2
Let $\mathbf{X} \succeq \mathbf{0}, \mathbf{S} \succeq \mathbf{0}, \mathbf{P} \succeq \mathbf{0}$ be given and suppose that Assumption 2 holds, then, a sufficient condition for systems (12.36) and (12.37) to have a unique solution is that $\hat{\mathbf{E}}\hat{\mathbf{F}} + \hat{\mathbf{F}}\hat{\mathbf{E}} \succ \mathbf{0}$. Moreover, the later condition holds if $\mathbf{H}(\hat{\mathbf{X}}\hat{\mathbf{S}}) = \mathbf{H_P}(\mathbf{XS}) \succeq \mathbf{0}$.

Several popular selections of $\mathbf{P}$ are: (a) $\mathbf{P} = \mathbf{I}$ proposed by Alizadeh, Haeberly, and Overton [4] (denoted as AHO), (b) $\mathbf{P} = \mathbf{S}^{1/2}$ proposed by Helmberg, Randl, Vanderbei, and Wolkowicz [52], Kojima, Shida, and Shindoh [73], and Menteiro [94] (denoted as HRVW/KSS/M), (c) $\mathbf{P} = \mathbf{X}^{-1/2}$ proposed by Kojima, Shida, and Shindoh [73], and Menteiro [94] (denoted as KSS/M), and (d) $\mathbf{P} = \mathbf{W}_{nt}^{1/2}$ proposed by Nesterov and Todd [102, 103] (denoted as NT), where

$$\mathbf{W}_{nt} := \mathbf{S}^{1/2}(\mathbf{S}^{1/2}\mathbf{X}\mathbf{S}^{1/2})^{-1/2}\mathbf{S}^{1/2} = \mathbf{X}^{-1/2}(\mathbf{X}^{1/2}\mathbf{S}\mathbf{X}^{1/2})^{1/2}\mathbf{X}^{-1/2}.$$

All these selections meet the definition of $\mathbf{H_P}(\mathbf{X}\dot{\mathbf{S}} + \dot{\mathbf{X}}\mathbf{S})$ and generate symmetric $\dot{\mathbf{X}}$ and $\dot{\mathbf{S}}$. To make the convergence analysis easier to handle, we want to select $\mathbf{P}$ such that the matrices $\hat{\mathbf{X}}$ and $\hat{\mathbf{S}}$ are commutative, i.e.,

$$\hat{\mathbf{X}}\hat{\mathbf{S}} = \hat{\mathbf{S}}\hat{\mathbf{X}}. \qquad (12.45)$$

We denote the set of **P** that meets the condition (12.45) as

$$\mathcal{P}(\mathbf{X},\mathbf{S}) = \{\mathbf{P} \in \mathbf{S}^n_{++} : \hat{\mathbf{X}}\hat{\mathbf{S}} = \hat{\mathbf{S}}\hat{\mathbf{X}}\}. \tag{12.46}$$

It is easy to see that this set is not empty as HRVW/KSS/M, KSS/M, and NT scaling matrices are all in the set defined in (12.46). However, AHO ($\mathbf{P} = \mathbf{I}$) scaling is not. Using HRVW/KSS/M scaling $\mathbf{P} = \mathbf{S}^{1/2}$ as an example, we have

$$\hat{\mathbf{X}} = \mathbf{S}^{1/2}\mathbf{X}\mathbf{S}^{1/2}, \quad \hat{\mathbf{S}} = \mathbf{S}^{-1/2}\mathbf{S}\mathbf{S}^{-1/2} = \mathbf{I},$$

therefore, $\hat{\mathbf{X}}\hat{\mathbf{S}} = \hat{\mathbf{S}}\hat{\mathbf{X}}$. The following lemma will be used later on.

Lemma 12.5
If $\hat{\mathbf{X}}\hat{\mathbf{S}} = \hat{\mathbf{S}}\hat{\mathbf{X}}$ holds, then, we have $\hat{\mathbf{E}}\hat{\mathbf{F}} = \hat{\mathbf{F}}\hat{\mathbf{E}}$.

Proof 12.4 It is straightforward to verify that

$$\begin{aligned}
\hat{\mathbf{E}}\hat{\mathbf{F}} &= \frac{1}{4}(\hat{\mathbf{S}} \otimes \mathbf{I} + \mathbf{I} \otimes \hat{\mathbf{S}})(\hat{\mathbf{X}} \otimes \mathbf{I} + \mathbf{I} \otimes \hat{\mathbf{X}}) \\
&= \frac{1}{4}(\hat{\mathbf{S}}\hat{\mathbf{X}} \otimes \mathbf{I} + \hat{\mathbf{S}} \otimes \hat{\mathbf{X}} + \hat{\mathbf{X}} \otimes \hat{\mathbf{S}} + \mathbf{I} \otimes \hat{\mathbf{S}}\hat{\mathbf{X}})
\end{aligned}$$

$$\begin{aligned}
\hat{\mathbf{F}}\hat{\mathbf{E}} &= \frac{1}{4}(\hat{\mathbf{X}} \otimes \mathbf{I} + \mathbf{I} \otimes \hat{\mathbf{X}})(\hat{\mathbf{S}} \otimes \mathbf{I} + \mathbf{I} \otimes \hat{\mathbf{S}}) \\
&= \frac{1}{4}(\hat{\mathbf{X}}\hat{\mathbf{S}} \otimes \mathbf{I} + \hat{\mathbf{S}} \otimes \hat{\mathbf{X}} + \hat{\mathbf{X}} \otimes \hat{\mathbf{S}} + \mathbf{I} \otimes \hat{\mathbf{X}}\hat{\mathbf{S}})
\end{aligned}$$

Since $\hat{\mathbf{X}}\hat{\mathbf{S}} = \hat{\mathbf{S}}\hat{\mathbf{X}}$ holds, we have $\hat{\mathbf{E}}\hat{\mathbf{F}} = \hat{\mathbf{F}}\hat{\mathbf{E}}$. ■

In the remaining discussion, we assume that NT scaling is used. Therefore, $\hat{\mathbf{E}}$ and $\hat{\mathbf{F}}$ are commutative. Now, we are ready to discuss arc-search technique for the aforementioned SDP problem. Let's consider the modified central path

$$\mathbf{A}_i \bullet \mathbf{X} = b_i, \quad i = 1, \ldots, m, \quad \mathbf{X} \succ \mathbf{0} \tag{12.47a}$$

$$\sum_{i=1}^{m} y_i \mathbf{A}_i + \mathbf{S} = \mathbf{C}, \quad \mathbf{S} \succ \mathbf{0}, \tag{12.47b}$$

$$\mathbf{H}_{\mathbf{P}}(\mathbf{X}\mathbf{S}) = \mu \mathbf{I}, \tag{12.47c}$$

The following lemma [173] shows that the central path (12.25) and the modified central path (12.47) are equivalent.

Lemma 12.6
For $\mathbf{M} \in \mathbb{R}^{n \times n}$ with real spectrum, nonsingular $\mathbf{P} \in \mathbb{R}^{n \times n}$ and a scalar μ,

$$\mathbf{H}_{\mathbf{P}}(\mathbf{M}) = \mu \mathbf{I} \quad \Leftrightarrow \quad \mathbf{M} = \mu \mathbf{I}. \tag{12.48}$$

Proof 12.5 Suppose that the equality holds on the left. We must have

$$\mathbf{PMP}^{-1} = \mu \mathbf{I} + \mathbf{G}$$

for some skew-symmetric matrix $\mathbf{G}$. Since the spectrum is real, we must have $\mathbf{G} = \mathbf{0}$, otherwise $\mathbf{M}$ would have an eigenvalue $\mu + \phi_k(\mathbf{G})$ for a nonzero pure imaginary number $\phi_k(\mathbf{M})$, contradicting the realness assumption for the spectrum of $\mathbf{M}$. The inverse is obvious. ■

The algorithm to be discussed restricts iterates to the negative infinity neighborhood of the central path, defined by

$$N_{-\infty}(\gamma) = \{(\mathbf{X}, \mathbf{y}, \mathbf{S}) \in \mathcal{F}^0 : \lambda_{\min}(\mathbf{XS}) \geq \gamma \mu\}, \tag{12.49}$$

where $\gamma \in (0,1)$ is a constant independent of n. Since

$$\lambda_{\min}(\mathbf{XS}) = \lambda_{\min}(\mathbf{PXPP}^{-1}\mathbf{SP}^{-1}) = \lambda_{\min}(\hat{\mathbf{X}}\hat{\mathbf{S}}),$$

this means that this negative infinity neighborhood definition is working for both original and modified central paths.

Following the idea in Chapter 5, we will use an ellipse $\mathcal{E}(\alpha)$ to approximate the central path $\mathcal{C}$ described by

$$\mathcal{E}(\alpha) = \{(\mathbf{X}(\alpha), \mathbf{y}(\alpha), \mathbf{S}(\alpha)) : (\mathbf{X}(\alpha), \mathbf{y}(\alpha), \mathbf{S}(\alpha)) = \cos(\alpha)\vec{\mathbf{a}} + \sin(\alpha)\vec{\mathbf{b}} + \vec{\mathbf{c}}\}, \tag{12.50}$$

where $\vec{\mathbf{a}}$, $\vec{\mathbf{b}} \in \mathcal{S}^n \times \mathbb{R}^m \times \mathcal{S}^n$ are the axes of the ellipse, and $\vec{\mathbf{c}} \in \mathcal{S}^n \times \mathbb{R}^m \times \mathcal{S}^n$ is the center of the ellipse. From the first two rows of the equations (12.36) and (12.37), it is obvious to see that

$$\dot{\mathbf{X}} \bullet \dot{\mathbf{S}} = 0, \ \ \ddot{\mathbf{X}} \bullet \dot{\mathbf{S}} = 0, \ \ \dot{\mathbf{X}} \bullet \ddot{\mathbf{S}} = 0, \ \ \ddot{\mathbf{X}} \bullet \ddot{\mathbf{S}} = 0. \tag{12.51}$$

Let $\alpha \in [0, \frac{\pi}{2}]$. The following lemma provides the formulas of the ellipsoidal approximation of the central path $\mathcal{C}$.

Lemma 12.7
Let $\mathcal{E}(\alpha)$ be the ellipse defined in (12.50) which passes the point $(\mathbf{X}, \mathbf{y}, \mathbf{S})$. Moreover, assume that the first and second derivatives of the central path $(\dot{\mathbf{X}}, \dot{\mathbf{y}}, \dot{\mathbf{S}})$ and $(\ddot{\mathbf{X}}, \ddot{\mathbf{y}}, \ddot{\mathbf{S}})$ at $(\mathbf{X}, \mathbf{y}, \mathbf{S})$ satisfy (12.36) and (12.37). Then, an ellipsoidal approximation of the (modified) central path is given by

$$\mathbf{X}(\alpha) = \mathbf{X} - \sin(\alpha)\dot{\mathbf{X}} + (1 - \cos(\alpha))\ddot{\mathbf{X}}, \tag{12.52}$$

$$\mathbf{S}(\alpha) = \mathbf{S} - \sin(\alpha)\dot{\mathbf{S}} + (1 - \cos(\alpha))\ddot{\mathbf{S}}, \tag{12.53}$$

$$\mathbf{y}(\alpha) = \mathbf{y} - \sin(\alpha)\dot{\mathbf{y}} + (1 - \cos(\alpha))\ddot{\mathbf{y}}. \tag{12.54}$$

Proof 12.6 Similar to the proof in Chapter 5, let

$$\mathbf{Z}(\alpha) = (\mathbf{X}(\alpha), \mathbf{y}(\alpha), \mathbf{S}(\alpha)) = \cos(\alpha)\vec{\mathbf{a}} + \sin(\alpha)\vec{\mathbf{b}} + \vec{\mathbf{c}}. \tag{12.55}$$

Then, taking derivative twice at $\mathbf{Z}(\alpha_0)$ gives

$$\dot{\mathbf{Z}}(\alpha) = (\dot{\mathbf{X}}(\alpha), \dot{\mathbf{y}}(\alpha), \dot{\mathbf{S}}(\alpha)) = -\sin(\alpha)\vec{\mathbf{a}} + \cos(\alpha)\vec{\mathbf{b}}, \tag{12.56}$$

$$\ddot{\mathbf{Z}}(\alpha) = (\ddot{\mathbf{X}}(\alpha), \ddot{\mathbf{y}}(\alpha), \ddot{\mathbf{S}}(\alpha)) = -\cos(\alpha)\vec{\mathbf{a}} - \sin(\alpha)\vec{\mathbf{b}}. \tag{12.57}$$

It is straightforward to verify the following relations from the results above:

$$\vec{\mathbf{a}} = -\sin(\alpha)\dot{\mathbf{Z}} - \cos(\alpha)\ddot{\mathbf{Z}}, \quad \vec{\mathbf{b}} = \cos(\alpha)\dot{\mathbf{Z}} - \sin(\alpha)\ddot{\mathbf{Z}}, \quad \vec{\mathbf{c}} = \mathbf{Z} + \ddot{\mathbf{Z}}. \tag{12.58}$$

Denote

$$\vec{\mathbf{a}} = (\mathbf{a}_X, \mathbf{a}_y, \mathbf{a}_S) = -\sin(\alpha)\dot{\mathbf{Z}} - \cos(\alpha)\ddot{\mathbf{Z}},$$

this gives

$$\begin{aligned}
\mathbf{a}_X &= -\sin(\alpha)\dot{\mathbf{X}} - \cos(\alpha)\ddot{\mathbf{X}}, \\
\mathbf{a}_y &= -\sin(\alpha)\dot{\mathbf{y}} - \cos(\alpha)\ddot{\mathbf{y}}, \\
\mathbf{a}_S &= -\sin(\alpha)\dot{\mathbf{S}} - \cos(\alpha)\ddot{\mathbf{S}}.
\end{aligned}$$

Similarly, denote

$$\vec{\mathbf{b}} = (\mathbf{b}_X, \mathbf{b}_y, \mathbf{b}_S) = \cos(\alpha)\dot{\mathbf{Z}} - \sin(\alpha)\ddot{\mathbf{Z}},$$

this gives

$$\begin{aligned}
\mathbf{b}_X &= \cos(\alpha)\dot{\mathbf{X}} - \sin(\alpha)\ddot{\mathbf{X}}, \\
\mathbf{b}_y &= \cos(\alpha)\dot{\mathbf{y}} - \sin(\alpha)\ddot{\mathbf{y}}, \\
\mathbf{b}_S &= \cos(\alpha)\dot{\mathbf{S}} - \sin(\alpha)\ddot{\mathbf{S}};
\end{aligned}$$

and denote

$$\vec{\mathbf{c}} = (\mathbf{c}_X, \mathbf{c}_y, \mathbf{c}_S) = \mathbf{Z} + \ddot{\mathbf{Z}},$$

this gives

$$\begin{aligned}
\mathbf{c}_X &= \mathbf{X} + \ddot{\mathbf{X}}, \\
\mathbf{c}_y &= \mathbf{y} + \ddot{\mathbf{y}}, \\
\mathbf{c}_S &= \mathbf{S} + \ddot{\mathbf{S}}.
\end{aligned}$$

Let $\mathbf{X}(\alpha)$ and $\mathbf{S}(\alpha)$ be the updated $\mathbf{X}$ and $\mathbf{S}$ after the search, since $\mathbf{X} = \mathbf{X}(\alpha_0) = \cos(\alpha_0)\mathbf{a}_X + \sin(\alpha_0)\mathbf{b}_X + \mathbf{c}_x$, we have

$$\mathbf{X}(\alpha) = \cos(\alpha_0 - \alpha)\mathbf{a}_X + \sin(\alpha_0 - \alpha)\mathbf{b}_X + \mathbf{c}_X$$

$$\begin{aligned}
&= [\cos(\alpha_0)\cos(\alpha)+\sin(\alpha_0)\sin(\alpha)]\mathbf{a}_X \\
&\quad +[\sin(\alpha_0)\cos(\alpha)-\cos(\alpha_0)\sin(\alpha)]\mathbf{b}_X \\
&\quad +\mathbf{c}_X-\cos(\alpha)\mathbf{c}_X+\cos(\alpha)\mathbf{c}_X \\
&= \cos(\alpha)\mathbf{X}+\sin(\alpha_0)\sin(\alpha)\mathbf{a}_X-\cos(\alpha_0)\sin(\alpha)\mathbf{b}_X+(1-\cos(\alpha))\mathbf{c}_X \\
&= \cos(\alpha)\mathbf{X}-[\sin(\alpha_0)\dot{\mathbf{X}}+\cos(\alpha_0)\ddot{\mathbf{X}}]\sin(\alpha_0)\sin(\alpha) \\
&\quad -[\cos(\alpha_0)\dot{\mathbf{X}}-\sin(\alpha_0)\ddot{\mathbf{X}}]\cos(\alpha_0)\sin(\alpha)+(1-\cos(\alpha))(\mathbf{X}+\ddot{\mathbf{X}}) \\
&= \mathbf{X}-[\sin^2(\alpha_0)\sin(\alpha)+\cos^2(\alpha_0)\sin(\alpha)]\dot{\mathbf{X}} \\
&\quad +[-\sin(\alpha_0)\cos(\alpha_0)\sin(\alpha)+\sin(\alpha_0)\cos(\alpha_0)\sin(\alpha)+(1-\cos(\alpha))]\ddot{\mathbf{X}} \\
&= \mathbf{X}-\sin(\alpha)\dot{\mathbf{X}}+(1-\cos(\alpha))\ddot{\mathbf{X}}. \qquad (12.59)
\end{aligned}$$

Similarly, it follows that

$$\mathbf{S}(\alpha)=\mathbf{S}-\sin(\alpha)\dot{\mathbf{S}}+(1-\cos(\alpha))\ddot{\mathbf{S}}$$

$$\mathbf{y}(\alpha)=\mathbf{y}-\sin(\alpha)\dot{\mathbf{y}}+(1-\cos(\alpha))\ddot{\mathbf{y}}.$$

This finishes the proof. ∎

Define

$$\sin(\hat{\alpha}) := \max\left\{\sin(\alpha):\ (\mathbf{X}(\alpha),\mathbf{y}(\alpha),\mathbf{S}(\alpha))\in N_{-\infty}(\gamma),\ \forall\, \alpha\in\left[0,\frac{\pi}{2}\right]\right\}. \qquad (12.60)$$

We are now in the position to describe the wide neighborhood interior-point algorithm with arc-search for SDP problem.

Algorithm 12.1
Data: $\varepsilon>0$, $\gamma\in(0,1)$, *and* $\sigma\in(0,1)$.
Initial point: $(\mathbf{X}^0,\mathbf{y}^0,\mathbf{S}^0)\in N_{-\infty}(\gamma)$, *and* $\mu_0=\frac{\mathbf{X}^0\bullet\mathbf{S}^0}{n}$.
for *iteration* $k=0,1,2,\ldots$

Step 1. If $\mathbf{X}^k\bullet\mathbf{S}^k\le\varepsilon$, *then stop.*

Step 2. Solve the systems (12.36) and (12.37) to get $(\dot{\mathbf{X}},\dot{\mathbf{y}},\dot{\mathbf{S}})$ *and* $(\ddot{\mathbf{X}},\ddot{\mathbf{y}},\ddot{\mathbf{S}})$.

Step 3. Compute $\sin(\hat{\alpha})$ *by (12.60).*

Step 4. Calculate $(\mathbf{X}^{k+1},\mathbf{y}^{k+1},\mathbf{S}^{k+1})=(\mathbf{X}(\hat{\alpha}^k),\mathbf{y}(\hat{\alpha}^k),\mathbf{S}(\hat{\alpha}^k))$ *and set* $\mu_{k+1}=\frac{\mathbf{X}^{k+1}\bullet\mathbf{S}^{k+1}}{n}$,

Step 5. $k:=k+1$. *Go back to step 1.*

end *(for)*

The following lemma indicates that if the initial point satisfies the equality constraints in $N_{-\infty}(\gamma)$, then a search along the ellipse will also meet the equality constraints in $N_{-\infty}(\gamma)$.

Lemma 12.8
Let $(\mathbf{X},\mathbf{y},\mathbf{S})$ *be a strictly feasible point of the primal SDP (12.19) and the dual SDP (12.20),* $(\dot{\mathbf{X}},\dot{\mathbf{y}},\dot{\mathbf{S}})$ *and* $(\ddot{\mathbf{X}},\ddot{\mathbf{y}},\ddot{\mathbf{S}})$ *satisfy (12.36) and (12.37), respectively,* $(\mathbf{X}(\alpha),\mathbf{y}(\alpha),\mathbf{S}(\alpha))$ *are calculated using Lemma 12.7, then*

$$\mathbf{A}_i \bullet \mathbf{X}(\alpha) = b_i, \qquad \sum_{i=1}^{m} y_i(\alpha)\mathbf{A}_i + \mathbf{S}(\alpha) = \mathbf{c}.$$

Proof 12.7 Since $(\mathbf{X},\mathbf{y},\mathbf{S})$ is a strictly feasible point, and from Lemma 12.7, (12.36) and (12.37), we have

$$\begin{aligned}
\mathbf{A}_i \bullet \mathbf{X}(\alpha) &= \mathbf{A}_i \bullet [\mathbf{X} - \sin(\alpha)\dot{\mathbf{X}} + (1-\cos(\alpha))\ddot{\mathbf{X}}] \\
&= \mathbf{A}_i \bullet \mathbf{X} - \sin(\alpha)\mathbf{A}_i \bullet \dot{\mathbf{X}} + (1-\cos(\alpha))\mathbf{A}_i \bullet \ddot{\mathbf{X}} \\
&= b_i.
\end{aligned}$$

Similarly, we obtain

$$\sum_{i=1}^{m} y_i(\alpha)\mathbf{A}_i + \mathbf{S}(\alpha) = \mathbf{c}.$$

This completes the proof. ■

The next two lemmas derive a relation between the duality measure and updated duality measure.

Lemma 12.9
Let $(\mathbf{X},\mathbf{y},\mathbf{S})$ *be a strictly feasible point of the primal SDP (12.19) and the dual (12.20),* $(\dot{\mathbf{X}},\dot{\mathbf{y}},\dot{\mathbf{S}})$ *and* $(\ddot{\mathbf{X}},\ddot{\mathbf{y}},\ddot{\mathbf{S}})$ *satisfy (12.36) and (12.37), respectively,* $(\mathbf{X}(\alpha),\mathbf{y}(\alpha),\mathbf{S}(\alpha))$ *are calculated using Lemma 12.7. Let* $g(\alpha) := 1-\cos(\alpha)$*, then*

$$\mathbf{H_P}(\mathbf{X}(\alpha)\mathbf{S}(\alpha)) = (1-\sin(\alpha))\mathbf{H_P}(\mathbf{XS}) + \sin(\alpha)\sigma\mu\mathbf{I} + \mathbf{H_P}(\chi(\alpha)), \tag{12.61}$$

where

$$\begin{aligned}
\mathbf{H_P}(\chi(\alpha)) := \; & -g^2(\alpha)\mathbf{H_P}(\dot{\mathbf{X}}\ddot{\mathbf{S}}) - \sin(\alpha)g(\alpha)\mathbf{H_P}(\dot{\mathbf{X}}\ddot{\mathbf{S}} + \ddot{\mathbf{X}}\dot{\mathbf{S}}) \\
& + g^2(\alpha)\mathbf{H_P}(\ddot{\mathbf{X}}\ddot{\mathbf{S}}).
\end{aligned} \tag{12.62}$$

Proof 12.8 From Lemma 12.7, we get

$$\begin{aligned}
& \mathbf{X}(\alpha)\mathbf{S}(\alpha) \\
= \; & [\mathbf{X} - \sin(\alpha)\dot{\mathbf{X}} + (1-\cos(\alpha))\ddot{\mathbf{X}}][\mathbf{S} - \sin(\alpha)\dot{\mathbf{S}} + (1-\cos(\alpha))\ddot{\mathbf{S}}] \\
= \; & \mathbf{XS} - \sin(\alpha)\dot{\mathbf{X}}\mathbf{S} + (1-\cos(\alpha))\ddot{\mathbf{X}}\mathbf{S} \\
& - \sin(\alpha)\mathbf{X}\dot{\mathbf{S}} + \sin^2(\alpha)\dot{\mathbf{X}}\dot{\mathbf{S}} - (1-\cos(\alpha))\sin(\alpha)\ddot{\mathbf{X}}\dot{\mathbf{S}} \\
& + (1-\cos(\alpha))\mathbf{X}\ddot{\mathbf{S}} - \sin(\alpha)(1-\cos(\alpha))\dot{\mathbf{X}}\ddot{\mathbf{S}} + (1-\cos(\alpha))^2\ddot{\mathbf{X}}\ddot{\mathbf{S}} \\
= \; & \mathbf{XS} - \sin(\alpha)(\dot{\mathbf{X}}\mathbf{S} + \mathbf{X}\dot{\mathbf{S}}) + g(\alpha)(\ddot{\mathbf{X}}\mathbf{S} + \mathbf{X}\ddot{\mathbf{S}}) + \sin^2(\alpha)(\dot{\mathbf{X}}\dot{\mathbf{S}}) \\
& - \sin(\alpha)g(\alpha)(\dot{\mathbf{X}}\ddot{\mathbf{S}} + \ddot{\mathbf{X}}\dot{\mathbf{S}}) + g^2(\alpha)(\ddot{\mathbf{X}}\ddot{\mathbf{S}}).
\end{aligned}$$

Applying Lemma 12.4 (the linearity of $\mathbf{H_P}(\cdot)$) to this equality, and using the last rows of (12.36) and (12.37) and the fact that

$$\begin{aligned}
\sin^2(\alpha) - 2g(\alpha) &= \sin^2(\alpha) - 2(1-\cos(\alpha)) \\
&= \sin^2(\alpha) - 1 + 2\cos(\alpha) - \sin^2(\alpha) - \cos^2(\alpha) \\
&= -(1-\cos(\alpha))^2 = -g^2(\alpha),
\end{aligned}$$

we obtain

$$\begin{aligned}
\mathbf{H_P}(\mathbf{X}(\alpha)\mathbf{S}(\alpha)) &= \mathbf{H_P}(\mathbf{XS}) - \sin(\alpha)\mathbf{H_P}(\dot{\mathbf{X}}\mathbf{S} + \mathbf{X}\dot{\mathbf{S}}) + g^2(\alpha)\mathbf{H_P}(\ddot{\mathbf{X}}\ddot{\mathbf{S}}) \\
&\quad + \sin^2(\alpha)\mathbf{H_P}(\dot{\mathbf{X}}\dot{\mathbf{S}}) - \sin(\alpha)g(\alpha)\mathbf{H_P}(\dot{\mathbf{X}}\ddot{\mathbf{S}} + \ddot{\mathbf{X}}\dot{\mathbf{S}}) \\
&\quad + g(\alpha)\mathbf{H_P}(\ddot{\mathbf{X}}\mathbf{S} + \mathbf{X}\ddot{\mathbf{S}}) \\
\text{[use (12.36)]} &= (1-\sin(\alpha))\mathbf{H_P}(\mathbf{XS}) + \sin(\alpha)\sigma\mu\mathbf{I} \\
&\quad - \sin(\alpha)g(\alpha)\mathbf{H_P}(\dot{\mathbf{X}}\ddot{\mathbf{S}} + \ddot{\mathbf{X}}\dot{\mathbf{S}}) \\
&\quad + \sin^2(\alpha)\mathbf{H_P}(\dot{\mathbf{X}}\dot{\mathbf{S}}) + g^2(\alpha)\mathbf{H_P}(\ddot{\mathbf{X}}\ddot{\mathbf{S}}) \\
&\quad + g(\alpha)\mathbf{H_P}(\ddot{\mathbf{X}}\mathbf{S} + \mathbf{X}\ddot{\mathbf{S}}) \\
\text{[use (12.37)]} &= (1-\sin(\alpha))\mathbf{H_P}(\mathbf{XS}) + \sin(\alpha)\sigma\mu\mathbf{I} \\
&\quad - \sin(\alpha)g(\alpha)\mathbf{H_P}(\dot{\mathbf{X}}\ddot{\mathbf{S}} + \ddot{\mathbf{X}}\dot{\mathbf{S}}) \\
&\quad + \sin^2(\alpha)\mathbf{H_P}(\dot{\mathbf{X}}\dot{\mathbf{S}}) - 2g(\alpha)\mathbf{H_P}(\dot{\mathbf{X}}\dot{\mathbf{S}}) \\
&\quad + g^2(\alpha)\mathbf{H_P}(\ddot{\mathbf{X}}\ddot{\mathbf{S}}) \\
&= (1 - \sin(\alpha))\mathbf{H_P}(\mathbf{XS}) + \sin(\alpha)\sigma\mu\mathbf{I} \\
&\quad - \sin(\alpha)g(\alpha)\mathbf{H_P}(\dot{\mathbf{X}}\ddot{\mathbf{S}} + \ddot{\mathbf{X}}\dot{\mathbf{S}}) \\
&\quad - g^2(\alpha)\mathbf{H_P}(\dot{\mathbf{X}}\dot{\mathbf{S}}) + g^2(\alpha)\mathbf{H_P}(\ddot{\mathbf{X}}\ddot{\mathbf{S}}) \\
&= (1-\sin(\alpha))\mathbf{H_P}(\mathbf{XS}) + \sin(\alpha)\sigma\mu\mathbf{I} + \mathbf{H_P}(\chi(\alpha)),
\end{aligned} \tag{12.63}$$

which completes the proof. ■

Lemma 12.10

Let $(\mathbf{X},\mathbf{y},\mathbf{S}) \in \mathcal{N}_{-\infty}(\gamma)$, $(\dot{\mathbf{X}},\dot{\mathbf{y}},\dot{\mathbf{S}})$ and $(\ddot{\mathbf{X}},\ddot{\mathbf{y}},\ddot{\mathbf{S}})$ be calculated from (12.36) and (12.37). Let $\mathbf{X}(\alpha)$ and $\mathbf{S}(\alpha)$ be defined as in Lemma 12.7. Then,

$$\mu(\alpha) = [1-(1-\sigma)\sin(\alpha)]\mu.$$

Proof 12.9 Using the identity $Tr(\mathbf{H_P}(\mathbf{M})) = Tr(\mathbf{M})$ in Lemma 12.4 and (12.61) in Lemma 12.9 and (12.51), we have

$$\begin{aligned}
\mathbf{X}(\alpha)\bullet\mathbf{S}(\alpha) &= Tr(\mathbf{X}(\alpha)\mathbf{S}(\alpha)) = Tr(\mathbf{H_P}(\mathbf{X}(\alpha)\mathbf{S}(\alpha))) \\
&= (1-\sin(\alpha))Tr(\mathbf{H_P}(\mathbf{XS})) + \sin(\alpha)\sigma\mu n + Tr(\mathbf{H_P}(\chi(\alpha))) \\
&= (1-\sin(\alpha))\mathbf{X}\bullet\mathbf{S} + \sin(\alpha)\sigma\mu n \\
&\quad - \sin(\alpha)g(\alpha)(\dot{\mathbf{X}}\bullet\ddot{\mathbf{S}} + \ddot{\mathbf{X}}\bullet\dot{\mathbf{S}})
\end{aligned}$$

$$-g^2(\alpha)(\dot{\mathbf{X}} \bullet \dot{\mathbf{S}}) + g^2(\alpha)(\ddot{\mathbf{X}} \bullet \ddot{\mathbf{S}}) \tag{12.64}$$

$$= [1-(1-\sigma)\sin(\alpha)]n\mu. \tag{12.65}$$

Dividing both sides by n yields

$$\mu(\alpha) = \frac{\mathbf{X}(\alpha) \bullet \mathbf{S}(\alpha)}{n} = [1-(1-\sigma)\sin(\alpha)]\mu.$$

This completes the proof. ∎

To proceed, we need several technical lemmas, which are derived by Monteiro and Zhang in [98]. We denote the eigenvalues of the matrix **XS** as $\lambda_1 \le \lambda_2 \le \ldots, \le \lambda_n$, and $\lambda[(\mathbf{XS})]$ be any eigenvalue of **XS**. Since $\lambda[(\mathbf{XS})] = \lambda[(\mathbf{XS})^{\mathrm{T}}] = \lambda[(\mathbf{S}^{\mathrm{T}}\mathbf{X}^{\mathrm{T}})] = \lambda[(\mathbf{SX})]$, and $\mathbf{S}^{1/2}\mathbf{X}\mathbf{S}^{1/2}$, $\mathbf{X}^{1/2}\mathbf{S}\mathbf{X}^{1/2}$, $\hat{\mathbf{X}}\hat{\mathbf{S}}$, and $\hat{\mathbf{S}}\hat{\mathbf{X}}$ are similar to either **XS** or **SX**, they all have the same eigenvalues. In addition, we let Λ denote the diagonal matrix

$$\Lambda = \mathrm{diag}(\lambda_1, \lambda_2, \ldots, \lambda_n).$$

Lemma 12.11
For any $\mathbf{P} \in \mathcal{P}(\mathbf{X},\mathbf{S})$*, there exists an orthogonal matrix* $\mathbf{Q_P}$ *and diagonal matrices* $\Lambda(\mathbf{X})$ *and* $\Lambda(\mathbf{S})$ *such that*

(i). $\hat{\mathbf{X}} := \mathbf{PXP} = \mathbf{Q_P}\Lambda(\hat{\mathbf{X}})\mathbf{Q_P^{\mathrm{T}}}$.

(ii). $\hat{\mathbf{S}} := \mathbf{P}^{-1}\mathbf{S}\mathbf{P}^{-1} = \mathbf{Q_P}\Lambda(\hat{\mathbf{S}})\mathbf{Q_P^{\mathrm{T}}}$.

(iii). $\Lambda = \Lambda(\hat{\mathbf{X}})\Lambda(\hat{\mathbf{S}})$*, and hence,* $\hat{\mathbf{X}}\hat{\mathbf{S}} = \hat{\mathbf{S}}\hat{\mathbf{X}} = \mathbf{Q_P}\Lambda\mathbf{Q_P^{\mathrm{T}}}$.

Proof 12.10 Noticing that $\hat{\mathbf{X}} \in \mathcal{S}^n_{++}$ and $\hat{\mathbf{S}} \in \mathcal{S}^n_{++}$, the commutativity of $\hat{\mathbf{X}}$ and $\hat{\mathbf{S}}$ ensures that the two matrices share a common set of orthogonal eigenvectors [54, Theorem 1.3.19], from which (i) and (ii) follow. Moreover, by (i) and (ii), we have $\hat{\mathbf{X}}\hat{\mathbf{S}} = \mathbf{Q_P}\Lambda(\hat{\mathbf{X}})\Lambda(\hat{\mathbf{S}})\mathbf{Q_P^{\mathrm{T}}}$. Since the spectra **XS** and $\hat{\mathbf{X}}\hat{\mathbf{S}}$ are the same, by permuting the columns of $\mathbf{Q_P}$, if necessary, we have $\Lambda = \Lambda(\hat{\mathbf{X}})\Lambda(\hat{\mathbf{S}})$, therefore, (iii) holds. ∎

Lemma 12.12
Let $\mathbf{P} \in \mathcal{P}(\mathbf{X},\mathbf{S})$ *and* $\mathbf{G} := \hat{\mathbf{E}}^{-1}\hat{\mathbf{F}}$*. Then,*

$$\|\mathbf{G}^{-1/2}\mathrm{vec}(\dot{\hat{\mathbf{X}}})\|^2 + \|\mathbf{G}^{1/2}\mathrm{vec}(\dot{\hat{\mathbf{S}}})\|^2 + 2\dot{\hat{\mathbf{X}}} \bullet \dot{\hat{\mathbf{S}}} = \sum_{i=1}^{n} \frac{(\sigma\mu - \lambda_i)^2}{\lambda_i}.$$

Moreover, if $\lambda_{\min}(\mathbf{XS}) \ge \gamma\mu$ *for some* $\gamma \in (0,1)$*, then*

$$\sum_{i=1}^{n} \frac{(\sigma\mu - \lambda_i)^2}{\lambda_i} \le (1 - 2\sigma + \sigma^2/\gamma)n\mu. \tag{12.66}$$

Proof 12.11 Pre-multiplying both sides of (12.42) by $(\hat{\mathbf{F}}\hat{\mathbf{E}})^{-\frac{1}{2}}$, taking 2-norm squared, noticing that $\hat{\mathbf{E}}$ and $\hat{\mathbf{F}}$ are positive definite and commutative, and using (12.12), we get

$$\begin{aligned} & \|(\hat{\mathbf{F}}\hat{\mathbf{E}})^{-\frac{1}{2}}\hat{\mathbf{E}}\text{vec}\dot{\hat{\mathbf{X}}}\|^2 + \|(\hat{\mathbf{F}}\hat{\mathbf{E}})^{-\frac{1}{2}}\hat{\mathbf{F}}\text{vec}\dot{\hat{\mathbf{S}}}\|^2 + 2\dot{\hat{\mathbf{X}}} \bullet \dot{\hat{\mathbf{S}}} \\ = & \|(\hat{\mathbf{F}}\hat{\mathbf{E}})^{-\frac{1}{2}}\text{vec}(\mathbf{H}(\hat{\mathbf{X}}\hat{\mathbf{S}}) - \sigma\mu\mathbf{I})\|^2. \end{aligned}$$

Since $\mathbf{P} \in \mathcal{P}(\mathbf{X}, \mathbf{S})$, $\hat{\mathbf{F}}$ and $\hat{\mathbf{E}}$ are commutative, which implies that

$$(\hat{\mathbf{F}}\hat{\mathbf{E}})^{-\frac{1}{2}}\hat{\mathbf{E}} = (\hat{\mathbf{E}}\hat{\mathbf{F}})^{-\frac{1}{2}}\hat{\mathbf{E}} = \hat{\mathbf{F}}^{-\frac{1}{2}}\hat{\mathbf{E}}^{-\frac{1}{2}}\hat{\mathbf{E}} = (\hat{\mathbf{E}}^{-1}\hat{\mathbf{F}})^{-\frac{1}{2}} = \mathbf{G}^{-\frac{1}{2}},$$

$$(\hat{\mathbf{F}}\hat{\mathbf{E}})^{-\frac{1}{2}}\hat{\mathbf{F}} = \hat{\mathbf{E}}^{-\frac{1}{2}}\hat{\mathbf{F}}^{-\frac{1}{2}}\hat{\mathbf{F}} = (\hat{\mathbf{E}}^{-1}\hat{\mathbf{F}})^{\frac{1}{2}} = \mathbf{G}^{\frac{1}{2}}.$$

Hence, for the proof of the first statement it remains to show that

$$\|(\hat{\mathbf{F}}\hat{\mathbf{E}})^{-\frac{1}{2}}\text{vec}(\mathbf{H}(\hat{\mathbf{X}}\hat{\mathbf{S}}) - \sigma\mu\mathbf{I})\|^2 = \sum_{i=1}^{n} \frac{(\sigma\mu - \lambda_i)^2}{\lambda_i}. \tag{12.67}$$

Using (12.43), Lemma 12.11 (ii), and (12.10), we find the spectral decomposition of $\hat{\mathbf{E}}$ to be

$$\hat{\mathbf{E}} = \frac{1}{2}(\hat{\mathbf{S}} \otimes \mathbf{I} + \mathbf{I} \otimes \hat{\mathbf{S}}) = \frac{1}{2}\hat{\mathbf{Q}}[\Lambda(\hat{\mathbf{S}}) \otimes \mathbf{I} + \mathbf{I} \otimes \Lambda(\hat{\mathbf{S}})]\hat{\mathbf{Q}}^{\mathrm{T}}.$$

where $\hat{\mathbf{Q}} = \mathbf{Q}_{\mathbf{P}} \otimes \mathbf{Q}_{\mathbf{P}}$ is an orthogonal matrix of dimension n^2. Similarly, by (12.43) and Lemma 12.11 (i), we have

$$\hat{\mathbf{F}} = \frac{1}{2}(\hat{\mathbf{X}} \otimes \mathbf{I} + \mathbf{I} \otimes \hat{\mathbf{X}}) = \frac{1}{2}\hat{\mathbf{Q}}[\Lambda(\hat{\mathbf{X}}) \otimes \mathbf{I} + \mathbf{I} \otimes \Lambda(\hat{\mathbf{X}})]\hat{\mathbf{Q}}^{\mathrm{T}}.$$

Therefore, using Lemma 12.11 (iii), we obtain

$$(\hat{\mathbf{F}}\hat{\mathbf{E}})^{-1} = 4\hat{\mathbf{Q}}[\Lambda \otimes \mathbf{I} + \mathbf{I} \otimes \Lambda + \Lambda(\mathbf{X}) \otimes \Lambda(\mathbf{S}) + \Lambda(\mathbf{S}) \otimes \Lambda(\mathbf{X})]^{-1}\hat{\mathbf{Q}}^{\mathrm{T}},$$

where, from Lemma 12.2, the matrix in the middle is diagonal with the property that its $((i-1)n+i)$-th diagonal element is equal to $1/(4\lambda_i)$, for $i = 1, \ldots, n$. For $i = 1$, the first diagonal element is equal to $1/(4\lambda_1)$. On the other hand, observe that

$$\text{vec}(\hat{\mathbf{X}}\hat{\mathbf{S}} - \sigma\mu\mathbf{I}) = \hat{\mathbf{Q}}\text{vec}(\Lambda - \sigma\mu\mathbf{I}),$$

where $\text{vec}(\Lambda - \sigma\mu\mathbf{I})$ is an n^2-vector having at most n nonzero components, namely, its $((i-1)n+i)$-th component is equal to $\lambda_i - \sigma\mu$ for $i = 1, \ldots, n$. The above two relations and a straightforward verification finally yield

$$\text{vec}(\hat{\mathbf{X}}\hat{\mathbf{S}} - \sigma\mu\mathbf{I})^{\mathrm{T}}(\hat{\mathbf{E}}\hat{\mathbf{F}})^{-1}\text{vec}(\hat{\mathbf{X}}\hat{\mathbf{S}} - \sigma\mu\mathbf{I}) = \sum_{i=1}^{n} \frac{(\sigma\mu - \lambda_i)^2}{\lambda_i},$$

which proves (12.67) and, hence, the first part of the lemma. To prove the second part of the lemma, we use the fact that $n\mu = Tr(\mathbf{XS}) = \sum_{i=1}^{n} \lambda_i$ in order to obtain

$$\sum_{i=1}^{n} \frac{(\sigma\mu - \lambda_i)^2}{\lambda_i} \leq \sigma^2\mu^2\frac{n}{\lambda_1} - 2\sigma n\mu + \sum_{i=1}^{n} \lambda_i \leq \frac{\sigma^2 n\mu}{\gamma} - 2\sigma n\mu + n\mu,$$

which completes the proof of the lemma. ∎

The next technical lemma is as follows:

Lemma 12.13
Suppose that $(\mathbf{X},\mathbf{y},\mathbf{S}) \in \mathcal{S}^n_{++} \times \mathbb{R}^m \times \mathcal{S}^n_{++}$, $\mathbf{P} \in \mathcal{S}^n_{++}$, *and* $\mathbf{Q} \in \mathcal{P}(\mathbf{X},\mathbf{S})$. *Then,*

$$\lambda_{\min}[\mathbf{H_P}(\mathbf{X},\mathbf{S})] \le \lambda_{\min}[\mathbf{XS}] = \lambda_{\min}[\mathbf{H_Q}(\mathbf{X},\mathbf{S})]. \tag{12.68}$$

Proof 12.12 Since $\mathbf{Q} \in \mathcal{P}(\mathbf{X},\mathbf{S})$, we have

$$\begin{aligned} \mathbf{H_Q}(\mathbf{XS}) &= \tfrac{1}{2}\left[\mathbf{QXQQ}^{-1}\mathbf{SQ}^{-1} + (\mathbf{QXQQ}^{-1}\mathbf{SQ}^{-1})^{\mathrm{T}}\right] \\ &= \tfrac{1}{2}\left[\hat{\mathbf{X}}\hat{\mathbf{S}} + \hat{\mathbf{S}}\hat{\mathbf{X}}\right] = \hat{\mathbf{X}}\hat{\mathbf{S}} = \mathbf{QXSQ}^{-1}. \end{aligned}$$

By similarity,

$$\lambda_{\min}[\mathbf{XS}] = \lambda_{\min}[\mathbf{QXSQ}^{-1}] = \lambda_{\min}[\mathbf{H_Q}(\mathbf{XS})].$$

Moreover,

$$\lambda_{\min}[\mathbf{XS}] = \lambda_{\min}[\mathbf{PXSP}^{-1}] \ge \lambda_{\min}[\mathbf{H}(\mathbf{PXSP}^{-1})] = \lambda_{\min}[\mathbf{H}(\mathbf{XS})],$$

where the inequality follows from the fact that the real part of the spectrum of a real matrix is contained between the largest and smallest eigenvalues of its Hermitian part (see p.187 of [54]), for example). We have, thus, shown that (12.68) holds. ∎

The last technical lemma is given as:

Lemma 12.14
For any $\mathbf{u},\mathbf{v} \in \mathbb{R}^n$ *and* $\mathbf{G} \in \mathcal{S}^n_{++}$, *we have*

$$\|\mathbf{u}\|\|\mathbf{v}\| \le \sqrt{cond(\mathbf{G})}\|\mathbf{G}^{-1/2}\mathbf{u}\|\|\mathbf{G}^{1/2}\mathbf{v}\| \le \frac{\sqrt{cond(\mathbf{G})}}{2}\left(\|\mathbf{G}^{-1/2}\mathbf{u}\|^2 + \|\mathbf{G}^{1/2}\mathbf{v}\|^2\right). \tag{12.69}$$

Proof 12.13 We have

$$\|\mathbf{u}\|^2 \le \frac{\mathbf{u}^{\mathrm{T}}\mathbf{G}^{-1}\mathbf{u}}{\lambda_{\min}(\mathbf{G}^{-1})} = \lambda_{\max}(\mathbf{G})\|\mathbf{G}^{-1/2}\mathbf{u}\|^2$$

and

$$\|\mathbf{v}\|^2 \le \frac{\mathbf{v}^{\mathrm{T}}\mathbf{G}\mathbf{v}}{\lambda_{\min}(\mathbf{G})} = \frac{\|\mathbf{G}^{1/2}\mathbf{v}\|^2}{\lambda_{\min}(\mathbf{G})}.$$

Combining these two inequality gives

$$\|\mathbf{u}\|\|\mathbf{v}\| \le \sqrt{cond(\mathbf{G})}\|\mathbf{G}^{-1/2}\mathbf{u}\|\|\mathbf{G}^{1/2}\mathbf{v}\| \le \frac{\sqrt{cond(\mathbf{G})}}{2}(\|\mathbf{G}^{-1/2}\mathbf{u}\|^2 + \|\mathbf{G}^{1/2}\mathbf{v}\|^2)$$

where the last inequality follows from $ab \leq \frac{1}{2}(a^2+b^2)$. ∎

The remaining proofs are provided by Zhang, Yuan, Zhou, Luo, and Huang in [170].

Lemma 12.15
Let a point $(\mathbf{X},\mathbf{y},\mathbf{S}) \in \mathcal{N}_{-\infty}(\gamma)$ *and a scaling matrix* $\mathbf{P} \in \mathcal{P}(\mathbf{X},\mathbf{S})$ *be given, and define* $\mathbf{G} = \hat{\mathbf{E}}^{-1}\hat{\mathbf{F}}$ *where* $\hat{\mathbf{E}}$ *and* $\hat{\mathbf{F}}$ *are given by equation (12.43). Then, the solution* $(\dot{\mathbf{X}},\dot{\mathbf{y}},\dot{\mathbf{S}})$ *of (12.36) satisfies*

$$||\mathbf{H_P}(\dot{\mathbf{X}}\dot{\mathbf{S}})||_F \leq \frac{\sqrt{cond(\mathbf{G})}}{2}\beta_1\mu n,$$

and

$$||\mathbf{G}^{-\frac{1}{2}}\mathrm{vec}\dot{\hat{\mathbf{X}}}|| \leq \sqrt{\beta_1\mu n}, \quad ||\mathbf{G}^{\frac{1}{2}}\mathrm{vec}\dot{\hat{\mathbf{S}}}|| \leq \sqrt{\beta_1\mu n},$$

where $\beta_1 \geq 1 \geq 1-2\sigma+\frac{\sigma^2}{\gamma}$.

Proof 12.14 Pre-multiplying the equation of (12.42) by $(\hat{\mathbf{E}}\hat{\mathbf{F}})^{-\frac{1}{2}}$ and taking norm-squared on both sides, we have

$$\begin{aligned} & ||(\hat{\mathbf{F}}\hat{\mathbf{E}})^{-\frac{1}{2}}\hat{\mathbf{E}}\mathrm{vec}\dot{\hat{\mathbf{X}}}||^2 + ||(\hat{\mathbf{F}}\hat{\mathbf{E}})^{-\frac{1}{2}}\hat{\mathbf{F}}\mathrm{vec}\dot{\hat{\mathbf{S}}}||^2 + 2\dot{\hat{\mathbf{X}}}\bullet\dot{\hat{\mathbf{S}}} \\ = \; & ||(\hat{\mathbf{F}}\hat{\mathbf{E}})^{-\frac{1}{2}}\mathrm{vec}(\mathbf{H}(\hat{\mathbf{X}}\hat{\mathbf{S}})-\sigma\mu\mathbf{I})||^2. \end{aligned}$$

Since $\mathbf{P} \in \mathcal{P}(\mathbf{X},\mathbf{S})$, $\hat{\mathbf{F}}$ and $\hat{\mathbf{E}}$ are commutative, which implies that

$$(\hat{\mathbf{F}}\hat{\mathbf{E}})^{-\frac{1}{2}}\hat{\mathbf{E}} = (\hat{\mathbf{E}}^{-1}\hat{\mathbf{F}})^{-\frac{1}{2}} = \mathbf{G}^{-\frac{1}{2}},$$

$$(\hat{\mathbf{F}}\hat{\mathbf{E}})^{-\frac{1}{2}}\hat{\mathbf{F}} = (\hat{\mathbf{E}}^{-1}\hat{\mathbf{F}})^{\frac{1}{2}} = \mathbf{G}^{\frac{1}{2}}.$$

It follows that

$$||\mathbf{G}^{-\frac{1}{2}}\mathrm{vec}\dot{\hat{\mathbf{X}}}||^2 + ||\mathbf{G}^{\frac{1}{2}}\mathrm{vec}\dot{\hat{\mathbf{S}}}||^2 + 2\dot{\hat{\mathbf{X}}}\bullet\dot{\hat{\mathbf{S}}} = ||(\hat{\mathbf{F}}\hat{\mathbf{E}})^{-\frac{1}{2}}\mathrm{vec}(\mathbf{H}(\hat{\mathbf{X}}\hat{\mathbf{S}})-\sigma\mu\mathbf{I})||^2.$$

From $\dot{\hat{\mathbf{X}}}\bullet\dot{\hat{\mathbf{S}}} = 0$ and Lemma 12.12, we have

$$||\mathbf{G}^{-\frac{1}{2}}\mathrm{vec}\dot{\hat{\mathbf{X}}}||^2 + ||\mathbf{G}^{\frac{1}{2}}\mathrm{vec}\dot{\hat{\mathbf{S}}}||^2 \leq \left(1-2\sigma+\frac{\sigma^2}{\gamma}\right)n\mu \leq \beta_1\mu n.$$

By Lemma 12.14, we can obtain

$$\begin{aligned} ||\dot{\hat{\mathbf{X}}}\dot{\hat{\mathbf{S}}}||_F & \leq ||\dot{\hat{\mathbf{X}}}||_F||\dot{\hat{\mathbf{S}}}||_F \\ & = ||\mathrm{vec}\dot{\hat{\mathbf{X}}}|| \; ||\mathrm{vec}\dot{\hat{\mathbf{S}}}|| \\ & \leq \frac{\sqrt{cond(\mathbf{G})}}{2}\left[||\mathbf{G}^{-\frac{1}{2}}\mathrm{vec}\dot{\hat{\mathbf{X}}}||^2 + ||\mathbf{G}^{\frac{1}{2}}\mathrm{vec}\dot{\hat{\mathbf{S}}}||^2\right] \\ & \leq \frac{\sqrt{cond(\mathbf{G})}}{2}\beta_1\mu n. \end{aligned}$$

Thus, we have

$$\begin{aligned}
\|\mathbf{H_P}(\dot{\mathbf{X}}\dot{\mathbf{S}})\|_F &= \|\mathbf{H}(\dot{\hat{\mathbf{X}}}\dot{\hat{\mathbf{S}}})\|_F \\
&= \left\| \tfrac{1}{2}\left[\dot{\hat{\mathbf{X}}}\dot{\hat{\mathbf{S}}} + \left(\dot{\hat{\mathbf{X}}}\dot{\hat{\mathbf{S}}}\right)^{\mathrm{T}}\right]\right\|_F \\
&\le \tfrac{1}{2}\left(\left\|\dot{\hat{\mathbf{X}}}\dot{\hat{\mathbf{S}}}\right\|_F + \left\|\left(\dot{\hat{\mathbf{X}}}\dot{\hat{\mathbf{S}}}\right)^{\mathrm{T}}\right\|_F\right) \\
&= \|\dot{\hat{\mathbf{X}}}\dot{\hat{\mathbf{S}}}\|_F \\
&\le \frac{\sqrt{cond(\mathbf{G})}}{2}\beta_1\mu n,
\end{aligned}$$

and

$$\|\mathbf{G}^{-\frac{1}{2}}\mathrm{vec}\dot{\hat{\mathbf{X}}}\| \le \sqrt{\beta_1\mu n}, \quad \|\mathbf{G}^{\frac{1}{2}}\mathrm{vec}\dot{\hat{\mathbf{S}}}\| \le \sqrt{\beta_1\mu n}.$$

This completes the proof. ■

Corollary 12.2
Let a point $(\mathbf{X},\mathbf{y},\mathbf{S}) \in \mathcal{N}_{-\infty}(\gamma)$, $\mathbf{G} \in \mathcal{S}^n_{++}$ and $\mathbf{P} \in \mathcal{P}(\mathbf{X},\mathbf{S})$ is the NT scaling matrix. Then,

$$\|\mathbf{H_P}(\dot{\mathbf{X}}\dot{\mathbf{S}})\|_F \le \frac{1}{2}\beta_1\mu n.$$

Proof 12.15 Since $\mathbf{P}$ is the NT scaling matrix, we have (see also [134, 98])

$$\mathbf{W}_{nt}\mathbf{X}\mathbf{W}_{nt} = \mathbf{X}^{-\frac{1}{2}}\underbrace{(\mathbf{X}^{\frac{1}{2}}\mathbf{S}\mathbf{X}^{\frac{1}{2}})^{\frac{1}{2}}\underbrace{\mathbf{X}^{-\frac{1}{2}}\mathbf{X}\mathbf{X}^{-\frac{1}{2}}}_{\mathbf{I}}(\mathbf{X}^{\frac{1}{2}}\mathbf{S}\mathbf{X}^{\frac{1}{2}})^{\frac{1}{2}}}_{\mathbf{X}^{\frac{1}{2}}\mathbf{S}\mathbf{X}^{\frac{1}{2}}}\mathbf{X}^{-\frac{1}{2}} = \mathbf{S},$$

which is equivalent to

$$\mathbf{W}_{nt}^{\frac{1}{2}}\mathbf{X}\mathbf{W}_{nt}^{\frac{1}{2}} = \mathbf{W}_{nt}^{-\frac{1}{2}}\mathbf{S}\mathbf{W}_{nt}^{-\frac{1}{2}},$$

which means $\hat{\mathbf{X}} = \mathbf{PXP} = \mathbf{P}^{-1}\mathbf{S}\mathbf{P}^{-1} = \hat{\mathbf{S}}$ and consequently $\hat{\mathbf{E}} = \hat{\mathbf{F}}$. This implies $cond(\mathbf{G}) = 1$. In view of Lemma 12.15, the claim follows. ■

Lemma 12.16
Let a point $(\mathbf{X},\mathbf{y},\mathbf{S}) \in \mathcal{N}_{-\infty}(\gamma)$ and $\mathbf{P} \in \mathcal{P}(\mathbf{X},\mathbf{S})$ be the NT scaling matrix, and define $\mathbf{G} = \hat{\mathbf{E}}^{-1}\hat{\mathbf{F}}$, where $\hat{\mathbf{E}}$ and $\hat{\mathbf{F}}$ are given by equation (12.43). Then, the solution $(\ddot{\mathbf{X}},\ddot{\mathbf{y}},\ddot{\mathbf{S}})$ of (12.37) satisfies

$$\|\mathbf{H_P}(\ddot{\mathbf{X}}\ddot{\mathbf{S}})\|_F \le \frac{1}{8}\beta_2\mu n^2,$$

and

$$\|\mathbf{G}^{-\frac{1}{2}}\mathrm{vec}\ddot{\hat{\mathbf{X}}}\| \le \frac{1}{2}\sqrt{\beta_2\mu}n, \quad \|\mathbf{G}^{\frac{1}{2}}\mathrm{vec}\ddot{\hat{\mathbf{S}}}\| \le \frac{1}{2}\sqrt{\beta_2\mu}n,$$

where $\beta_2 = \frac{\beta_1^2}{\gamma} \ge 1$.

Proof 12.16 Pre-multiplying $(\hat{\mathbf{F}}\hat{\mathbf{E}})^{-\frac{1}{2}}$ to the equation (12.44), and taking the squared norm on both sides, we have

$$\|\mathbf{G}^{-\frac{1}{2}}\text{vec}\ddot{\mathbf{X}}\|^2 + \|\mathbf{G}^{\frac{1}{2}}\text{vec}\ddot{\mathbf{S}}\|^2 + 2\ddot{\mathbf{X}} \bullet \ddot{\mathbf{S}} = 4\|(\hat{\mathbf{F}}\hat{\mathbf{E}})^{-\frac{1}{2}}\text{vec}(\mathbf{H}(\dot{\mathbf{X}}\dot{\mathbf{S}}))\|^2.$$

Since NT scaling is used, $\ddot{\mathbf{X}} = \ddot{\mathbf{S}}$ and, consequently, $\hat{\mathbf{E}} = \hat{\mathbf{F}}$, which implies $cond(\mathbf{G}) = 1$. In addition, it has been proven in Lemma 12.12 that $\rho((\hat{\mathbf{F}}\hat{\mathbf{E}})^{-1}) \le \frac{1}{4\lambda_1}$. Using Lemmas 12.14, 12.15, we obtain

$$\begin{aligned}
4\|(\hat{\mathbf{F}}\hat{\mathbf{E}})^{-\frac{1}{2}}\text{vec}(\mathbf{H}(\dot{\mathbf{X}}\dot{\mathbf{S}}))\|^2 &\le 4\|(\hat{\mathbf{F}}\hat{\mathbf{E}})^{-\frac{1}{2}}\|^2\|\text{vec}(\dot{\mathbf{X}}\dot{\mathbf{S}})\|^2 \\
&= 4\rho((\hat{\mathbf{F}}\hat{\mathbf{E}})^{-1})\|(\dot{\mathbf{X}}\dot{\mathbf{S}})\|_F^2 \\
\text{since } \rho((\hat{\mathbf{F}}\hat{\mathbf{E}})^{-1}) \le \tfrac{1}{4\lambda_1} \quad &\le 4\tfrac{1}{4\lambda_1}\tfrac{1}{4}\beta_1^2\mu^2n^2 \\
&\le \tfrac{1}{4\gamma\mu}\beta_1^2\mu^2n^2 \\
&= \tfrac{1}{4}\beta_2\mu n^2.
\end{aligned}$$

Since $\ddot{\mathbf{X}} \bullet \ddot{\mathbf{S}} = 0$, we have

$$\|\mathbf{G}^{-\frac{1}{2}}\text{vec}\ddot{\mathbf{X}}\|^2 + \|\mathbf{G}^{\frac{1}{2}}\text{vec}\ddot{\mathbf{S}}\|^2 \le \frac{1}{4}\beta_2\mu n^2.$$

From Lemma 12.14 and $cond(\mathbf{G}) = 1$, we get

$$\begin{aligned}
\|\ddot{\mathbf{X}}\ddot{\mathbf{S}}\|_F &\le \|\ddot{\mathbf{X}}\|_F\|\ddot{\mathbf{S}}\|_F = \|\text{vec}\ddot{\mathbf{X}}\|\ \|\text{vec}\ddot{\mathbf{S}}\| \\
&\le \frac{\sqrt{cond(\mathbf{G})}}{2}\left(\|\mathbf{G}^{-1/2}\text{vec}\ddot{\mathbf{X}}\|^2 + \|\mathbf{G}^{1/2}\text{vec}\ddot{\mathbf{S}}\|^2\right) \\
&\le \frac{1}{2}\left(\frac{1}{4}\beta_2\mu n^2\right) = \frac{1}{8}\beta_2\mu n^2.
\end{aligned}$$

Hence, it follows that

$$\begin{aligned}
\|\mathbf{H_P}(\ddot{\mathbf{X}}\ddot{\mathbf{S}})\|_F &= \|\mathbf{H}(\ddot{\mathbf{X}}\ddot{\mathbf{S}})\|_F \\
&= \left\|\frac{1}{2}\left[\ddot{\mathbf{X}}\ddot{\mathbf{S}} + \left(\ddot{\mathbf{X}}\ddot{\mathbf{S}}\right)^{\mathrm{T}}\right]\right\|_F \\
&\le \frac{1}{2}\left(\left\|\ddot{\mathbf{X}}\ddot{\mathbf{S}}\right\|_F + \left\|\left(\ddot{\mathbf{X}}\ddot{\mathbf{S}}\right)^{\mathrm{T}}\right\|_F\right) \\
&= \|\ddot{\mathbf{X}}\ddot{\mathbf{S}}\|_F \\
&\le \frac{1}{8}\beta_2\mu n^2,
\end{aligned}$$

and

$$\|\mathbf{G}^{-\frac{1}{2}}\text{vec}\ddot{\mathbf{X}}\| \le \frac{1}{2}\sqrt{\beta_2\mu}n, \quad \|\mathbf{G}^{\frac{1}{2}}\text{vec}\ddot{\mathbf{S}}\| \le \frac{1}{2}\sqrt{\beta_2\mu}n.$$

The proof is completed. ∎

Lemma 12.17
Assume that NT scaling is used. Let $\beta_3 = \sqrt{\beta_1\beta_2} \geq 1$, *then*

$$\|\mathbf{H_P}(\dot{\mathbf{X}}\ddot{\mathbf{S}})\|_F \leq \frac{1}{2}\beta_3\mu n^{\frac{3}{2}}, \quad \|\mathbf{H}_P(\ddot{\mathbf{X}}\dot{\mathbf{S}})\|_F \leq \frac{1}{2}\beta_3\mu n^{\frac{3}{2}}.$$

Proof 12.17 Using Lemma 12.14 and $cond(\mathbf{G}) = 1$, some straightforward computation yields

$$\begin{aligned}
\|\mathbf{H_P}(\dot{\mathbf{X}}\ddot{\mathbf{S}})\|_F &= \|\mathbf{H}(\hat{\dot{\mathbf{X}}}\hat{\ddot{\mathbf{S}}})\|_F \\
&= \left\|\frac{1}{2}\left[\hat{\dot{\mathbf{X}}}\hat{\ddot{\mathbf{S}}} + \left(\hat{\dot{\mathbf{X}}}\hat{\ddot{\mathbf{S}}}\right)^{\mathrm{T}}\right]\right\|_F \\
&\leq \|\hat{\dot{\mathbf{X}}}\hat{\ddot{\mathbf{S}}}\|_F \\
&= \|\mathrm{vec}\hat{\dot{\mathbf{X}}}\| \, \|\mathrm{vec}\hat{\ddot{\mathbf{S}}}\| \\
&\leq \|\mathbf{G}^{-\frac{1}{2}}\mathrm{vec}\hat{\dot{\mathbf{X}}}\| \, \|\mathbf{G}^{\frac{1}{2}}\mathrm{vec}\hat{\ddot{\mathbf{S}}}\|,
\end{aligned}$$

From lemmas 12.15 and 12.16, it follows that

$$\|\mathbf{H_P}(\dot{\mathbf{X}}\ddot{\mathbf{S}})\|_F \leq \frac{1}{2}\beta_3\mu n^{\frac{3}{2}}.$$

Similarly,

$$\|\mathbf{H_P}(\ddot{\mathbf{X}}\dot{\mathbf{S}})\|_F \leq \frac{1}{2}\beta_3\mu n^{\frac{3}{2}}.$$

This proof is completed. ∎

We need a result in [54, Theorem 4.3.27, page 194] in our proof, which is stated as the following lemma.

Lemma 12.18
Let $\mathbf{A}, \mathbf{B} \in \mathcal{S}^n$, *and let* $\lambda_i(\mathbf{A})$, $\lambda_i(\mathbf{B})$, $\lambda_i(\mathbf{A}+\mathbf{B})$ *denote the ith eigenvalues of* $\mathbf{A}$, $\mathbf{B}$, *and* $\mathbf{A}+\mathbf{B}$ *arranged in increasing order, i.e.,* $\lambda_1(\mathbf{A}+\mathbf{B}) = \lambda_{\min}(\mathbf{A}+\mathbf{B})$. *Then*

$$\lambda_1(\mathbf{A}+\mathbf{B}) \geq \lambda_1(\mathbf{A}) + \lambda_1(\mathbf{B}).$$

Let $\sin(\hat{\alpha}_0) = \frac{1}{\beta n^{\frac{3}{4}}}$ where $\beta = \frac{\beta_1+\beta_2+2\beta_3}{\sigma(1-\gamma)}$, which will be used in the following lemmas.

Lemma 12.19
Let $\mathbf{H_P}(\chi(\alpha))$ *be defined as (12.62),* $\alpha \in (0, \hat{\alpha}_0]$ *and* $\mathbf{P}$ *be a NT scaling matrix. Then, we have*

$$\lambda_{min}(\mathbf{H_P}(\chi(\alpha))) \geq -\frac{1}{2}\sin(\alpha)(1-\gamma)\mu\sigma.$$

Proof 12.18 We will use Lemmas 12.9, 12.15, 12.16, 12.17, the simple fact of $g(\alpha) \le \sin^2(\alpha)$ and $\lambda_{\min}(\cdot)$ is a homogeneous concave function on the space of symmetric matrices. From (12.62) and Lemma 12.18, we have

$$\begin{aligned}
\lambda_{\min}(\mathbf{H_P}(\chi(\alpha))) &= \lambda_{\min}\left[-g^2(\alpha)(\mathbf{H_P}(\dot{\mathbf{X}}\dot{\mathbf{S}})) \right.\\
&\quad -\sin(\alpha)g(\alpha)(\mathbf{H_P}(\mathbf{XS}+\mathbf{X}\dot{\mathbf{S}}))\\
&\quad \left.+g^2(\alpha)(\mathbf{H_P}(\ddot{\mathbf{X}}\ddot{\mathbf{S}}))\right]\\
&\ge -g^2(\alpha)\lambda_{\max}(\mathbf{H_P}(\dot{\mathbf{X}}\dot{\mathbf{S}}))\\
&\quad -\sin(\alpha)g(\alpha)\lambda_{\max}(\mathbf{H_P}(\dot{\mathbf{X}}\ddot{\mathbf{S}}+\ddot{\mathbf{X}}\dot{\mathbf{S}}))\\
&\quad +g^2(\alpha)\lambda_{\min}(\mathbf{H_P}(\ddot{\mathbf{X}}\ddot{\mathbf{S}}))\\
&\ge -g^2(\alpha)\|\mathbf{H_P}(\dot{\mathbf{X}}\dot{\mathbf{S}})\|_F\\
&\quad -\sin(\alpha)g(\alpha)\|\mathbf{H_P}(\dot{\mathbf{X}}\ddot{\mathbf{S}}+\ddot{\mathbf{X}}\dot{\mathbf{S}})\|_F\\
&\quad -g^2(\alpha)\|\mathbf{H_P}(\ddot{\mathbf{X}}\ddot{\mathbf{S}})\|_F\\
&\ge -g^2(\alpha)\|\mathbf{H_P}(\dot{\mathbf{X}}\dot{\mathbf{S}})\|_F-\sin(\alpha)g(\alpha)\|\mathbf{H_P}(\dot{\mathbf{X}}\ddot{\mathbf{S}})\|_F\\
&\quad -\sin(\alpha)g(\alpha)\|\mathbf{H_P}(\ddot{\mathbf{X}}\dot{\mathbf{S}})\|_F-g^2(\alpha)\|\mathbf{H_P}(\ddot{\mathbf{X}}\ddot{\mathbf{S}})\|_F\\
&\ge -\frac{1}{2}\left[\sin^4(\alpha)\beta_1\mu n+2\sin^3(\alpha)\beta_3\mu n^{\frac{3}{2}}+\frac{1}{4}\sin^4(\alpha)\beta_2\mu n^2\right]\\
&\ge -\frac{1}{2}\sin(\alpha)\mu\left[\sin^3(\hat{\alpha}_0)\beta_1 n+2\sin^2(\hat{\alpha}_0)\beta_3 n^{\frac{3}{2}}+\frac{1}{4}\sin^3(\hat{\alpha}_0)\beta_2 n^2\right]\\
&= -\frac{1}{2}\sin(\alpha)\mu\left[\frac{\beta_1}{\beta^3 n^{\frac{5}{4}}}+\frac{2\beta_3}{\beta^2}+\frac{\beta_2}{4\beta^3 n^{\frac{1}{4}}}\right]\\
&\ge -\frac{1}{2}\sin(\alpha)\mu\frac{1}{\beta}\left[\beta_1+\beta_2+2\beta_3\right]\\
&= -\frac{1}{2}\sin(\alpha)(1-\gamma)\mu\sigma, \qquad (12.70)
\end{aligned}$$

where the last inequality follows from $\beta = \frac{\beta_1+\beta_2+2\beta_3}{\sigma(1-\gamma)}$. This finishes the proof. ■

Now, we are ready to present the main result of the analysis.

Lemma 12.20
Let $(\mathbf{X},\mathbf{y},\mathbf{S}) \in \mathcal{N}_{-\infty}(\gamma)$, $\alpha \in (0,\hat{\alpha}_0]$, *and* $\sin(\hat{\alpha})$ *be defined by* (12.60), *then* $(\mathbf{X}(\alpha),\mathbf{y}(\alpha),\mathbf{S}(\alpha)) \in \mathcal{N}_{-\infty}(\gamma)$ *and* $\sin(\hat{\alpha}) \ge \sin(\hat{\alpha}_0)$.

Proof 12.19 Combining (12.61), (12.62), Lemmas 12.10, 12.13, 12.18, 12.19, one has

$$\begin{aligned}
&\lambda_{\min}(\mathbf{X}(\alpha)\mathbf{S}(\alpha))\\
[\text{use Lemma 12.13}] \ &\ge \lambda_{\min}(\mathbf{H_P}(\mathbf{X}(\alpha)\mathbf{S}(\alpha)))\\
[\text{use (12.61)}] \ &\ge \lambda_{\min}\left[(1-\sin(\alpha))\mathbf{H_P}(\mathbf{XS})+\sigma\mu\sin(\alpha)\mathbf{I}+\mathbf{H_P}(\chi(\alpha))\right]
\end{aligned}$$

$$\begin{aligned}
\text{[use Lemma 12.18]} &\geq (1-\sin(\alpha))\lambda_{\min}(\mathbf{H_P}(\mathbf{XS}))+\sigma\mu\sin(\alpha)+\lambda_{\min}(\mathbf{H_P}(\chi(\alpha))) \\
\text{[use Lemma 12.13]} &= (1-\sin(\alpha))\lambda_{\min}(\hat{\mathbf{X}}\hat{\mathbf{S}})+\sigma\mu\sin(\alpha)+\lambda_{\min}(\mathbf{H_P}(\chi(\alpha))) \\
\text{[use Lemma 12.19]} &\geq (1-\sin(\alpha))\gamma\mu+\sigma\mu\sin(\alpha)-\frac{1}{2}\sin(\alpha)(1-\gamma)\sigma\mu \\
\text{[use Lemma 12.10]} &= \gamma\mu(\alpha)+\sin(\alpha)(1-\gamma)\sigma\mu-\frac{1}{2}\sin(\alpha)(1-\gamma)\sigma\mu \\
&\geq \gamma\mu(\alpha)>0. \qquad (12.71)
\end{aligned}$$

This reveals that $\mathbf{X}(\alpha)\mathbf{S}(\alpha)$ is nonsingular for all $\alpha\in[0,\hat{\alpha}]$. By using continuity of the eigenvalues of a symmetric matrix [173] and $\mathbf{X}\succ\mathbf{0}$, $\mathbf{S}\succ\mathbf{0}$, it follows that $\mathbf{X}(\alpha)\succ\mathbf{0}$, $\mathbf{S}(\alpha)\succ\mathbf{0}$ for all $\alpha\in[0,\hat{\alpha}_0]$. From (12.49) and Lemma 12.8, we have $(\mathbf{X}(\alpha),\mathbf{S}(\alpha))\in\mathcal{N}_{-\infty}(\gamma)$. By (12.60), we obtain $\sin(\hat{\alpha})\geq\sin(\hat{\alpha}_0)$. This completes the proof. ∎

In the following theorem, we give an upper bound for the required number of iterations for Algorithm 12.1 to obtain an ε-approximate solution of (12.19) and (12.20).

Theorem 12.1
Let $(\mathbf{X}^0,\mathbf{y}^0,\mathbf{S}^0)\in\mathcal{N}_{-\infty}(\gamma)$, $\sin(\hat{\alpha}_0)=\frac{1}{\beta n^{\frac{3}{4}}}$ and $\beta=\frac{\beta_1+\beta_2+2\beta_3}{\sigma(1-\gamma)}$. Then, the Algorithm 12.1 terminates in at most $O(n^{\frac{3}{4}}\log\frac{\mathbf{X}^0\bullet\mathbf{S}^0}{\varepsilon})$ iterations.

Proof 12.20 Due to Lemma 12.10 and the inequality $\sin(\hat{\alpha})\geq\sin(\hat{\alpha}_0)$, we have

$$\mu(\hat{\alpha})=[1-(1-\sigma)\sin(\hat{\alpha})]\mu\leq[1-(1-\sigma)\sin(\hat{\alpha}_0)]\mu=\left(1-\frac{1-\sigma}{\beta n^{\frac{3}{4}}}\right)\mu.$$

Thus, we obtain

$$\mathbf{X}^k\bullet\mathbf{S}^k\leq\left(1-\frac{1-\sigma}{\beta n^{\frac{3}{4}}}\right)^k(\mathbf{X}^0)\bullet\mathbf{S}^0.$$

Since we need to have $\mathbf{X}^k\bullet\mathbf{S}^k\leq\varepsilon$, it suffices to have

$$\left(1-\frac{1-\sigma}{\beta n^{\frac{3}{4}}}\right)^k(\mathbf{X}^0)\bullet\mathbf{S}^0\leq\varepsilon. \qquad (12.72)$$

Take logarithms, we obtain

$$k\log\left(1-\frac{1-\sigma}{\beta n^{\frac{3}{4}}}\right)\leq\log\frac{\varepsilon}{\mathbf{X}^0\bullet\mathbf{S}^0}.$$

Using $-\log(1-\theta)\geq\theta$ for $0<\theta<1$, which is equivalent to Lemma 1.2, then we have that (12.72) holds, if $k\geq\frac{\beta n^{\frac{3}{4}}}{1-\sigma}\log\frac{\mathbf{X}^0\bullet\mathbf{S}^0}{\varepsilon}$. This completes the proof. ∎

12.3 Concluding Remarks

Monteiro and Zhang [98] analyzed several popular interior-point algorithms, such as AHO [4], HRVW/KSS/M [52, 73, 94], KSS/M [73, 94], and NT [103]. They derived polynomial bounds, as summarized here:

(a) $O(n^{\frac{3}{2}} \log \frac{\mathbf{X}^0 \bullet \mathbf{S}^0}{\varepsilon})$ for Algorithm HRVW/KSS/M.

(b) $O(n^{\frac{3}{2}} \log \frac{\mathbf{X}^0 \bullet \mathbf{S}^0}{\varepsilon})$ for Algorithm KSS/M.

(c) $O(n \log \frac{\mathbf{X}^0 \bullet \mathbf{S}^0}{\varepsilon})$ for Algorithm NT.

Clearly, the algorithm discussed in this chapter has a better polynomial complexity bound, which is $O(n^{\frac{3}{4}} \log \frac{\mathbf{X}^0 \bullet \mathbf{S}^0}{\varepsilon})$.

Still, there are problems to be investigated and questions to be answered. For example, can we develop infeasible interior-point arc-search algorithms for SDP problem? Can we find infeasible interior-point arc-search algorithms for SDP problem with lower polynomial bound? Computationally, is arc-search infeasible interior-point algorithm(s) more efficient than traditional infeasible interior-point algorithm(s) for SDP? To answer the last question, a software package has to be developed and compared to the ones such as SDPT5 [136], SeduMi [129], and SDPA [146], etc.

Finally, There are quite a few papers on interior-point algorithms for nonlinear programming problems, for example, [13, 16, 17, 32, 33, 45, 78, 105, 131, 137, 140]. There should be many opportunities to extend the arc-search techniques to this broad research area. A very recent effort in this direction is described in [76].

References

[1] P. A. Absil, R. Mahony and R. Sepulchre, *Optimization algorithms on matrix manifolds*, Princeton University Press, Princeton, 2008.

[2] F. Alizadeh, *Combinatorial optimization with interior-point methods and semi-definite matrices*, Ph.D. thesis, Department of Computer Science, University of Minnesota, Minneapolis, MN, 1993.

[3] F. Alizadeh, *Interior-point methods in semidefinite programming with applications to combinatorial optimization*, SIAM Journal on Optimization, 5 (1993), pp. 13-51.

[4] F. Alizadeh, J. P. A. Haeberly and M. L. Overton, *Complementarity and nondegeneracy in semidefinite programming*, Mathematical Programming, 77 (1997), pp. 111-128.

[5] A. Altman and J. Gondzio, *Regularized symmetric indefinite systems in interior point methods for linear and quadratic optimization*, Optimization Methods and Software, 11 (1999), pp. 275-302.

[6] E. D. Andersen, *Finding all linearly dependent rows in large-scale linear programming*, Optimization methods and software, 6 (1995), pp. 219-227.

[7] E. D. Andersen and K. D. Andersen, *Presolving in linear programming*, Mathematical Programming, 71 (1993), pp. 221-245.

[8] D. A. Bayer and J. C. Lagaris, *The nonlinear geometry of linear programming, I. Affine and projective scaling trajectories*, Transactions of the American Mathematical Society, 314 (1989), pp. 499-526.

[9] D. P. Bertsekas, *Projected Newton methods for optimization problems with simple constraints*, SIAM Journal on Control and Optimization, 20 (1982), pp. 221-246.

[10] A. B. Berkelaar, K. Roos and T. Terlaky, *The optimal set and optimal partition approach to linear and quadratic programming*, in Recent Advances in Sensitivity Analysis and Parametric Programming, H. Greenberg and T. Gal, eds., Kluwer Publishers Berlin, 1997.

[11] R. Bland, D. Goldfarb and M. Todd, *The ellipsoid method: A survey*, Operations Research, 29 (1981), pp. 1039-1091.

[12] A. L. Brearley, G. Mitra and H. P. Williams, *Analysis of mathematical programming problems prior to applying the simplex algorithm*, Mathematical Programming, 8 (1975), pp. 54-83.

[13] S. Browne, J. Dongarra, E. Grosse and T. Rowan, *The Netlib mathematical software repository*, DLib magazine, http://www.dlib.org/dlib/september95/netlib/09browne.html, Accessed 2 January, 2020.

[14] S. Boyd, L. El Ghaoui, E. Feron and V. Balakrishnan, *Linear matrix inequalities in system and control theory*, Society for Industrial and Applied Mathematics, Philadelphia, PA, 1994.

[15] S. Boyd and L. Vandenberghe, *Convex optimization*, Cambridge University Press, Cambridge, UK, 1994.

[16] R. H. Byrd, J. C. Gilbert and J. Nocedal, *A trust region method based on interior point techniques for nonlinear programming*, Mathematical Programming, 89 (2000), pp. 149-185.

[17] R. H. Byrd, M. E. Hribar and J. Nocedal, *An interior point algorithm for large-scale nonlinear programming*, SIAM Journal on Optimization, 9 (1999), 877-900.

[18] M. P. Do Carmo, *Differential geometry of curves and surfaces*, Prentice-Hall, New Jersey, 1976.

[19] C. Cartis, *Some disadvantages of a Mehrotra-type primal-dual corrector interior point algorithm for linear programming*, Applied Numerical Mathematics, 59 (2009), pp. 1110-1119.

[20] C. Cartis and N. I. M. Could, *Finding a point in the relative interior of a polyhedron*, Technical Report NA-07/01, Computing Laboratory, Oxford University, Oxford, UK, 2007.

[21] R. W. Cottle, J. S. Pang and R. E. Stone, *The linear complementarity problem*, New York, Academic Press, 1992.

[22] A. R. Curtis and J. K. Reid, *On the automatic scaling of matrices for Gaussian elimination*, IMA Journal of Applied Mathematics, 10 (1972), pp. 118-124.

[23] J. Czyzyk, S. Mehrotra, M. Wagner and S. J. Wright, *PCx user guide (version 1.1)*, Technical Report OTC 96/01, Optimization Technology Center, 1997.

[24] G. B. Dantzig, *Programming in a linear structure*, Econometrica, 17 (1949), pp. 73-74.

[25] G. B. Dantzig, *Linear programming and extension*, Princeton University Press, New Jersey, 1963.

[26] T. A. Davis, *Multifrontal multithreaded rank-revealing sparse QR factorization*, Technical Report, Department of Computer and information Science and engineering, University of Florida, Florida, 2008.

[27] I. I. Dikin, *Iterative solution of problems of linear and quadratic programming*, Soviet Mathematics Doklady, 8 (1967), pp. 674-675.

[28] J. Dobes, *A modified Markowitz criterion for the fast modes of the LU factorization*. Proceedings of 48th Midwest Symposium on Circuits and Systems, (2005), pp. 955-959.

[29] E. D. Dolan and J. J. More, *Benchmarking optimization software with performance profiles*, Mathematical Programming, 91 (2002), pp. 201-213.

[30] J. F. Duff, A. M. Erisman and J. K. Reid, *Direct method for sparse matrices*, Oxford University Press, New York, 1989.

[31] A. Edelman, T. A. Arias and S. T. Smith, *The Geometry of algorithms with orthogonality of constraints*, SIAM J. Matrix Anal. Appl, 20 (1998), 303-353.

[32] A. S. El-Bakry, R. A. Tapia, T. Tsuchiya and Y. Zhang, *On the formulation and theory of the Newton interior-point method for nonlinear programming*, Journal of Optimization Theory and Applications, 89 (1996), pp. 507-541,

[33] A. Forsgren and P. E. Gill, *Primal-dual interior methods for nonconvex nonlinear programming*, SIAM Journal on Optimization, 8 (1998), pp. 1132-1152.

[34] J. Ekefer, *Sequential minimax search for a maximum*, Proceedings of the American Mathematical Society, 4 (1953), pp. 502-506.

[35] D. Herbison-Evans, *Solving quartics and cubics for graphics*, Technical Report TR94-487, Basser Department of Computer Science, University of Sydney, Sydney, Australia, 1994.

[36] A. V. Fiacco and G. P. McCormick, *Nonlinear programming: Sequential unconstrined minimization techniques*, Wiley, New York, 1968.

[37] O. Friedmann, *A subexponential lower bound for Zadeh's pivoting rule for solving linear programs and games*, in Integer Programming and Combinatoral Optimization 2011, Lecture Notes in Computer Science 6655, O. Gunluk and G. J. Woeginger, eds., Springer, Berlin, 2011, pp. 192-206.

[38] K. R. Frisch, *The logarithmic potential method of convex programming*, Technical Report, University Institute of Economics, Oslo, Norway, 1955.

[39] D. Gabay, *Minimizing a differentiable function over a differentiable manifold*, Journal of Optimization Theory and Applications, 37 (1982), pp. 177-219.

[40] A. Goldman and A. Tucker, *Theory of linear programming*, in Linear equalities and related systems, H. Kuhn and A. Tucker, eds., Princeton University Press, Princeton, 1956, pp. 53-97.

[41] J. Gondzio, *Multiple centrality corrections in a primal-dual method for linear programming*, Computational Optimization and Applications, 6 (1994), pp. 137-156

[42] C. C. Gonzaga, *Polynomial affine algorithms for linear programming*, Mathematical Programming 49 (1990), pp. 7-21.

[43] C. T. L. S. Ghidini, A. R. L. Oliveira, J. Silvab and M. I. Velazco, *Combining a hybrid preconditioner and a optimal adjustment algorithm to accelerate the convergence of interior point methods*, Linear Algebra and its Applications, 436 (2012), pp. 1267-1284.

[44] P. E. Gill, W. Murray, M. A. Saunders, J. A. Tomlin and M. H. Wright, *On projected Newton barrier methods for linear programming and an equivalence of Karmarkar's projective method*, Mathematical Programming, 36 (1986), pp. 183-209.

[45] N. I. M. Gould, D. Orban and P. L. Toint, *CUTEst: A constrained and unconstrained testing environment with safe threads for mathematical optimization*, Computational Optimization and Applications, 60 (2015), pp. 545–557.

[46] N. I. M. Gould and J. Scott, *A note on performance profiles for benchmarking software*, ACM Transactions on Mathematical Software, 43 (2016), pp. 15.

[47] I. Griva, D. F. Shanno, R. J. Vanderbei and H. Y. Benson, *Global convergence of a primal-dual interior-point method for nonlinear programming*, Algorithmic Operations Research, 3 (2008), pp. 27-52.

[48] O. Guler, D. den Hertog, C. Roos, T. Terlaky and Tsuchiya, *Degeneracy in interior-point methods for linear programming: A survey*, Annals of Operations Research, 46 (1993), pp. 107-138.

[49] O. Guler and Y. Ye, *Convergence behavior of interior-point algorithms*, Mathematical Programming, 60, (1993), pp. 215-228.

[50] http://users.clas.ufl.edu/hager/coap/format.htm.

[51] C. G. Han, P. Pardalos and Y. Ye, *Computational aspects of an interior point algorithm for quadratic programming problem with box constraints*, in Large-Scale Numerical Optimization, T. F. Coleman and Y. Li, eds. SIAM Publications, Philadelphia, PA, 1990, pp. 92-112.

[52] C. Helmberg, F. Rendl, R. J. Vanderbei and H. Wolkowicz, *An interior-point method for semidefinite programming*, SIAM Journal on Optimization, 6 (1996), pp. 342-361.

[53] W. Hock and K. Schittkowski, *Test examples for nonlinear programming codes*, Lecture Notes in Economics and Mathematical Systems, 187, Springer-Verlag, Berlin, 1981.

[54] R. A. Horn and C. R. Johnson, *Topics in matrix analysis*, Cambridge University Press, Cambridge, UK, 1991.

[55] P. Hung and Y. Ye, *An asymptotical $O(\sqrt{n}L)$-iteration path-following linear programming algorithm that use wide neighborhoods*, SIAM Journal on Optimization, 6 (1996), pp. 570-586.

[56] B. Jansen, C. Roos and T. Terlaky, *Improved complexity using higher-order correctors for primal-dual Dikin affine scaling*, Mathematical Programming, 76 (1996), pp. 117-130.

[57] N. Karmarkar, *A new polynomial-time algorithm for linear programming*, Combinatorics, 4 (1984), pp. 373-395.

[58] W. Karush, *Minima of functions of several variables with inequalities as side constraints*, M.Sc. Dissertation. Dept. of Mathematics, Univ. of Chicago, Chicago, Illinois, 1939.

[59] L. Khachiyan, *A polynomial algorithm in linear programming*, Doklady Akademiia Nauk SSSR, 224 (1979), pp. 1093-1096.

[60] B. Kheirfam, *An arc-search interior point method in the $N^{-\infty}$ neighborhood for symmetric optimization*, Fundamenta Informaticae, 146 (2016), pp. 255-269.

[61] B. Kheirfam, *An arc-search infeasible interior-point algorithm for horizontal linear complementarity problem in the $N^{-\infty}$ neighbourhood of the central path*, International Journal of Computer Mathematics, 94 (2017), pp. 2271-2282.

[62] B. Kheirfam and M. Moslemi, *On the extend of an arc-search interior-point algorithm for semidefinite optimization*, Numerical Algebra, Control, and Optimization, 2 (2018), pp. 261-275.

[63] B. Kheirfam, K. Ahmadi and F. Hasani, *A modified full-Newton step infeasible interior-point algoirhm for linear optimization*, Asia-Pacific Journal of Operational Research, 30 (2013), pp. 11-23.

[64] B. Kheirfam and M. Chitsaz, *A corrector-predictor arc-search interior point algorithm for $P^*(k)$-LCP acting in a wide neighborhood of the central path*, Iranian Journal of Operations Research, 6 (2015), pp. 1-18.

[65] V. Klee and G. J. Minty, *How good is the simplex algorithm*? in Inequalities, O. Shisha, eds., Academic Press, Providence, RI, 1972, pp. 159-175.

[66] E. de Klerk, *Aspects of semidefinite programming: Interior point algorithms and selected applications*, Kluwer Academic Publishers, Dordrecht, The Netherlands. 2002.

[67] M. Kojima, *Basic lemmas in polynomial-time infeasible-interior-point methods for linear programs*, Annals of Operations Research, 62 (1996), pp. 1-28.

[68] M. Kojima, N. Megiddo and S. Mizuno, *A primal-dual infeasible interior-point algorithm for linear programming*, Mathematical Programming, 61 (1993), pp. 261-280.

[69] M. Kojima, N. Megiddo, T. Noma and A. Yoshise, *A unified approach to interior point algorithms for linear complementarity problems: A summary*, Operations Research Letters, 10, (1991), pp. 247254.

[70] M. Kojima, S. Mizuno and A. Yoshise, *A polynomial-time algorithm for a class of linear complementarity problem*, Mathematical Programming, 44 (1989), pp. 1039-1091.

[71] M. Kojima, S. Mizuno and A. Yoshise, *A primal-dual interior point algorithm for linear programming*, in Mathematical Programming: Interior-point and related methods, N. Megiddo, eds., Springer-Verlag, New York, 1989, pp. 29-47.

[72] M. Kojima, S. Mizuno and A. Yoshise, *A $\mathcal{O}(\sqrt{n}L)$ iteration potential reduction algorithm for linear complementarity programming*, Mathematical Programming, 50 (1991), pp. 331-342.

[73] M. Kojima, S. Shindoh and S. Hara, *Interior-point methods for the monotone semidefinite linear complementarity problem in symmetric matrices*, SIAM journal on Optimization, 7 (1997), pp. 86-125.

[74] M. H. Koulaei and T. Terlaky, *On the extension of a Mehrotra-type algorithm for semidefinit optimization*, Technical Report 2007/4, Advanced optimization Lab., Department of Computing and Software, McMaster University, Hamilton, Ontario, 2007.

[75] H. W. Kuhn and A. W. Tucker, *Nonlinear programming*, Proceedings of 2nd Berkeley Symposium, Berkeley, University of California Press, pp. 481-492, 1951.

[76] E. Lida, Y. Yang and M. Yamashita, *An infeasible interior-point arc-search algorithm for nonlinear constrained optimization*, arXiv:1909.10706[math.OC], 2019.

[77] J. W. Liu, *Modification of the minimum degree algorithm by multiple elimination*, ACM Transactions on Mathematical Software, 11 (1985), pp. 141-153.

[78] T. T. Lu and S. H. Shiou, *Inverses of 2 $\times$ 2 block matrices*, Computers and Mathematics with Applications, 43 (2002), pp. 119-129.

[79] D. G. Luenberger, *Introduction to linear and nonlinear programming*, Addison Wesley, Massachusetts, 1972.

[80] D. Luenberger, *Linear and nonlinear programming*, Second Edition, Addison-Wesley Publishing Company, Menlo Park, 1984.

[81] D. G. Luenberger and Y. Ye, *Linear and nonlinear programming*, Springer, New York, 2008.

[82] I. Lustig, R. Marsten and D. Shannon, *Computational experience with a primal-dual interior-point method for linear programming*, Linear Algebra and Its Applications, 152, 1991, pp. 191-222.

[83] I. Lustig, R. Marsten and D. Shannon, *On implementing Mehrotra's predictor-corrector interior-point method for linear programming*, SIAM journal on Optimization, 2 (1992), pp. 432-449.

[84] A. Mahajan, *Presolving mixed-integer linear programs*, Preprint ANL/MCS-P1752-0510, Argonne National Laboratory, 2010.

[85] H. Mansouri, M. Pirhaji and M. Zangiabadi, *An arc search infeasible interior-point algorithm for symmetric optimization using a new wide neighborhood*, Acta Applicandae Mathematicae, 157 (2018), pp. 75-91.

[86] H. Markowitz, *Portfolio selection*, The Journal of Finance, 7 (1952), pp. 77-91.

[87] G. P. McCormick, *A modification of Armijo's step-size rule for negative curvature*, Mathematical Programming, 13 (1977), pp. 111–115.

[88] N. Megiddo, *Pathway to the optimal set in linear proprramming*, in Program in mathematical programming: interior point and related methods, N. Megiddo, eds., Springer Verlag, New York, 1989, pp. 131-158.

[89] S Mehrotra, *On the implementation of a primal-dual interior point method*, SIAM Journal on Optimization, 2 (1992), pp. 575-601.

[90] J. Miao, *Two infeasible interior-point predictor-corrector algorithms for linear programming*, SIAM journal on Optimization, 6 (1996), pp. 587-599.

[91] S. Mizuno, *A new polynomial time method for a linear complementarity problem*, Mathematical Programming, 56 1(992), pp. 31-43.

[92] S. Mizuno, *Polynomiality of infeasible-interior-point algorithms for linear programming*, Mathematical Programming, 67 (1994), pp. 109-119.

[93] S. Mizuno, M. Todd and Y. Te, *On adaptive step primal-dual interior-point algorithms for linear programming*, Mathematics of Operations Research, 18 (1993), pp. 964-981.

[94] R. Monteiro, *Primal-dual path-following algorithms for semidefinite programming*, SIAM Journal on Optimization, 7 (1997), pp. 663-678.

[95] R. Monteiro and I. Adler, *Interior path following primal-dual algorithm, Part I: linear programming*, Mathematical Programming, 44 (1989), pp. 27-41.

[96] R. Monteiro and I. Adler, *Interior path following primal-dual algorithms. Part II: convex quadratic programming*, Mathematical Programming, 44 (1989), pp. 43-66.

[97] R. Monteiro, I. Adler and M. Resende, *A polynominal-time primal-dual affine scaling algorithm for linear and convex quadratic programming and its power series extension*, Mathematics of Operations Research, 15 (1990), pp. 191-214.

[98] R. D. C. Monteiro and Y. Zhang, *A unified analysis for a class of long-step primal-dual path-following interior-point algorithms for semidefinite programming*, Mathematical Programming, 81 (1998), pp. 281-299.

[99] Y. Nesterov, *Long-step strategies in interior-point primal-dual methods*, Mathematical Programming, 76 (1996), pp. 47-94.

[100] Y. Nesterov, *Lectures on convex optimization*, Springer, Gewerbestrasse, Switzerland, 2018.

[101] Y. Nesterov and A. Nemirovskii, *Interior-point polynomial methods in convex programming*, Society for Industrial and Applied Mathematics, Philadelphia, PA, 1994.

[102] Y. Nesterov and M. Todd, *Self-scaled barriers and interior-point methods for convex programming*, Mathematics of Operations Research, 22 (1997), pp. 1-42.

[103] Y. Nesterov and M. Todd, *Primal-dual interior-point methods for self-scaled cones*, SIAM Journal on Optimization, 8 (1998), pp. 256-268.

[104] E. Ng and B.W. Peyton, *Block sparse Cholesky algorithm on advanced uniprocessor computers*, SIAM Journal on Scientific Computing, 14 (1993), pp. 1034-1056.

[105] J. Nocedal, A. Wachter and R. A. Waltz, *Adaptive barrier update strategies for nonlinear interior methods*, SIAM Journal on Optimization, 19 (2009), pp. 1674-1693.

[106] J. Nocedal and S. J. Wright, *Numerical optimization*, Springer-Verlag, New York, 1999.

[107] , M. L. Overton, *Large-scale optimization of eigenvalues*, SIAM Journal on Optimization, 2 (1992), pp. 88-120.

[108] K. Paparrizos, N. Samaras and D. Zissopoulos, *Linear programming: Klee–Minty examples*, In Encyclopedia of Optimization, C. Floudas and P. Pardalos, eds, Springer, Boston, MA, 2008, pp. 17-36.

[109] K. B. Petersen and M. S. Pedersen, *The matrix cookbook*, https://www.math.uwaterloo.ca/ hwolkowi/matrixcookbook.pdf, 2012.

[110] M. Pirhaji, M. Zangiabadi and H. Mansouri, *An ℓ_2-neighborhood infeasible interior-point algorithm for linear complementarity problems*, 4OR, 15 (2017), pp. 111-131.

[111] M. Pirhaji, M. Zangiabadi and H. Mansouri, *A corrector-predictor arc search interior-point algorithm for symmetric optimization*, Acta Mathematica Scientia, 38 (2018), pp. 1269-1284.

[112] M. Pirhaji, M. Zangiabadi, H. Mansouri and S. H. Amin, *An arc-search interior-point algorithm for monotone linear complementary problem over symmetric cones*, Mathematical Modeling and Analysis, 23 (2018), pp. 1-16.

[113] A. D. Polyanin and Manzhirov, *Handbook of mathematics for engineers and scientist*, Chapman & Hall/CRC, Noca Raton,2007.

[114] J. Renegar, *A polynomial-time algorithm, based on Newton's method, for linear programming*, Mathematical Programming, 40 (1988), pp. 59-93.

[115] R. T. Rockafellar, *Convex analysis*, Princeton University Press, New Jersey, 1970.

[116] C. Roos, *A full-Newton step $O(n)$ infeasible interior-point algorithm for linear optimization*, SIAM Journal on Optimization, 16 (2006), pp. 1110-1136.

[117] C. Roos, T. Terlaky and J-Ph. Vial, *Theory and algorithms for linear optimization: An interior-point approach*, John Wiley and Sons, Chichester, 1997.

[118] C. Roos, T. Terlaky and J-Ph. Vial, *Interior-point methods for linear optimization*, Springer, New York, 2006.

[119] J. B. Rosen, *Pattern separation by convex programming*, Analysis and Applications, 10 (1965), pp. 123-134.

[120] M. Salahi, J. Peng and T. Terlaky, *On Mehrotra-type predictor-corrector algorithms*, SIAM Journal on Optimization, 18 (2007), pp. 1377-1397.

[121] F. Santos, *A counterexample to the Hirsch conjecture*, Annals of Mathematics, 176 (2012), pp. 383-412.

[122] M. S. Shahraki and A. Delavarkhalafi, *An arc-search predictor-corrector infeasible-interior-point algorithm for $P_*(\kappa)$-SCLCPs*, Numerical Algorithms, Accepted, doi:10.1007/s11075-019-00736-4 (2019).

[123] M. S. Shahraki, H. Mansouri and M. Zangiabadi, *A wide neighborhood infeasible-interior-point method with arc-search for SCLCPs*, Optimization, 67 (2018), pp. 409-425.

[124] M. Shida, S. Shindoh and M. Kojima, *Existence and uniqueness of search directions in interior-point algorithms for the SDP and the monotone SDLCP*, SIAM Journal on Optimization, 8 (1998), pp. 387-396.

[125] S. L. Shmakov, *A universal method of solving quartic equations*, International Journal of Pure and Applied Mathematics, 71 (2011), pp. 251-259.

[126] S. Smale, *Mathematical problems for the next century*, in V. I. Arnold, M. Atiyah, P. Lax and B. Mazur, Mathematics: Frontiers and perspectives, American Mathematical Society, 1999, pp. 271294.

[127] S. T. Smith, *Geometric optimization methods for adaptive filtering*, Ph.D. thesis, Department of Applied Mathematics, Harvard University, Cambridge, MA, 1993.

[128] P. M. Stoltz, S. Sivapiragasam and T. Anthony, *Satellite orbit-raising using LQR control with fixed thrusters*, Advances in the Astronautical Sciences, 98 (1998), pp. 109-120.

[129] J. F. Sturm, *Using SeDuMi 1.02, A Matlab toolbox for optimization over symmetric cone*, Optimization Methods and Software, 11 (1999), pp. 625-653.

[130] K. Tanabe, *Centered Newton method for mathematical programming*, in System modeling and optimization: Proceedings of the 13th IFIP conference, Lecture Notes in Control and Information Systems 113, Berlin, 1987, Springer-verlag, New York, 1988, pp. 197-206.

[131] A. L. Tits, A. Wachter, S. Bakhtiarl, T. J. Urban and C. T. Lawrence, *A primal-dual method for nonlinear programming with strong global and local convergence properties*, Mathematical Programming, 8 (1998), pp. 1132-1152.

[132] A. L. Tits and Y. Yang, *Globally convergent algorithms for robust pole assignment by state feedback*, IEEE transactions on Automatic Control, 41 (1996), pp. 1432-1452.

[133] M. J. Todd, *The many facets of linear programming*, Mathematical Programming, Ser. B, 91 (2002), pp. 417-436.

[134] M. J. Todd, K. C. Toh and R. H. Tutuncu, *On the Nesterov–Todd direction in semidefinite programming*, SIAM Journal on Optimization, 8(3) (1999), pp. 769796.

[135] M. J. Todd and Y. Ye, *A centered projective algorithm for linear programming*, Mathematics of Operations Research, 15 (1990), pp. 508-529.

[136] K. C. Toh, M. J. Todd and R. H. Tutuncu, *SDPT3 A Matlab software package for semidefinite programming, Version 1.3*, Optimization Methods and Software, 11 (1999), pp. 545-581.

[137] M. Ulbrich, S. Ulbrich and L. N. Vicente, *A globally convergent primal-dual interior-point filter method for nonlinear programming*, Mathematical Programming, 100 (2004), pp. 379-410.

[138] L. Vandenberghe and S. Boyd, *A primal-dual potential reduction method for problems involving matrix inequalities*, Mathematical Programming, 69 (1995), pp. 205-236.

[139] R. J. Vanderbei. *LOQO: An interior point code for quadratic programming*, Optimization Methods and Software, 12 (1999), pp. 451-484.

[140] R. J. Vanderbei and D. F. Shanno, *An interior-point algorithm for nonconvex nonlinear programming*, Computational Optimization and Applications, 13 (1999), pp. 231-252.

[141] S. A. Vavasis and Y. Ye, *A primal dual interior-point method whose running time depends on the constraint matrix*, Mathematical Programming, 74 (1996), pp .79-120.

[142] L. B. Winternitz, A. L. Tits and P.-A. Absil, *Addressing rank degeneracy in constraint-reduced interior-point methods for linear optimization*, Journal of Optimization Theory and Applications, 160 (2014), pp. 127-157.

[143] P. Wolfe, *The simplex method for quadratic programming*, Econometrica, 27, (1959), pp. 382-398.

[144] S. Wright, *Primal-Dual Interior-Point Methods*, Society for Industrial and Applied Mathematics, Philadelphia, 1997.

[145] X. Yang, *Study on wide neighborhood in interior-point method for symmetric cone programming*, Ph.D thesis, Xidian University, Xian, 2017.

[146] M. Yamashita, K. Fujisawa, M. Fukuda, K. Nakata and M. Nakata, *A high-performance software package for semidefinite programs: SDPA 7*, Research Report B-463, Dept. of Mathematical and Computing Science, Tokyo Institute of Technology, Tokyo Japan, 2010.

[147] X. Yang, H. Liu and Y. Zhang, *An arc-search infeasible-interior-point method for symmetric optimization in a wide neighborhood of the central path*, Optimization Letters, 11 (2017), pp. 135-152.

[148] X. Yang and Y. Zhang, *A Mizuno-Todd-Ye predictor-corrector infeasible-interior-point method for symmetric optimization with the arc-search strategy*, Journal of Inequalities and Applications, 2017, 2017:291.

[149] X. Yang, Y. Zhang and H. Liu, *A wide neighborhood infeasible-interior-point method with arc-search for linear programming*, Journal of Applied Mathematics and Computing, 51 (2016), pp. 209-225.

[150] Y. Yang, *Robust system design: Pole assignment approach*, Ph.D. thesis, Department of Electrical and Computer Engineering, University of Maryland, College Park, MD, 1996.

[151] Y. Yang, *Arc-search path-following interior-point algorithm for linear programming*, Optimization Online, 2009.

[152] Y. Yang, *A polynomial arc-search interior-point algorithm for convex quadratic programming*, European Journal of Operational Research, 215 (2011), pp. 25-38.

[153] Y. Yang, *A polynomial arc-search interior-point algorithm for linear programming*, Journal of Optimization Theory and Applications, 158 (2013), pp. 859-873.

[154] Y. Yang, *An efficient polynomial interior-point algorithm for linear programming*, arXiv:1304.3677[math.OC], 2013.

[155] Y. Yang, *Constrained LQR design using interior-point arc-search method for convex quadratic programming with box constraints*, arXiv:1304.4685[math.OC], 2013.

[156] Y. Yang, *Arc-search infeasible interior-point algorithm for linear programming*, arXiv:1406.4539 [math.OC], 2014.

[157] https://www.mathworks.com/matlabcentral/fileexchange/53911-curvelp.

[158] Y. Yang, *CurveLP-a MATLAB implementation of an infeasible interior-point algorithm for linear programming*, Numerical Algorithms, 74 (2017), pp. 967-996.

[159] Y. Yang, *Two computationally efficient polynomial-iteration infeasible interior-point algorithms for linear programming*, Numerical Algorithms, 79 (2018), pp. 957-992.

[160] Y. Yang, *Spacecraft modeling, attitude determination, and control: Quaternion-based method*, CRC Press, Boca Raton, 2019.

[161] Y. Yang and M. Yamashita, *An arc-search $\mathcal{O}(nL)$ infeasible-interior-point algorithm for linear programming*, Optimization Letters, 12 (2018), pp. 781-798.

[162] Y. Yang and Z. Zhou, *An analytic solution to Wahba's problem*, Aerospace Science and Technology, 30 (2013), pp. 46-49.

[163] Y. Ye, *An $\mathcal{O}(n^3L)$ potential reduction algorithm for linear programming*, Mathematical programming, 50 (1991), pp. 239-258.

[164] Y. Ye, *Interior point algorithms: Theory and analysis*, John Wiley & Son, Inc., New York, 1997.

[165] Y. Ye, *Conic linear programming*, Preprint available on internet http://web.stanford.edu/class/msande314/sdpmain.pdf, 2017.

[166] Y. Ye and E. Tse, *An extension of Karmarkar's projective algorithm for convex quadratic programming*, Mathematical Programming, 44 (1989), pp. 157-179.

[167] B. Yuan, M. Zhang and Z. Huang, *A wide neighborhood primal-dual interior-point algorithm with arc-search for linear complementarity problem*, Journal of Nonlinear Functional Analysis, Article ID 31, (2018).

[168] B. Yuan, M. Zhang and Z. Huang, *A wide neighborhood interior-point algorithm with arc-search for $P^*(k)$ linear complementarity problem*, Applied Numerical Mathematics, 136 (2019), pp. 293-304.

[169] N. Zadeh, *What is the worst case behavior of the simplex algorithm?* Technical Report No. 27, Department of Operations Research, Stanford University, Stanford, California, 1980.

[170] M. Zhang, B. Yuan, Y. Zhou, X. Luo and Z. Huang, *A primal-dual interior-point algorithm with arc-search for semidefinite programming*, Optimization Letters, 13 (2019), pp. 1157-1175.

[171] Y. Zhang, *On the convergence of a class of infeasible interior-point methods for the horizontal linear complementarity problem*, SIAM Journal on Optimization, 4 (1994), pp. 208-227.

[172] Y. Zhang, *Solving large-scale linear programs by interior-point methods under the matlab environment*, Technical Report TR96-01, Department of Mathematics and Statistics, University of Maryland, Baltimore, MD, 1996.

[173] Y. Zhang, *On extending some primal-dual interior-point algorithms from linear programming to semi-definite programming*, SIAM Journal on Optimization, 8 (1998), pp. 365-386.

[174] Y. Zhang and D. T. Zhang, *On polynomiality of the Mehrotra-type predictor-corrector interior-point algorithms*, Mathematical Programming, 68 (1995), pp. 303-318.

Index